Animal Models in Eye Research

Animal Models in Eye Research

Edited by
Panagiotis A. Tsonis
University of Dayton,
Dayton, OH 45469-2320, USA

ELSEVIER

AMSTERDAM • BOSTON • HEIDELBERG • LONDON • NEW YORK • OXFORD
PARIS • SAN DIEGO • SAN FRANCISCO • SINGAPORE • SYDNEY • TOKYO
Academic Press is an imprint of Elsevier

Academic Press is an imprint of Elsevier

525 B Street, Suite 1900, San Diego, CA 92101-4495, USA
30 Corporate Drive, Suite 400, Burlington, MA 01803, USA
32 Jamestown Road, London NW1 7BY, UK
Radarweg 29, PO Box 211, 1000 AE Amsterdam, The Netherlands

First edition 2008

British Library Cataloguing in Publication Data
A catalogue record for this book is available from the British Library

Library of Congress Cataloging-in-Publication Data
A catalog record for this book is available from the Library of Congress

ISBN: 978-0-12-374169-1

For information on all Academic Press publications
visit our web site at books.elsevier.com

Typeset by Charon Tec Ltd., A Macmillan Company.
(www.macmillansolutions.com)

Printed and bound in the United Kingdom
Transferred to Digital Print 2011

Contents

Preface

The eye is a complex sensory organ, which enables visual perception of the world. Thus the eye has several tissues that do different tasks. One of the most basic aspects of eye function is the sensitivity of cells to light and its transduction though the optic nerve to the brain. Different organisms use different ways to achieve these tasks. In this sense, the eye function becomes a very important evolutionary aspect as well and different animal models provide unique accessibility to eye experimentation.

It is largely accepted that vision originated in the early Cambrian about half a billion years ago. During the important evolutionary event that is known as the Cambrian explosion, it seems that an incredible number of phyla that gave rise to modern species had come into existence within a few million years. During that period compound eyes appeared in species, such as tribolites and arthropods. Other invertebrates possessed uncomplicated eye designs made up of simple visual organs mainly composed of photoreceptor cells protected by a pigment cell. Throughout evolution as well as in modern species, there are many different types of eyes. The grouping depends on the type of photoreceptors that the eye uses and of the eye architecture (compound or single-chambered). One of the greatest debates in the eye field is how many times eyes have evolved independently during evolution and if there is a common ancestor. Based on the different types, it is obvious that eyes have evolved more than once. Also embryology teaches us that in different species eyes derive from different tissues. However, the identification of *pax-6* as the master gene in the development of different eye types testifies for a common ancestry. Given the number of different visual devises that animals have come up with it is obvious that depending on the eye type, we can acquire distinct knowledge from each one. This eventually will help to clarify the issues pertaining to eye evolution, development, and diseases. Thus, in this book we have assembled a series of chapters that address the uniqueness of different animal models in eye research. The reader will navigate through animal models spanning from bacteria to primates. Each animal has something unique to contribute to our understanding of how vision was evolved and how we can approach issues that affect it.

Panagiotis A. Tsonis
Dayton, OH, USA

List of Contributors

J. Bronwyn Bateman, Ophthalmology and Pediatrics, Rocky Mountain Lions Eye Institute, The Children's Hospital, University of Colorado Health Sciences Center at Denver and Aurora, CO 80262, USA

Renata Batistoni, Dipartimento di Biologia, Università di Pisa, Pisa, Italy

Teri Belecky-Adams, Department of Biology and Center for Regenerative Biology and Medicine, Indiana University-Purdue University Indianapolis, Indianapolis, IN 46202, USA

Elke K. Buschbeck, Department of Biological Sciences, University of Cincinnati, Cincinnati, OH 45221-0006, USA

Claude Desplan, Department of Biology, Center for Developmental Genetics, New York University, New York, NY 10003-6688, USA

Nicholas S. Foulkes, Institute of Toxicology and Genetics, Forschungszentrum Karlsruhe, Hermann-von-Helmholtz-Platz 1, Eggenstein-Leopoldshafen 76344, Germany

Lisa Fukui, Department of Cell and Developmental Biology (formerly Department of Cell and Structural Biology), University of Illinois, Urbana, IL 61801, USA

Arlene Gwon, Advanced Medical Optics, 1700 E. St. Andrew Place, Santa Ana, CA 92799-5162, USA

Tracy Haynes, Department of Zoology, Miami University, Oxford, OH 45056, USA

Jonathan J. Henry, Department of Cell and Developmental Biology (formerly Department of Cell and Structural Biology), University of Illinois, Urbana, IL 61801, USA

Andor J. Kiss, Laboratory for Ecophysiological Cryobiology, Department of Zoology, Miami University, Oxford, OH 45056, USA

Richard A. Lang, Division of Pediatric Ophthalmology and Developmental Biology, Children's Hospital Research Foundation; Department of Ophthalmology, and Graduate Program of Molecular and Developmental Biology, College of Medicine, University of Cincinnati, Cincinnati, OH 45229, USA

F. J. Lovicu, Department of Anatomy and Histology, Save Sight Institute, and Discipline of Ophthalmology, The Vision Cooperative Research Centre, University of Sydney, Sydney Eye Hospital Campus, Sydney, NSW, Australia

Robert J. Lucas, Faculty of Life Sciences, University of Manchester, Manchester M13 9PT, UK

Satori A. Marchitti, Molecular Toxicology and Environmental Health Sciences Program, Departments of Pharmaceutical Sciences, The Children's Hospital, University of Colorado Health Sciences Center at Denver and Aurora, CO 80262, USA

John W. McAvoy, Department of Anatomy and Histology, Save Sight Institute and Discipline of Ophthalmology, the Vision Cooperative Research Centre, University of Sydney, Sydney Eye Hospital Campus, Sydney, NSW, Australia

J. Mark Petrash, Department of Ophthalmology and Visual Science, Washington University School of Medicine, St. Louis, MO 63110, USA

Katia Del Rio-Tsonis, Department of Zoology, Miami University, Oxford, OH 45056-1400, USA

Meagan Roddy, Department of Biology and Center for Tissue Regeneration and Engineering, University of Dayton, Dayton, OH 45469-2320, USA

Emili Saló, Departament de Genètica, Facultat de Biologia, Universitat de Barcelona, Barcelona, Spain

Ni Song, Divisions of Pediatric Ophthalmology and Developmental Biology, Children's Hospital Research Foundation; Department of Ophthalmology, and Graduate Program of Molecular and Developmental Biology, College of Medicine, University of Cincinnati, Cincinnati, OH 45229, USA

Simon G. Sprecher, Department of Biology, Center for Developmental Genetics, New York University, New York, NY 10003-6688, USA

John L. Spudich, Center for Membrane Biology, Departments of Biochemistry and Molecular Biology, and Microbiology and Molecular Genetics, University of Texas Medical School, Houston, TX 77030, USA

Elena N. Spudich, Department of Biochemistry and Molecular Biology, Center for Membrane Biology, University of Texas Medical School, Houston, TX 77030, USA

Panagiotis A. Tsonis, Department of Biology and Center for Tissue Regeneration and Engineering, University of Dayton, Dayton, OH 45469-2320, USA

Daniela Vallone, Institute of Toxicology and Genetics, Forschungszentrum Karlsruhe, Hermann-von-Helmholtz Platz 1, Eggenstein-Leopoldshafen 76344, Germany

Vasilis Vasiliou, Molecular Toxicology and Environmental Health Sciences Program, Department of Pharmaceutical Sciences, University of Colorado Health Sciences Center (UCHSC) at Denver and Aurora, CO 80262, USA

M. Natalia Vergara, Department of Zoology, Miami University, Oxford, OH 45056, USA

Jason M. Wever, Department of Cell and Developmental Biology (formerly Department of Cell and Structural Biology), University of Illinois, Urbana, IL 61801, USA

Jonathan Wilson, Department of Biology and Center for Regenerative Biology and Medicine, Indiana University-Purdue University Indianapolis, Indianapolis, IN 46202, USA

CHAPTER

1

Anatomical and Functional Diversity of Animal Eyes

Elke K. Buschbeck
Department of Biological Sciences, University of Cincinnati, Cincinnati, OH 45221-0006, USA

OUTLINE

Animal eyes come in many different shapes and sizes, comprising a great example of the tension between adaptive fitness and physical constraints. Since vision is the preeminent sense in human perceptual experience, it is easy for us to relate to the importance of eyes in animals, the diversity of which can be viewed as a collection of physical filters for electromagnetic waves. Each type of filter allows its bearer to extract particularly relevant information from its surroundings. Studying eyes in one animal alone can therefore never shed full light on the functional design of eyes in general. Rather, investigating a variety of animals is necessary to take advantage of the laboratory of nature, and to analyze diversely crafted specific solutions.

Looking at different eyes, it is tempting to be judgmental, and to label different designs as "better" or "worse". However, it is important to understand that the elaboration of an eye does not necessarily lead to an adaptive advantage for that specific animal. Evolution can only improve an eye if it is developmentally possible, and if its bearer gains overall fitness benefits from doing so. Therefore any advantage from better eye performance must exceed any costs arising from the new design. These include costs from underlying brain structures as much as from the eye itself; a sophisticated eye can only evolve in conjunction with sophisticated visual processing centers.

There indeed is support for the fact that the presence of eyes is costly. For example, there is a substantial energetic cost that simply arises from maintaining photoreceptor cells. In *Drosophila* it has been estimated that the ATP used to maintain illuminated photoreceptor cells accounts for 8% of the total energy consumed at rest (Laughlin *et al.*, 1998). Note that these calculations do not account for any costs arising from the constant renewal of the photoactive membrane, or from the neural activity needed to process the visual information (Laughlin, 2001). Other evidence for the cost of eyes derives from the fact that they tend to be reduced (Tan *et al.*, 2005) or lost (Fong *et al.*, 1995) relatively rapidly if they are not needed. For example many cave dwelling animals tend to be blind (see Chapter 8 in this volume). Some light has been shed on the loss of eyes in the fish *Astyanax mexicanus* through comparison of cave populations with surface populations. Being part of the same species, these fish interbreed and the populations are thought to have diverged within the last 10,000 years. Surface and cave forms start developing eyes early on; however, the cave dwelling form looses its eye before adulthood. Interestingly, it has been recently shown (Yamamoto *et al.*, 2004) that this loss is correlated with an expansion in expression of the midline genes *sonic hedgehog* and *tiggy-winkle hedgehog*, suggesting that it could be due to an active repression of eye development rather than to loss of function mutants in eye genes.

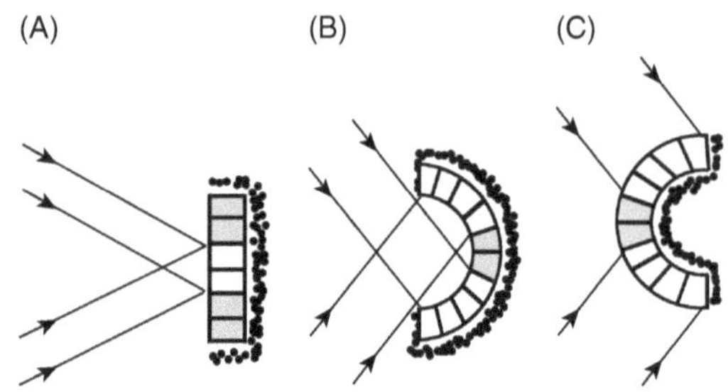

FIGURE 1.1 Spatial resolution can be achieved by folding a flat epithelium (A) either in (B) or out (C). In each case the curvature limits the directions from which light can be absorbed. Modified after Oyster (1999) and (Land, 1981).

PHYSICAL LIMITS OF EYE DESIGNS

While it may be difficult to judge if a particular eye design is "better" or "worse" for its bearer, it is relatively easy to judge its quality against the limits of physics (Land, 1981; Land and Nilsson, 2002). As photographers know well, specific camera designs allow for maximizing properties such as spatial resolution, photon capture under dim light conditions, or the capture of specific wavelengths. Since improvement of one property often degrades other properties, the definition of "good" needs to take into consideration what needs to be achieved. A camera can be designed to be particularly good at capturing one specific aspect of the visual environment, or it could balance multiple optical properties. The same is true for animal eyes.

Some of the simplest visual organs are limited to the detection of major changes in light levels, failing to resolve spatial information which arguably is the most important visual quality of eyes (Land and Nilsson, 2002). A flat piece of photosensitive epithelium that is shielded on one side by pigment will only respond to light from the non-shielded direction, however all cells of the epithelium have the same large visual field of nearly 180° (Fig. 1.1A). Spatial resolution hence is not possible, and the benefit of such an eye spot is primarily to detect dramatic changes in light levels such as could result from the shadow of a predator, or from altering the eye's orientation. In order to achieve spatial resolution, the epithelium needs to acquire curvature that limits the field of view of individual receptors. In principle there are two ways to do so, namely to form the epithelium into a concave (Fig. 1.1B) or a convex surface (Fig. 1.1C). The former is thought to have occurred during the evolution of single chamber camera type eyes (Nilsson and Pelger, 1994), whereas the latter is the basis for spatial resolution in compound eyes. The extent of spatial resolution depends on several key factors such as the degree of curvature, if and what optical support structures (such as a lens or iris) are present, the layout of the photoreceptors, and the absolute size of the eye. Ultimately the level of spatial resolution that can be achieved is frequently limited by diffraction, and indeed it is not uncommon for animal eyes that function in bright daylight to operate near that limit (Land and Nilsson, 2002). However, not all eyes operate close to the diffraction limit, and in fact it can be a disadvantage to do so if the eye needs to function at low light levels. One way to look at this is that an eye of a given size can either split up the captured light between many receptor cells (high spatial resolution), or it can use the light to activate fewer receptor cells more strongly (high sensitivity). Considering that a minimum of light needs to be captured in order to reliably be detected, it is not surprising that eyes of nocturnal animals tend to be optimized for sensitivity rather than spatial resolution (Land and Nilsson, 2002).

Although the limits of spatial resolution and light sensitivity are major physical constraints of eyes, there are several other limits that influence vision. Just to mention a few, the dynamic range of light levels over which an eye can operate, and the need to differentiate contrast irrespective of the light level have to be considered. Eyes furthermore may be optimized to detect color or polarization information, possessing different receptor classes that are tuned to specific wavelengths or polarization planes, respectively (the latter is primarily found among invertebrates). A detailed examination of such limits is beyond the scope of this introduction, but see Land and Nilsson (2002) for a comprehensive account.

THE EVOLUTIONARY ORIGIN OF EYES

In the previous section we discussed constraints in regards to the physical design momentarily ignoring the fact that each eye must have originated through evolution, and that a specific optical design might not be possible for a given bearer. Specifically, once eye evolution has started to follow a certain path, developmental constraints may favor staying on that path. This is because evolution only can act on the existing genes that contribute to the developmental machine, and unless pre-existing machinery can be recruited simultaneously, eye evolution only can proceed through small incremental improvements. Considering that our planet is inhabited by thousands of millions of animal species, it is not too surprising that there are numerous animals that have evolved eyes that are in some ways optimized near the boundaries of physics, despite costs. Indeed, comparisons suggest that a variety of complex optical solutions have evolved multiple times.

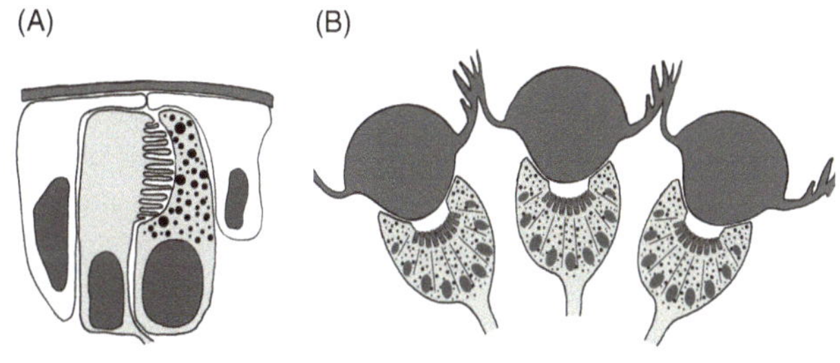

FIGURE 1.2 (A) Example of a simple visual organ consisting of one photoreceptive as well as one pigment cell as is found in polychaete worms. Modified after (Rhode, 1992). (B) The strepsipteran eye has evolved a series of camera eyes from a compound eye ancestor, resulting in an overall convex surface with concave receptor arrays.

It is largely accepted that proper vision originated in the early Cambrian about half a billion years ago. During the important evolutionary event that is known as the Cambrian explosion, nearly all phyla that gave rise to modern species came into existence within just five to forty million years. Most of our knowledge about this time period is derived from the Canadian Burgess shale fossils (Briggs *et al.*, 1994), that are extraordinarily well preserved in regards to soft body parts. Other fossilized faunas were discovered in China, Greenland, and Russia. During the Cambrian explosion, animals started to become mobile, requiring increased information about their surroundings. Consequently the evolution of sensory organs was favored and many of the early Cambrian species likely carried eyes, some of them presumably quite large (Land and Nilsson, 2002). Among the most common animals were arthropods, including trilobites, suggesting that compound eyes evolved particularly early. Other invertebrates possessed uncomplicated eye designs made up of simple visual organs. These were mainly composed of a photoreceptor cell and a pigment cell, and were possibly similar to those that still can be found in many basal groups such as certain annelids (Fig. 1.2A). The evolution of single chamber eyes in chordates took place slightly later.

One of the greatest debates in the eye field is how many times eyes have evolved independently during the history of life on earth, and if there is a single common ancestor for all eye types. Much of this question depends on how an eye is defined, or which component of an eye is under investigation. For example, if one accepts a simple visual organ consisting of one photoreceptive cell with adjacent pigment cell as an eye (Gehring and Ikeo, 1999), then this organ potentially could have been the common ancestor of the major lines of eyes. This has been suggested by Arendt and Wittbrodt (2001) who performed an extensive comparative analysis of the most basal bilateran animal lines. On the other hand, if an eye is defined as complex image-forming visual organ (Nilsson, 1996; Land and Nilsson, 2002) then one has to conclude that eyes evolved many times. Although in all major lines of eyes phototransduction is based on rhodopsins, a phylogenetic analysis of the rhodopsin gene suggests that major subtypes evolved prior to the Cambrian explosion (Arendt and Wittbrodt, 2001; Nilsson, 2005). In addition, there are profound differences in eye organization between major lines. Those include the presence of rhabdomeric receptor cells in arthropods and ciliary receptor cells in vertebrates, as well as the presence of different G-proteins that mediate the transduction mechanism. This latter aspect results in a division between hyperpolarizing current in response to light as is typical for vertebrates, and a depolarizing current that is found in almost all invertebrates (but see Gorman and McReynolds, 1969; Arendt *et al.*, 2004). Major differences between key groups exist in regards to the formation of the lens (Fernald, 2006). In addition to the independent evolution of the complex single chamber and compound eyes, there is evidence that each of those types evolved several times. Perhaps the most amazing example here is the convergence between the eyes of certain cephalopods and those of fish, which arrived at strikingly similar physical configurations. However, when investigated further, it becomes clear that while the former evolved from a combination of neural and epidermal epithelium, the latter is exclusively formed by the epidermis (Nilsson, 1996). Consequently the retina of the former has inverted ciliary receptors, and the latter is characterized by upright microvillar receptors. Similarly, a close comparison of the compound eyes of polychaete worms and those of insect or crustacean compound eyes suggest independent evolution of similarly organized compound eyes.

There has been relative consensus about eye evolution prior to the early 1990s, when the identification of *Pax6* as a master control gene in the development of different eye types (Quiring *et al.*, 1994) renewed the debate. Perhaps the most striking challenge to the multiple eye origin hypothesis was the discovery that the mouse *Pax6* gene, a homolog of the *Drosophila* gene *eyeless*, can initiate the development of a compound eye in the fruit fly when misexpressed. This is the case not only at the usual place for an eye, but also at unconventional positions such as the leg, wing, or antennae (Halder *et al.*, 1995). In addition to *Pax6*, other major homologous genes have been identified in the developmental pathways of various animal eyes. While these findings are striking, their interpretation is confounded by the fact that *Pax* genes are not exclusive to eye development, and are also found in animals the ancestors of which likely never possessed eyes such as certain echinoderms (Arendt and Wittbrodt, 2001). The same is true for other important genes in eye development. It is therefore likely, as Nilsson (1996) pointed out, that the same old genes and developmental pathways are being independently recruited into "new" eyes (see also Fernald, 2006). Evidence for the independent evolution of the two major photoreceptor types also derives from the fact that in many organisms there is growing evidence for the presence of receptor elements of both types.

DIVERSITY OF EYE TYPES

One key point in understanding eye evolution is that it is possible to evolve eyes through a sequence of small steps, each of which represents an improvement in regards to vision. This has been demonstrated for the evolution of spatial resolution in a camera eye design (Nilsson and Pelger, 1994). Using the initial conditions of a flat piece of epithelium, covered with a transparent cell layer on top, and a layer of opaque pigment cells at the bottom, Nilsson and Pelger (1994) showed that spatial resolution first can gradually be improved through an increasingly concave surface, and then through the evolution of a lens with rising refractive power. The key condition for this evolutionary sequence to occur is that each of the steps results in a fitness benefit, such as might be expected by the improvement of spatial vision. A similar argument for gradual improvement could be made for the compound eye. Here too spatial resolution first can be improved by increasing curvature, and then through the addition of refractive or reflective components. Alternatively, compound eyes could evolve through duplications of existing convex visual organs. Regardless, this scenario implies that once an eye starts being convex, it is more likely to gain further resolution by becoming more convex, or by evolving arrays of lenses or mirrors, than by turning into a single chamber camera-type eye. Interestingly, there is nevertheless evidence for several incidences of evolution of single chamber lens eyes from an ancestor with compound eyes. Those include examples from the three major arthropod lines, such as the eyes of spiders (Paulus, 1979), certain larvae of holometabolous insects (Paulus, 1986; Liu and Friedrich, 2004; Sbita *et al.*, 2007), and mysid shrimp (Nilsson and Modlin, 1994). Among insects, scale insects (Duelli, 1978) and twisted wing insects (Buschbeck *et al.*, 1999; Maksimovic *et al.*, 2007) are noteworthy. Except for mysid shrimp, common to all these examples is that they possess eyes with extended retinas that are covered by corneal lenses, together resolving relatively large visual fields. In Strepsiptera, for example, spatial resolution is achieved by an overall convex surface of the compounded eye, with units that are formed by individual concave retinae (Fig. 1.2B). For all these examples special circumstances must have facilitated the transition between these fundamentally different eye types.

Given the number of different, specialized visual devices that animals have come up with, it is expected that different questions can be best addressed within different animals. While rhodopsins are present even in bacteria (Chapter 2), fundamental mechanisms of relatively simple concave eyes can be studied in several of the most basal lines of bilateria. The flatworm *Planaria* (Chapter 3) for example makes a particularly interesting model system since it can fully regenerate major parts of its body including its eyes, which regenerate by expressing a *Pax6* ortholog. The best studied compound eye is undoubtedly *Drosophila* (Chapter 4) which offers a battery of genetic tools, and an accessibility of individually identified neurons; a combination that is available in no other organism. Within the vertebrates there are major differences and specific strength, even though individual model systems follow the same evolutionary line of single chamber eyes. Those range from the accessibility of early eye development in *Xenopus* (Chapter 6) and the power of regeneration in newts, chicks, and mice (Chapters 7, 8, and 9) to unique adaptations to life in caves and water (Chapter 5). Much vision research revolves around mice (Chapters 11 and 12) which offer the availability of unique genetic tools in combination with relative proximity to our own eyes. The latter quality renders vision research in primates particularly important

(Chapter 14). Thus, this book contains a series of chapters that address the uniqueness of different animal models in eye research.

ACKNOWLEDGMENTS

I thank Drs. Ilya Vilinsky, John E. Layne, and Dan-Eric Nilsson for helpful discussions and comments on this chapter, and the National Science Foundation for funding our research.

REFERENCES

Arendt D, Tessmar-Raible K, Snyman H, Dorresteijn AW, Wittbrodt J (2004). Ciliary photoreceptors with a vertebrate-type opsin in an invertebrate brain. *Science* 306:869–871.

Arendt D, Wittbrodt J (2001). Reconstructing the eyes of Urbilateria. *Philos T R Soc B* 356:1545–1563.

Briggs DEG, Erwin DH, Collier F (1994). *The Fossils of the Burgess Shale*. Smithsonian.

Buschbeck E, Ehmer B, Hoy R (1999). Chunk versus point sampling: visual imaging in a small insect. *Science* 286:1178–1180.

Duelli P (1978). An insect retina without microvilli in the male scale insect, *Eriococcus sp* (Eriococcidae, Homoptera). *Cell Tissue Res* 187:417–427.

Fernald RD (2006). Casting a genetic light on the evolution of eyes. *Science* 313:1914–1918.

Fong DW, Kane TC, Culver DC (1995). Vestigialization and loss of nonfunctional characters. *Annu Rev Ecol Syst* 26:249–268.

Gehring WJ, Ikeo K (1999). Pax 6: mastering eye morphogenesis and eye evolution. *Trends Genet* 15:371–377.

Gorman ALF, McReynolds JS (1969). Hyperpolarizing and depolarizing receptor potentials in scallops eye. *Science* 165:309.

Halder G, Callaerts P, Gehring WJ (1995). Induction of ectopic eyes by targeted expression of the eyeless gene in *Drosophila. Science* 267:1788–1792.

Land MF (1981). Optics and vision in invertebrates. In: Autrum H (ed.), *Comparative Physiology and Evolution of Vision in Invertebrates, Handbook of Sensory Physiology*. Springer, Berlin, pp. 471–592.

Land MF, Nilsson DE (2002). *Animal Eyes*. Oxford University Press, Oxford. pp. 1–221

Laughlin SB (2001). Energy as a constraint on the coding and processing of sensory information. *Curr Opin Neurobiol* 11:475–480.

Laughlin SB, van Steveninck RRD, Anderson JC (1998). The metabolic cost of neural information. *Nat Neurosci* 1:36–41.

Liu ZY, Friedrich M (2004). The *Tribolium* homologue of glass and the evolution of insect larval eyes. *Dev Biol* 269:36–54.

Maksimovic S, Layne JE, Buschbeck EK (2007). Behavioral evidence for within-eyelet resolution in twisted-winged insects (Strepsiptera). *J Exp Biol* 210:2819–2828.

Nilsson DE (1996). Eye ancestry: old genes for new eyes. *Curr Biol* 6:39–42.

Nilsson DE (2005). Photoreceptor evolution: ancient siblings serve different tasks. *Curr Biol* 15:R94–R96.

Nilsson DE, Modlin RF (1994). A mysid shrimp carrying a pair of binoculars. *J Exp Biol* 189:213–236.

Nilsson DE, Pelger S (1994). A pessimistic estimate of the time required for an eye to evolve. *Proc R Soc B* 256:53–58.

Oyster CW (1999). *The Human Eye: Structure and Function*. Sinauer Associates, Inc, Sunderland, MA.

Paulus HF (1979). Eye structure and the monophyly of the Arthropoda. In: Gupta AP (ed.), *Arthropod Phylogeny*. Van Nostrad Reinhold, New York, pp. 299–384.

Paulus HF (1986). Evolutionswege zum Larvalauge der Insekten – Ein Modell für die Entstehung und Ableitung von ozellären Lateralaugen der Myriapoden von Facettenaugen. *Zool Jb Syst* 113:353–371.

Quiring R, Walldorf U, Kloter U, Gehring WJ (1994). Homology of the eyeless gene of *Drosophila* to the small eye gene in mice and aniridia in humans. *Science* 265:785–789.

Rhode B (1992). Development and differentiation of the eye in *Platynereis dumerilii* (Annelida, Polychaeta). *J Morphol* 212:71–85.

Sbita SJ, Morgan RC, Buschbeck EK (2007). Eye and optic lobe metamorphosis in the sunburst diving beetle, *Thermonectus marmoratus* (Coleoptera: Dytiscidae). *Arthropod Struct Dev* 37:449–462.

Tan SJ, Amos W, Laughlin SB (2005). Captivity selects for smaller eyes. *Curr Biol* 15:R540–R542.

Yamamoto Y, Stock DW, Jeffery WR (2004). Hedgehog signalling controls eye degeneration in blind cavefish. *Nature* 431:844–847.

CHAPTER

2

The Simplest Eyes: Rhodopsin-mediated Phototaxis Reception in Microorganisms

John L. Spudich, Elena N. Spudich

Center for Membrane Biology, Department of Biochemistry and Molecular Biology, University of Texas Medical School, Houston, TX 77030, USA

OUTLINE

INTRODUCTION

If we accept a broad definition of vision as the capability of an organism to sense spatial patterns of light in its environment and to use this information to modulate its behavior, then unicellular microorganisms contain the simplest of visual systems. Phototaxis, which entails altered motility of an organism in response to changes in light intensity, color, or direction, is a widespread capability in the microbial world, since sunlight is as important a component of microscopic niches as it is in our macroscopic world. Phototaxis requires light sensors which are able to convert photon energy into a chemical signal, transmission of this signal to a motility apparatus, and a motility apparatus response appropriate for positioning the cell in a preferred region of illumination.

A priori, the phototaxis machinery from photoreception to response might have been related to our visual system only in the very broadest of terms. However, 25 years ago when the first light-sensing receptor was discovered in a microorganism, namely a phototaxis receptor in the haloarchaeal prokaryote *Halobacterium salinarum* (Spudich and Spudich, 1982; Bogomolni and Spudich, 1982), it turned out to be a membrane-embedded protein using photoisomerization of a retinylidene chromophore as the photoreception mechanism, like visual pigments in our retina. Furthermore, this first photoreceptor, now called sensory rhodopsin I, in later studies (reviewed in Hoff *et al.*, 1997) was found to consist of seven transmembrane helices forming a pocket in which retinal is attached in a protonated Schiff base (PSB) linkage to a mid-membrane lysyl residue in the seventh helix, like in mammalian visual pigments. Twenty years later, phototaxis receptors in a phylogenetically distant organism, the unicellular eukaryote *Chlamydomonas reinhardtii* which uses a signal transmission and motility system very different from that of prokaryotic cells, were found to have as part of their structure seven-helix retinylidene protein components homologous to SRI (Sineshchekov *et al.*, 2002). Thus, the photoreceptors themselves are the components of microbial phototaxis systems having the most direct bearing on animal visual systems, and are the focus of this brief review.

MICROBIAL RHODOPSINS, A LARGE FAMILY WITH DIVERSE PHOTOTRANSDUCING FUNCTIONS

The first 35 years of research on microbial rhodopsins concerned exclusively four proteins in the cytoplasmic membranes of halophilic Archaea. Prior to genome sequencing projects starting in 1999, four haloarchaeal types sharing the *H. salinarum* cytoplasmic membrane were the only microbial retinylidene proteins known, namely: the light-driven proton transporter bacteriorhodopsin [BR (Oesterhelt and Stoeckenius, 1973)], the light-driven chloride transporter halorhodopsin [HR (Schobert and Lanyi, 1982)], and the phototaxis receptors sensory rhodopsin I [SRI (Bogomolni and Spudich, 1982)], and sensory rhodopsin II [SRII (Takahashi *et al.*, 1985)]. Studies of the haloarchaeal rhodopsins by the most incisive biophysical and biochemical tools available produced a wealth of information making them one of the best understood membrane-embedded protein families in terms of their structure–function relationships. Crystal structures of three [BR (Grigorieff *et al.*, 1996; Essen *et al.*, 1998; Luecke *et al.*, 1999), HR (Kolbe *et al.*, 2000), and SRII (Fig. 2.1) plus one of a later found eubacterial homologs Anabaena sensory rhodopsin (Vogeley *et al.*, 2004)] revealed seven-transmembrane α-helical structures with nearly identical helix positions in the membrane, despite their differing functions and identity in only ~25% of their residues.

The helix positions (Fig. 2.1) differ from those of visual pigments, as shown by the crystal structure of bovine rod rhodopsin (Palczewski *et al.*, 2000), but their overall topologies are similar, namely the seven helices form an interior binding pocket in the hydrophobic core of the membrane for the retinal chromophore. In both the microbial and visual pigments, the retinal is attached by a PSB linkage and

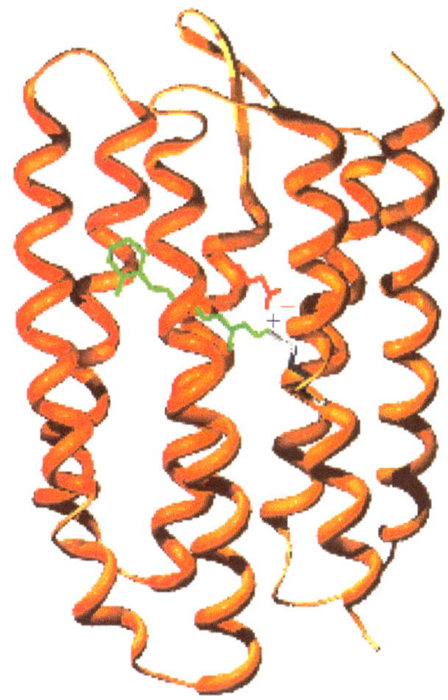

FIGURE 2.1 Structure of sensory rhodopsin II (SRII) from the haloarchaeon *Natronomonas pharaonis* (redrawn from Spudich, 2006). Crystal structures of two microbial sensory rhodopsins, SRII (Kunji *et al.*, 2001; Luecke *et al.*, 2001; Royant *et al.*, 2001) and *Anabaena* sensory rhodopsin (ASR; Vogeley *et al.*, 2004) show seven transmembrane α-helical structures with helix positions in the membrane closely similar to those of the transport rhodopsins, bacteriorhodopsin, and halorhodopsin (see text), despite their differing functions and identity in only ~25% of their residues. The helix positions differ from those of the visual pigment bovine rod rhodopsin (Palczewski *et al.*, 2000), but rhodopsins from prokaryotes to humans share common structural and mechanistic features: (i) their seven transmembrane α-helices form a binding pocket for a retinal molecule; (ii) the retinal is attached by a PSB linkage to a lysyl ε-amino group in the middle of the seventh helix; and (iii) retinal photoisomerization initiates their photochemical reactions, which (iv) include, in general, transfer of the Schiff base proton from the chromophore to a carboxylate on the third helix of the protein in both the microbial (type 1 rhodopsins) and mammalian visual pigments (type 2 rhodopsins).

chromophore photoisomerization initiates their photochemical reactions. In microbial rhodopsins retinal photoisomerization is from all-*trans* to 13-*cis*, the two thermodynamically most stable isomers of retinal, whereas in visual pigments it is 11-*cis* to all-*trans*. A notable similarity between nearly all microbial rhodopsins and mammalian rhodopsins is that the PSB on the seventh helix forms a salt bridge with a carboxylate anion (aspartate or glutamate) on the third helix. As discussed below, the disruption of this salt bridge by light-induced proton transfer from the PSB to the carboxylate counterion is functionally important in both microbial and mammalian rhodopsin activation.

Starting in 1999, genome sequencing of cultivated microorganisms began to reveal the previously unsuspected presence of archaeal rhodopsin homologs in several organisms in the other two domains of life, namely Bacteria and Eukarya (Bieszke *et al.*, 1999; Sineshchekov *et al.*, 2002; Jung *et al.*, 2003). Further in 2000, "environmental genomics" of populations of uncultivated microorganisms in ocean plankton showed the presence of a homolog in marine proteobacteria [hence given the name proteorhodopsin (Béjà *et al.*, 2001)], which has swiftly expanded so far to >4500 relatives identified throughout the world's oceans by a number of laboratories (as summarized by Frigaard *et al.*, 2006) with the greatest number found by high-throughput shotgun sequencing (Venter *et al.*, 2004; Rusch *et al.*, 2007). Nearly all of the marine proteorhodopsins have sequence features of proton transporters, although a small number appear to be sensory rhodopsins with transducer protein partners (Spudich, 2006).

MODES OF SIGNALING BY THE VERSATILE MICROBIAL SENSORY RHODOPSINS

The conservation of residues in microbial rhodopsins, especially in the retinal-binding pocket, define a large phylogenetic class called type 1 rhodopsins to distinguish them from the visual pigments and related retinylidene proteins in higher organisms (type 2 rhodopsins) (Spudich *et al.*, 2000). The sequences of the newly found type 1 rhodopsins, their heterologous expression and study, and in some cases study of the photosensory physiology of the organisms containing them, have shown that the newfound pigments fulfill both ion transport and sensory functions, the latter with a variety of signal-transduction mechanisms (Fig. 2.2).

Microbial rhodopsins functioning as light-driven proton pumps are widespread in prokaryotic and eukaryotic species. Phylogenetic analysis strongly suggests that microbial rhodopsin photosensors evolved from the proton pumps and that this evolutionary event occurred multiple times in different lineages independently (Sharma *et al.*, 2006). The reasoning is that the pumps, which have a single protein function without need for interaction with other proteins, readily undergo lateral gene transfer, followed by duplication and modification to develop a functional interaction with signal transduction machinery of the new host. Such parallel evolution of sensory rhodopsins fits the experimental observation that their signaling mechanisms are strikingly different in different branches of the phylogenetic tree. The three best studied are the haloarchaeal phototaxis receptors, the cyanobacterial Anabaena sensory rhodopsin, and the phototaxis receptors in *C. reinhardtii* (Fig. 2.2).

Signaling to a Membrane-embedded Transducer in Haloarchaeal Prokaryotic Phototaxis

The haloarchaeal sensory rhodopsins SRI and SRII are photoactive subunits of molecular 2:2 complexes containing homodimers of their cognate transducers, HtrI and HtrII respectively (Hoff *et al.*, 1997; Klare *et al.*, 2004). The Htr transducer subunits are homologous to prokaryotic chemotaxis receptors, consisting of a homodimer in which each monomer contains two transmembrane segments and a mostly α-helical rod-like extension into the cytoplasm (Falke and Hazelbauer, 2001; Parkinson *et al.*, 2005; Baker *et al.*, 2006). Like the chemotaxis transducers, Htr proteins bind a CheW–CheA pair at their distal end responsible for phosphorylation of the cytoplasmic CheY regulator protein that controls flagellar motor switching behavior (Rudolph *et al.*, 1995), and the Htr cytoplasmic domain contains adaptive methylation sites (Spudich *et al.*, 1989; Perazzona and Spudich, 1999). Photoactivation of the SR subunits modulate the CheA kinase activity, thereby controlling the extent of phosphorylation of CheY. Orange light activation of SRI elicits an attractant response by transiently inhibiting kinase activity. The resulting decrease in phospho–CheY concentration reduces the probability of motor switching (i.e. changes in direction of rotation of the flagella), and therefore the cells continue to swim in the direction of increasing orange light. Blue light activation of SRII has the opposite effect, transiently activating CheA. The increased phosphorylation of CheY increases the probability of a reversal in swimming direction, and therefore the cells swimming path is biased toward lower intensities of blue light. A novel aspect is that SRI also mediates strong repellent responses when it is activated by two sequential photons, i.e. orange followed by near-UV light (Spudich and Bogomolni, 1984).

The adaptive value of this color-sensitive two-receptor signaling system seems clear. The repellent receptor SRII is produced by cells when their respiration activity is high and they seek the dark thereby minimizing the danger of photooxidative damage. In relatively anoxic conditions, the electrogenic proton and chloride pumps BR and HR are induced in cells

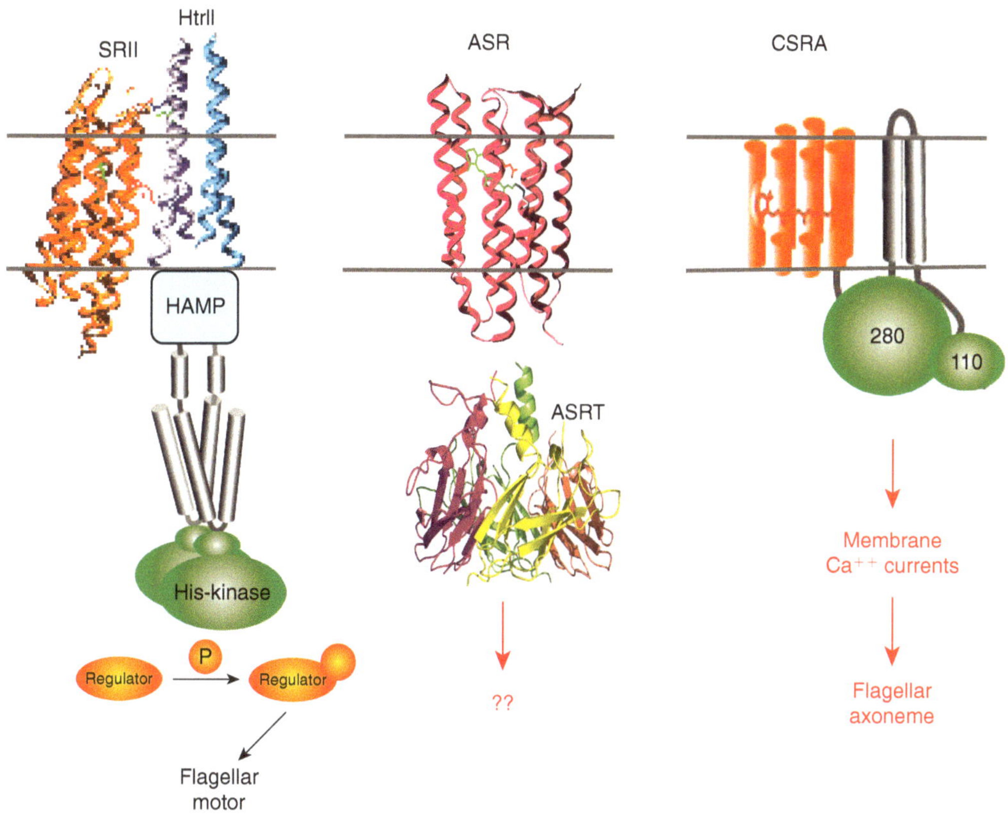

FIGURE 2.2 Three different modes of signaling by microbial sensory rhodopsins. Shown from the left are the SRII–HtrII receptor-transducer phototaxis signaling complex from *Natronomonas pharaonis,* the sensory rhodopsin from the cyanobacterium *Anabaena* sp. PCC7120 with interacting protein ASRT, and the phototaxis receptor CSRA (also called channelrhodopsin-1) from *C. reinhardtii.* Structures shown are from X-ray crystallography (see text). Domains of CSRA are based on secondary structure predictions and numbers in the cytoplasmic domains indicates numbers of amino acid residues.

enabling proton motive force to be generated with light while maintaining pH homeostasis. SRI is also produced in such low oxygen conditions and its color-sensitive attractant and repellent signals enable the cells to seek orange light effective for the pumps while avoiding regions containing high intensities of near-UV light (Hoff *et al.*, 1997).

Notable progress on the SRII–HtrII signaling complex in particular creates an opportunity to resolve a first example of signal relay between membrane proteins at the atomic level. Molecular events in the photoactive site of the receptor have been elucidated by the combined efforts of several groups merging insights from X-ray crystallography and optical and molecular spectroscopy coupled with the strong genetics and behavioral physiology of *H. salinarum* (reviewed in Hoff *et al.*, 1997; Kamo *et al.*, 2001; Klare *et al.*, 2004; Spudich, 2006). These efforts have identified specific chemical transitions and hydrogen-bond alterations in SRII responsible for its activation, and these chemical events have been demonstrated to be both necessary and sufficient for signaling by engineering the responsible structures into the proton pump bacteriorhodopsin (Sudo and Spudich, 2006).

In SRI and in SRII, two key chemical reactions in the photoactive site contribute to formation of the receptor signaling state: (i) a "steric trigger", i.e. a steric interaction between the photoisomerizing retinal and the protein, both in SRI (Yan *et al.*, 1991) and SRII (Sudo and Spudich, 2006; Sudo *et al.*, 2006), and (ii) proton transfer from the PSB to the protein (Yan and Spudich, 1991; Spudich *et al.*, 1997). Analogous chromophore/protein steric interactions (Koch and Gärtner, 1997; Shieh *et al.*, 1997) and PSB to counterion proton transfer (Bennett *et al.*, 1982; Longstaff *et al.*, 1986; Arnis and Hofmann, 1993) are also key contributors to the activation of mammalian rod rhodopsin (reviewed in Hofmann, 1999; Abduleav and Ridge, 2005; Palczewski, 2006).

How the photoactivated SRII relays the signal to HtrII is a current focus in the field. The SRII and HtrII subunits exhibit extensive contacts starting from the periplasmic membrane surface, continuing through the membrane-embedded domain, and into the cytoplasmic membrane-proximal domain (Spudich, 2006). An X-ray crystal structure of the membrane-embedded domain of the complex shows that the interface is comprised of tight van der Waals interaction within the membrane and two hydrogen-bonded regions between the two HtrII transmembrane helices and SRII helices F and G (Gordeliy *et al.*, 2002). One hydrogen-bonded region is at the periplasm-membrane border and the second is buried deep in the hydrophobic interior. An atomic structure of the hydrophilic membrane-proximal domain of the transducer is not available; however, fluorescent probe accessibility and Förster resonance energy transfer measurements (Yang *et al.*, 2004), EPR of spin-labels (Bordignon *et al.*, 2005), and *in vitro* binding of HtrII peptides to SRII (Hippler-Mreyen *et al.*, 2003; Sudo *et al.*, 2005) demonstrate interaction of the HtrII membrane-proximal domain with the cytoplasmic helix F and the E-F loop of the receptor.

Disulfide cross-linking (Yang and Spudich, 2001), site-directed spin labeling and EPR (Wegener *et al.*, 2001), time-resolved FTIR (Bergo *et al.*, 2005), fluorescent probes (Yang *et al.*, 2004; Taniguchi *et al.*, 2007), and illumination of X-ray diffracting crystals (Moukhametzianov *et al.*, 2006) show that light-induced structural changes occur throughout the long SRII–HtrII interface. Specifically, disulfide cross-linking shows light-induced changes in the transmembrane region, and FTIR light–dark difference spectra of the complex in proteoliposomes show both hydrogen-bonded regions undergo major structural changes consistent with light-induced disruption of the hydrogen bonds present in the dark (Bergo *et al.*, 2005). EPR spectra of proteoliposomes reveal a rotatory motion of the cytoplasmic end of the HtrII second transmembrane helix (TM2) in the membrane (Wegener *et al.*, 2001); a small (0.9 Å) displacement and rotation of TM2 has also been reported in illuminated crystals of the complex (Moukhametzianov *et al.*, 2006). In addition fluorescent probes show that light-induced structural changes also occur in the cytoplasmic membrane proximal region of the complex (Yang *et al.*, 2004; Taniguchi *et al.*, 2007).

The structural changes in all regions of the interface raise the challenge of distinguishing which of the structural changes is (are) responsible for signal relay from those that are non-essential consequences of the activation of the complex. Some authors emphasize the membrane domain changes and others the changes in the cytoplasmic region of the complex, leading to two prevailing models for SRII–HtrII signal relay: (i) the "steric trigger-transmembrane coupling model" which proposes that retinal isomerization directly signals HtrII through a steric conflict with the Tyr124–Thr204 hydrogen bonded pair through adjacent residues in contact with the mid-membrane SRII–HtrII interface, and (ii) the "linker-switch" model which proposes signal relay by light-induced alteration of the demonstrated receptor E-F loop/transducer HAMP domain (a molecular switch) contact sites in the cytoplasmic membrane-proximal domain.

Recent results strongly favor transmembrane coupling rather than membrane-proximal domain coupling (Sasaki *et al.*, 2007). The study localized the transmission of signals from SRII receptors constitutively activated by mutagenic disruption of the PSB-counterion salt bridge to homologous HtrII proteins from different species that exhibit different sensitivities to the constitutively active receptor signal. Chimeric transducers localized the transducer sensitivity differences to the transmembrane segments; i.e. swapping transducer membrane domains, but not cytoplasmic domains altered the sensitivity of the transducers to the constitutive signal. The study concludes that the transmembrane domains (TM) of the transducers detect the conformational change of SRII induced by counterion mutations. The implication is that also light-induced conformational changes of the receptor are transmitted to the transducer through the TM domain of the transducer. This stronger conclusion assumes that the constitutive signals caused by mutation are transmitted to the transducer through the same interactions that photostimulus-induced signals are transmitted. Such an assumption is very reasonable, and often implicitly made in studies of receptor constitutive activity, and in the cited study is supported by a strong reciprocal correlation between extents of constitutive activity and decreased light-induced responses by the mutants. Deletion constructs lacking the known contact region in the cytoplasmic domain of the NpSRII–NpHtrII complex further support this conclusion, since the cytoplasmic interactions, although known to exist and to be altered by light activation, were found to be not necessary for light-induced phototaxis responses (Sasaki *et al.*, 2007).

Signaling to a Cytoplasmic Transducer by a Cyanobacterial Sensory Rhodopsin

A eubacterial sensory rhodopsin (ASR) in the freshwater cyanobacterium *Anabaena* sp. PCC7120, interacts with a protein very different from the haloarchaeal taxis transducers (Fig. 2.2). The receptor is encoded in an operon containing a second gene which encodes a small soluble cytoplasmic protein, recently named ASRT for "ASR transducer" (Vogeley *et al.*, 2007). Physical interaction between the two proteins has been shown by *in vitro* binding measurements (Jung *et al.*, 2003). More recently isothermal titration calorimetry with purified ASRT and ASR in detergent measured moderate affinity binding (K_d 8μM) with an ASRT:ASR stoichiometry of 4:1, and ASRT crystals show a tight tetramer (Vogeley *et al.*, 2007). Thus, the binding data fit quantitative binding of the tetramer to a single ASR molecule.

Atomic resolution structures of both proteins are available from X-ray crystallography (Vogeley *et al.*, 2004; Vogeley *et al.*, 2007). The overall membrane-embedded seven-helical structure of ASR is very similar to those of the haloarchaeal rhodopsins; however, it differs significantly in its cytoplasmic-side structure and in its chromophore configuration. First, its interior on the cytoplasmic side, highly hydrophobic in the archaeal rhodopsins, contains numerous hydrophilic residues networked by water molecules, providing a hydrophilic connection from the photoactive site to the cytoplasmic surface, the expected region of ASRT binding. Second, it exhibits both a stable all-*trans* and a stable 13-*cis* retinal isomeric form.

The physiological function of the ASR–ASRT pair is not known, but a second unusual property of ASR suggests its function involves color discrimination. Initially suggested by the crystal structure and confirmed by chromophore extraction and spectroscopic analysis, ASR exists in a mix of stable all-*trans* and 13-*cis* isomeric configurations (Vogeley *et al.*, 2004), with different absorption maxima (550nm and 537nm, respectively). The pigment is photochromic, i.e. each of the forms exhibits efficient light-induced conversion to the other (Sineshchekov *et al.*, 2005a,b; Kawanabe *et al.*, 2007). Therefore the ratio of the *cis*- and *trans*-chromophore forms depends on the wavelength of illumination, providing a potential mechanism for single-pigment color sensing. Its two distinct ground-state species thermally interconvert with halftimes of ~100min and ~300min for the *trans* and *cis* forms, respectively. Such relatively long-lasting color sensitivity is similar to that of the red/far-red photochromic states of phytochrome and may be used, in the *Anabaena* cell in analogy to the plant photosensor phytochrome, to control expression of different proteins (e.g. phycobilisome pigments) required under orange-light versus blue-light illumination in the photosynthesis system.

Light-gated Channel Activity in Chlamydomonas Phototaxis

The Chlamydomonas photomotility receptors are the only identified eukaryotic microbial rhodopsins whose function in the cell has been established (Sineshchekov *et al.*, 2002). *Chlamydomonas reinhardtii* and other algal species exhibit two types of motility responses to light: phototaxis, in algae defined as the oriented swimming of cells along the direction of a light beam, and the photophobic response, a reorientation of swimming direction induced by an abrupt increase in light intensity. Both photobehaviors result from the generation of Ca^{2+} currents in the plasma membrane that have been well characterized electrophysiologically (Sineshchekov and Govorunova, 1999). Two type 1 rhodopsins identified by genomic analysis were demonstrated to mediate the currents responsible for phototaxis orientation (Sineshchekov *et al.*, 2002) as well as the photophobic response (Govorunova *et al.*, 2004) by electrophysiological measurements of transformants in which their cellular concentrations were selectively reduced by RNA*i*. The proteins were therefore named Chlamydomonas sensory rhodopsins A and B (CSRA and CSRB). They each consist of a rhodopsin domain which is part of a larger protein (712 and 737 residues, respectively) (CSRA in Fig. 2.2). The electrophysiological measurements revealed several differences between the two receptors. CSRA and CSRB mediate two kinetically different photoreceptor currents, a fast and a slow current, at high and low light intensities, respectively. The absorption maximum deduced from action spectra of CSRA (500–510nm) is red-shifted with respect to that of CSRB (460–470nm). CSRA generates a current with no latency at 3μs resolution suggesting the protein is pre-associated with a Ca^{2+} channel or itself is a light-gated channel. On the other hand a 2-ms latency prior to onset of CSRB-mediated currents suggest biochemical steps between the receptor and channel activity.

Heterologously expressed CSRA and CSRB are both light-gated ion channels. CSRA and CSRB genes were expressed in Xenopus oocytes and found to mediate light-gated channel activity for protons (CSRA) and other cations (CSRB) (Nagel *et al.*, 2002, 2003). On this basis, the authors named the proteins channelrhodopsin-1 and channelrhodopsin-2, which correspond to CSRA and CSRB, respectively. The channel opening is mediated by the seven helices of the rhodopsin domains and does not require the remaining parts of the proteins. The relationship of the channel activities of the heterologously expressed proteins to their Ca^{2+} flux regulation in Chlamydomonas is not clear. Nevertheless, the CSRB rhodopsin domain is finding use in interesting heterologous expression experiments as a tool to induce membrane depolarization and action potentials in *Caenorhabditis elegans*, and avian and mammalian excitable cells (Nagel *et al.*, 2005; Li *et al.*, 2005; Bi *et al.*, 2006).

The general scheme of photoreception and signal transduction based on two rhodopsin-mediated Ca^{2+} currents, tuned to different intensity and spectral ranges, appears to be universal for green algae (Sineshchekov and Spudich, 2005), and similar signaling properties have been identified also in a phylogenetically distant group of cryptophyte algae (Sineshchekov *et al.*, 2005a,b).

Of particular interest to the subject of models for eye research, Chlamydomonas phototaxis reception involves a specialized organelle, the "eyespot" or stigma, which together with the receptor proteins, forms a spatially fixed photoreceptive structure reminiscent of animal eyes. This asymmetrically positioned apparatus is used for phototactic orientation (Foster and Smyth, 1980; Dieckmann, 2003). It consists of the multilayered pigmented eyespot, which serves as an accessory device, and a portion of the plasma membrane containing the receptor proteins, which underlies the eyespot. Illumination of the photoreceptor membrane during the cell's helical swimming path is modulated by the eyespot and the rest of the cell. When the axis of the helical swimming path of the cell deviates from the axis of incident light, the periodic changes in photoreceptor illumination during the rotation cycle give rise to unbalanced responses of the two flagella, which lead to a correction of the swimming path with respect to light direction. When the direction of the cell's movement becomes parallel with that of light, illumination of the photoreceptor becomes constant, and no corrective motor responses occur. A more detailed description of this mechanism can be found in several reviews (Foster and Smyth, 1980; Witman, 1993; Hegemann, 1997; Kreimer, 2001). More recently, proteomic analysis of the eyespot and associated proteins reveal complex signal transduction machinery in the algal "eye" which contains >200 different proteins (Schmidt *et al.*, 2006).

EVOLUTIONARY RELATIONSHIP BETWEEN MICROBIAL RHODOPSINS AND VISUAL PIGMENTS

The detailed similarities of the microbial pigments (type 1 rhodopsins) and animal visual and related pigments (type 2 rhodopsins) have long-driven speculation as to whether both derive from a seven-helix retinylidene pigment in a common ancestor or, alternatively, are stunning examples of convergence at the molecular level. The minimal identity in primary sequence between known type 1 and type 2 rhodopsins has raised the possibility that nature discovered use of retinal as a chromophore twice, and both times found it useful, when solvated with seven helices and linked as a PSB, for photosensory signaling as well as other phototransduction functions. Such a two-progenitor hypothesis would require that microbial sensory rhodopsins and animal visual pigments have converged on remarkably similar mechanisms of receptor photoactivation, but such similarity could result from "likely reinvention" determined by the inherent chemical properties of retinal. On the other hand, a common origin may exist but be obscured in the examples that we know, since they are from evolutionarily very distant organisms, all type 1 rhodopsins so far identified in unicellular microorganisms and all type 2 in multicellular animals. The single-progenitor hypothesis will be tested in future genome projects and may be confirmed if a missing link were to be found, i.e. a gene encoding a retinylidene protein with both type 1 and type 2 sequence identity.

ACKNOWLEDGMENTS

Research findings by the authors were supported primarily by National Institutes of Health grant R37GM27750 and the Robert A. Welch Foundation.

REFERENCES

Abduleav NG, Ridge KD (2005). Structural and functional aspects of the mammalian rod-cell photoreceptor rhodopsin. In: Briggs W, Spudich JL (eds), *Handbook of Photosensory Receptors*. Wiley-VCH, Weinheim, pp. 77–92.

Arnis S, Hofmann KP (1993). Two different forms of metarhodopsin II: schiff base deprotonation precedes proton uptake and signaling state. *P Natl Acad Sci USA* 90:7849–7853.

Baker MD, Wolanin PM, Stock JB (2006). Systems biology of bacterial chemotaxis. *Curr Opin Microbiol* 9:187–192.

Béjà O, Spudich EN, Spudich JL, Leclerc M, DeLong EF (2001). Proteorhodopsin phototrophy in the ocean. *Nature* 411:786–789.

Bennett N, Michel-Villaz M, Kuhn H (1982). Light-induced interaction between rhodopsin and the GTP-binding protein Metarhodopsin II is the major photoproduct involved. *Eur J Biochem* 127:97–103.

Bergo VB, Spudich EN, Rothschild KJ, Spudich JL (2005). Photoactivation perturbs the membrane-embedded contacts between sensory rhodopsin II and its transducer. *J Biol Chem* 280:28365–28369.

Bi A, Cui J, Ma YP, Olshevskaya E, Pu M, Dizhoor AM, Pan ZH (2006). Ectopic expression of a microbial-type rhodopsin restores visual responses in mice with photoreceptor degeneration. *Neuron* 50:23–33.

Bieszke JA, Spudich EN, Scott KL, Borkovich KA, Spudich JL (1999). A eukaryotic protein, NOP-1, binds retinal to form an archaeal

rhodopsin-like photochemically reactive pigment. *Biochemistry* 38:14138–14145.
Bogomolni RA, Spudich JL (1982). Identification of a third rhodopsin-like pigment in phototactic Halobacterium halobium. *P Natl Acad Sci USA* 79:6250–6254.
Bordignon E, Klare JP, Doebber M, Wegener AA, Martell S, Engelhard M, Steinhoff HJ (2005). Structural analysis of a HAMP domain: the linker region of the phototransducer in complex with sensory rhodopsin II. *J Biol Chem* 280:38767–38775.
Dieckmann CL (2003). Eyespot placement and assembly in the green alga Chlamydomonas. *BioEssays* 25:410–416.
Essen L, Siegert R, Lehmann WD, Oesterhelt D (1998). Lipid patches in membrane protein oligomers: crystal structure of the bacteriorhodopsin–lipid complex. *P Natl Acad Sci USA* 95:11673–11678.
Falke JJ, Hazelbauer GL (2001). Transmembrane signaling in bacterial chemoreceptors. *Trends Biochem Sci* 26:257–265.
Foster KW, Smyth RD (1980). Light antennas in phototactic algae. *Microbiol Rev* 44:572–630.
Frigaard NU, Martinez A, Mincer TJ, DeLong EF (2006). Proteorhodopsin lateral gene transfer between marine planktonic bacteria and archaea. *Nature* 439:847–850.
Gordeliy VI, Labahn J, Moukhametzianov R, Efremov R, Granzin J, Schlesinger R, Buldt G, Savopol T, Scheidig AJ, Klare JP, Engelhard M (2002). Molecular basis of transmembrane signalling by sensory rhodopsin II-transducer complex. *Nature* 419:484–487.
Govorunova EG, Jung KH, Sineshchekov OA, Spudich JL (2004). Chlamydomonas sensory rhodopsins A and B: cellular content and role in photophobic responses. *Biophys J* 86:2342–2349.
Grigorieff N, Ceska TA, Downing KH, Baldwin JM, Henderson R (1996). Electron-crystallographic refinement of the structure of bacteriorhodopsin. *J Mol Biol* 259:393–421.
Hegemann P (1997). Vision in microalgae. *Planta* 203:265–274.
Hippler-Mreyen S, Klare JP, Wegener AA, Seidel R, Herrmann C, Schmies G, Nagel G, Bamberg E, Engelhard M (2003). Probing the sensory rhodopsin II binding domain of its cognate transducer by calorimetry and electrophysiology. *J Mol Biol* 330:1203–1213.
Hoff WD, Jung K-H, Spudich JL (1997). Molecular mechanism of photosignaling by archaeal sensory rhodopsins. *Annu Rev Biophys Biomol Struct* 26:223–258.
Hofmann KP (1999). Signalling states of photoactivated rhodopsin. *Novart Found Symp* 224:158–175. discussion 175–180.
Jung KH, Trivedi VD, Spudich JL (2003). Demonstration of a sensory rhodopsin in eubacteria. *Mol Microbiol* 47:1513–1522.
Kamo N, Shimono K, Iwamoto M, Sudo Y (2001). Photochemistry and photoinduced proton-transfer by pharaonis phoborhodopsin. *Biochemistry* 66:1277–1282.
Kawanabe A, Furutani Y, Jung KH, Kandori H (2007). Photochromism of Anabaena sensory rhodopsin. *J Am Chem Soc* 129:8644–8649.
Klare JP, Gordeliy VI, Labahn J, Buldt G, Steinhoff HJ, Engelhard M (2004). The archaeal sensory rhodopsin II/transducer complex: a model for transmembrane signal transfer. *FEBS Lett* 564:219–224.
Koch D, Gärtner W (1997). Steric hindrance between chromophore substituents as the driving force of rhodopsin photoisomerization: 10-methyl-13-demethyl retinal containing rhodopsin. *Photochem Photobiol* 65:181–186.
Kolbe M, Besir H, Essen LO, Oesterhelt D (2000). Structure of the light-driven chloride pump halorhodopsin at 1.8 Å resolution. *Science* 288:1390–1396.
Kreimer G (2001). Light perception and signal modulation during photoorientation of flagellate green algae. In: Häder DP , Jori G (eds), *Comprehensive Series in Photosciences, Photomovement*, Vol. 1. Elsevier, Amsterdam, pp. 193–227.
Kunji ER, Spudich EN, Grisshammer R, Henderson R, Spudich JL (2001). Electron crystallographic analysis of two-dimensional crystals of sensory rhodopsin II: a 6.9 Å projection structure. *J Mol Biol* 308:279–293.
Li X, Gutierrez DV, Hanson MG, Han J, Mark MD, Chiel H, Hegemann P, Landmesser LT, Herlitze S (2005). Fast noninvasive activation and inhibition of neural and network activity by vertebrate rhodopsin and green algae channelrhodopsin. *P Natl Acad Sci USA* 102:17816–17821.
Longstaff C, Calhoon RD, Rando RR (1986). Deprotonation of the Schiff base of rhodopsin is obligate in the activation of the G protein. *P Natl Acad Sci USA* 83:4209–4213.
Luecke H, Schobert B, Lanyi JK, Spudich EN, Spudich JL (2001). Crystal structure of sensory rhodopsin II at 2.4 angstroms: insights into color tuning and transducer interaction. *Science* 293:1499–1503.
Luecke H, Schobert B, Richter HT, Cartailler JP, Lanyi JK (1999). Structure of bacteriorhodopsin at 1.55 Å resolution. *J Mol Biol* 291:899–911.
Moukhametzianov R, Klare JP, Efremov R, Baeken C, Goppner A, Labahn J, Engelhard M, Buldt G, Gordeliy VI (2006). Development of the signal in sensory rhodopsin and its transfer to the cognate transducer. *Nature* 440:115–119.
Nagel G, Brauner M, Liewald JF, Adeishvili N, Bamberg E, Gottschalk A (2005). Light activation of channelrhodopsin-2 in excitable cells of Caenorhabditis elegans triggers rapid behavioral responses. *Curr Biol* 15:2279–2284.
Nagel G, Ollig D, Fuhrmann M, Kateriya S, Musti AM, Bamberg E, Hegemann P (2002). Channelrhodopsin-1: a light-gated proton channel in green algae. *Science* 296:2395–2398.
Nagel G, Szellas T, Huhn W, Kateriya S, Adeishvili N, Berthold P, Ollig D, Hegemann P, Bamberg E (2003). Channelrhodopsin-2, a directly light-gated cation-selective membrane channel. *P Natl Acad Sci USA* 100:13940–13945.
Oesterhelt D, Stoeckenius W (1973). Functions of a new photoreceptor membrane. *P Natl Acad Sci USA* 70:2853–2857.
Palczewski K (2006). G protein-coupled receptor rhodopsin. *Annu Rev Biochem* 75:743–767.
Palczewski K, Kumasaka T, Hori T, Behnke CA, Motoshima H, Fox BA, Le Trong I, Teller DC, Okada T, Stenkamp RE, Yamamoto M, Miyano M (2000). Crystal structure of rhodopsin: a G protein-coupled receptor. *Science* 289:739–745.
Parkinson JS, Ames P, Studdert CA (2005). Collaborative signaling by bacterial chemoreceptors. *Curr Opin Microbiol* 8:116–121.
Perazzona B, Spudich JL (1999). Identification of methylation sites and effects of phototaxis stimuli on transducer methylation in Halobacterium salinarum. *J Bacteriol* 181:5676–5683.
Royant A, Nollert P, Edman K, Neutze R, Landau EM, Pebay-Peyroula E, Navarro J (2001). X-ray structure of sensory rhodopsin II at 2.1-Å resolution. *P Natl Acad Sci USA* 98:10131–10136.
Rudolph J, Tolliday N, Schmitt C, Schuster SC, Oesterhelt D (1995). Phosphorylation in halobacterial signal transduction. *EMBO J* 14:4249–4257.
Rusch DB, Halpern AL, Sutton G, Heidelberg KB, Williamson S, Yooseph S, Wu D, Eisen JA, Hoffman JM, Remington K et al. (2007). The sorcerer II global ocean sampling expedition: northwest Atlantic through eastern tropical Pacific. *PLoS Biol* 5:e77.
Sasaki J, Nara T, Spudich EN, Spudich JL (2007). Constitutive activity in chimeras and deletions localize sensory rhodopsin II/HtrII signal relay to the membrane-inserted domain. *Mol Microbiol* 66:1321–1330.
Schmidt M, Gessner G, Luff M, Heiland I, Wagner V, Kaminski M, Geimer S, Eitzinger N, Reissenweber T, Voytsekh O et al. (2006). Proteomic analysis of the eyespot of Chlamydomonas reinhardtii

provides novel insights into its components and tactic movements. *Plant Cell* 18:1908–1930.

Schobert B, Lanyi JK (1982). Halorhodopsin is a light-driven chloride pump. *J Biol Chem* 257:10306–10313.

Sharma AK, Spudich JL, Doolittle WF (2006). Microbial rhodopsins: functional versatility and genetic mobility. *Trends Microbiol* 14:463–469.

Shieh T, Han M, Sakmar TP, Smith SO (1997). The steric trigger in rhodopsin activation. *J Mol Biol* 269:373–384.

Sineshchekov OA, Govorunova EG (1999). Rhodopsin-mediated photosensing in green flagellated algae. *Trends Plant Sci* 4:201.

Sineshchekov OA, Govorunova EG, Jung KH, Zauner S, Maier UG, Spudich JL (2005). Rhodopsin-mediated photoreception in cryptophyte flagellates. *Biophys J* 89:4310–4319.

Sineshchekov OA, Jung KH, Spudich JL (2002). Two rhodopsins mediate phototaxis to low- and high-intensity light in Chlamydomonas reinhardtii. *P Natl Acad Sci USA* 99:8689–8694.

Sineshchekov OA, Spudich JL (2005a). Sensory rhodopsin signaling in green flagellate algae. In: Briggs W, Spudich JL (eds), *Handbook of Photosensory Receptors*. Wiley-VCH, Weinheim, pp. 25–42.

Sineshchekov OA, Trivedi VD, Sasaki J, Spudich JL (2005b). Photochromicity of Anabaena sensory rhodopsin, an atypical microbial receptor with a cis-retinal light-adapted form. *J Biol Chem* 280:14663–14668.

Spudich EN, Spudich JL (1982). Control of transmembrane ion fluxes to select halorhodopsin-deficient and other energy transduction mutants of Halobacterium halobium. *P Natl Acad Sci USA* 79:4308–4312.

Spudich EN, Takahashi T, Spudich JL (1989). Sensory rhodopsins I and II modulate a methylation/demethylation system in Halobacterium halobium phototaxis. *P Natl Acad Sci USA* 86:7746–7750.

Spudich EN, Zhang W, Alam M, Spudich JL (1997). Constitutive signaling by the phototaxis receptor sensory rhodopsin II from disruption of its protonated Schiff base-Asp73 intrahelical salt bridge. *P Natl Acad Sci USA* 94:4960–4965.

Spudich JL (2006). The multitalented microbial sensory rhodopsins. *Trends Microbiol* 14:480–487.

Spudich JL, Bogomolni RA (1984). Mechanism of colour discrimination by a bacterial sensory rhodopsin. *Nature* 312:509–513.

Spudich JL, Yang CS, Jung KH, Spudich EN (2000). Retinylidene proteins: structures and functions from archaea to humans. *Annu Rev Cell Dev Biol* 16:365–392.

Sudo Y, Furutani Y, Kandori H, Spudich JL (2006). Functional importance of the interhelical hydrogen bond between Thr204 and Tyr174 of sensory rhodopsin II and its alteration during the signaling process. *J Biol Chem* 281:34239–34245.

Sudo Y, Okuda H, Yamabi M, Fukuzaki Y, Mishima M, Kamo N, Kojima C (2005). Linker region of a halobacterial transducer protein interacts directly with its sensor retinal protein. *Biochemistry* 44:6144–6152.

Sudo Y, Spudich JL (2006). Three strategically placed hydrogen-bonding residues convert a proton pump into a sensory receptor. *P Natl Acad Sci USA* 103:16129–16134.

Takahashi T, Tomioka H, Kamo N, Kobatake Y (1985). A photosystem other than PS370 also mediates the negative phototaxis of *Halobacterium halobium*. *FEMS Microbiol Lett* 28:161–164.

Taniguchi Y, Ikehara T, Kamo N, Yamasaki H, Toyoshima Y (2007). Dynamics of light-induced conformational changes of the phoborhodopsin/transducer complex formed in the n-dodecyl beta-D-maltoside micelle. *Biochemistry* 46:5349–5357.

Venter JC, Remington K, Heidelberg JF, Halpern AL, Rusch D, Eisen JA, Wu D, Paulsen I, Nelson KE, Nelson W et al. (2004). Environmental genome shotgun sequencing of the Sargasso Sea. *Science* 304:66–74.

Vogeley L, Sineshchekov OA, Trivedi VD, Sasaki J, Spudich JL, Luecke H (2004). Anabaena sensory rhodopsin: a photochromic color sensor at 2.0 A. *Science* 306:1390–1393.

Vogeley L, Trivedi VD, Sineshchekov OA, Spudich EN, Spudich JL, Luecke H (2007). Crystal structure of the Anabaena sensory rhodopsin transducer. *J Mol Biol* 367:741–751.

Wegener AA, Klare JP, Engelhard M, Steinhoff HJ (2001). Structural insights into the early steps of receptor-transducer signal transfer in archaeal phototaxis. *Embo J* 20:5312–5319.

Witman GB (1993). Chlamydomonas phototaxis. *Trends Cell Biol* 3:403–408.

Yan B, Nakanishi K, Spudich JL (1991). Mechanism of activation of sensory rhodopsin-I: evidence for a steric trigger. *P Natl Acad Sci USA* 88:9412–9416.

Yan B, Spudich JL (1991). Evidence that the repellent receptor form of sensory rhodopsin I is an attractant signaling state. *Photochem Photobiol* 54:1023–1126.

Yang CS, Sineshchekov O, Spudich EN, Spudich JL (2004). The cytoplasmic membrane-proximal domain of the HtrII transducer interacts with the E-F loop of photoactivated Natronomonas pharaonis sensory rhodopsin II. *J Biol Chem* 279:42970–42976.

Yang CS, Spudich JL (2001). Light-induced structural changes occur in the transmembrane helices of the Natronobacterium pharaonis HtrII transducer. *Biochemistry* 40:14207–14214.

CHAPTER

3

The Planarian Eye: A Simple and Plastic System with Great Regenerative Capacity

Emili Saló[1], *Renata Batistoni*[2]

[1]Departament de Genètica, Facultat de Biologia, Universitat de Barcelona, Barcelona, Spain

[2]Dipartimento di Biologia, Università di Pisa, Pisa, Italy

OUTLINE

INTRODUCTION

A number of adult animals display an intrinsic potential to regrow missing body parts. Cellular and molecular studies of these natural models of regeneration are beginning to reveal the secrets of this fascinating phenomenon. Regeneration occurs through a more or less complex mixture of strategies that make use of stem cells or specialized cells that can dedifferentiate to assume a stem cell-like state, and also includes remodeling of existing tissue to restore body symmetry and proportions (Sanchez-Alvarado and Tsonis, 2006). Planarians, free-living members of the phylum Platyhelminthes (Lophotrochozoa), occupy a special place in regeneration research. Planarian worms possess the remarkable capability to regenerate an entire animal from only a small body fragment in a short period of time (Saló, 2006 and references therein). Their regenerative ability is dependent on small, undifferentiated cells present in the parenchyma, the neoblasts. Neoblasts are considered true stem cells from which all planarian cell types can be derived. The presence of pluripotent stem cells that can generate all cell types, including the germ line, is a unique characteristic of the Platyhelminthes phylum. Planarian cells are continuously turning over at different rates, depending on the animal's size, to maintain the form and functionality of the organism. This process, which requires precise control of cell renewal and differentiation rates, leads to an extraordinary plasticity in which the size of the adult animal changes according to the availability of food. When neoblasts are destroyed by X-ray irradiation (Dubois, 1949), planarians do not regenerate and die in a short time. Injecting non-irradiated neoblasts into an irradiated host leads to recovery of the regenerative capacity, whereas injection of differentiated cells does not, suggesting that neoblasts are the regenerative cells and that dedifferentiation processes are not available for planarian regeneration in such specific circumstances (Baguñà *et al.*, 1989). On the other hand, there is evidence for neoblast transdetermination (Gremigni *et al.*, 1980) and, recently, studies of autophagy have suggested the possibility of transdetermination of neoblasts during regeneration and starvation (Gonzàlez-Estèvez *et al.*, 2007a,b; Tettamanti *et al.*, 2008). Although proliferation, differentiation and migration of neoblasts occur as part of a physiological homeostatic mechanism of cell renewal in intact worms, the same processes are activated by regeneration. As a result, a new unpigmented tissue, the blastema, is obtained. Blastemal regeneration represents a fascinating example of pattern formation and a valuable source of information for understanding the cellular and molecular events that direct the morphogenesis of different structures. In the blastema the new tissue reorganizes itself in an anteroposterior morphogenetic sequence (Bayascas *et al.*, 1998), while morphallactic remodeling respecifies the pre-existing tissue (Morgan, 1898; Saló and Baguñà, 2002; Oviedo *et al.*, 2003; Saló, 2006). Unlike classical model systems used to study development, such as *Drosophila, C. elegans*, zebrafish, chicken and mouse, which cannot replicate the phases of their development until the next generation, planarian regeneration is a very plastic process that can be continually repeated during adult life. In fact a blastema develops anew after wounding of an already regenerating piece, due to the presence of neoblasts.

Besides their regenerative capability, planarians display the typical body plan of a bilaterian animal, including a well-defined cephalization of the nervous system (CNS) and sensory organs (Fig. 3.1 A–C). Despite their simple morphology, the neural structures of these organisms display a high level of molecular complexity and well-organized functional activity (Agata *et al.*, 1998; Umesono *et al.*, 1999; Cebrià *et al.*, 2002; Mineta *et al.*, 2003; Nakazawa *et al.*, 2003; Okamoto *et al.*, 2005; Cebrià, 2007; Inoue *et al.*, 2007). Nevertheless, when a planarian is beheaded, a perfect head can be regenerated from a small piece of tissue. The eye represents a simple neural structure that has emerged as a model of particular interest to shed light on the molecular regulation of morphogenesis during blastemal regeneration. In the remainder of this chapter we will focus on planarian eyes and review current knowledge regarding how these structures form during regeneration.

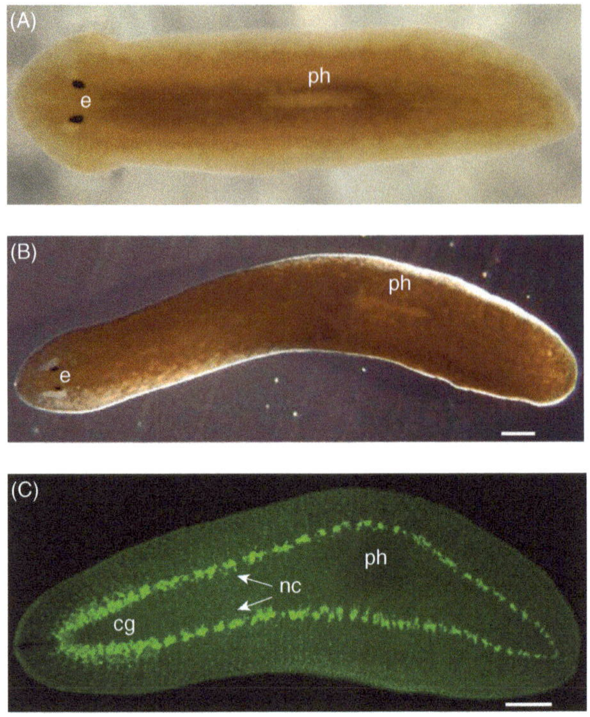

FIGURE 3.1 Dorsal view of two planarians commonly used for molecular studies: (A) *Dugesia japonica*, (B) *Schmidtea mediterranea*. A pair of eyes can be seen close to the anterior body margin, dorsal to the brain. (C) Ventral view of the central nervous system in *S. mediterranea* visualized by whole mount staining with the anti-synapsin antibody 3C11. Perfect bilateral symmetry can be observed, with two cephalic ganglia (cg) connected with two ventral nerve cords (nc). e, eyes; ph, pharynx. Scale bar: 1 mm.

PLANARIAN EYES

Light perception occurs in planarians through specialized visual cells that are clustered together into eyespots. The light-perceptive cells only register the brightness of the environment. The eyes do not form images in planarians but allow these lower invertebrates to determine the direction from which light is coming and to avoid direct exposure by sheltering on the underside of rocks and leaves. Although there is variety in terms of eye number, size and location, most planarian species possess two eyespots on the dorsal side of cephalic ganglia (Agata *et al.*, 1998; Tazaki *et al.*, 1999; Sakai *et al.*, 2000; Saló *et al.*, 2002; Inoue *et al.*, 2004). Photoreceptors are bipolar neurons with a rhabdomeric structure in which the microvillar dendrites are enclosed in a monolayered cup of melanosome-containing pigment cells that allows detection of light direction. Pigment cells also act as phagocytes to remove photoreceptor membranes during the daily turnover of the visual cells (Tamamaki, 1990). Biochemical studies demonstrate that light radiation induces extracellular changes in Ca^{+} and Na^{+} concentrations around the photoreceptors (Azuma, 1998). The microvillar area of the photoreceptors represents the photoreceptive component, where

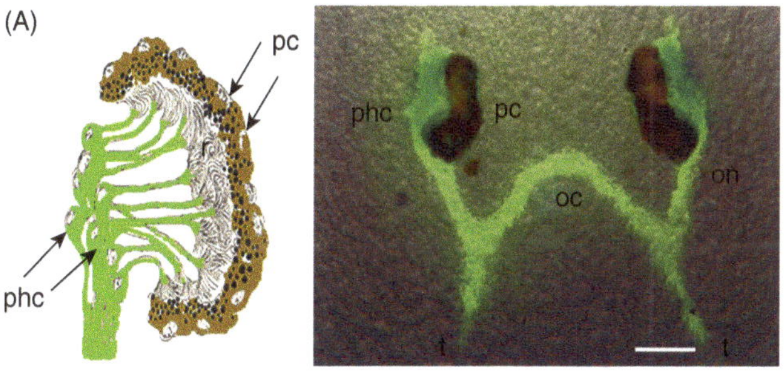

FIGURE 3.2 The planarian eye. (A) Diagram of the planarian eyespot or ocellus showing a section through the eye cup. (B) Planarian photoreceptors stained with the MA-VC1 antibody, showing the optic chiasm and the optic nerve pathway to the cephalic ganglia, located more ventrally. pc, multicellular pigment cup; phc, bipolar photoreceptor cells; r, rhabdomeric dendritic ends; oc, optic chiasm; on, optic nerve; t, optic nerve terminals. Scale bar: 0.1 mm.

opsin accumulates (Orii *et al.*, 1998), triggering the β-arrestin-mediated phototransduction cascade (Mineta *et al.*, 2003; Nakazawa *et al.*, 2003; Inoue *et al.*, 2004). The photoreceptor cell bodies are located just outside the pigmented eyecup and their axons project to the brain (Fig. 3.2A,B). Studies involving immunostaining with an antibody specific for the visual cells (anti-arrestin monoclonal antibody VC-1) and tracing with fluorescent dyes revealed the topographical mapping of visual axons (Sakai *et al.*, 2000; Agata *et al.*, 2003; Okamoto *et al.*, 2005). These studies showed that some axons directly connect to the opposite eye, whereas others project sensory connections to the ipsilateral side or cross to the contralateral side of the dorsal-medial brain, producing an optic chiasm. Connecting the eye with the brain allows photosensory inputs to be integrated and processed, suggesting that complex neuronal circuits regulate planarian behavior. Further evidence that planarian eye cells integrate in a complex neural circuit has been obtained from recent studies of the netrin/DCC and ROBO (Roundabout) systems, which provide attractive and repulsive cues to guide axons, including photoreceptors axons, to their proper targets. In particular, it was observed in the planarian *Schmidtea mediterranea* that functional ablation of a member of DCC family of netrin receptors, Smed-netR, or its potential ligand, Smed-netrin2, led to an abnormal photophobic response (Cebrià and Newmark, 2005). In both cases, ectopic projection of photoreceptor axons was observed as a consequence of RNA interference (RNAi). RNAi-mediated knockdown of the planarian *robo* homolog, *Smed roboA*, also affected neuronal connectivity, including that of visual neurons (Cebrià and Newmark, 2007). A precise correlation has also been observed between the pattern of projections and the anteroposterior functional subdivisions of photoreceptors (Okamoto *et al.*, 2005). However, no morphological or molecular differences in the visual cells have yet been reported. The presence of only two cell types (photoreceptors and pigment cells) identifies the planarian eye as one of the most simple rhabdomeric eye types (Callaerts *et al.*, 1999; Gehring and Ikeo, 1999; Gehring, 2005).

TECHNOLOGICAL ADVANCES IN PLANARIAN STUDIES

Building on classical experimental approaches, the repertoire of the methods available today for use in planarians has allowed considerable progress to be made in the understanding of how planarians regenerate their body parts, including the eyes (Saló *et al.*, 2008). Importantly, emerging technologies have been successfully applied in these organisms. For example, a large number of gene sequences have been identified in expressed sequence tag (EST) projects (more than 10,000 unique ESTs) and by genome sequencing. These genome-scale resources now make it easier to clone candidate genes, and an ever-increasing number of molecular markers are becoming available. Furthermore, high-throughput expression profiling can now be performed by in situ hybridization or microarrays (Sanchez-Alvarado *et al.*, 2002; Mineta *et al.*, 2003; Rossi *et al.*, 2007). The screening criteria have also been changed by the rapid progress of methods for reverse genetics, such as large-scale RNAi-mediated functional screens (Newmark *et al.*, 2003; Reddien *et al.*, 2005) and transgenesis (Gonzàlez-Estèvez *et al.*, 2003). In the absence of naturally occurring planarian mutants, these methods now give us the opportunity to functionally explore genes of interest through analysis of mutant phenotypes that can be monitored by morphological observation or marker-guided screens, as well as by evaluation of behavioral responses. In particular, the use of systematic RNAi-mediated gene knockdown has been instrumental in identifying a variety of genes that, on the basis of phenotypic abnormalities of the eyes, may play key roles in regeneration of these structures (Newmark *et al.*, 2003; Reddien *et al.*, 2005). Further characterization of the genes identified by this approach has the potential to significantly contribute to elucidating the mechanism by which an eye-cell fate is specified in some neoblasts and also how these cells reorganize to form a new functional visual system.

PLANARIAN EYE REGENERATION: A UNIQUE MODEL FOR THE STUDY OF EYE ORGANOGENESIS

Regeneration provides a unique opportunity for tracing the fate of the cells that make up the eyes in planarians. The time at which morphogenesis of two bilaterally symmetric eyespots becomes apparent in the cephalic blastema depends on the temperature at which regeneration proceeds (2–3 days at 23°C or 4–5 days at 17°C). Following their initial appearance in the blastema, the eyespots grow to their normal size by aggregation of newly differentiated cells. Sakai *et al.* (2000) suggested that in these organisms the eyes originate early in the dorsal anterior part of the blastema from a single eye-regeneration field (i.e. a domain of blastemal cells competent to develop visual cells) that soon resolves into two lateral eye primordia. Although the molecular mechanism required for eye field resolution is still unknown, classical transplantation experiments demonstrated that the planarian brain plays a key role in eye regeneration (Lender, 1950, 1951), and it is noteworthy that brain rudiments form within the blastema as two bilateral cell clusters (Umesono *et al.*, 1999; Cebrià *et al.*, 2002; Inoue *et al.*, 2007). Consequently, it is possible that neural induction and/or neural patterning events may be involved in regenerating two physically separate eyes. Consistent with this possibility, it has recently been found that after RNAi-mediated functional ablation of the planarian *Slit* homolog *Smed-slit*, planarians regenerate a cyclopic eye at the midline rather than forming a bilateral pair of eyes. The Slit proteins play a conserved role as major determinants of axonal pathway formation inside the CNS, secreting repulsive signals from the midline. This role, which is mediated by different Robo receptors, determines the distance of axonal projections from the midline (Englund *et al.*, 2002; Dickson and Gilestro, 2006). Smed-slit-mediated signaling prevents axon crossing at the midline (Cebrià *et al.*, 2007). However, neither of the two *robo*-related genes identified in *S. mediterranea* appears to encode the receptors for *Smed-slit*, and additional *robo* genes probably exist in planarians (Cebrià *et al.*, 2007). The progression of eye regeneration in planarians also involves other signaling pathways, including the Wnt (*DjWntA*) (Kobayashi *et al.*, 2007), fibroblast growth factor (FGF-R) (*Djnou darake*) (Cebrià *et al.*, 2002) and Dpp/transforming growth factor-β (TGF-β) (Molina *et al.*, 2007; Orii and Watanabe, 2007; Reddien *et al.*, 2007) and wnt-beta-catenin (Adell *et al.*, 2008; Gurley *et al.*, 2008; Petersen and Reddien, 2008; Iglesias *et al.*, 2008) pathways. When the function of *DjWntA* or *Djnou darake* is interfered with RNAi, an anteroposterior expansion of the brain is produced, whereas inhibition of the canonical wnt pathway modifies the antero-posterior axis (Guo *et al.*, 2008; Petersen and Reddien, 2008) the stronger phenotype produces radial-like hypercephalized planarians with circular brain and ectopic eyes all around the circular border (Iglesias *et al.*, 2008). Finally inhibition of *DjBMP*-related pathway causes dorsoventral duplication of the brain; in the absence of these signals, ectopic eyes are also generated (Cebrià *et al.*, 2002; Kobayashi *et al.*, 2007; Molina *et al.*, 2007; Orii and Watanabe, 2007; Reddien *et al.*, 2007). These data clearly support the possibility that dynamic integration of different signaling pathways is involved in regulating the intrinsic activity of the transcription factors involved in eye-cell specification. A better understanding of the molecular mechanisms that orchestrate these signaling pathways will be crucial to elucidating the spatio-temporal control of eye regeneration.

By using a combination of molecular markers, RNAi and a phototaxis assay system, Inoue *et al.* (2004) subdivided the process of planarian eye regeneration into three steps. The first step occurs early during cephalic regeneration (2 days at 23°C) with the formation of two bilaterally symmetric visual cell clusters in the dorsal blastema. The first molecular evidence of eye-cell specification in planarians coincides with the regeneration of a regionally defined brain primordium, visualized by enhanced expression of planarian *Otx/Otd*-related genes. Members of the *Otx/Otd* gene family encode homeobox transcription factors with a conserved role in early specification of brain and eye cells of different animal groups (Vandendries *et al.*, 1996; Furukawa *et al.*, 1997; Takayato *et al.*, 2003; Viczian *et al.*, 2003; Plouhinec *et al.*, 2005), strongly supporting the notion that a genetic ground plan for brain and eye development was already present in a common urbilaterian ancestor (reviewed in Lichtneckert and Reichert, 2005). Planarians possess three *Otx/Otd*-related genes expressed in distinct brain regions (Umesono *et al.*, 1997, 1999; Stornaiuolo *et al.*, 1998). Transcripts of one of these genes, *DjotxA*, are present in the eye cells of *Dugesia japonica* and also appear restricted to a medial region of the cephalic ganglia, considered the putative visual center to which the photoreceptor axons project (Umesono *et al.*, 1999). Although the strong activation of these genes observed during early stages of cephalic regeneration is consistent with their involvement in brain patterning, it is still not clear what role is played by *DjotxA* in the eyes, for example, in the specification of pigment eye cells, as occurs in vertebrate eyes (Martinez-Morales *et al.*, 2004), because *DjotxA* RNAi

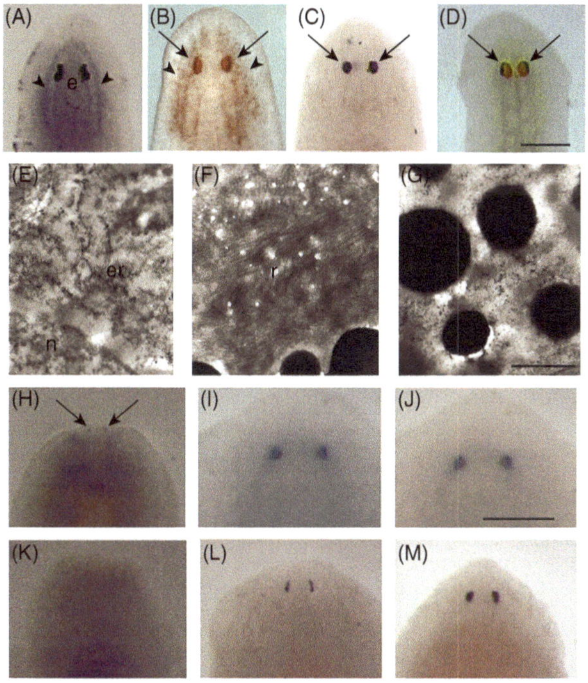

FIGURE 3.3 Expression pattern of the eye genetic network genes *Pax6*, *eya*, *six-1* and *opsin* by whole mount in situ hybridization in intact and regenerating planarians. (A) *Pax6* expression in the adult planarian cephalic ganglia (arrowheads). (B) *Eya* expression in the adult planarian cephalic ganglia (arrowheads) and eyes (arrows). (C) *six-1* expression in the adult planarian eyes (arrows). (D) *opsin* expression in the adult planarian eyes (arrows). (E)–(G) Transmission electron microscopy in situ hybridization of *Pax6*; (E)–(F) photoreceptor cell; (E) gold particles in the endoplasmic reticulum; (F) no staining is observed in the rhabdomeric region; (G) eye pigment cell with gold particles in the cytoplasmic region between the pigment granules. (H)–(J) *six-1* expression at 3, 6 and 9 days of head regeneration, respectively; (H) two spots of *six-1* expression in the two eye field precursor cells (arrows); (I) during early differentiation of the eyes, differentiated pigment cells containing brown pigment can be observed close to the positive blue signal of *six-1* expression; (J) differentiated eyes maintain *six-1* expression. (K)—(M) *opsin* expression during the same regenerative stages as (H)–(J); (K) before eye differentiation no *opsin* expression can be observed; (L) *opsin* expression starts during early eye differentiation; (M) *opsin* expression is maintained at late differentiation. e, eyes; n, nucleus; r, rhabdomeres; er, endoplasmic reticulum. Scale bar: 1 mm, except (E)–(G) (0.5 μm).

experiments did not produce any relevant defects in these organisms.

The process of photoreceptor differentiation coincides with the initiation of *opsin* expression. A gene coding for a rhabdomeric opsin type (r-opsin) has been isolated in three planarian species *(Girardia tigrina, Gtops; S. mediterranea, Smedops;* and *D. japonica, Djops)* and represents a useful molecular marker for differentiated photoreceptors, as it is expressed only in these cells (Fig. 3.3D; K-M). Interestingly, transient expression of a gene related to the *nanos* gene family (*Smednos*)

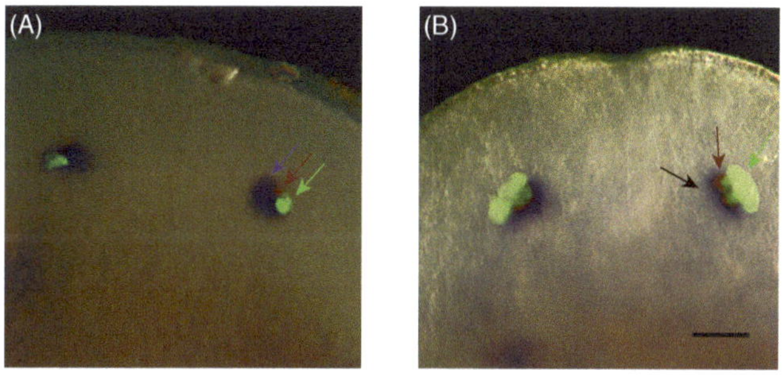

FIGURE 3.4 Expression of *Smednos* during late stages of eye regeneration visualized by a combination of whole-mount in situ hybridization (in blue) and immunostaining of photoreceptor cells with the monoclonal antibody VC-1 (in green). The images are dorsal views with anterior to the top. (A) At 7 days of regeneration, a small overlap can be observed between the *Smednos*-positive cells (blue arrow) and the photoreceptor cells (green arrow). The brown pigment cells are now visible (brown arrow). (B) An almost fully regenerated eye at 10 days of regeneration. The three described cell types can be observed in a close but not overlapping pattern. The pigment cells (brown arrow) are located in between the photosensitive cells (green arrow) and the *Smednos*-positive presumptive eye precursor cells (black arrow). Scale bar: 100 μm.

has recently been described at this stage of eye formation in *S. mediterranea* by a combination of staining with the eye-specific antibody VC-1 (Sakai *et al.*, 2000) and *Smednos* in situ hybridization (Handberg-Thorsager and Saló, 2007). At later stages, when the eyes are completely differentiated, *Smednos* expression declines to undetectable levels. At 10 days of regeneration, three areas can be distinguished in the eye region: the photosensitive cells labeled with VC-1, the pigment cells, easily identified by their brown pigment granules, and finally, the *Smednos*-positive cells, situated next to the pigment cells (Fig. 3.4). *Smednos* transcripts can also be observed during planarian development. The final development and maturation of the eyes takes place at stage 8 of embryogenesis in planarians and, upon hatching, the juvenile possesses a pair of completely differentiated eyes (Cardona *et al.*, 2005). At this stage, *Smednos* expression was found in differentiating eye cells and in cells adjacent to the eye that could correspond to eye precursor cells. *Smednos* transcripts could no longer be detected in newly hatched planarians (Handberg-Thorsager and Saló, 2007). It has been suggested that the *Smednos* gene product may play a role in the process of proliferation and maintenance of these precursors to prevent their early differentiation as eye cells during regeneration (Handberg-Thorsager and Saló, 2007).

Differentiated photoreceptors may transduce physical properties of light, but planarians cannot process them until neural connections are re-established. As regeneration proceeds, visual axons project between the two eye centers. Visual system regeneration is then completed by the formation of axonal projections from the photoreceptors to the cephalic ganglia. The planarian *clathrin heavy chain* (*DjCHC*) gene is required for CNS regeneration, where it is essential for neurite outgrowth and maintenance and for neuronal cell survival. Projection of left and right visual neurons to the opposite sides and the formation of connections between the optic nerves and the brain are also inhibited by silencing of *DjCHC* (Inoue *et al.*, 2007).

Recent studies have shown that innexin-based gap junctions are essential for the development of normal synaptic connections in the *Drosophila* visual system (Curtin *et al.*, 2002a,b). Innexins represent fundamental components of gap junctions in invertebrates (Phelan *et al.*, 1998). Characterization of several members of the *innexin* gene family in *D. japonica* and *S. mediterranea* has provided evidence that regeneration and maintenance of planarian body pattern are mediated by innexin-based gap-junctional communication (Nogi and Levin, 2005; Oviedo and Levin, 2007). Characterization of the expression of different *innexin* mRNAs in adult and regenerating planarians by in situ hybridization clearly shows that *Djinx4* is expressed in regenerating photoreceptors and in the medial visual region of the brain, the area to which the photoreceptors project (Umesono *et al.*, 1999). *Djinx4* expression was found to appear only 4 days after cutting. These results were interpreted as suggesting that, as in *Drosophila*, Djinx4-mediated gap-junctional communication plays a role in transmitting visual information from adult photoreceptors rather than in eye-cell specification (Nogi and Levin, 2005).

Interestingly, it has been found that functional recovery of the characteristic negative phototaxis behavior in planarians does not correspond to the conclusion of this process, but depends on the expression of other late-expressed genes – i.e. genes upregulated within the cephalic blastema only after 4–5 days of regeneration – that are probably involved in the functional reorganization of the neuronal circuitry through the re-establishment of connections between the photoreceptors and the appropriate targets in the brain (Cebrià *et al.*, 2002). Using behavioral screening for the loss of negative phototaxis, two late-expressed genes named *eye53* and *1020HH*, which encode novel secreted proteins, have been identified as essential for the functional recovery of negative phototaxis in regenerated animals (Inoue *et al.*, 2004). Moreover, using a similar approach, a synaptosome-associated protein 25-related gene, *Djsnap-25*, was also found to modulate functional brain recovery in planarians (Takano *et al.*, 2007).

Although the studies reviewed above provide evidence of the remarkable progress in the identification and functional characterization of a large number of genes critical for regeneration of the visual system in planarians (Table 3.1), we are only just beginning to

TABLE 3.1 Summary of genes that affect regeneration of the visual system in planarians

Planarian gene	Vertebrate homolog Expression/function	*Drosophila* homolog Expression/function	Planarian expression	Eye-related planarian RNAi phenotypes	References in planarians
Djnou darake	FGF-receptor-like 1 (FGFRL1)	FGF-receptor	Head region	Ectopic brain and eye tissues	(Cebrià *et al.*, 2002)
DjWntA	Wnt-A	Wnt-A	Posterior brain	Ectopic eyes	(Kobayashi *et al.*, 2007)
DjBMP *SmedBMP*	BMP2-4 Ventral determination	BMP2-4 Dorsal determination	Dorsal midline	Dorsoventral eye duplication	(Orii *et al.*, 1998) (Orii and Watanabe, 2007) (Molina *et al.*, 2007) (Reddien *et al.*, 2007)
Smed-βcatenin 1	βcatenin	βcatenin	Ubiquitous	Ectopic eyes	Iglesias et al., 2008)
Smed-netR *Smed-netrin2*	netrin/DCC Axon guidance	netrin/frazzled Axon guidance	Central nervous system	Defects in visual axon targeting and abnormal photophobic behaviour	(Cebrià and Newmark, 2005)
Smed-slit	Slit Axon guidance	Slit Axon guidance	Dorsal and ventral midline	Cyclopic eye	(Cebrià *et al.*, 2007)
Smed roboA	roundabout Axon guidance	robo Axon guidance	Central nervous system	Aberrant visual projections	(Cebrià and Newmark, 2007)
Djβ-arrestin	phototransduction	Visual cells	Visual cells		(Nakazawa *et al.*, 2003)
Djeye53 *Dj1020HH*			Brain region where the optic nerves project	Impaired negative phototaxis	(Inoue *et al.*, 2004)

(Continued)

TABLE 3.1 *Continued*

Planarian gene	Vertebrate homolog	*Drosophila* homolog	Planarian expression	Eye-related planarian RNAi phenotypes	References in planarians
DjCHC	Clathrin heavy chain Membrane trafficking	Clathrin heavy chain Membrane trafficking	Ubiquitous, but higher expression in CNS	Perturbation of CNS and eye regeneration	(Inoue *et al.*, 2007)
Djsnap-25	Synaptosome-associated protein (Snap-25) Regulate exocytosis	Synaptosome-associated protein (Snap-25) Regulate exocytosis	Nervous system	Impaired negative phototaxis	(Takano *et al.*, 2007)
DjotxA	Crx Neural retina	Otd Photoreceptor cells	Photoreceptor cells		(Umesono *et al.*, 1999)
Dj/Gt/Smed Pax-6A *Dj/Gt/Smed Pax-6B*	Pax6 Lens placode, optic vesicle	eyeless twin of eyeless Eye imaginal discs	Low expression in photoreceptor and pigmented eye cells	No phenotype	(Callaerts *et al.*, 1999) (Rossi *et al.*, 2001) (Pineda *et al.*, 2001) (Pineda *et al.*, 2002) Work in progress
Dj/Gt/Smedsix-1	Six 1 -2 No eye cell expression	sine oculis Eye disc epithelium, photoreceptor cells and optic lobes	Eye precursor cells, photoreceptor cells and pigmented cells	No eyes	(Pineda *et al.*, 2000) (Pineda *et al.*, 2001) (Mannini *et al.*, 2004) Work in progress
Dj/Smed eya	Eye absent 1 to 4	Eyes absent Eye imaginal discs	Eye precursor cells, photoreceptor cells and pigmented cells	No eyes	(Mannini *et al.*, 2004) Work in progress
Smeddac	dach	Dachshund Eye imaginal discs	Work in progress	Work in progress	Work in progress
Gt/Smed/Dj ops	Opsin Photoreceptor cells	Opsin Photoreceptor cells	Photoreceptor cells	Loss of negative phototaxis	(Sanchez Alvarado and Newmark, 1999) (Pineda *et al.*, 2000) (Pineda *et al.*, 2001) (Saló *et al.*, 2002)
Smednos	Nanos Germ line precursor cells	Nanos Germ line precursor cells	Germ line, neoblasts and eye precursor cells	No phenotype	(Handberg-Thorsager and Saló, 2007)

understand their function and interactions. After blastema formation, early neural patterning events and different signaling systems appear to play key roles in the formation of the eyes.

EYE CELL SPECIFICATION IN PLANARIANS: IDENTIFYING PLANARIAN MEMBERS OF THE RETINAL DETERMINATION GENE NETWORK

Despite great eye morphological differences, the increasing amount of data from invertebrates and vertebrates have provided evidence that early morphogenesis of animal eyes requires the regulatory activity of a gene network known to as retinal determination gene network (RDGN) in *Drosophila* (Treisman, 1999; Silver and Rebay, 2005). This network includes genes belonging to the *Pax*, *Six*, *Eyes absent* (*Eya*) and *Dachshund* (*Dach*) families. In particular, the almost universal use of *Pax6*, the most evolutionarily conserved member of the *Pax* gene family among metazoans (Callaerts *et al.*, 1997), has been considered as evidence for a monophyletic origin of all eye types (Gehring and Ikeo, 1999; Gehring, 2005). However, in addition to homologous genes such as Pax6, the RDGN of vertebrates also contains non-orthologous genes from the same family. For example, vertebrate RDGN includes the *sine oculis*-related gene *Six3*, but not *Six1/2* gene characterized in invertebrates. The most parsimonious interpretation is that some members of different gene families formed an ancestral gene regulatory network that later diversified in the different RDGNs during evolution, co-opting closely related transcription factors for the construction of different eye types (Niwa *et al.*, 2004; Davidson and

Erwin, 2006). As a first step toward a mechanistic understanding of the involvement of the RDGN in the early specification of a primitive eye type, such as the planarian eye (Saló *et al.*, 2002), we analyzed the role of the Pax6 homolog in planarians. The first *Pax6*-related gene characterized in these organisms was *GtPax6B*, isolated in *G. tigrina* (Callaerts *et al.*, 1999). A second *Pax6* gene, *GtPax6A*, was later identified in the same species (Pineda *et al.*, 2001, 2002). Two genes, *DjPax6A* and *DjPax6B*, were also found in the phylogenetically distant planarian species *D. japonica* (Rossi *et al.*, 2001; Pineda *et al.*, 2002). These two genes are very similar to those found in *G. tigrina*, even outside the DNA-binding domains, suggesting that *Pax-6* gene duplication events occurred in the Platyhelminthes lineage. Moreover, comparative sequence analysis demonstrated that *Pax6* duplication in planarians occurred independently of that originating the *Pax-6*-related genes *eyeless (ey)* (Quiring *et al.*, 1994) and *twin of eyeless (toy)* (Czerny *et al.*, 1999) in *Drosophila* and other holometabolous insects (Pineda *et al.*, 2002). While expression of the two planarian genes *Pax6A* and *Pax6B* was clearly detected by whole mount in situ hybridization in the central nervous system of regenerating and intact animals (Fig. 3.3A) (Pineda *et al.*, 2002), the presence of specific transcripts could be visualized only by electron microscopy in situ hybridization in adult and regenerating eyes (Fig. 3.3E–G) (Callaerts *et al.*, 1999; Pineda *et al.*, 2002). We also generated transgenic planarians by injection and subsequent electroporation of transposon-derived vectors, using an artificial P3 opsin promoter, repeated three times in tandem, driving EGFP expression in photoreceptor cells under the control of Pax6. Our results with the 3xP3-EGFP marker confirmed the presence of Pax6 or Pax6-related activity in the planarian photoreceptors (Fig. 3.5C) (Pineda *et al.*, 2002; Gonzàlez-Estèvez *et al.*, 2003). To clarify whether *Pax6* expression is functional in the eye, we used RNAi to generate *Pax6A/Pax6B* double loss-of-function planarians. As was the case for the specimens injected with *Pax6A*- or *Pax6B* dsRNA, no morphological abnormalities were observed during eye regeneration or eye maintenance after *Pax6A/Pax6B* RNAi (Pineda *et al.*, 2002). These data support the possibility that more molecular pathways can be utilized to regenerate and/or maintain functional eye cells in planarians. Analyses of the eye genetic network in other species such as the polychaete *Platynereis* or the protochordate amphioxus show other cases of dissociation involving *Pax-6*. Interestingly, it has been found that Pax6 and Eya act synergistically to induce eye development in *Drosophila* (Bonini, 1997), and a possible genetic interaction between EYA-1 and PAX-6 has also been demonstrated in *C. elegans* (Furuya *et al.*, 2005). *Eya* genes encode nuclear proteins that function as transcriptional co-factors through interaction with Six family members and/or the retinal determination protein dachshund (Chen *et al.*, 1997; Pignoni *et al.*, 1997; Ikeda *et al.*, 2002). In addition, Eya may regulate the phosphorylation state of either itself or its transcriptional co-factors through its protein tyrosine phosphatase activity (Rayapureddi *et al.*, 2003). Due to their dual-function, Eya proteins are probably more important for development than previously supposed. We characterized planarian *eya* homologs in *D. japonica* (*Djeya*) and *S. mediterranea* (*Smedeya*). These genes are expressed in brain and eye cells of intact planarians (Fig. 3.3B) and during regeneration, and *eya* RNAi during regeneration leads to an eyeless phenotype (Mannini *et al.*, 2004 and work in progress). Since planarian Pax6 and Eya are co-expressed in eye cells, they could have both distinct functions and act co-operatively on common targets. The next challenge will be to identify the molecules that mediate the effects of these factors during eye formation. Recently, we demonstrated that planarian Eya protein participates in eye formation and maintenance by modulating the action of Six-1, a member of the Six1/2 family, through functional synergy. This finding indicates that the Eya-Six1/2 regulatory cassette may have an ancient, conserved role in protostome eye formation, including formation of the simple planarian eye (Saló *et al.*, 2002). Various members of the *Six* gene family have been isolated in planarians. First we isolated *Gtsix-1* from *G. tigrina* and showed it to be related to *Drosophila sine oculis* and *C. elegans Ceh-33* and *Ceh-34* (Pineda *et al.*, 2000). Subsequently, *Gtsix-3*, a *Six-3* ortholog closely related to *Drosophila optix* and *C. elegans Ceh-32*, was isolated in the same species (Pineda and Saló, 2002). *Gtsix-1*, like its homolog in *D. japonica*, *Djsix1* (Mannini *et al.*, 2004), is specifically expressed in both adult and regenerating eyes (Fig. 3.3C), whereas *Gtsix-3* transcripts are found in the cephalic branches that connect the brain with the head sensors, but have not been detected in the eyes (Pineda and Saló, 2002). Detection of *six-1* transcripts has been used to assess the appearance of eye precursor cells before their differentiation during regeneration (Saló *et al.*, 2002; Mannini *et al.*, 2004). Two to three days after amputation, depending on the temperature, two small spots of hybridization signal were detected in the dorsal anterior region of the cephalic blastema, where no visible structure had yet formed (Fig. 3.3H). Subsequently, *six-1* expression was seen in the eye regions identified by the presence of pigment cells (Fig. 3.3I). Loss of function of *six-1* by dsRNA caused an eyeless phenotype in planarians regenerating a head (Pineda *et al.*, 2000; Mannini *et al.*, 2004). Morphologically normal heads, with cephalic

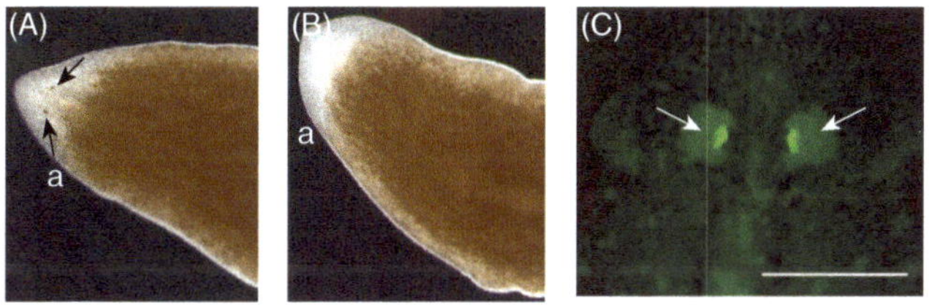

FIGURE 3.5 Modification of eye genetic expression. (A)–(B) Inhibition of eye regeneration by *Smedsix-1* dsRNA in *Schmidtea mediterranea*. Bright field images of dorsal views of living regenerated planarian heads. (A) Control with a regenerated head with differentiated eyes (black arrows). (B) *Smedsix-1* dsRNA-fed organism; although the head is normal with a complete brain (not shown) and auricles, no eyes are observed. a, auricles. (C) Fluorescent image of a transgenic *Girardia tigrina* head transformed by electroporation with the EGFP *Hermes*-derived vector construct (Gonzàlez-Estèvez *et al.*, 2003). Homogeneous fluorescent signal is observed in the photoreceptor cells of both eyes (white arrows). Scale bars: 1 mm.

ganglia and other sense organs, but lacking eyes, could also be detected 5–30 days after injection of *six1* dsRNA (Fig. 3.5B). In intact planarians, *Gtsix-1* RNAi rapidly reduced endogenous *Gtsix-1* expression within the first 24 h, and *opsin* (*Gtops*) expression decreased progressively in the following week (Pineda *et al.*, 2001). Compared with the eye size of water-injected controls, adult planarians also had small eyes with fewer photoreceptors following *Gtsix-1* RNAi (Saló *et al.*, 2002). As we had succeeded in demonstrating that Djsix-1 and Djeya can physically interact through their evolutionarily conserved domains, and *Djsix-1* and *Djeya* double RNAi was substantially more effective in producing regenerating planarian heads without eyes (Mannini *et al.*, 2004), we also asked whether a planarian homolog of Dachshund was included in the Eya-Six network. *Dachshund* (*Dach*) genes encode RDGN members with DNA-binding capacities that interact with the Eya domain of Eya in different organisms (Silver and Rebay, 2005). To extend our knowledge regarding the conservation of the RDGN in planarians, we are now characterizing a planarian homolog of *Dachshund* (*Dach*) and a partial sequence of a *Dachshund* ortholog has been detected *in silico* and amplified from *S. mediterranea* (work in progress).

CONCLUSIONS AND FUTURE PROSPECTS

The structural simplicity of the eye in planarians, combined with the regenerative abilities of these organisms, provides a unique system for dissecting the genetic mechanisms that allow a simple visual structure to be built and may provide insights into the morphogenesis of more complex eyes. Information emerging from the past few years of research provides evidence of a conserved use of regulatory genes, intercellular signaling (TGF-b, Wnt) and axon guidance (DCC/netrin, SLIT/ROBO) mechanisms for establishing early head patterning, supporting the view that bilaterian animals share common regulatory mechanisms to specify anterior structures both during development and regeneration. Most genes related to the RDGN have also been studied functionally in adult planarians and during regeneration. One of the most fascinating observations in these studies is that the two planarian *Pax6*-related genes do not affect the regeneration or maintenance of the eye. On the other hand, analysis of reported genes for planarian *Pax6* confirms its expression in the photoreceptor cells. Why Pax6 appears not to play a role in planarian eye specification is not understood. Keeping in mind that regulatory genes are typically involved in multiple distinct developmental processes and that changes in interactions between regulatory genes and their targets often underlie evolutionary or functional constraints, it is tempting to speculate that adult planarians may make use of alternative molecular pathways for eye specification. A still unexplored possibility is that *Pax6* may be recruited only during planarian embryonic development, while unknown constraints during regeneration might lead to the selection of alternative gene interactions. This condition does not appear generally applicable to all planarian RDGN components, as demonstrated by the conserved role of the Eya-Six1/2 regulatory cassette. Eye regeneration may therefore represent an exciting developmental context in which novelty (emergence of new characteristics) and conservation (conserved features) can be combined to rebuild a new structure in an adult organism.

Although several genes linked to the formation of the planarian visual system have now been characterized, we are still a long way from completely understanding how the fate of planarian stem cells is specified during the formation of eye cells and many questions remain to be answered before a model for this fascinating process can be proposed. Consequently, a comprehensive search for new factors that are involved in eye formation is of fundamental importance. We recently exploited the possibility of coupling the ability to specifically inhibit planarian eye regeneration by RNAi-induced knockdown of *six-1* with a subtractive library approach (Batistoni *et al.*, 2006 and work in progress). Using a similar experimental strategy to generate two sets of probes that hybridize differentially, we intend to screen for coding regions with a custom-designed microarray using NimbleGen maskless photolithography technologies (Nuwaysir *et al.*, 2002 and work in progress). Bioinformatics will play an important role in the integration of all the data to define annotation hypotheses to be validated or refuted later at the bench, as well as in the analysis of the results obtained with high-throughput technologies. With such strategies we expect to characterize new genes and genetic networks involved in regulating planarian eye regeneration and maintenance that can be extrapolated to more complex visual systems.

ACKNOWLEDGMENTS

We are grateful to Dr. Iain Patten for critical comments and corrections, Kay Eckelt for sharing Figures 3.2 and 3.5A,B and Mette Handberg-Thorsager for sharing Figure 3.4. Work summarized in this chapter was supported by grants BFU2005-00422 from the Ministerio de Educación y Ciencia, Spain, and grant 2005SGR00769 from AGAUR (Generalitat de Catalunya) and grants from MURST-Italy (Cofinanziamento Programmi di Ricerca di Interesse Nazionale) to R.B.

REFERENCES

Adell T, Marsal M, Saló E (2008). Planarian GSK3s are involved in neural regeneration. *Dev Genes Evol* 218:89–103.

Agata K, Soejima Y, Kato K, Kobayashi C, Umesono Y, Watanabe K (1998). Structure of the planarian central nervous system (CNS) revealed by neuronal cell markers. *Zool Sci* 15:433–440.

Agata K, Tanaka T, Kobayashi C, Kato K, Saitoh Y (2003). Intercalary regeneration in planarians. *Dev Dynam* 226:308–316.

Azuma K (1998). Light-induced extracellular changes of calcium and sodium concentrations in the Planarian ocellus. *Comparat Biochem Physiol Molecul Integrat Physiol* 119:321–325.

Baguñà J, Saló E, Auladell C (1989). Regeneration and pattern formation in planarians III. Evidence that neoblasts are totipotent stem cells and the source of blastema cells. *Development* 107:77–86.

Batistoni R, Mannini L, Salvetti A, Rossi L, Gremigni V, Deri P (2006). Genetic regulation of planarian head morphogenesis during regeneration. *Ital J Zool* 73:295–301.

Bayascas JR, Castillo E, Saló E (1998). Platyhelminthes have a hox code differentially activated during regeneration, with genes closely related to those of spiralian protostomes. *Dev Genes Evol* 208:467–473.

Bonini NM (1997). Surviving Drosophila eye development. *Cell Death Diff* 4:4–11.

Callaerts P, Halder G, Gehring WJ (1997). Pax-6 in development and evolution. *Annu Rev Neurosci* 20:483–532.

Callaerts P, Munoz-Marmol AM, Glardon S, Castillo E, Sun H, Li WH, Gehring WJ, Salo E (1999). Isolation and expression of a Pax-6 gene in the regenerating and intact Planarian Dugesia(G)tigrina. *Proc Natl Acad Sci USA* 96:558–563.

Cardona A, Hartenstein V, Romero R (2005). The embryonic development of the triclad Schmidtea polychroa. *Dev Genes Evol* 215:109–131.

Cebrià F (2007). Regenerating the central nervous system: how easy for planarians! *Dev Genes Evol* 217:733–748.

Cebrià F, Guo T, Jopek J, Newmark PA (2007). Regeneration and maintenance of the planarian midline is regulated by a slit orthologue. *Dev Biol* 307:394–406.

Cebrià F, Kobayashi C, Umesono Y, Nakazawa M, Mineta K, Ikeo K, Gojobori T, Itoh M, Taira M, Sanchez Alvarado A et al. (2002). FGFR-related gene nou-darake restricts brain tissues to the head region of planarians. *Nature* 419:620–624.

Cebrià F, Newmark PA (2005). Planarian homologs of netrin and netrin receptor are required for proper regeneration of the central nervous system and the maintenance of nervous system architecture. *Development* 132:3691–3703.

Cebrià F, Newmark PA (2007). Morphogenesis defects are associated with abnormal nervous system regeneration following roboA RNAi in planarians. *Development* 134:833–837.

Curtin KD, Zhang Z, Wyman RJ (2002a). Gap junction proteins are not interchangeable in development of neural function in the Drosophila visual system. *J Cell Sci* 115:3379–3388.

Curtin KD, Zhang Z, Wyman RJ (2002b). Gap junction proteins expressed during development are required for adult neural function in the Drosophila optic lamina. *J Neurosci* 22:7088–7096.

Czerny T, Halder G, Kloter U, Souabni A, Gehring WJ, Busslinger M (1999). twin of eyeless, a second Pax-6 gene of Drosophila, acts upstream of eyeless in the control of eye development. *Mol Cell* 3:297–307.

Chen SM, Wang QL, Nie ZQ, Sun H, Lennon G, Copeland NG, Gilbert DJ, Jenkins NA, Zack DJ (1997). Crx, a novel Otx-like paired-homeodomain protein, binds to and transactivates photoreceptor cell-specific genes. *Neuron* 19:1017–1030.

Davidson EH, Erwin DH (2006). Gene regulatory networks and the evolution of animal body plans. *Science* 311:796–800.

Dickson BJ, Gilestro GF (2006). Regulation of commissural axon pathfinding by slit and its robo receptors. *Annu Rev Cell Dev Biol* 22:651–675.

Dubois F (1949). Contribution á l'ètude de la migration des cellules de régénération chez les Planaires dulcicoles. *Bull Biol Fr Belg* 93:213–283.

Englund C, Steneberg P, Falileeva L, Xylourgidis N, Samakovlis C (2002). Attractive and repulsive functions of Slit are mediated by different receptors in the Drosophila trachea. *Development* 129:4941–4951.

Furukawa T, Morrow EM, Cepko CL (1997). Crx, a novel otx-like homeobox gene, shows photoreceptor-specific expression and regulates photoreceptor differentiation. *Cell* 91:531–541.

Furuya M, Qadota H, Chisholm AD, Sugimoto A (2005). The C-elegans eyes absent ortholog EYA-1 is required for tissue differentiation and plays partially redundant roles with PAX-6. *Dev Biol* 286:452–463.

Gehring WJ (2005). New perspectives on eye development and the evolution of eyes and photoreceptors. *J Hered* 96:171–184.

Gehring WJ, Ikeo K (1999). Pax 6 – mastering eye morphogenesis and eye evolution. *Trends Genet* 15:371–377.

Gonzàlez-Estèvez C, Felix DA, Aboobaker AA, Saló E (2007a). Gtdap-1 promotes autophagy and is required for planarian remodeling during regeneration and starvation. *Proc Natl Acad Sci USA* 104:13373–13378.

Gonzàlez-Estèvez C, Felix DA, Aboobaker AA, Saló E (2007b). Gtdap-1 and the role of Autophagy during planarian regeneration. *Autophagy* 3(6):640–642.

Gonzàlez-Estèvez C, Momose T, Gehring WJ, Saló E (2003). Transgenic planarian lines obtained by electroporation using transposon-derived vectors and an eye-specific GFP marker. *Proc Natl Acad Sci USA* 100:14046–14051.

Gremigni V, Miceli C, Puccinelli I (1980). On the role of germ cells in planarian regeneration. I. A karyological investigation. *J Embryol Exp Morphol* 55:53–63.

Gurley KA, Rink C, Sáncez-Alavarado S (2008). B-catenin defines head versus tail identity during planarian regeneration and homeostasis. *Science* 319:323–327.

Handberg-Thorsager M, Saló E (2007). The planarian nanos-like gene Smednos is expressed in germline and eye precursor cells during development and regeneration. *Dev Genes Evol* 217: 403–411.

Iglesias M, Gomez-Skarmeta JL, Saló E, Adell T (2008). Silencing of *Smed-β-catenin1* generates radial-like hypercephalized planarians. *Development* 135:1215–1221.

Ikeda K, Watanabe Y, Ohto H, Kawakami K (2002). Molecular interaction and synergistic activation of a promoter by Six, Eya, and Dach proteins mediated through CREB binding protein. *Mol Cell Biol* 22:6759–6766.

Inoue T, Hayashi T, Takechi K, Agata K (2007). Clathrin-mediated endocytic signals are required for the regeneration of, as well as homeostasis in, the planarian CNS. *Development* 134:1679–1689.

Inoue T, Kumamoto H, Okamoto K, Umesono Y, Sakai M, Sanchez Alvarado A, Agata K (2004). Morphological and functional recovery of the planarian photosensing system during head regeneration. *Zool Sci* 21:275–283.

Kobayashi C, Saito Y, Ogawa K, Agata K (2007). Wnt signaling is required for antero-posterior patterning of the planarian brain. *Dev Biol* 306:714–724.

Lender T (1950). [Demonstration of the organizing role of the brain in regeneration of the eyes in planaria Polycelis nigra by grafting methods.]. *CR Seances Soc Biol Fil* 144:1407–1409.

Lender T (1951). [Inductive capacity of eye organizor in the regeneration of the planarian Polycelis nigra (Ehr.); action of the brain in degeneration and in heteroplastic grafts.]. *CR Seances Soc Biol Fil* 145:1378–1381.

Lichtneckert R, Reichert H (2005). Insights into the urbilaterian brain: conserved genetic patterning mechanisms in insect and vertebrate brain development. *Heredity* 94:465–477.

Mannini L, Rossi L, Deri P, Gremigni V, Salvetti A, Salo E, Batistoni R (2004). Djeyes absent (Djeya) controls prototypic planarian eye regeneration by cooperating with the transcription factor Djsix-1. *Dev Biol* 269:346–359.

Martinez-Morales JR, Rodrigo I, Bovolenta P (2004). Eye development: a view from the retina pigmented epithelium. *Bioessays* 26: 766–777.

Mineta K, Nakazawa M, Cebria F, Ikeo K, Agata K, Gojobori T (2003). Origin and evolutionary process of the CNS elucidated by comparative genomics analysis of planarian ESTs. *Proc Natl Acad Sci USA* 100:7666–7671.

Molina MD, Saló E, Cebrià F (2007). The BMP pathway is essential for re-specification and maintenance of the dorsoventral axis in regenerating and intact planarians. *Dev Biol* 311:79–94.

Morgan TH (1898). Experimental studies of the regeneration of Planaria maculata. *Archiv für Entwickelungsmechanik der organismen* 7:364–397.

Nakazawa M, Cebria F, Mineta K, Ikeo K, Agata K, Gojobori T (2003). Search for the evolutionary origin of a brain: planarian brain characterized by microarray. *Mol Biol Evol* 20:784–791.

Newmark PA, Reddien PW, Cebrià F, Sanchez Alvarado A (2003). Ingestion of bacterially expressed double-stranded RNA inhibits gene expression in planarians. *Proc Natl Acad Sci USA* 100(Suppl 1):11861–11865.

Niwa N, Hiromi Y, Okabe M (2004). A conserved developmental program for sensory organ formation in Drosophila melanogaster. *Nat Genet* 36:293–297.

Nogi T, Levin M (2005). Characterization of innexin gene expression and functional roles of gap-junctional communication in planarian regeneration. *Dev Biol* 287:314–335.

Nuwaysir EF, Huang W, Albert TJ, Singh J, Nuwaysir K, Pitas A, Richmond T, Gorski T, Berg JP, Ballin J et al. (2002). Gene expression analysis using oligonucleotide arrays produced by maskless photolithography. *Genom Res* 12:1749–1755.

Okamoto K, Takeuchi K, Agata K (2005). Neural projections in planarian brain revealed by fluorescent dye tracing. *Zool Sci* 22:535–546.

Orii H, Katayama T, Sakurai T, Agata K, Watanabe K (1998). Immunohistochemical detection of opsins in turbellarians. *Hydrobiologia* 383:183–187.

Orii H, Watanabe K (2007). Bone morphogenetic protein is required for dorso-ventral patterning in the planarian Dugesia japonica. *Dev Growth Differ* 49:345–349.

Oviedo NJ, Levin M (2007). Smedinx-11 is a planarian stem cell gap junction gene required for regeneration and homeostasis. *Development* 134:3121–3131.

Oviedo NJ, Newmark PA, Sanchez Alvarado A (2003). Allometric scaling and proportion regulation in the freshwater planarian Schmidtea mediterranea. *Dev Dyn* 226:326–333.

Petersen C, Reddien PW (2008). Smed-betacatenin-1 is required for anteroposterior blastema polarity in planarian regeneration. *Science* 319:327–330.

Phelan P, Bacon JP, Davies JA, Stebbings LA, Todman MG, Avery L, Baines RA (1998). Innexins: a family of invertebrate gap-junction proteins. *Trend Genet* 14:348–349.

Pignoni F, Hu BR, Zipursky SL (1997). Identification of genes required for Drosophila eye development using a phenotypic enhancer-type. *Proc Natl Acad Sci USA* 94:9220–9225.

Pineda D, Gonzalez J, Callaerts P, Ikeo K, Gehring WJ, Salo E (2000). Searching for the prototypic eye genetic network: Sine oculis is essential for eye regeneration in planarians. *Proc Natl Acad Sci USA* 97:4525–4529.

Pineda D, Gonzalez J, Marsal M, Salo E (2001). Evolutionary conservation of the initial eye genetic pathway in planarians. *Belg J Zool* 131:77–82.

Pineda D, Rossi L, Batistoni R, Salvetti A, Marsal M, Gremigni V, Falleni A, Gonzalez-Linares J, Deri P, Salo E (2002). The genetic network of prototypic planarian eye regeneration is Pax6 independent. *Development* 129:1423–1434.

Pineda D, Salo E (2002). Planarian Gtsix3, a member of the Six/so gene family, is expressed in brain branches but not in eye cells. *Mech Dev* 119(Suppl 1):S167–S171.

Plouhinec JL, Leconte L, Sauka-Spengler T, Bovolenta P, Mazan S, Saule S (2005). Comparative analysis of gnathostome Otx gene expression patterns in the developing eye: implications for the functional evolution of the multigene family. *Dev Biol* 278:560–575.

Quiring R, Walldorf U, Kloter U, Gehring WJ (1994). Homology of the eyeless gene of Drosophila to Small eye gene in mice and Aniridia in humans. *Science* 265:785–789.

Rayapureddi JP, Kattamuri C, Steinmetz BD, Frankfort BJ, Ostrin EJ, Mardon G, Hegde RS (2003). Eyes absent represents a class of protein tyrosine phosphatases. *Nature* 426:295–298.

Reddien PW, Bermange AL, Murfitt KJ, Jennings JR, Alvarado AS (2005). Identification of genes needed for regeneration, stem cell function, and tissue homeostasis by systematic gene perturbation in planaria. *Dev Cell* 8:635–649.

Reddien PW, Bermange AL, Kicza AM, Sánchez-Alavarado A (2007). BMP signaling regulates the dorsal planarian midline and is needed for asymmetric regeneration. *Development* 134:4043–4051.

Rossi L, Batistoni R, Salvetti A, Deri P, Bernini F, Andreoli I, Falleni A, Gremigni V (2001). Molecular aspects of cell proliferation and neurogenesis in planarians. *Belg J Zool* 131:83–87.

Rossi L, Salvetti A, Marincola FM, Lena A, Deri P, Mannini L, Batistoni R, Wang E, Gremigni V (2007). Deciphering the molecular machinery of stem cells: a look at the neoblast gene expression profile. *Genom Biol* 8:R62.

Sakai F, Agata K, Orii H, Watanabe K (2000). Organization and regeneration ability of spontaneous supernumerary eyes in planarians – Eye regeneration field and pathway selection by optic nerves. *Zool Sci* 17:375–381.

Saló E (2006). The power of regeneration and the stem-cell kingdom: freshwater planarians (platyhelminthes). *Bioessays* 28:546–559.

Saló E, Baguñà J (2002). Regeneration in Planarians and other worms: New findings, new tools, and new perspectives. *J Exp Zool* 292:528–539.

Saló E, Pineda D, Marsal M, Gonzalez J, Gremigni V, Batistoni R (2002). Genetic network of the eye in Platyhelminthes: expression and functional analysis of some players during planarian regeneration. *Gene* 287:67–74.

Saló E, Abril JF, Adell T, Cebrià F, Eckelt K, Fernández-Taboada E, Handberg-Thorsager M, Iglesias M, Molina MD, Rodríguez-Esteban G (2008). Planarian regeneration: achievements and future directions after 20 years of research. *Int J Dev Biol* 52: in press.

Sanchez-Alvarado A, Newmark PA, Robb SMC, Juste R (2002). The Schmidtea mediterranea database as a molecular resource for studying platyhelminthes, stem cells and regeneration. *Development* 129:5659–5665.

Sanchez-Alvarado A, Tsonis PA (2006). Bridging the regeneration gap: genetic insights from diverse animal models. *Nat Rev Genet* 7:873–884.

Sanchez Alvarado A, Newmark PA (1999). Double-stranded RNA specifically disrupts gene expression during planarian regeneration. *Proc Natl Acad Sci USA* 96:5049–5054.

Silver SJ, Rebay L (2005). Signaling circuitries in development: insights from the retinal determination gene network. *Development* 132:3–13.

Stornaiuolo A, Bayascas JR, Salo E, Boncinelli E (1998). A homeobox gene of the orthodenticle family is involved in anteroposterior patterning of regenerating planarians. *Int J Dev Biol* 42:1153–1158.

Takano T, Pulvers JN, Inoue T, Tarui H, Sakamoto H, Agata K, Umesono Y (2007). Regeneration-dependent conditional gene knockdown (Readyknock) in planarian: demonstration of requirement for Djsnap-25 expression in the brain for negative phototactic behavior. *Dev Growth Differ* 49:383–394.

Takayato A, Sonneville R, Pichaud F, Wernet M, Papatsenko D, Beaufils P, Cook T, Desplan C (2003). Otd/Crx, a dual regulator for the specification of ommatidia subtypes in the Drosophila retina. *Dev Cell* 5:391–402.

Tamamaki N (1990). Evidence for the Phagocytotic Removal of Photoreceptive Membrane by Pigment-Cells in the Eye of the Planarian, Dugesia-Japonica. *Zool Sci* 7:385–393.

Tazaki A, Gaudieri S, Ikeo K, Gojobori T, Watanabe K, Agata K (1999). Neural network in planarian revealed by an antibody against planarian synaptotagmin homologue. *Biochem Biophys Res Commun* 260:426–432.

Tettamanti G, Saló E, González-Estevéz C, Felix AD, Grimaldi A, De Eguileor M (2008). Autophagy in Invertebrates: Insights into Development, Regeneration and Body Remodeling. *Current Pharmaceutical Design* 14:116–125.

Treisman JE (1999). A conserved blueprint for the eye? *Bioessays* 21:843–850.

Umesono Y, Watanabe K, Agata K (1997). A planarian orthopedia homolog is specifically expressed in the branch region of both the mature and regenerating brain. *Dev Growth Differ* 39:723–727.

Umesono Y, Watanabe K, Agata K (1999). Distinct structural domains in the planarian brain defined by the expression of evolutionarily conserved homeobox genes. *Dev Genes Evol* 209:31–39.

Vandendries E, Johnson D, Reinke R (1996). Orthodenticle is required for photoreceptor cell development in the Drosophila eye. *Dev Biol* 173:243–255.

Viczian AS, Vignali R, Zuber ME, Barsacchi G, Harris WA (2003). XOtx5b and XOtx2 regulate photoreceptor and bipolar fates in the Xenopus retina. *Development* 130:1281–1294.

CHAPTER

4

Development of the *Drosophila melanogaster* Eye: from Precursor Specification to Terminal Differentiation

Simon G. Sprecher, Claude Desplan
Department of Biology, New York University, New York, NY 10003-6688, USA

OUTLINE

INTRODUCTION

Drosophila as A Model System

The fruit fly *Drosophila melanogaster* was first introduced as a genetic model system about a century ago (Fig. 4.1A). Already then, the striking morphological characteristics of the fly's eye were used to build the basis of modern genetics, several decades before the discovery of the double helix DNA structure or the genetic code. In 1910, Thomas Hunt Morgan discovered a fly in which the typical red pigmentation of the eye was missing (Fig. 4.1B,C). The description of the *white* mutation was part of the demonstration of hereditary transmission in *Drosophila* for which T.H. Morgan was awarded the Nobel Prize in Physiology and Medicine in 1933. Over the last 100 years, *Drosophila* has emerged as a precious model system to study various fields

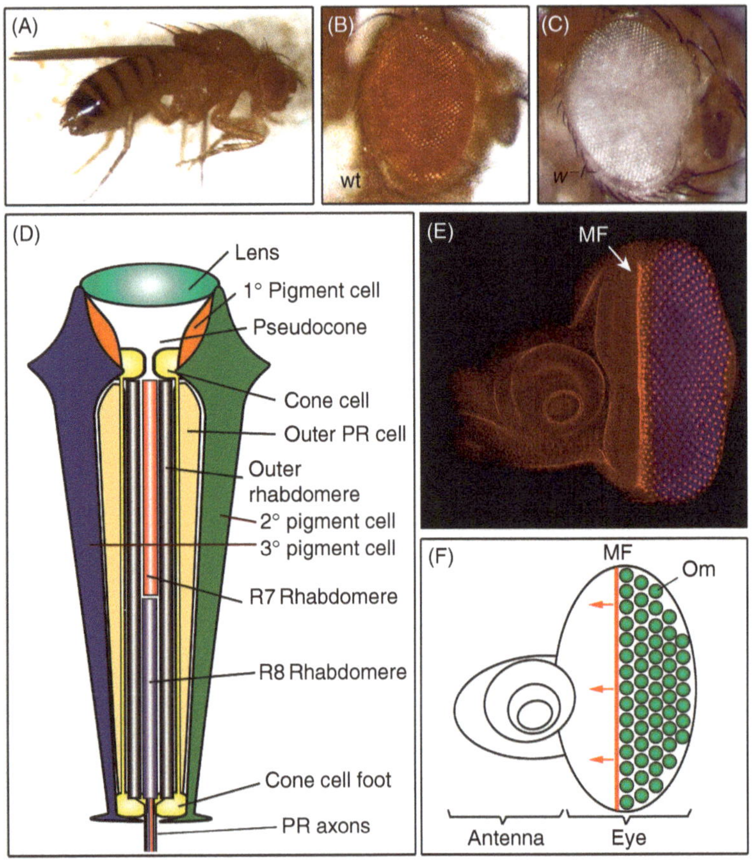

FIGURE 4.1 The *Drosophila* compound eye and anatomy of the ommatidium. The adult *Drosophila* fly is comparably small (A; female, lateral view). The red pigmentation of the adult compound eye is a characteristic (B), is lost in *white* mutant animals (C). The *Drosophila* eye consists of about 800 ommatidial units (D) each comprised of 19 cells. In the center of the ommatidium are the eight photoreceptors. The outer PRs (R1–R6) surround the two inner PRs (R7 and R8). The photoreceptors are surrounded by pigment cells acting as light insulators. Distal to the photoreceptors are the cone cells, the pseudocone and lens acting establishing the correct optical path for light detection in the underlying photoreceptors. During late larval stages individual ommatidia develop in the eye-antennal imaginal disc (E, schematic representation in F). The morphogenetic furrow (MF) passes the disc from posterior to anterior (red arrows) patterning the tissue. Photoreceptors (blue in E) develop posterior to the furrow (adherens junctions of the epithelium shown in red). The eye-antennal disc consists of an anterior "antenna-part" and a posterior "eye-part" (F). Wt, wildtype; $w^{-/-}$, *white* mutant; MF, morphogenetic furrow; Om, ommatidium.

of modern biology, well beyond its role in studying genetic transmission. The discovery of homeotic genes and transcriptional networks that establish the body segmentation, signal transduction pathways or genes controlling eye formation are just a few examples of the major contributions of *Drosophila* molecular genetics. Of particular notice is the constant effort to establish increasingly elaborate genetic techniques to study the function of genes in various biological contexts, such as specific cell types or life stages. This has maintained the status of *Drosophila* as a most powerful genetic model system.

In contrast to vertebrate model systems such as the mouse, *Drosophila* is easy and economical to keep. It is small, has a relatively short generation cycle of about 10 days and large quantities of fly stocks can be used and maintained at low cost. Therefore, large classical genetic screens, either making use of mutagens such as chemical reagents or X-ray, have led to the discovery of a wide range of mutations affecting almost every biological process. In 1982, Allan Spradling and Gerry Rubin published a method which revolutionized *Drosophila* genetics: P-element mediated transformation. Modifying an existing transposable element, they established a system in which almost any piece of DNA (up to several 10's of Kb) can be randomly inserted into the fly genome in a stable and inheritable manner. This technique provided the basis of *Drosophila* genetics for a novel area and its modern modified versions still represent one of the most essential toolkits of *Drosophila*. Since the introduction of transgenics, a large number of modifications have been made to this technique. Using reporter genes, such as *lacZ*, enhancer-trap screens have led to the identification of genes based on their expression patterns. A next major breakthrough was made when Andrea Brand and Norbert Perrimon developed the binary Gal4/UAS-system in 1993. The system makes use of the yeast transcription activator gene Gal4, which binds to specific DNA target sequences, the UAS (upstream-activating-sequence). The Gal4/UAS-system enables temporal or developmental stage- and tissue-specific genetic manipulations (Brand and Perrimon, 1993). During the same time period several laboratories worked on the development of the FLP/FRT system, which uses the yeast site-specific Flip recombinase. The FLP/FRT-system in combination with cell markers enables to study mosaic marked clones to label mutant tissue for instance in the developing eye (for review see Theodosiou and Xu, 1998). During the following years a large variety modifications of both systems led to the development of novel techniques. The MARCM (**M**osaic **A**nalysis with a **R**epressible **C**ell **M**arker) makes use of both systems and is largely used to understand the developmental mode of different parts of the nervous system (for review see (Lee and Luo, 2001).

But perhaps the most significant accomplishment was the sequencing in 2000 of the complete *Drosophila* genome, providing an important step to identify genomic networks and identifying novel genes (Adams *et al.*, 2000). The genome sequences of more than 15 fly species have since become available. The development of all these techniques and ongoing invention of sophisticated molecular and genetic methods provides a powerful basis to study the development of the *Drosophila* compound eye.

Anatomy and Morphology of the *Drosophila* Eye

The *Drosophila* compound eye is an assembly of about 750–800 light-sensing unites, termed ommatidia. Externally, the eye appears as a regular hexagonal array of facets, which are the lenses of each ommatidium. This highly organized repetitive array of facets in the compound eye is a direct consequence of the underlying cellular lattice of cells in the retina. If the cells at the surface are organized in a virtually crystalline lattice, this reflects the underlying architecture of photoreceptors, pigment cells, and other accessory cells in the retina. Each ommatidium represents an independent eye-unit and consists of an assembly of 19 cells. The anatomical concentric architecture of the ommatidium stems from its origin in the epithelium of the eye-antennal imaginal disc (see below; Fig. 4.1E, F). The 8 photoreceptors can be classified as outer photoreceptors (named R1–R6) or inners photoreceptors (R7 and R8). In the adult fly retina, the eight photoreceptors sit in the core of the ommatidium and are surrounded by an array of accessory cone and pigment cells. R7 lies distal (on top) of the proximal R8 photoreceptor (for details see Wolff and Ready, 1993).

The pigment cells form a hexagonal net around the photoreceptors and are responsible for keeping each ommatidium optically insulated from the neighboring ommatidia (Fig. 4.1D). Above the photoreceptors lie four cone cells, the equatorial/polar and the anterior/posterior pair. The two sets of cone cells have their cell bodies apically above the photoreceptors. These cells secrete the overlaying pseudocone and lens material that builds the physical basis for a correct optical system within each ommatidium. There are three sets of pigment cells present. Two primary pigment cells encircle the cone cells. The corneal lens consists of layered laminae of chitinous material which is secreted by the cone cells and primary pigment cells building the lens. Below the corneal lens is a thin layer, called pseudocone of secreted non-chitinous material (Fig. 4.1D). An exact hexagonal mesh of shared pigment cells is arranged surrounding the core of the ommatidium (photoreceptors, cone cells, and primary pigment cells). The edges of each corneal facet at the surface are an exact representation of the hexagonal mesh of secondary and tertiary pigment cells. Beside the photoreceptor neurons there, are small mechanosensory bristles distributed between ommatidia, normally at the anterior end of each horizontal face (for details see Wolff and Ready, 1993).

EARLY DEVELOPMENT AND SPECIFICATION OF EYE PRECURSOR CELLS

During embryogenesis, before the larva hatches, an array of about 20 so-called eye precursor cells are set aside, which will later develop during larval life into the eye-antennal imaginal disc (Garcia-Bellido and Merriam, 1969). Imaginal discs are monolayered epithelial sacs from which most parts of the adult fly will be formed during metamorphosis in the pupa. The eye-antennal imaginal disc gives rise to the eye, the antenna and most of the head capsule as well as the photoreceptor of the ocelli, three eyes located at the posterior medial part of the fly head (Pichaud and Casares,

2000). During the first and second larval instar, cells of the eye-antennal imaginal disc proliferate giving rise to a pool of progenitors which will later develop into most of the external adult head structures. During terminal larval stages (mid-late third larval instar) a wave of differentiation, the morphogenetic furrow, crosses the eye-disc providing the basis of pattern formation of the adult fly retina (Fig. 4.1E, F). During this process, cells of the fly's eye are generated and specified to adopt particular fates.

Embryonic Origin of the Eye

All structures that form the adult fly are either already generated during embryogenesis and transformed during metamorphosis for re-utilization in the adult, or develop from imaginal discs during late larval stages and metamorphosis. Initially, the embryo consists of a monolayered epithelium called blastoderm. Fatemap studies and lineage tracing have placed the origin of the eye imaginal disc to about 5–20 cells located at the anterior dorsolateral region of the early embryo. These cells will be part of the ectodermal region which has been described as the eye field (Fig. 4.2A), containing not only the eye-antennal imaginal disc but also most parts of the larval and adult visual system (Jurgens and Hartenstein, 1993). The cells giving rise to the eye-antennal imaginal disc are within close vicinity to the precursors of the larval eye (Bolwig's organ, BO) and the optic lobe primordium (Fig. 4.2A). After invagination of the optic lobe epithelium, the eye-precursors remain anterior to the optic lobe as a part of the dorsal pouch (Jurgens and Hartenstein, 1993). During head involution, cells of the eye disc form an epithelial sack of about 70 cells whose lumen remains open to the pharynx (Wolff and Ready, 1993).

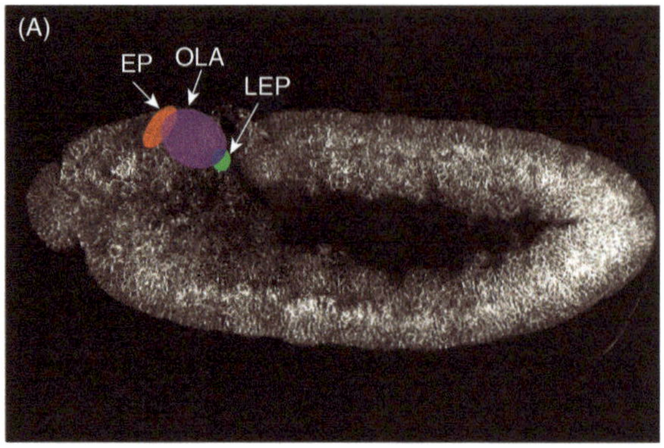

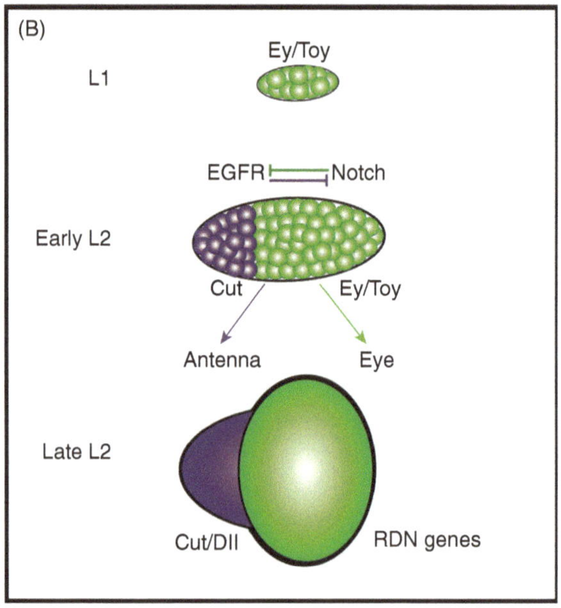

FIGURE 4.2 Development of the eye-antennal imaginal disc. Precursors of the eye-antennal disc develop as part of the eye field at the anterior dorsal part of the embryonic head region (A; membranes of neurogenic region shown in white). In the neurogenic epithelium eye precursors cells (red) locate anterior to the optic lobe anlage (blue). Cells giving rise to the larval eye are at the ventral lateral tip of the optic lobe anlage (green). During second larval instar EGFR versus Notch signaling specify eye versus antennal part of the eye-antennal disc (B). The initial uniform expression of Ey/Toy gets restricted to the posterior 2/3 of the disc by EGFR signaling. The anterior third requires Cut and Dll to develop into the antenna-part, whereas the posterior part requires RDN gene action to develop into the eye-part. EP, eye precursors; OLA, optic lobe anlage; LEP, larval eye precursors; L2, second larval instar.

Early Development of the Eye-antennal Disc: Specification of Eye Versus Antenna

Even though most processes in eye formation occur during late larval and pupa stages, initial steps of eye disc patterning take place already during early larval life. Clonal analyses of the first instar eye-antennal disc indicate that about 20 cells will give rise to the eye, although clone size varies between individuals (for review see Wolff and Ready, 1993). During early stages, cells of the eye disc proliferate leading to disc growth. During early second larval instar, changes in gene expression occur to subdivide the eye-antennal disc into an "eye-part"and "antennal part". Expression of the Pax genes *eyeless* (*ey*) and *twin of eyeless* (*toy*), two major factors in eye development (see below) which were expressed in the entire disc (Fig. 4.2B), retracts to the posterior two-thirds while the homeodomain transcription factor Cut starts to be expressed in the anterior third. Ey/Toy expression marks the prospective eye-part, whereas Cut expression defines the antenna-part (Kenyon *et al.*, 2003). Two other homeodomain transcription factors Distal-less (Dll) and Homothorax (Hth) in the Cut domain specify the antennal fate (Fig. 4.2B). Interestingly removal of Ey in combination of ectopic activation of Dll in the eye-part results in the formation of an ectopic antenna (when apoptosis is prevented) (Punzo *et al.*, 2004). Therefore, Ey might act as an activator of eye disc fate but also as a repressor for antennal fate. Another key event in the distinction between eye and antenna development is provided by antagonistic action of Notch and EGFR signaling. EGFR signaling actively represses Ey in the antennal part, whereas the Notch pathway antagonizes EGFR in the eye-part thereby enabling eye development to proceed (Fig. 4.2B). Ectopic Notch activation is sufficient to induce Ey expression (Kumar and Moses, 2001a).

It is to mention that a large number of steps in the developmental program of eye organogenesis which will be discussed in the following require the interaction of the EGFR- and Notch-signaling pathways. However, the way these two signaling pathways interact varies and largely depends on the context and the developmental stage as antagonistic or cooperating interactions of these pathways may occur (For review see Sundaram, 2005).

THE RETINAL DETERMINATION NETWORK (RDN) PROVIDES THE BASIS FOR EYE FORMATION

A genetic network of evolutionary conserved transcription factors provides the basis for the development of the eye in vertebrates and invertebrates. In *Drosophila,* the specification of the compound eye is controlled by the genes *ey, toy, eye gone (eyg), sine oculis (so), optix, eyes absent (eya),* and *dachshund (dac)* which encode nuclear factors (for review see Kumar, 2001; Silver and Rebay, 2005). Recent studies suggest that these nuclear factors act as a regulatory network, termed retinal determination network (RDN), rather than in a linear manner (Fig. 4.3E). Inactivation of any of RDN genes in the developing eye primordium leads to the loss or severe reduction of the adult compound eye (Fig. 4.3B). The discovery of the *Drosophila* Pax6 homologs Ey and Toy has led to the realization that these genes are sufficient to induce ectopic eyes in other body parts, such as the antenna, legs, and wings (Fig. 4.3C,D, Halder *et al.*, 1995). This ability to induce ectopic eyes is not restricted to Ey/Toy. All RDN genes, with the exception of *So*, are sufficient to induce ectopic eyes. Moreover the genetic interaction of RDN genes is not only required for the establishment of the eye, but also for the development of a number of other tissues including the brain (Kurusu *et al.*, 2000; Noveen *et al.*, 2000; Kammermeier *et al.*, 2001; Anderson *et al.*, 2006).

Function of the RDN

The most prominent members of the RDN are the Pax-family genes *ey* and *toy*. Pax genes are defined by the presence of a paired-box, which codes for a highly conserved DNA binding domain (Wilson *et al.*, 1993, 1995). They also often contain a paired-class homeobox (which is characterized by the presence of serine at position 50). Both genes have been placed at the top of the RDN hierarchy as they are required for the expression of the other members of the network. The second Pax6 homolog in *Drosophila, toy,* is only present in holometabolous insects, where it appears to act upstream of *ey* (Czerny *et al.*, 1999). Loss of Ey/Toy results in the absence of expression of downstream RDN genes, whereas Ey/Toy misexpression is sufficient to induce expression of RDN genes (Fig. 4.3E). Since other RDN members are also able to induce ectopic eyes, the hierarchy is not absolute, and *eyg, eya,* and *dac* can also induce *ey* expression. As Ey/Toy acts to initiate the genetic program underlying eye formation in *Drosophila,* they have been termed "master control genes" (Gehring and Ikeo, 1999).

Another Pax gene, with some similarities to Pax6, is *eyg* and its paralogue *twin of eyegone* (*toe*) which acts in parallel to *ey* (Fig. 4.3E). *eyg* null mutant flies have no eyes (no *toe* mutants have been described so far), as *ey/toy, eyg,* and *toe* code for transcription factors containing a paired domain and a homeodomain. However, the paired domain of Eyg and Toe exhibit a major alteration as compared to the Ey/Toy paired domain. The canonical bipartite 128 residue paired domain consists of an N-terminal "PAI" part and an adjacent C-terminal "RED" part, which both can bind DNA independently or synergistically. The "PAI" and "RED" parts have different DNA binding specificities, with the "PAI" domain being predominant. In contrast to the canonical paired domain of Pax6 and Ey/Toy, the Eyg/Toe paired domain shows similarities to the human Pax6(5a) splicing isoform where the "PAI" domain is absent, and therefore displays an altered binding specificity. Initially identified by enhancer traps, *eyg* is expressed early as a stripe along the anteroposterior axis along the equator (see below).

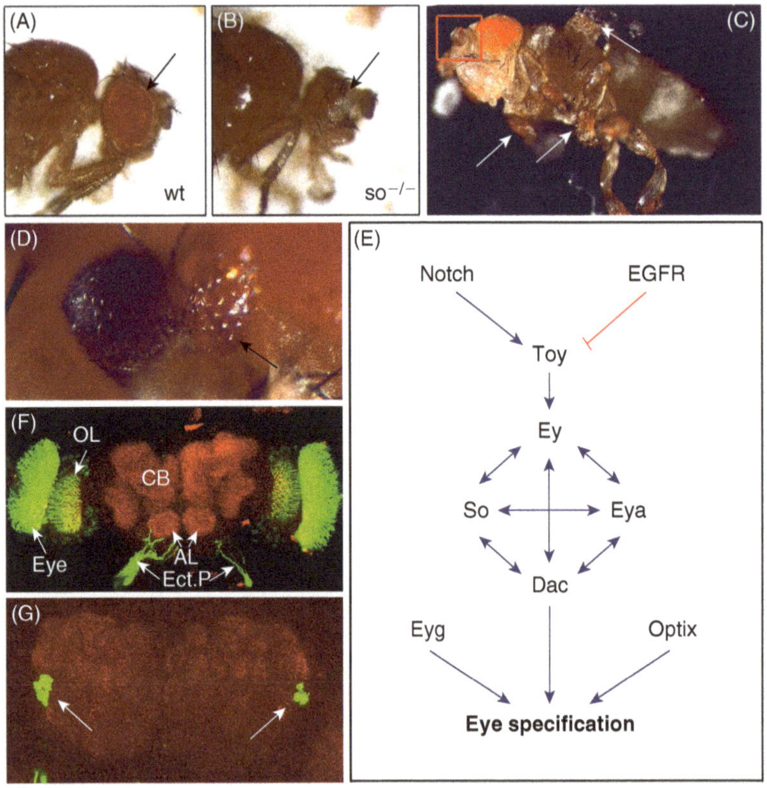

FIGURE 4.3 The (RDN) in the formation of *Drosophila* eyes. In flies mutant for any RDN gene, such as *sine oculis* (*so*), the compound eyes are absent (compare A and B). Ectopic activation of the Pax6 gene *ey* in imaginal discs leads to the induction of ectopic eyes in antenna, legs, and wings (C, D). The ectopic eyes formed in the antenna do not project to the optic lobe like normal photoreceptors (F), but instead extend their axons into the antennal lobe, a center in the brain for olfactory information processing (F, G: photoreceptor and their axonal projections in Green; Brain neuropile in Red). Like the olfactory receptor neurons of the antenna, these ectopic eyes contribute the glomeruli of the antennal lobe (G, arrows). Genetic interactions of genes involved in eye formation (E). Initially EGFR and Notch signaling act to promote and inhibit expression of the Ey/Toy. So, Eya, and Dac are sufficient to induce Ey expression, however act genetically downstream of Ey. Optix and Eyg act genetically, partly independently, in Eye specification (modified after Kumar, 2001). Wt, wildtype; $so^{-/-}$, *sine oculis* mutant; OL, optic lobe; CB, central brain; AL, antennal lobe; Ect.P, ectopic PR projections.

Eyg acts genetically downstream of other RDN genes, since the expression of Ey, Dac, and So is unaffected in *eyg* mutant tissue (although *toe* is still present in this situation). However, Ey/Toy is not able to rescue the loss of *eyg* function, indicating that the differences in the paired domain are essential. Moreover Eyg/Toe is not able to induce ectopic eyes by themselves (Jang *et al.*, 2003). However, *eyg* can lead to an increase in eye size of *ey*-induced ectopic eyes, suggesting that it acts to control growth of the eye disc. Indeed *eyg* acts downstream of Notch to control growth of the early eye disc in the equatorial region. Removing *eyg* function outside this territory has no effect. Thus, in contrast to the canonical Pax6 homologs Ey/Toy, Eyg acts to control eye tissue growth. Eyg and Toe in *Drosophila* seems to act in a comparable fashion to the Pax6–5a splice isoforms in vertebrates, whose genome does not contain an *eyg* homolog (Dominguez *et al.*, 2004; Rodrigues and Moses, 2004).

so and *optix* belong to the Six gene family characterized by a Six domain that mediates protein–protein interactions, and has a DNA binding homeodomain. The vertebrate *so* homolog Six1 is involved in regulating cell proliferation, but does not seem to affect eye development, whereas the *optix* homologs Six3/Six6 act as transcriptional repressors that are important for eye and brain development. *so* is required for the development of the entire visual system but no *optix* mutants have been published so far. Expression of *so* initially covers the entire eye-field and is also required for the generation of the larval eye and optic lobe primordium. In strong *so* alleles, the optic lobe primordium does not invaginate and precursors of the larval eye are not specified. In the eye-antennal imaginal disc, *so* mutations lead to apoptosis anterior to the furrow (Cheyette *et al.*, 1994). *Optix* is expressed anterior to the furrow, whereas *so* expression is much broader (Seimiya and Gehring, 2000). Interestingly, the two

Six genes seem to conduct very distinct functions in *Drosophila. so* only has weak ability to induce ectopic eyes, whereas *Optix* ectopic expression is more efficient and occurs in *ey*-independent manner, in contrast to *so*. However, the molecular basis of this alternative pathway for eye development remains largely unknown (Seimiya and Gehring, 2000). Studies using chimeric proteins between Optix and So showed that specificity lies in the C-terminal part of the protein. An engineered So protein containing the C-terminus of Optix is able to induce ectopic eyes, whereas the C-terminus of So prevents the Optix protein from inducing ectopic eyes (Weasner *et al.*, 2007).

dac encodes a novel DNA binding protein with two conserved domains, the DachBox-N and the DachBox-C. Even though the molecular function of Dac remains largely unknown, it seems likely that the Dac protein interacts with So and Eya to form a complex (Chen *et al.*, 1997). Dac is acting downstream of *ey*, since *dac* is not required for *ey* expression (Fig. 4.3E). Furthermore, ectopic *ey* induces *dac* expression and *dac* function is required for ectopic eye formation by *ey*.

The Eya family members are characterized by the presence of a conserved C-terminal EYA domain and an N-terminal EYA domain 2. The EYA domain is essential for protein–protein interactions with other RDN members such as So and Dac (Chen *et al.*, 1997; Pignoni *et al.*, 1997), whereas the N-terminal regions are important for its co-activator function. Surprisingly the EYA domain appears to be a protein phosphatase, a role unusual for co-activator of transcription. Eya and Dac act synergistically in ectopic eye formation. Interestingly Eya and Dac also seem to regulate each other's expression in the mushroom body (Noveen *et al.*, 2000), a part of the *Drosophila* brain involved in learning and memory formation, suggesting that the RDN gene network is also required in other developmental contexts.

Evolutionary Conservation of the RDN

One of the most striking findings concerning the function of RDN genes is the degree of evolutionary conservation of those genes. The two Pax6 homologs *ey* and *toy* were initially been identified by their "eyeless" phenotype. The first *eyeless* mutant was described in 1915 by Hodge, but it is the cloning of the gene that showed the similarity to the vertebrate *Pax6* gene (Quiring *et al.*, 1994), which is also involved in eye development. Mutation of one copy of the vertebrate *Pax6* gene leads to the *small eyes* phenotype in mouse and aniridia in humans, a genetic eye disorder which results in an underdeveloped iris and retina. Loss of both copies leads to the absence of eyes and nose in both species. In *Drosophila*, ectopic expression of the *ey* gene is sufficient to induce ectopic eyes in other tissues such as wings, antennae, and legs (Halder *et al.*, 1995). Interestingly apart from their mislocation these eyes appear normal in their morphology, extend axonal projections to the central nervous system (CNS), and seem to be functional according to electrical retinogram (ERG) recordings. Ectopic eyes in the antenna, where odorant receptor neurons are located, project to the antennal lobe in the brain, the center where odor information is processed (Fig. 4.3F,G; Sprecher and Desplan, unpublished).

Pax6 homologs are expressed in the eyes of a large number of other species, including lower chordates such as *Phallusia*, and the mollusk cephalopod *Loligo*. Interestingly induction of ectopic eyes in *Drosophila* can be achieved not only by the two endogenous genes *ey* and *toy*, but also by mouse Pax6, the Pax6 homologs of the lancet *Amphioxus*, the mollusk *Loligo* or even the nematode *C. elegans*, a species that has no eyes (for review see Gehring, 2005). Conversely ectopic expression of *Drosophila ey* and *toy* is sufficient to induce the formation of an ectopic lens and retina in the frog *Xenopus* (Onuma *et al.*, 2002). Even though *Pax6* genes seem to play a pivotal role in eye development in a large variety of species, there are also examples where eyes develop in a *Pax6*-independent manner. For instance in planarians, *Dugesia japonica* and *Girardia tigrina*, the *Pax6* genes are neither expressed in the eye, nor required for eye regeneration (Pineda *et al.*, 2002). In contrast, the *so* homolog is expressed in the eye of this species, and appears to be required for eye regeneration (Pineda *et al.*, 2000). In the larval eye, Bolwig's organ, neither *ey* or *toy* is expressed, nor are they required for its formation (Daniel *et al.*, 1999). Other RDN genes such as *so* or *eya* are, however, required for the development of the larval eye (Suzuki and Saigo, 2000; Sprecher and Desplan, unpublished).

Thus, at least in some cases, eyes can develop without *Pax6* genes, although RDN genes seem to execute major functions in eye development in all currently investigated phylae. The debate whether eyes, which are so different in shape and development, are of monophyletic origin or have been invented several times in different animal phyla is continuing (Gehring, 2001; Pichaud and Desplan, 2002; Gehring, 2005). Studies from the cnidarian jellyfish *Podocoryne* indicate that PaxC, which seems to be the closest related Pax gene of *Podocoryne*, is expressed in developing neurons (Groger *et al.*, 2000). Pax6 is expressed and seems to play a major role in the developing nervous system of a large variety of species in vertebrates and invertebrates. The ancestral function of Pax6 is therefore likely in controlling neuronal development, and it was later recruited for eye development (for review see Pichaud and Desplan, 2002). Similarly, RDN genes seem to be evolutionary conserved and act in a variety of other processes including muscle development or in otic placode development in flies and mice (for review see Silver and Rebay, 2005).

THE MORPHOGENETIC FURROW AND THE SPECIFICATION OF PHOTORECEPTOR AND ACCESORY CELLS

The actual determination and differentiation of photoreceptors as well as other cells of the adult retina occur as a result of highly dynamic developmental events initiated at the morphogenetic furrow. Prior to the initiation of the morphogenetic furrow, the presumptive eye consists of a monolayer epithelium containing undifferentiated cells which proliferate with no apparent pattern (Fig. 4.4A). After initiation at the posterior edge of the imaginal disc, the morphogenetic furrow sweeps anteriorly across the eye imaginal disc. The "furrow" itself is the morphological consequence of constriction of apical actin cytoskeleton rings, which coincides with cell-cycle arrest in G1 (Fig. 4.4B). Directly posterior to the furrow are precisely spaced rows of ommatidial founder cells, each row being specified roughly for every 2 h. Once the R8 precursor cell is specified, it then recruits the other photoreceptors in a stereotyped manner, first the R2/R5 pair, then R3/R4 pair, and finally the R1/R6 pair (Fig. 4.4B). The last cell to be recruited is R7. Cells within the ommatidium are not clonally related as shown by clonal analysis (Lawrence *et al.*, 1979).

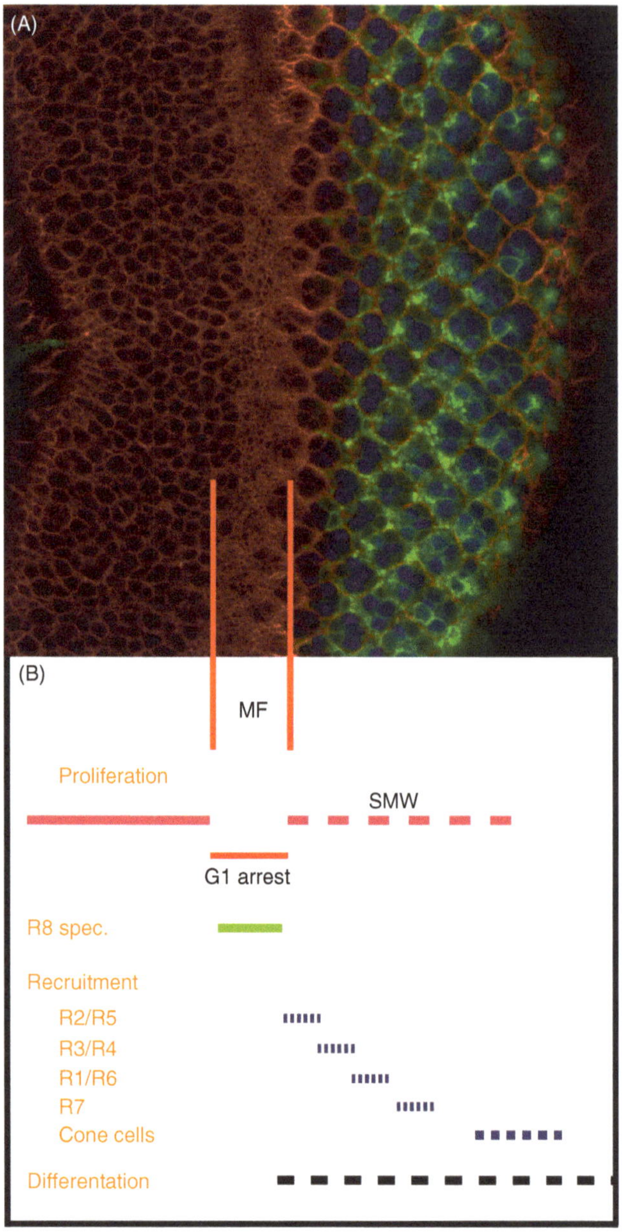

FIGURE 4.4 The morphogenetic furrow and processes in ommatidia development. Passing of the morphogenetic furrow across the epithelium initiates the formation of ommatidia and photoreceptor formation and specification (A, B). The initially uniform population of proliferating cells (right of the furrow in A) enters cell cycle arrest (in the furrow in B). The first neuron to get specified is the R8 photoreceptor (B). Subsequently all remaining photoreceptors are specified and recruited (R8, R2/R5) or undergo an additional round of cell division (second mitotic wave). Posterior to the furrow as cells get specified as photoreceptors they initiate their terminal differentiation (A: adherens junctions of the epithelium shown in red; differentiating neurons are shown in blue; Cell membranes of differentiation photoreceptors are shown in green). EP, eye precursors; OLA, optic lobe anlage; LEP, larval eye precursors; L2, second larval instar; MF, morphogenetic furrow; SMW, second mitotic wave; R8 spec, R8 specification.

Initiation of the Morphogenetic Furrow

The generation of immature photoreceptors in the eye imaginal disc starts early in third larval instar. Movement of the morphogenetic furrow can be subdivided into two phases: First, initiation of the furrow, followed by furrow progression from posterior to anterior across the eye disc. Regularly spaced ommatidia form posterior to the furrow (see below). The Hedgehog (Hh), Wingless (Wg), Decapentaplegic (Dpp), Notch, and EGFR signaling pathways provide the basis for initiation of the furrow (Curtiss and Mlodzik, 2000; Kumar and Moses, 2001b). Birth of the furrow occurs at the posterior end of the eye disc at the optic stalk when the first row of retinal cells forms

(Fig. 4.5A). Since the eye disc has a round shape, the width of the furrow dorso-ventrally increases as it progresses (Fig. 4.5B). Therefore, furrow formation has to be reinitiated repeatedly at the rim of the disc, a process termed furrow reincarnation (Kumar and Moses, 2001b). Hh and Dpp act to promote furrow initiation and reincarnation, whereas Wg acts as an inhibitor (Treisman and Rubin, 1995). Prior to morphogenetic furrow initiation, *hh* is expressed at the posterior margin. Lack of the *hh* signal results in the complete inhibition of pattern formation while ectopic activation of the *hh* signal ahead of the furrow leads to ectopic furrow formation and retinal development (Ma *et al.*, 1993; Treisman and Rubin, 1995). Dpp expression is found along the posterior and lateral margins of the disc. Ectopic Dpp signal leads to new precocious furrow formation along the anterior margin. Loss of Dpp signal in contrast leads to the absence of the morphogenetic furrow, comparable to the function of Hh (Fig. 4.5B). Furthermore Dpp acts genetically to activate *eya* (as well as *so* and *dac*), thereby controlling the expression of RDN genes (Curtiss and Mlodzik, 2000). Wg is expressed along the lateral margins just anterior to the morphogenetic furrow. Since Wg acts as an inhibitor, ectopic *wg* signal stops morphogenetic furrow progression, whereas the loss of *wg* leads to ectopic furrow formation. The Jak/STAT signaling pathway acts to repress *wingless* and thereby allows birth and reincarnation of the morphogenetic furrow (Ekas *et al.*, 2006; Tsai *et al.*, 2007). Beside the action of these secreted signaling molecules which tightly control the formation of the morphogenetic furrow, Notch and EGFR signaling are also core components of the initiation of the morphogenetic furrow. The antagonistic action of Notch and EGFR is required to specify the antennal and eye part in the eye-antennal imaginal disc (see above). In the context of furrow formation Notch and EGFR signaling act synergistically to promote furrow initiation (Kumar and Moses, 2001b). Temperature shift experiments indicate that Dpp, EGFR, and Hh signaling are required for birth of the morphogenetic furrow.

The progressive movement of the morphogenetic furrow across the epithelium is a reiterative process, which also uses Hh, Dpp, and Notch signaling. After the furrow is initiated, differentiating photoreceptors behind the furrow express *hh*, which signals to cells at the anterior edge of the furrow to express Dpp (Dominguez and Hafen, 1997). Hh, as well as Dpp in conjunction propagate the progression of the "pre-proneural" state to the "proneural" state (Fig. 4.5C, Greenwood and Struhl, 1999). Initially *hairy* (*h*) is expressed in a broad stripe (pre-proneural), which subsequently ceases and *ato* expression comes up (proneural state). Repression of *h* is essential to adopt the correct cell fate (see below).

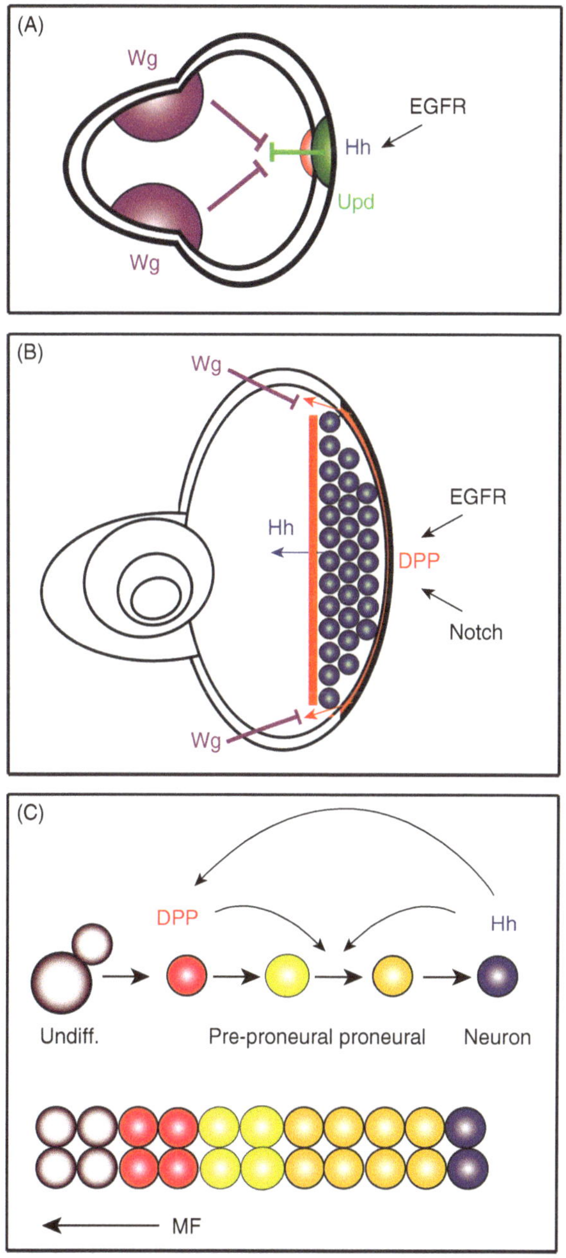

FIGURE 4.5 Model for the initiation and propagation of the morphogenetic furrow. Birth of the morphogenetic furrow depends on the interactions of the Wg, Hh, EGFR, and Jak/STAT signaling pathways (A). Upd is required to repress wg, and thereby allows birth and reincarnation of the morphogenetic furrow (A). Reincarnation of the morphogenetic furrow depends on Wg, Hh, EGFR, Dpp, and Notch signaling pathways (B). EGFR and Notch act upstream of Hh for initiation of the furrow, whereas Wg acts negatively on furrow formation. Dpp propagates furrow reincarnation at the lateral margins (B). The reiterative propagation of the furrow depends on the interaction of Hh and Dpp. Differentiating photoreceptors behind the furrow express *hh*, which signals to cells at the anterior edge of the furrow to express Dpp. Hh, as well as Dpp induce the maturation of the "pre-proneural" state to "proneural" state thereby propagating the progression of the furrow. MF, morphogenetic furrow.

Hh and Notch repress expression of a negative regulator of photoreceptor differentiation, *h*. Hh, Dpp, and with Notch signaling are required for the expression of the proneural gene *atonal* (see below). As cells start to express *ato* and later *hh* propagating the signal for furrow progression. The loss of *dpp* expression of cells emerging posterior of the furrow is essential for subsequent development and differentiation. This down regulation of Dpp expression depends on the cAMP-dependent protein kinase (PKA) (Pan and Rubin, 1995; Strutt *et al.*, 1995). The consequence of these events includes the steady movement of the furrow across the eye disc, providing the basis of photoreceptor development.

The Starting Point in Ommatidia Development: Specification of the R8 Photoreceptor Precursor

The development of individual photoreceptors is tightly linked with the development of ommatidial units as individual photoreceptors are recruited into ommatidia. As the morphogenetic furrow progresses, a new column forms about every 2 h. The first neuron to be specified is the R8 founder cell (Fig. 4.6). Founder cells in each column are not specified simultaneously as R8 at the equator are formed about 20 min before polar cells. Subsequently R8 recruits other photoreceptors into the ommatidal cluster. If R8 cells are absent, none of the other photoreceptors can develop. The formation and specification of this first neuron in the ommatidium require a complex molecular mechanism with positive and negative interactions involved to specify R8 precursors (for review see Frankfort and Mardon, 2002; Hsiung and Moses, 2002). R8 cells have to be evenly spaced in the epithelium to ensure the crystalline architecture of the eye.

The proneural gene *atonal* (*ato*) is required for the development of the peripheral nervous systems. It is also required for the selection of the R8 founder cell. Ato acts together with its heterodimeric partner Daughterless (Da), and if either of these genes is mutated, R8 cells do not develop and the other PRs are not recruited into the ommatidium. Expression of Ato is highly dynamic, first covering a broad stripe just anterior to the furrow. It becomes more and more restricted to smaller clusters until it is only expressed in a single cell per cluster (for review see Frankfort and Mardon, 2002). The broad band of *ato* expression (stage 1) splits into clusters of about 10 cells (stage 2; named intermediate group). Two to three nuclei then move apically and form an R8 equivalence group (stage 3), in which all cells are equipotent to give rise to R8. Only one of these 2–3 cells maintains *ato* expression and finally becomes

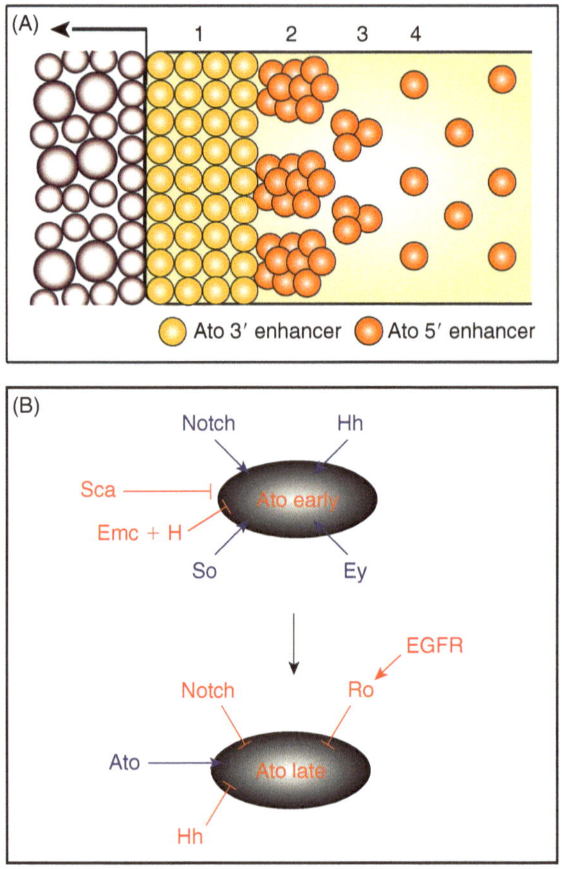

FIGURE 4.6 Specification of the R8 cell is an initial step in ommatidium development. The first cell to be specified in each ommatidium is the R8 photoreceptor which subsequently recruits the other cells into the ommatidium. Specification of the R8 cells depends on the proneural gene *atonal* (A). Initial *ato* is expressed in a broad band (stage 1) and subsequently gets restricted into intermediate groups (stage 2). About 2–3 cells then move form an R8 equivalence group (stage 3), only one of these 2–3 cells maintains *ato* expression and finally becomes the R8 photoreceptor (stage 4). Initial *ato* expression depends on a 3′ enhancer region, whereas later expression is regulated by a 5′ enhancer. The tight control of *ato* expression ensures proper spacing of ommatidia and that only one cell per ommatidium develops as R8 cell (B). During the initial broad *ato* expression (stage 1) Notch, Hh signaling as well as RDN genes (So and Ey) act to promote *ato* expression, whereas Emc in conjunction with H as well as Sca represses *ato* expression (B). During later stages Notch and Hh signaling repress *ato* (B). Ro acts genetically downstream of EGFR to repress *ato* expression. During this period *ato* auto-regulation is required to maintain proper development.

the R8 photoreceptor (stage 4). Therefore the tight control of *ato* expression is essential for the specification of a single R8 precursor per cluster, as well as the proper spacing of ommatidia (Fig. 4.6A). Interestingly the regulatory elements controlling *ato* expression changes during R8 precursor selection. The broad initial *ato* expression is under the control of a 3′ enhancer, whereas later expression is controlled by a distinct 5′ element (Fig. 4.6A), which requires auto-regulatory feedback from Ato (Sun *et al.*, 1998). The interaction of the Notch, Hh, and EGFR signaling pathways is required to ensure that only one cell is specified per cluster. Initially Hh and Notch signaling promote Ato expression (Fig. 4.6B). However, later Hh act negatively on *ato* expression (Fig. 4.6B). This might be due to the Hh gradient, with low level in the furrow promoting *ato* expression, whereas high levels of Hh anterior to the furrow repress *ato* (Dominguez, 1999).

The two transcriptional repressors H and Extra macrochaetae (Emc) act in conjunction to repress *ato* expression. Notch signaling first promotes *ato* expression by repressing *h* and *emc*. Notch acts during later stages to repress Ato expression. The fibronectin-like secreted protein Scabrous (Sca) interacts with Notch and is required to control proper *ato* expression (Fig. 4.6B). Sca is expressed in a subset of Ato cells in the intermediate group and remains expressed at high levels in R8. In *sca* mutants, there is an excess of R8 cells with incorrect spacing (Mlodzik *et al.*, 1990a). Similarly, the inactivation of EGFR results in too closely spaced R8 cells (Baonza *et al.*, 2001). The homeodomain transcription factor Rough (Ro) is activated in non-R8 cells due to lateral inhibition. Ro inhibits R8 development by repressing *ato*, and in *rough* mutants, the other cells of the R8 equivalence group adopt the R8 fate (Frankfort *et al.*, 2001). Conversely, the transcription factors *senseless* (*sens*) act in R8 to repress Ro. *sens* is also required to repress the expression of *pointed*, a nuclear mediator of the EGFR pathway, thereby inhibiting autocrine Spitz signaling in R8 (Frankfort *et al.*, 2004). Even though R8 selection occurs normally in *sens* mutants, R8 cells fail to differentiate and adopt an R2/R5-like fate, expressing Ro (Frankfort and Mardon, 2002).

Recruitment and Specification of R1–R7

The recruitment of the photoreceptors R1–R7 is controlled by R8. This progressive process follows a strictly stereotypical order: first R2 and R5, then R3 and R4, then R1 and R6, then R7 (Fig. 4.7A). After R8 has been specified, the recruitment of R2/R5 and Rh3/R4 occurs rapidly and forms a five-cell pre-cluster. The precursors of the remaining cells of the ommatidium, including R1/R6, R7, and other accessory cells undergo a further cell division before they become specified. This post-furrow proliferation is termed the "second mitotic wave" (see below). The recruitment of R1–R6 requires the expression of the EGFR ligand Spitz in the R8 cell

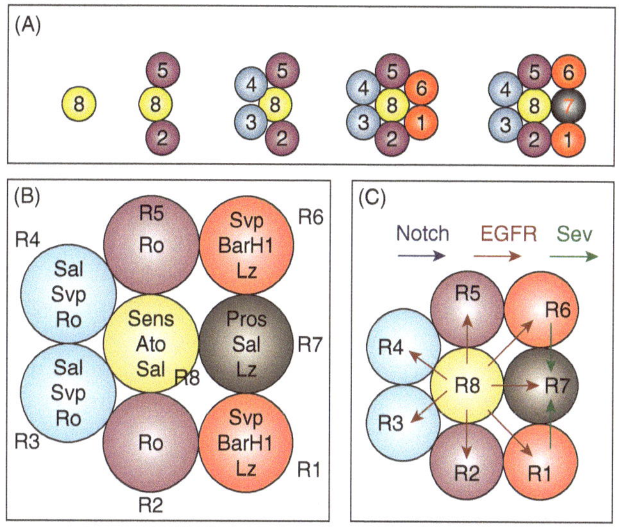

FIGURE 4.7 Recruitment of R1–R7 into the ommatidium. Recruitment by R8 follows a strictly stereotypical order: first R2 and R5, then R3 and R4, then R1 and R6, then R7 (A). Specification of the different photoreceptor requires the combinatorial code of several transcription factors (B) as well as Notch-, EGFR-, and Sev-signaling (C). The R3/R4 pair expresses Sal, Svp, and Ro; the R2/R5 pair Ro; the R1/R6 pair expresses Svp and BarH1; R7 expresses Pros, Sal, and Lz; R8 expresses Sens, Ato, and Sal (for details and function see text). The recruitment of R1–R6 depends on EGFR signaling from the R8 photoreceptor, R7 specification requires Sev-signaling from the R8 photoreceptor as well as Notch signal from R1 and R6 (C).

(Fig. 4.7C). The series of photoreceptor cell recruitment has provided major insights into the basis of molecular interactions of the Ras-pathway (for review see Dickson and Hafen, 1993). A number of cell–cell signaling events (Fig. 4.7C) as well as transcription factor action (Fig. 4.7B) is essential for proper specification and development of individual PRs. The first pair to be specified is the R2 and R5 pair, which requires the expression of Ro for proper specification (Tomlinson *et al.*, 1988). *rough* is expressed in R2/R5 and R3/R4 pairs (Fig. 4.7B). In *rough* mutants R2 and R5 are misspecified and express the R1/R3/R4/R6 marker *seven-up*. Furthermore ectopic expression of Svp in R2/R5 abolishes differentiation of photoreceptors, indicating that Ro acts to specify R2/R5 by repressing Svp (Kramer *et al.*, 1995). Furthermore, if *rough* is mutant, R3 and R4 are not properly recruited into the developing ommatidium. It has been proposed that R3/R4 partly depend upon a signal from R2/R5 for their development (Heberlein *et al.*, 1991; Dickson and Hafen, 1993).

The orphan nuclear receptor Seven-up (Svp) is required for the specification of R3/R4 and R1/R6 (Fig. 4.7B). In *svp* mutant clones, R3/R4 and R1/R6 are transformed into an R7-like cell (Mlodzik *et al.*, 1990b).

Later specification of the R3/R4 pair requires the interaction of Wingless and Notch signaling, which is a major determinant for the chirality of the ommatidium (see below). R3/R4 depends on the *spalt* gene complex, which encodes the two zinkfinger transcription factors *spalt major* and *spalt related.* In R3/R4 Sal is essential for the establishment of planar cell polarity and the activation of Svp expression (Domingos *et al.*, 2004). The transcription factor Lozenge (Lz) is essential for the development of R1 and R6 fates, which promotes the expression of the homeodomain transcription factor BarH1 (Daga *et al.*, 1996), which is required in R1 and R6 for their proper differentiation (Higashijima *et al.*, 1992). *lozenge* is furthermore required in R7 and cone cells to repress *svp* (Daga *et al.*, 1996). The last photoreceptor to be specified within each ommatidium is R7. In addition to EGFR a second receptor tyrosine kinase, Sevenless (Sev) is specifically required for R7 development (Fig. 4.7C), even though it is also expressed in R3/R4, cone cell precursors, and weakly in R1/R6 (Tomlinson *et al.*, 1987). The ligand for Sev, Bride of Sevenless (Boss) is found specifically in R8 photoreceptors. Even though R3/R4 is exposed to the R8 cell, and therefore to the Boss ligand, they do not adopt the R7 fate. Ectopic activation of Sev transforms cone cells into R7, but not R3/R4 or R1/R6 (Basler *et al.*, 1991), likely because these latter cells express *svp*. Removal of *svp* function in R3/R4 and R1/R6 transforms those cells into R7 cells. Conversely the expression of high levels of Svp is sufficient to transform R7 and cone cells into outer photoreceptors (Hiromi *et al.*, 1993; Begemann *et al.*, 1995; Kramer *et al.*, 1995). Thus, *svp* acts in a context-dependent manner to promote outer photoreceptor fate. In addition of Sev signaling, R7 development also depends on Notch and EGFR signaling (Fig. 4.7C). R7 receives the Notch signal from its neighboring cells R1 and R6; if this signal is absent, R7 differentiates into an R1/R6-like cell expressing BarH1 (Cooper and Bray, 1999; Tomlinson and Struhl, 2001). A further gene expressed in R7 and cone cells development is the homeodomain transcription factor Prospero (Pros), which is required for proper specification of R7 (see below Kauffmann *et al.*, 1996; Cook *et al.*, 2003).

Cell-cycle Control and Apoptosis in the Eye

A tight control of proliferation as well as the elimination of supernumerous cells by controlled cell death, or apoptosis, is of great importance for proper eye formation. Initially, all eye imaginal disc cells proliferate in a seemingly uncoordinated manner before cells are patterned by the passage of the morphogenetic furrow. This proliferative growth increases cell number from 70 to about 1300–1600 by the beginning of third larval instar (Wolff and Ready, 1993). EGFR signaling is essential for correct proliferation, as lack of EGFR signaling results in severe proliferation deficits, whereas high levels of EGFR (and Ras) triggers terminal neuronal differentiation (Dominguez and de Celis, 1998; Kumar *et al.*, 1998; Halfar *et al.*, 2001). As the morphogenetic furrow progresses, cells ahead of the furrow stop proliferation in G1. This cell cycle arrest is initially controlled by Dpp, but then becomes independent of Dpp. Since Dpp expression is strongest in the furrow and gradually ceases, leaving a morphogenetic gradient, it is thought that Dpp acts as an initial signal to induce cell cycle arrest. Ectopic Dpp anterior to the furrow is sufficient to stop the cell cycle (Penton *et al.*, 1997; Horsfield *et al.*, 1998). Dpp acts in conjunction with Hh, which arrests cells that do not respond to Dpp (Firth and Baker, 2005).

Subsequently to the recruitment of R8 and R2/R5 and R3/R4, cells of the five-cell pre-cluster start neuronal differentiation. Surroundings cells re-enter cell cycle in the second mitotic wave. During this stage, non-differentiating cells are in S-phase and mitosis. Apart from the 5 cells which are already differentiating, the remaining 14 cells of the ommatidium are born during this phase. Posterior to the furrow, most cells progress form G1 to S-phase and express cyclin D and cyclin E. The expression of Ato in R8 and EGFR signaling in R2/R5 and R3/R4 is required to maintain these cells in G1 arrest (Baker and Yu, 2001). The transition of G1 to S-phase in the second mitotic wave depends on Notch. The decision to re-enter S-phase or to remain in G1 is tightly regulated by the levels of cyclin E (Baonza and Freeman, 2001; Firth and Baker, 2005).

The major function of apoptosis, which occurs relatively late during development, is to remove excess cells of the eye. About 20 h after formation of the pupa, the specification of photoreceptors, primary pigment cells, and cone cells is completed. Interommatidial cells, however, are not yet patterned and remain undifferentiated. Cells start to organize, resulting in the final highly symmetric architecture of the ommatidium. Initially, two or more layers of interommatidial cells might separate individual ommatidia. The future lattice cells seem to prefer the contact with primary pigment cells compared to other interommatidial cells so that finally each lattice cell will contact at least two primary pigment cells (Cagan and Ready, 1989). Ommatidia are then reorganized in a tighter configuration separated by a single layer of interommatidial cells. The number of cells surrounding an ommatidium will subsequently decrease through apoptosis of excess cells, such that each ommatidium ends up being surrounded by exactly nine cells (Wolff and Ready, 1991). The two genes *irregular chiasm-C-roughest* (*irreC-rst*) and *echinus* (*ec*) are required for apoptosis of interommatidial cells and *irreC-rst* or *ec* mutants have an excess lattice cells. The two phenotypes, however, are different since in *irreC-rst* extra cells are piled up and stretched out along each other, whereas in *ec*, cells are correctly aligned (Wolff and Ready, 1991). The mechanism that control the correct number of interommatidial cells is also controlled by Notch, which acts to promote cell death, whereas EGFR signaling acts to suppress apoptosis (Miller and Cagan, 1998; Yu *et al.*, 2002). Primary pigment cells express the EGFR ligand spitz to promote survival of surrounding cells (Miller and Cagan, 1998).

Planar Cell Polarity

The eight photoreceptors of each ommatidium form trapezoids which are aligned in two mirror-image fields as compared to the equator. Therefore, each ommatidium is oriented in a two-dimensional grid, anterior–posterior and dorsal–ventral. The genetic program underlying the generation of these mirror-imaged fields requires several processes most of which are not completely understood at the molecular level. After its formation, the ommatidial pre-cluster is bilaterally symmetric along the anteroposterior axis. However, only a few columns later, the ommatidia start to rotate. The rotation occurs in a two-step process, first to 45° and then to 90° so that R7 is lying closest to the equator (Fig. 4.8A). Cells dorsal to the midline turn clockwise, whereas cells ventral to the midline counterclockwise (in the left eye). The midline has been proposed to act as an organizer to establish polarity, which finally results in the equator of the eye (Fig. 4.8B). For instance the gene *four-jointed* and a number of enhancer trap lines are expressed along the midline prior to the passage of the morphogenetic furrow. The secreted morphogen Wg, is expressed at both polar margins, if expressed in the ventral margin can either inhibit furrow formation or dorsalize the eye field, so that furrow initiation occurs more ventrally (Wehrli and Tomlinson, 1998). Interestingly generating mosaic clones of the Wg pathway genes including ectopic Wg, or *armadillo* (*arm*), and *arrow* (*arr*) loss-of-function reorient adjacent ommatidia and form a new equator. Loss of arrow or armadillo, reorient the ommatidia at the equatorial side of the clone, whereas ectopic Wg expression has the opposite effect. The homeodomain transcription factor *mirror* is expressed in the dorsal half of the developing retina. *mirror* (*mirr*) mosaic clones are sufficient to induce ectopic equators either by loss-of-function in

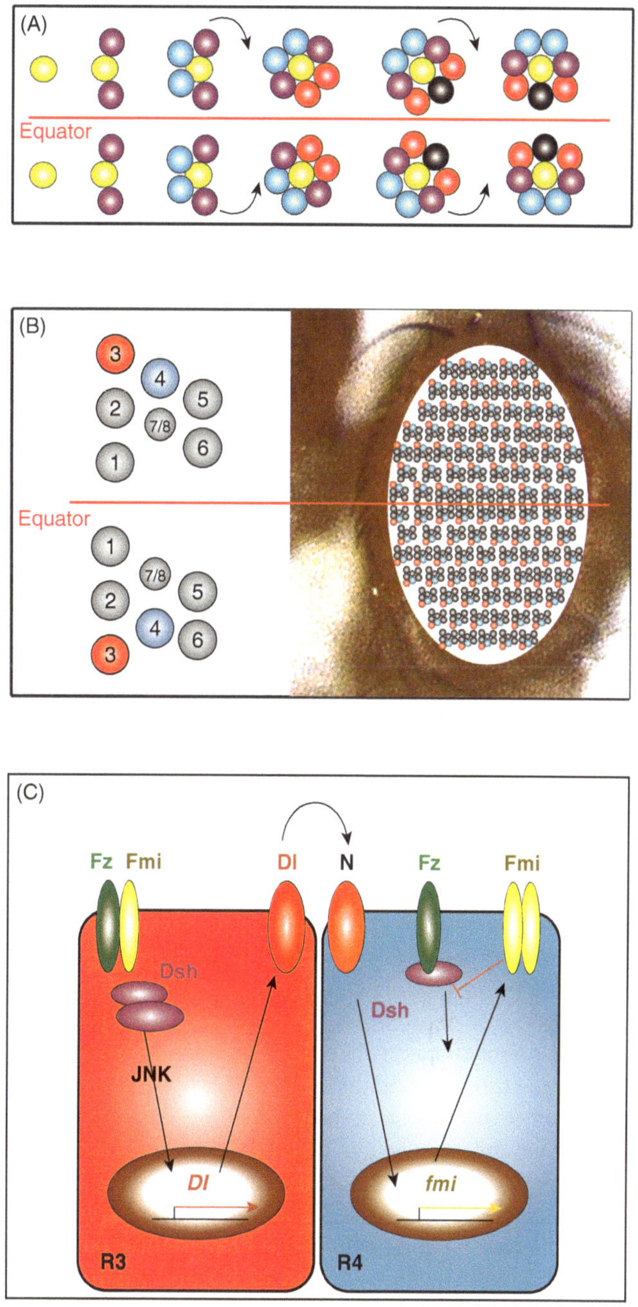

FIGURE 4.8 Establishing planar cell polarity in the retina. The ommatidium rotation is a two-step process, first to 45° and then to 90° so that R7 ends up lying closest to the equator (A). The eight photoreceptor of the ommatidium in the fly retina form a trapezoid. The trapezoids ventral and dorsal to the equator are aligned in two mirror-image fields (B). Ommatidial polarity is determined by the interaction of the R3/R4 pair. In the proto R3 cell, Dsh activates the JNK-cascade which results in the expression of Dl in R3. This leads to the activation of Notch in the neighboring R4 cell which in turn activates Fmi expression. Fmi acts as an antagonist of the Fz-signaling pathway, thereby ensuring the decision of R3 and R4 development (C modified after Mlodzik, 2002).

the dorsal half, or by gain-of-function in the ventral half (McNeill *et al.*, 1997). Fringe (Fng) a secreted protein is expressed in the ventral domain, where it promotes Notch activation by its ligand Serrate, whereas in the dorsal half *mirror* promotes elevated levels of Delta (another Notch ligand). Thereby abutting Fng and Mirr domains at the midline lead to a high level of Notch activity (Cho and Choi, 1998; Dominguez and de Celis, 1998; Papayannopoulos *et al.*, 1998).

Signaling between R3 and R4 photoreceptors is an essential step to induce chirality of the ommatidium. The last two cells to join the five-cell pre-cluster are the future R3 and R4. The cell on the polar side will become R4 and the cell at the equatorial side will become R3. The communication between R3 and R4 as well as their ability to read the dorso-ventral information has been studied extensively. Mutations affecting planar polarity in the eye include *frizzled* (*fz*), *strabismus/Van Gogh* (*stbm/Vang*), *disheveled* (*dsh*), *shaggy* (*sgg*), and *RhoA*. The transmembrane protein Fz acts as a Wnt receptor (for review see Mlodzik, 2002). Fz mutant clones display disrupted ommatidial polarity and R3/R4 differentiation is affected. If only one of the proto-R3/R4 cells expresses Fz, this cell will become R4, while the mutant cell will become the R3 and the ommatidium will rotate in the wrong direction. Thus Fz activity in this five-cell pre-cluster is not only defining which cell will become R3 or R4, but also determine the subsequent chirality and rotation (Fig. 4.8C). The frizzled receptor pathway uses the *dsh*, *sgg*, *stbm/Vang*, and *RhoA* to mediate non-canonical Wnt signaling. Even though Fz and Dsh also act in canonical Wnt signaling, the other components of the canonical Wnt pathway do not affect planar cell polarity. The Notch pathway also acts to distinguish member of the R3/R4 pair. In the proto R3 cell, Dsh activates the JNK-cascade which results in the expression of Dl (Fig. 4.8C). This leads to the activation of Notch in the neighboring R4 cell which in turn will promotes *flamingo* (*fmi*) expression. Fmi acts as an antagonist of the Fz-signaling pathway, thereby ensuring the decision of R3 and R4 to adopt the proper cell fate (for review see Mlodzik, 2002).

TERMINAL DIFFERENTIATION AND SUBTYPE SPECIFICATION OF PHOTORECEPTORS

The basic subdivision of the eight photoreceptors into outers and inners is not only characterized by their size and orientation within the ommatidium, but also reflects their functional role. The outer photoreceptors, R1–R6, are the fly equivalent of the vertebrate rods and have been implicated in motion detection, dim light vision, and image formation. Outer photoreceptors contain the broad spectrum photopigment *rhodopsin 1* and display an ommatidium of large diameter that spans the entire thickness of the retina, thus containing an increased volume of membranous structures in their rhabdomere, which enables them to capture photons with high efficiency.

The inner photoreceptors, R7 and R8, have been proposed to function like vertebrate cones in color vision (Morante and Desplan, 2004). In contrast to outer photoreceptors which span the whole length of the ommatidium, inners photoreceptors span only half the length, being located on top of each other. R7 is located on top of R8, thus sharing the same light path which allows them to compare their sensory outputs. Their diameter is considerably smaller than that of the outer photoreceptors. The first stage of sensory integration occurs in the optic lobes. The outer photoreceptors project into the lamina neuropile where they are primarily contacted by neurons localized in the lamina cortex. Inner photoreceptors extend their axonal projections to the medulla neuropile where they are contacted by neurons located in the medulla cortex.

Different Ommatidia Subtypes

The external morphology and appearance of all ommatidia of the *Drosophila* eye appear the same, ommatidia that can be subdivided into three distinct classes: the ommatidia of the dorsal rim area (DRA), the yellow ommatidia (y), and the pale ommatidia (p) (Fig. 4.9A–C). Initially p- and y-ommatidia have been identified by fluoroscopy and are randomly distributed throughout the eye. However the ratio of p- versus y-ommatidia is 70% yellow and 30% pale. The differences in their spectral absorbance lead to speculation that they might be involved in discrimination of different wavelengths of light. Indeed, the inner PRs of p- and y-ommatidia express different *rhodopsins* (Wernet and Desplan, 2004; Mikeladze-Dvali *et al.*, 2005a). In yellow ommatidia R7 expresses the UV-sensitive Rh4 and R8 expresses the green-sensitive Rh6, whereas in pale ommatidia R7 expresses the UV-sensitive Rh3 and R8 expresses the blue-sensitive Rh5. The coupling between Rh3 and Rh5 or Rh4 and Rh6, respectively, is kept in a stringent manner. The combination of a UV-rhodopsin and a green/blue-rhodopsin has been thought to be required for color discrimination. p-ommatidia discriminate between shorter wavelength light (UV and blue), whereas y-ommatidia

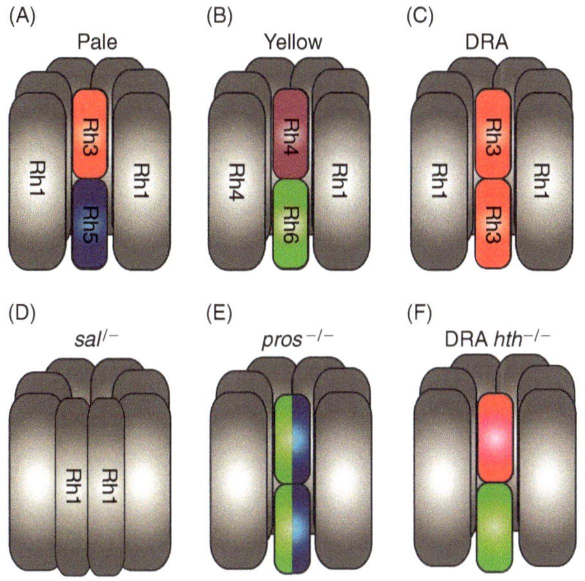

FIGURE 4.9 Different ommatidia types and inner PR versus outer PRs specification and R7 versus R8 specification. There are three types of ommatidia: pale (A; expressing Rh3 in R7, and Rh5 in R8), yellow (B; expressing Rh4 in R7, and Rh6 in R8), and DRA (C; expressing Rh3 in R7, and Rh3 in R8). Outer PRs express Rh1. Sal is required of the specification of inner PRs, in *sal* mutants inner PRs develop like outers and express Rh1 (D). Pros is required for R7 development, in *pros* mutants R7 displays R8 characteristics and expresses the R8 opsins Rh5 and Rh6 (E). If *hth* is lacking DRA ommatidia lack DRA-specific morphology and R7 and R8 express the combination of Rh3/Rh5 or Rh4/Rh6 (F).

discriminate between longer wavelength light (UV and green). Inner PRs of the third type of ommatidia in the DRA expresses Rh3 in both R7 and R8. The DRA ommatidia have been implicated in the perception of polarized light.

Specification of Inner Versus Outer Photoreceptors

The molecular basis distinguishing inner from outer PRs depends on the *spalt* gene complex. The *spalt* genes are expressed specifically in R7 and R8. In *spalt* mutants the two inner PRs display morphological characteristics of outer PRs (Fig. 4.9D), such as the rhabdomere morphology and rh1 expression (Mollereau *et al.*, 2001). In addition they lose inner PRs specific expression of Rh3, Rh4, Rh5, and Rh6. However the axonal projections of the transformed PRs are still terminating in the medulla, therefore initial specification of R7 and R8 in the eye-antennal imaginal disc does not seem to be affected but later only in terminal differentiation adopt an outer PR-identity. This suggests that there are two independent genetic programs in photoreceptor development; during an early-phase the establishment of neural and general photoreceptor identity and later the terminal differentiation into distinct photoreceptor types and subtypes. The *spalt* gene complex is therefore necessary to maintain inner photoreceptor terminal differentiation, in otherwise an "outer-PR" ground state. Therefore in a first step the *spalt* genes establish the specification of inner PRs, which might explain why the originally diverse R1–R6 all adopt an outer-PR fate and start to express *rh1*. It may further have implication on the divergent functions of inner PRs versus outer PRs. In an initial step they get specified as photoreceptors and a neuronal cell type which establish appropriate neuronal connectivity in the optic lobe, subsequently they get specified into specific photoreceptor types. Currently it is not known how the fine tuning of photoreceptor axon targeting is achieved both in a temporal as well as in a subtype specific manner (yR7versus yR8 and, pR7versus pR8, respectively).

Making Inner Photoreceptors to R7 and R8 cells

In addition to the differences to the outer photoreceptors, R7 and R8 display several morphological differences. First R7 is located on top of R8; second the nucleus of R7 is located distal, whereas the R8 nucleus is located proximal; third the projections of R7 terminate in a deeper layer than R8 projections. In addition to morphological criteria, R7 and R8 differ in the expression of *rhodopsins* (see above). The homeodomain transcription factor Pros plays a major role in distinguishing R7 fate from R8 fate (Fig. 4.9E). In the development of the central nervous system and peripheral nervous system Pros acts as a major determinant in asymmetric cell division of neuronal precursor cells (Egger *et al.*, 2008). Pros is expressed specifically in R7 photoreceptors in response to EGFR, Notch, and Sevenless signaling, and absent in R8 cells (and outer PRs) (Cook *et al.*, 2003). Initially *pros* has been identified in a screen for factors binding conserved regions upstream of rhodopsin promoters. More precisely prospero acts by directly binding to enhancer regions of R8 specific *rhodopsins* (*rh5* and *rh6*) and thereby repressed them in R7. Loss of *pros* leads to the de-repression of *rh5* and *rh6* in R7 cells, as well as nuclear mislocation therefore resulting in a second R8-like cell in each ommatidium. Conversely, the misexpression of Pros leads to the repression of *rh5* and *rh6* in R8 photoreceptors (Cook *et al.*, 2003). Axonal projections in the medulla of R7 however remain unaltered, again indicating that initial R7 specification is not affected.

Stochastic Specification of Yellow Versus Pale Ommatidia

A major question in the development of different ommatidia types is what determines an ommatidium to develop into either the p-type or the y-type, and which cell is controlling this fate. The bHLH-PAS (basic helix-loop-helix-Period-Arnt-Single-minded) transcription factor *spineless* (*ss*) is the major determinant in this process (Wernet *et al.*, 2006). *Ss* is expressed during metamorphosis in a subset of R7 cells in a stochastic manner. About 60–80% of all R7 express Ss, suggesting that *ss* is acting in y-ommatidia specification. Ss is both necessary and sufficient for the yR7 fate. R7 photoreceptors in *spineless* mutants all express Rh3, therefore adopt the pR7 fate (Fig. 4.10C). Conversely the ectopic misexpression of Ss results in the transformation of all R7 into yR7 (all R7 express Rh4) (Fig. 4.10D). Moreover Ss misexpression is also sufficient to induce R4 expression in outer PRs. How do R8 cells behave if R7 are transformed by lack of ectopic Ss expression? In *ss* mutants, most R8 photoreceptors express Rh5, whereas the ectopic expression of *ss* makes R8 photoreceptors express Rh6. Therefore spineless is acting at multiple levels in y-ommatidium specification (Wernet *et al.*, 2006). First, in R7 where it promotes Rh4 expression and represses Rh3. Secondly, in the underlying R8 photoreceptor where spineless is non-autonomously required for the expression of Rh6. This suggests that in yR7 Ss is controlling a signal to R8, which is required in this cell to adopt the yR8 fate.

Developmental Choice to Specify Yellow Versus Pale R8 Photoreceptors

The stochastic choice of a given ommatidium to adopt a pale or yellow fate is made in R7 by Ss. Depending on the decision of R7, the underlying R8 cell will adopt the same fate as R7, thereby leading to a strict coupling of inner *rhodopsins* (Mikeladze-Dvali *et al.*, 2005b). The genetic bistable loop of the growth regulator *warts* (*wts*) and the tumor suppressor *melted* (*melt*) leads to an unambiguous decision in R8 whether to adopt an yR8 (expressing Rh6) or a pR8 (expressing Rh5) fate. Wts is expressed in yR8 where it promotes Rh6 expression and represses Rh5 (Fig. 4.10G,H). Conversely, Melt is expressed in pR8 and promotes Rh5 expression and represses Rh6 (Fig. 4.10E,F). Both *wts* and *melt* are expressed in a mutually exclusive fashion. Wts represses *melt* and vise versa. In *wts* mutants all R8 PRs express Melt and therefore Rh5. In *melt* mutants all R8 PRs express *wts* and therefore Rh6. Ectopic activation of *wts* or *melt* leads to the repression of the opponent. In other tissues, both players are involved in distinct pathways. Hippo (Hpo) and Salvador (Sav) are two molecular partners of *wts* in the tumor suppression pathway and display identical phenotypes in R8-subtype specification as *wts*. Growth pathway components such as TOR and insulin receptor on the other hand are not involved in R8 cellfate specification (Mikeladze-Dvali *et al.*, 2005b). The novel role growth regulator and the tumor suppressor pathways in post-mitotic cells and their specification is rather surprising.

Specification of Inners Photoreceptors in the Dorsal Rim Area (DRA)

The dorsal half of the eye contains a set of highly specialized ommatidia which are involved in the detection of polarized light. The DRA is composed of an array of one to two rows of ommatidia directly adjacent to the head cuticle. In DRA ommatidia, inner PRs are highly adapted in their morphology and configuration to act as polarized light sensors. The ability to sense polarized light is dependent on strict alignment of microvilli which forms the rhabdomere. Other PRs are insensitive to polarized light due to the misalignment of microvilli caused by rhabdomere twisting. Furthermore the diameter of inner PR rhabdomeres is significantly enlarged. One factor involved in the development of DRA ommatidia is the secreted morphogen Wg, which is expressed in the head cuticle surrounding the eye. Ectopic activation of the Wg pathway in the eye transforms ommatidia in the dorsal half into DRA ommatidia, suggesting that these ommatidia are competent to respond to Wg signaling. Members of the Iroquois-complex (IRO-C) are expressed in the dorsal half of the eye (Tomlinson, 2003; Wernet *et al.*, 2003). Ectopic activation of any of the three IRO-C genes *auracan*, *caupolican*, and *mirror* in the whole eye leads to the expansion of DRA to the ventral margin. Therefore the combination of Wg and IRO-C functions leads to the specification of DRA ommatidia in the right place (Tomlinson, 2003; Wernet *et al.*, 2003). The major factor for DRA ommatidia development is the homeodomain transcription factors and Hox-protein co-factor Hth. In the eye, Hth is specifically expressed in R8 and R7 cells of DRA ommatidia (Fig. 4.9F). If *hth* is lacking in DRA ommatidia leads to the absence of DRA specific morphology

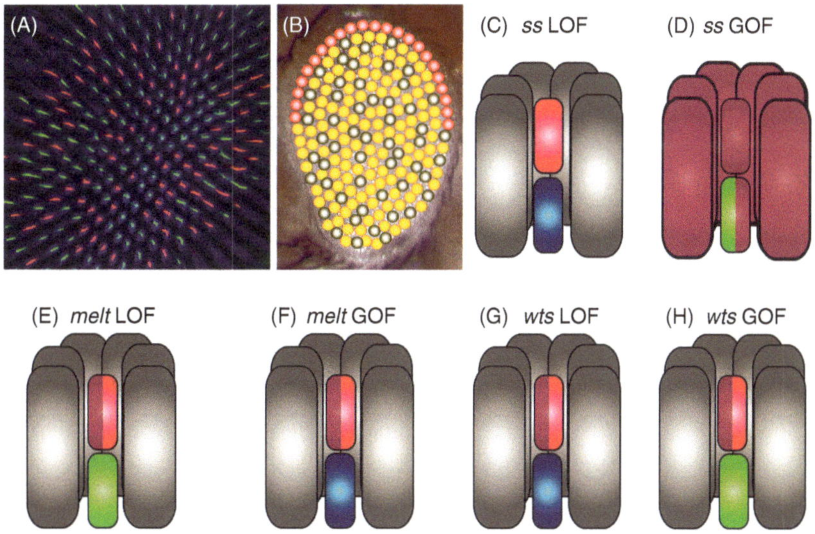

FIGURE 4.10 The stochastic choice of pale versus yellow ommatidia specification and developmental program to specify yellow versus pale R8 photoreceptors. R8 photoreceptors express Rh5 and Rh6 in a 30:70 ratio (A), which is consistent with the distribution of pale and yellow ommatidia (B; DRA photoreceptors are specified by Hth, see Fig 4.9). In *ss* mutants all R7 photoreceptors express Rh3 and therefore adopt the pR7 fate, while R8 PRs express Rh5 (C). Ectopic misexpression of Ss results in the transformation of all R7 into yR7 (all R7 express Rh4). SS is also sufficient to induce R4 expression in outer PRs (D). The choice to develop as yR8 or pR8 depends on *wts* and *melt*. In *melt* mutants all R8 photoreceptors express Rh6, therefore adopt the yR8 fate (E), whereas ectopic activation of *melt* results in Rh5 expression in R8, therefore adopt the pR8 fate (F). In *wts* mutants all R8 photoreceptors express Rh5, therefore adopt the pR8 fate (G), whereas ectopic activation of *wts* results in Rh6 expression in R8, therefore adopt the yR8 fate (H).

such as the increased rhabdomere diameter and the expression of Rh3 in R7 and Rh8. Furthermore, R7 and R8 express the untypical combination of Rh3/Rh6 (Wernet *et al.*, 2003). Ectopic expression of Hth is sufficient to transform all ommatidia into DRA ommatidia, with an increased inner PR diameter, expression of Rh3 and lack of normal coupling of Rh3/Rh5 or Rh4/Rh6. Interestingly, the ectopic activation of Ss is not sufficient to induce Rh4 expression in DRA inner PRs. Thus Hth seems to be sufficient to antagonize Ss function in the DRA. *hth* is both necessary and sufficient for the specification of polarized light sensors by coordinating R7 and R8 terminal differentiation.

DEVELOPMENT AND SPECIFICATION OF THE LARVAL EYE

The life cycle of all holometabolous insects, such as *Drosophila*, is bipartite. During larval stages the animal makes use of a set of sensory organs which develop during embryonic stages, as compared to the sensory organs of the adult fly which only get terminally specified during pupation. The eyes of the *Drosophila* larva, also termed as Bolwig's organ (BO), are comparably simple and consist of a paired structure each eye containing only about 12 photoreceptors. There are two distinct subtypes: about four PRs express the blue-sensitive *rh5* and about eight express the green-sensitive *rh6* (Sprecher *et al.*, 2007). The lateral-ventral tip of the optic lobe primordium gives rise to precursors cells of the larval eye. Development of larval PRs occurs in a two-step process (Fig. 4.11A,B). First, primary BO precursor cells get specified which requires the proneural gene *ato* and the RDN genes *so* and *eya*, as well as *hh* signaling (Green *et al.*, 1993; Daniel *et al.*, 1999; Suzuki and Saigo, 2000). Second, primary precursors signal to adjacent cells via EGFR signaling to develop as secondary precursors. The orphan nuclear receptor *tailless* (*tll*) is expressed in the rest of the optic lobe anlage and inhibits cells to develop as secondary precursors (Daniel *et al.*, 1999). Primary precursors will give rise to the Rh5-subtype, whereas secondary precursors give rise to the Rh6-subtype (Fig. 4.11C). The use of EGFR pathway in larval and ommatidal photoreceptors seems to be quite different, since in the adult EGFR together with Sev is required to recruit individual cells into the ommatidal cluster (see above). Even though both larval PR-subtypes express R8 specific *rhodopsins*, their specification does not depend on *wts* and *melt* as compared to the adult retina. Instead a set of three transcription

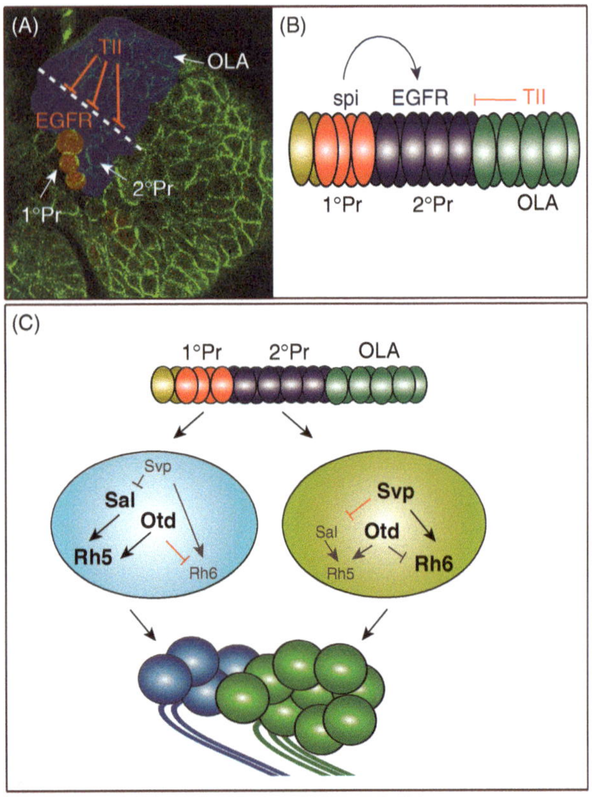

FIGURE 4.11 Development and subtype specification of larval PRs. Primary precursors (expressing Ato-red) signal to the adjacent cells to develop as secondary precursors (blue, region giving rise to secondary precursors; green, membranes in neurogenic region) (A). Tll antagonizes EGFR signaling inhibiting secondary precursor formation (B). Primary precursors give rise to the Rh5-PRs whereas secondary precursors give rise to Rh6-PRs (C). Rh5-PRs, Sal is required for Rh5 expression, whereas Otd represses Rh6 and promotes Rh5 expression. In Rh6 PRs, Svp represses Sal and promotes Rh6 expression (modified after Sprecher *et al.*, 2007). OLA, optic lobe anlage; 1° Pr, primary precursors; 2° Pr, secondary precursors.

factors orchestrate subtype specification. The transcription factor Sal is expressed in the Rh5-subtype, where it is required for expression of *rh5*. The role of Sal is quite different from the function in adult ommatidia development where Sal specifies inner PR cell fate (see above). The orphan nuclear receptor Svp acts in the opposite manner. Svp is only expressed in the Rh6-subtype, where it is required to repress Sal and to promote Rh6-expression (Fig. 4.11C). In *svp* mutants all PRs express Rh6. As with the role of Sal, the function of Svp is surprisingly different when compared to the adult retina (see above). A third player acting in larval PR-subtype specification is the homeodomain transcription factor Orthodenticle (Otd). The expression of *otd* is not restricted to a specific subtype, but is only required in the Rh5-PRs (Sprecher *et al.*, 2007). In *otd* mutants all PRs express Rh6. Therefore the combinatorial action of Svp, Sal, and Otd are required to specify larval PRs. Even though larval PRs and adult R8 express the same set of *rhodopsins*, the genetic mechanisms underlying PR subtypes specification are distinct.

ACKNOWLEDGMENTS

We would like to thank our colleagues at the Department of Biology at New York University, especially Dr. Robert Johnston and Dr. Daniel Vasiliauskas, for fruitful discussions and comments on the manuscript.

REFERENCES

Adams MD, Celniker SE, Holt RA, Evans CA, Gocayne JD, Amanatides PG, Scherer SE, Li PW, Hoskins RA, Galle RF, George RA, Lewis SE, Richards S, Ashburner M, Henderson SN, Sutton GG, Wortman JR, Yandell MD, Zhang Q, Chen LX, Brandon RC, Rogers YH, Blazej RG, Champe M, Pfeiffer BD, Wan KH, Doyle C, Baxter EG, Helt G, Nelson CR, Gabor GL, Abril JF, Agbayani A, An HJ, Andrews-Pfannkoch C, Baldwin D, Ballew RM, Basu A, Baxendale J, Bayraktaroglu L, Beasley EM, Beeson KY, Benos PV, Berman BP, Bhandari D, Bolshakov S, Borkova D, Botchan MR, Bouck J, Brokstein P, Brottier P, Burtis KC, Busam DA, Butler H, Cadieu E, Center A, Chandra I, Cherry JM, Cawley S, Dahlke C, Davenport LB, Davies P, de Pablos B, Delcher A, Deng Z, Mays AD, Dew I, Dietz SM, Dodson K, Doup LE, Downes M, Dugan-Rocha S, Dunkov BC, Dunn P, Durbin KJ, Evangelista CC, Ferraz C, Ferriera S, Fleischmann W, Fosler C, Gabrielian AE, Garg NS, Gelbart WM, Glasser K, Glodek A, Gong F, Gorrell JH, Gu Z, Guan P, Harris M, Harris NL, Harvey D, Heiman TJ, Hernandez JR, Houck J, Hostin D, Houston KA, Howland TJ, Wei MH, Ibegwam C, Jalali M, Kalush F, Karpen GH, Ke Z, Kennison JA, Ketchum KA, Kimmel BE, Kodira CD, Kraft C, Kravitz S, Kulp D, Lai Z, Lasko P, Lei Y, Levitsky AA, Li J, Li Z, Liang Y, Lin X, Liu X, Mattei B, McIntosh TC, McLeod MP, McPherson D, Merkulov G, Milshina NV, Mobarry C, Morris J, Moshrefi A, Mount SM, Moy M, Murphy B, Murphy L, Muzny DM, Nelson DL, Nelson DR, Nelson KA, Nixon K, Nusskern DR, Pacleb JM, Palazzolo M, Pittman GS, Pan S, Pollard J, Puri V, Reese MG, Reinert K, Remington K, Saunders RD, Scheeler F, Shen H, Shue BC, Siden-Kiamos I, Simpson M, Skupski MP, Smith T, Spier E, Spradling AC, Stapleton M, Strong R, Sun E, Svirskas R, Tector C, Turner R, Venter E, Wang AH, Wang X, Wang ZY, Wassarman DA, Weinstock GM, Weissenbach J, Williams SM, Woodage T, Worley KC, Wu D, Yang S, Yao QA, Ye J, Yeh RF, Zaveri JS, Zhan M, Zhang G, Zhao Q, Zheng L, Zheng XH, Zhong FN, Zhong W, Zhou X, Zhu S, Zhu X, Smith HO, Gibbs RA, Myers EW, Rubin GM, Venter JC (2000). The genome sequence of *Drosophila melanogaster*. *Science* 287:2185–2195.

Anderson J, Salzer CL, Kumar JP (2006). Regulation of the retinal determination gene dachshund in the embryonic head and developing eye of *Drosophila*. *Dev Biol* 297:536–549.

Baker NE, Yu SY (2001). The EGF receptor defines domains of cell cycle progression and survival to regulate cell number in the developing *Drosophila* eye. *Cell* 104:699–708.

Baonza A, Casci T, Freeman M (2001). A primary role for the epidermal growth factor receptor in ommatidial spacing in the *Drosophila* eye. *Curr Biol* 11:396–404.

Baonza A, Freeman M (2001). Notch signalling and the initiation of neural development in the *Drosophila* eye. *Development* 128:3889–3898.

Basler K, Christen B, Hafen E (1991). Ligand-independent activation of the sevenless receptor tyrosine kinase changes the fate of cells in the developing *Drosophila* eye. *Cell* 64:1069–1081.

Begemann G, Michon AM, Voorn L vd, Wepf R, Mlodzik M (1995). The *Drosophila* orphan nuclear receptor seven-up requires the Ras pathway for its function in photoreceptor determination. *Development* 121:225–235.

Brand AH, Perrimon N (1993). Targeted gene expression as a means of altering cell fates and generating dominant phenotypes. *Development* 118:401–415.

Cagan RL, Ready DF (1989). The emergence of order in the *Drosophila* pupal retina. *Dev Biol* 136:346–362.

Chen R, Amoui M, Zhang Z, Mardon G (1997). Dachshund and eyes absent proteins form a complex and function synergistically to induce ectopic eye development in *Drosophila* [see comments]. *Cell* 91:893–903.

Cheyette BN, Green PJ, Martin K, Garren H, Hartenstein V, Zipursky SL (1994). The *Drosophila* sine oculis locus encodes a homeodomain-containing protein required for the development of the entire visual system. *Neuron* 12:977–996.

Cho KO, Choi KW (1998). Fringe is essential for mirror symmetry and morphogenesis in the *Drosophila* eye. *Nature* 396:272–276.

Cook T, Pichaud F, Sonneville R, Papatsenko D, Desplan C (2003). Distinction between color photoreceptor cell fates is controlled by Prospero in *Drosophila*. *Dev Cell* 4:853–864.

Cooper MT, Bray SJ (1999). Frizzled regulation of Notch signalling polarizes cell fate in the *Drosophila* eye. *Nature* 397:526–530.

Curtiss J, Mlodzik M (2000). Morphogenetic furrow initiation and progression during eye development in *Drosophila*: the roles of decapentaplegic, hedgehog and eyes absent. *Development* 127:1325–1336.

Czerny T, Halder G, Kloter U, Souabni A, Gehring WJ, Busslinger M (1999). *twin* of eyeless, a second *Pax-6* gene of *Drosophila*, acts upstream of eyeless in the control of eye development. *Mol Cell* 3:297–307.

Daga A, Karlovich CA, Dumstrei K, Banerjee U (1996). Patterning of cells in the *Drosophila* eye by Lozenge, which shares homologous domains with AML1. *Genes Dev* 10:1194–1205.

Daniel A, Dumstrei K, Lengyel JA, Hartenstein V (1999). The control of cell fate in the embryonic visual system by atonal, tailless and EGFR signaling. *Development* 126:2945–2954.

Dickson B, Hafen E (1993). Genetic dissection of eye development in *Drosophila*. In: Bate M, Arias AM (eds), *The Development of Drosophila melanogaster*, Vol. 2. Cold Spring Harbor Laboratory Press, New York, pp. 1327–1363.

Domingos PM, Mlodzik M, Mendes CS, Brown S, Steller H, Mollereau B (2004). Spalt transcription factors are required for R3/R4 specification and establishment of planar cell polarity in the *Drosophila* eye. *Development* 131:5695–5702.

Dominguez M (1999). Dual role for Hedgehog in the regulation of the proneural gene atonal during ommatidia development. *Development* 126:2345–2353.

Dominguez M, de Celis JF (1998). A dorsal/ventral boundary established by Notch controls growth and polarity in the *Drosophila* eye. *Nature* 396:276–278.

Dominguez M, Ferres-Marco D, Gutierrez-Avino FJ, Speicher SA, Beneyto M (2004). Growth and specification of the eye are controlled independently by Eyegone and Eyeless in *Drosophila melanogaster*. *Nat Genet* 36:31–39.

Dominguez M, Hafen E (1997). Hedgehog directly controls initiation and propagation of retinal differentiation in the *Drosophila* eye. *Genes Dev* 11:3254–3264.

Egger B, Chell JM, Brand AH (2008). Insights into neural stem cell biology from flies. *Philos Trans R Soc Lond B Biol Sci* 363(1489):39–56.

Ekas LA, Baeg GH, Flaherty MS, Ayala-Camargo A, Bach EA (2006). JAK/STAT signaling promotes regional specification by negatively regulating wingless expression in *Drosophila*. *Development* 133:4721–4729.

Firth LC, Baker NE (2005). Extracellular signals responsible for spatially regulated proliferation in the differentiating *Drosophila* eye. *Dev Cell* 8:541–551.

Frankfort BJ, Mardon G (2002). R8 development in the *Drosophila* eye: a paradigm for neural selection and differentiation. *Development* 129:1295–1306.

Frankfort BJ, Nolo R, Zhang Z, Bellen H, Mardon G (2001). Senseless repression of rough is required for R8 photoreceptor differentiation in the developing *Drosophila* eye. *Neuron* 32:403–414.

Frankfort BJ, Pepple KL, Mamlouk M, Rose MF, Mardon G (2004). Senseless is required for pupal retinal development in Drosophila. *Genesis* 38:182–194.

Garcia-Bellido A, Merriam JR (1969). Cell lineage of the imaginal discs in *Drosophila* gynandromorphs. *J Exp Zool* 170:61–75.

Gehring WJ (2001). The genetic control of eye development and its implications for the evolution of the various eye-types. *Zoology-Jena* 104:171–183.

Gehring WJ (2005). New perspectives on eye development and the evolution of eyes and photoreceptors. *J Hered* 96:171–184.

Gehring WJ, Ikeo K (1999). Pax 6: mastering eye morphogenesis and eye evolution. *Trends Genet* 15:371–377.

Green P, Hartenstein AY, Hartenstein V (1993). The embryonic development of the *Drosophila* visual system. *Cell Tissue Res* 273:583–598.

Greenwood S, Struhl G (1999). Progression of the morphogenetic furrow in the *Drosophila* eye: the roles of Hedgehog, Decapentaplegic and the Raf pathway. *Development* 126:5795–5808.

Groger H, Callaerts P, Gehring WJ, Schmid V (2000). Characterization and expression analysis of an ancestor-type Pax gene in the hydrozoan jellyfish *Podocoryne carnea*. *Mech Dev* 94:157–169.

Halder G, Callaerts P, Gehring WJ (1995). Induction of ectopic eyes by targeted expression of the eyeless gene in *Drosophila* [see comments]. *Science* 267:1788–1792.

Halfar K, Rommel C, Stocker H, Hafen E (2001). Ras controls growth, survival and differentiation in the *Drosophila* eye by different thresholds of MAP kinase activity. *Development* 128:1687–1696.

Heberlein U, Mlodzik M, Rubin GM (1991). Cell-fate determination in the developing *Drosophila* eye: role of the rough gene. *Development* 112:703–712.

Higashijima S, Kojima T, Michiue T, Ishimaru S, Emori Y, Saigo K (1992). Dual Bar homeo box genes of *Drosophila* required in two photoreceptor cells, R1 and R6, and primary pigment cells for normal eye development. *Gene Dev* 6:50–60.

Hiromi Y, Mlodzik M, West SR, Rubin GM, Goodman CS (1993). Ectopic expression of seven-up causes cell fate changes during ommatidial assembly. *Development* 118:1123–1135.

Horsfield J, Penton A, Secombe J, Hoffman FM, Richardson H (1998). Decapentaplegic is required for arrest in G1 phase during *Drosophila* eye development. *Development* 125:5069–5078.

Hsiung F, Moses K (2002). Retinal development in *Drosophila*: specifying the first neuron. *Hum Mol Genet* 11:1207–1214.

Jang CC, Chao JL, Jones N, Yao LC, Bessarab DA, Kuo YM, Jun S, Desplan C, Beckendorf SK, Sun YH (2003). Two Pax genes, eye gone and eyeless, act cooperatively in promoting *Drosophila* eye development. *Development* 130:2939–2951.

Jurgens G, Hartenstein V (1993). The terminal regions of the body pattern. In: Bate M, Arias AM (eds), *The Development of Drosophila melanogaster*, Vol. 1. Cold Spring Harbor Laboratory Press, Cold Spring Harbor, pp. 687–745.

Kammermeier L, Leemans R, Hirth F, Flister S, Wenger U, Walldorf U, Gehring WJ, Reichert H (2001). Differential expression and function of the *Drosophila* Pax6 genes eyeless and twin of eyeless in embryonic central nervous system development. *Mech Dev* 103:71–78.

Kauffmann RC, Li S, Gallagher PA, Zhang J, Carthew RW (1996). Ras1 signaling and transcriptional competence in the R7 cell of *Drosophila*. *Genes Dev* 10:2167–2178.

Kenyon KL, Ranade SS, Curtiss J, Mlodzik M, Pignoni F (2003). Coordinating proliferation and tissue specification to promote regional identity in the *Drosophila* head. *Dev Cell* 5:403–414.

Kramer S, West SR, Hiromi Y (1995). Cell fate control in the *Drosophila* retina by the orphan receptor seven-up: its role in the decisions mediated by the ras signaling pathway. *Development* 121:1361–1372.

Kumar JP (2001). Signalling pathways in *Drosophila* and vertebrate retinal development. *Nat Rev Genet* 2:846–857.

Kumar JP, Moses K (2001a). EGF receptor and Notch signaling act upstream of Eyeless/Pax6 to control eye specification. *Cell* 104:687–697.

Kumar JP, Moses K (2001b). The EGF receptor and notch signaling pathways control the initiation of the morphogenetic furrow during *Drosophila* eye development. *Development* 128:2689–2697.

Kumar JP, Tio M, Hsiung F, Akopyan S, Gabay L, Seger R, Shilo BZ, Moses K (1998). Dissecting the roles of the *Drosophila* EGF receptor in eye development and MAP kinase activation. *Development* 125:3875–3885.

Kurusu M, Nagao T, Walldorf U, Flister S, Gehring WJ, Furukubo-Tokunaga K (2000). Genetic control of development of the mushroom bodies, the associative learning centers in the *Drosophila* brain, by the eyeless, twin of eyeless, and Dachshund genes. *P Natl Acad Sci USA* 97:2140–2144.

Lawrence PA, Struhl G, Morata G (1979). Bristle patterns and compartment boundaries in the tarsi of *Drosophila*. *J Embryol Exp Morphol* 51:195–208.

Lee T, Luo L (2001). Mosaic analysis with a repressible cell marker (MARCM) for *Drosophila* neural development. *Trends Neurosci* 24:251–254.

Ma C, Zhou Y, Beachy PA, Moses K (1993). The segment polarity gene hedgehog is required for progression of the morphogenetic furrow in the developing *Drosophila* eye. *Cell* 75:927–938.

McNeill H, Yang CH, Brodsky M, Ungos J, Simon MA (1997). *mirror* encodes a novel PBX-class homeoprotein that functions in the definition of the dorsal-ventral border in the *Drosophila* eye. *Gene Dev* 11:1073–1082.

Mikeladze-Dvali T, Desplan C, Pistillo D (2005a). Flipping coins in the fly retina. *Curr Top Dev Biol* 69:1–15.

Mikeladze-Dvali T, Wernet MF, Pistillo D, Mazzoni EO, Teleman AA, Chen YW, Cohen S, Desplan C (2005b). The growth regulators warts/lats and melted interact in a bistable loop to specify opposite fates in *Drosophila* R8 photoreceptors. *Cell* 122:775–787.

Miller DT, Cagan RL (1998). Local induction of patterning and programmed cell death in the developing *Drosophila* retina. *Development* 125:2327–2335.

Mlodzik M (2002). Planar cell polarization: do the same mechanisms regulate *Drosophila* tissue polarity and vertebrate gastrulation? *Trends Genet* 18:564–571.

Mlodzik M, Baker NE, Rubin GM (1990a). Isolation and expression of scabrous, a gene regulating neurogenesis in *Drosophila*. *Gene Dev* 4:1848–1861.

Mlodzik M, Hiromi Y, Weber U, Goodman CS, Rubin GM (1990b). The *Drosophila* seven-up gene, a member of the steroid receptor gene superfamily, controls photoreceptor cell fates. *Cell* 60:211–224.

Mollereau B, Dominguez M, Webel R, Colley NJ, Keung B, de Celis JF, Desplan C (2001). Two-step process for photoreceptor formation in *Drosophila*. *Nature* 412:911–913.

Morante J, Desplan C (2004). Building a projection map for photoreceptor neurons in the *Drosophila* optic lobes. *Semin Cell Dev Biol* 15:137–143.

Noveen A, Daniel A, Hartenstein V (2000). Early development of the *Drosophila* mushroom body: the roles of eyeless and dachshund. *Development* 127:3475–3488.

Onuma Y, Takahashi S, Asashima M, Kurata S, Gehring WJ (2002). Conservation of Pax 6 function and upstream activation by Notch signaling in eye development of frogs and flies. *P Natl Acad Sci USA* 99:2020–2025.

Pan D, Rubin GM (1995). cAmp-dependent protein kinase and hedgehog act antagonistically in regulating decapentaplegic transcription in *Drosophila* imaginal discs. *Cell* 80:543–552.

Papayannopoulos V, Tomlinson A, Panin VM, Rauskolb C, Irvine KD (1998). Dorsal-ventral signaling in the *Drosophila* eye. *Science* 281:2031–2034.

Penton A, Selleck SB, Hoffmann FM (1997). Regulation of cell cycle synchronization by decapentaplegic during *Drosophila* eye development. *Science* 275:203–206.

Pichaud F, Casares F (2000). homothorax and iroquois-C genes are required for the establishment of territories within the developing eye disc. *Mech Dev* 96:15–25.

Pichaud F, Desplan C (2002). Pax genes and eye organogenesis. *Curr Opin Genet Dev* 12:430–434.

Pignoni F, Hu B, Zavitz KH, Xiao J, Garrity PA, Zipursky SL (1997). The eye-specification proteins So and Eya form a complex and regulate multiple steps in *Drosophila* eye development. *Cell* 91:881–891.

Pineda D, Gonzalez J, Callaerts P, Ikeo K, Gehring WJ, Salo E (2000). Searching for the prototypic eye genetic network: sine oculis is essential for eye regeneration in planarians. *P Natl Acad Sci USA* 97:4525–4529.

Pineda D, Rossi L, Batistoni R, Salvetti A, Marsal M, Gremigni V, Falleni A, Gonzalez-Linares J, Deri P, Salo E (2002). The genetic network of prototypic planarian eye regeneration is Pax6 independent. *Development* 129:1423–1434.

Punzo C, Plaza S, Seimiya M, Schnupf P, Kurata S, Jaeger J, Gehring WJ (2004). Functional divergence between eyeless and twin of eyeless in *Drosophila melanogaster*. *Development* 131:3943–3953.

Quiring R, Walldorf U, Kloter U, Gehring WJ (1994). Homology of the eyeless gene of *Drosophila* to the Small eye gene in mice and Aniridia in humans [see comments]. *Science* 265:785–789.

Rodrigues AB, Moses K (2004). Growth and specification: fly Pax6 homologs eyegone and eyeless have distinct functions. *Bioessays* 26:600–603.

Seimiya M, Gehring WJ (2000). The *Drosophila* homeobox gene optix is capable of inducing ectopic eyes by an eyeless-independent mechanism. *Development* 127:1879–1886.

Silver SJ, Rebay I (2005). Signaling circuitries in development: insights from the retinal determination gene network. *Development* 132:3–13.

Sprecher SG, Pichaud F, Desplan C (2007). Adult and larval photoreceptors use different mechanisms to specify the same rhodopsin fates. *Gene Dev* 21:2182–2195.

Strutt DI, Wiersdorff V, Mlodzik M (1995). Regulation of furrow progression in the *Drosophila* eye by cAMP-dependent protein kinase A [see comments]. *Nature* 373:705–709.

Sun Y, Jan LY, Jan YN (1998). Transcriptional regulation of atonal during development of the *Drosophila* peripheral nervous system. *Development* 125:3731–3740.

Sundaram MV (2005). The love-hate relationship between Ras and Notch. *Gene Dev* 19:1825–1839.

Suzuki T, Saigo K (2000). Transcriptional regulation of atonal required for *Drosophila* larval eye development by concerted action of eyes absent, sine oculis and hedgehog signaling independent of fused kinase and cubitus interruptus. *Development* 127:1531–1540.

Theodosiou NA, Xu T (1998). Use of FLP/FRT system to study *Drosophila* development. *Methods* 14:355–365.

Tomlinson A (2003). Patterning the peripheral retina of the fly: decoding a gradient. *Dev Cell* 5:799–809.

Tomlinson A, Bowtell DD, Hafen E, Rubin GM (1987). Localization of the sevenless protein, a putative receptor for positional information, in the eye imaginal disc of *Drosophila*. *Cell* 51:143–150.

Tomlinson A, Kimmel BE, Rubin GM (1988). Rough, a *Drosophila* homeobox gene required in photoreceptors R2 and R5 for inductive interactions in the developing eye. *Cell* 55:771–784.

Tomlinson A, Struhl G (2001). Delta/Notch and Boss/Sevenless signals act combinatorially to specify the *Drosophila* R7 photoreceptor. *Mol Cell* 7:487–495.

Treisman JE, Rubin GM (1995). wingless inhibits morphogenetic furrow movement in the *Drosophila* eye disc. *Development* 121:3519–3527.

Tsai YC, Yao JG, Chen PH, Posakony JW, Barolo S, Kim J, Sun YH (2007). Upd/Jak/STAT signaling represses wg transcription to allow initiation of morphogenetic furrow in *Drosophila* eye development. *Dev Biol* 306:760–771.

Weasner B, Salzer C, Kumar JP (2007). Sine oculis, a member of the SIX family of transcription factors, directs eye formation. *Dev Biol* 303:756–771.

Wehrli M, Tomlinson A (1998). Independent regulation of anterior/posterior and equatorial/polar polarity in the *Drosophila* eye; evidence for the involvement of Wnt signaling in the equatorial/polar axis. *Development* 125:1421–1432.

Wernet MF, Desplan C (2004). Building a retinal mosaic: cell-fate decision in the fly eye. *Trends Cell Biol* 14:576–584.

Wernet MF, Labhart T, Baumann F, Mazzoni EO, Pichaud F, Desplan C (2003). Homothorax switches function of *Drosophila* photoreceptors from color to polarized light sensors. *Cell* 115:267–279.

Wernet MF, Mazzoni EO, Celik A, Duncan DM, Duncan I, Desplan C (2006). Stochastic spineless expression creates the retinal mosaic for colour vision. *Nature* 440:174–180.

Wilson D, Sheng G, Lecute T, Dostatni N, Desplan C (1993). Cooperative dimerization of paired class homeo domains on DNA. *Gene Dev* 7:2120–2134.

Wilson DS, Guenther B, Desplan C, Kuriyan J (1995). High resolution crystal structure of a paired (Pax) class cooperative homeodomain dimer on DNA. *Cell* 82:709–719.

Wolff T, Ready DF (1991). Cell death in normal and rough eye mutants of *Drosophila*. *Development* 113:825–839.

Wolff T, Ready DF (1993). Pattern formation in the *Drosophila* retina. In: Bate M, Arias AM (eds), *The Development of Drosophila melanogaster*. Cold Spring Harbor Laboratory Press, New York, pp. 1277–1325.

Yu SY, Yoo SJ, Yang L, Zapata C, Srinivasan A, Hay BA, Baker NE (2002). A pathway of signals regulating effector and initiator caspases in the developing *Drosophila* eye. *Development* 129:3269–3278.

CHAPTER

5

The Antarctic Toothfish: A New Model System for Eye Lens Biology

Andor J. Kiss

Laboratory for Ecophysiological Cryobiology, Department of Zoology, Miami University, Oxford, OH 45056, USA

OUTLINE

The giant Antarctic toothfish *Dissostichus mawsoni* is a large perciform fish belonging to the suborder Notothenioidei endemic to the inshore waters of the Southern Ocean encircling Antarctica (Eastman, 1993). It is very large fish often reaching 1 m in length and weighing between 34 and 55 kg (Dewitt *et al.*, 1990), and living up to 50 years (Horn *et al.*, 2003). The mean water temperature in the Southern Ocean year-round is −1.86°C, which is at or near the freezing point of seawater (Hunt *et al.*, 2003). The toothfish is a well-studied animal displaying a number of biochemical and physiological adaptations to the cold (Williams *et al.*, 1985; Eastman, 1993; Chen *et al.*, 1997; Metcalf *et al.*, 1999; Pointer *et al.*, 2005). As a large fish, it has a large prominent eye (Fig. 5.1). Interestingly, no reports of lens cataracts have been noted in the literature during the approximately 40 years of study. Because of unique habitat of the toothfish, it is worth placing this animal in context of existing model systems, both from the perspective of its environment and its relevant general biology if we are to properly exploit the toothfish as a new model system for lens biology.

THE ANTARCTIC ENVIRONMENT

The Antarctic continent covers approximately 14 million km^2 and is the Southern-most of the earth's land masses. It is oceanographically isolated from all other continental land masses by large amounts of open water. This isolation has allowed for the development of the Antarctic circumpolar current (ACC). The ACC has the largest volume of any oceanic current and extends from the ocean surface to the sea-floor (2000 ~ 4000 m). The ACC varies from 200 km to 1000 km wide (slightly deflected and compressed by the tip of South America) thereby representing a true physical barrier separating the fishes of the Southern Ocean from those of the rest of the world (Eastman, 1993).

The temperature on the Antarctic continent itself is at or below 0°C, with the world's lowest recorded temperature of −89.6°C reported at Vostok Station (USSR). In addition to being extremely cold year-round there is very little yearly precipitation, 40~100 cm per year on the coast with trace amounts at the South Pole

(Eastman, 1993). As thousands of years of snowfall have accumulated on the Antarctic continent, they have been compressed and formed a large continental sized ice sheet which averages about 2 km in depth. This massive ice sheet accounts for approximately 70% of the world's fresh water. The tremendous weight of the Antarctic ice sheet has depressed the continental plate several hundred meters with the resulting pressure on the underside being so great as to cause melting. This small amount of melt-water flows outwards from underneath the ice sheet off the continental plate and into the Southern Ocean. This melting is usually replaced by new precipitation each year, thereby preserving the overall size and nature of the ice sheet. Global warming upsets this balance and in recent years the melting has become greater than the contribution from precipitation resulting in an overall shrinking of the ice sheet.

Partially due to the melt-waters flowing down into the Southern Ocean from the underside of ice sheet, and partially due to annual sea ice formation, the inshore waters about the Antarctic continental shelf are extremely cold averaging a year-round constant of −1.86°C (Hunt *et al.*, 2003). The temperature of the Southern Ocean has been at or near current temperatures (−2°C) for about 10 million years (Chen *et al.*, 1997). On the 200 km wide Antarctic continental shelf, there is a diverse community of life surviving on the basal communities of phytoplankton and zooplankton communities (Brierley and Thomas, 2002). This diversity of life is due to in part the very cold water, with its elevated amount of dissolved oxygen (70~105%); about 1.6 times higher than dissolved oxygen content at 20°C. Where there are plankton, there are usually fish, and the Antarctic is no exception. The fishes of the Southern Ocean have evolved and adapted to this polar niche and they have done so over several millenia. Thus the stability of their eye lens is based on long-term adaptations, and not short-term seasonal acclimations.

TOOTHFISH BIOLOGY

The majority of the fishes that inhabit the subzero Antarctic waters are from the teleost suborder Nototheniodei. As ectothermic animals, the Antarctic fishes have developed a number of biochemical adaptations to the cold including catalytically efficient enzymes, cold stable structural proteins, cold adapted membrane fluidity, and cold adapted protein translocation (Williams *et al.*, 1985; Fields and Somero, 1998; Cossins *et al.*, 2002; Hochachka and Somero, 2002; Romisch *et al.*, 2003). Paramount among all of these adaptations was the evolution of a blood-borne antifreeze glycoprotein (AFGP), which is believed to inhibit ice crystal growth thus allowing the notothenioid fishes to inhabit the ice laden waters of the Southern Ocean without freezing (DeVries, 1983; Knight *et al.*, 1991). These adaptations have enabled the notothenioids to become extremely successful in the Antarctic environment.

FIGURE 5.1 Photograph of Antarctic toothfish *D. mawsoni* partially in a recirculating seawater tank in McMurdo Sound aquarium. The size of this fish is typical of the size caught

One of the best studied Antarctic fishes is the giant Antarctic toothfish *Dissostichus mawsoni*[1]. This fish is especially well-suited to the study of eye biology as it is a large fish with a very large eye lens comparable to that of the well-characterized cow *Bos taurus* lens (Fig. 5.1). Recently, three papers have been published describing the basic biochemical properties of the cold-adapted toothfish eye lens (Kiss *et al.*, 2004), the cDNA sequences of the lens crystallins (Kiss *et al.*, 2008) and one describing the retinal organization and spectral properties of photoreceptors from several Antarctic fish, including the toothfish (Pointer *et al.*, 2005).

In the first of these three papers, the basic properties of the overall lens stability were investigated by whole lens cooling experiments as well as the basic

[1]The toothfish has unfortunately begun to be commercially fished (legally and illegally) and is turning up in restaurants and seafood shops as "Chilean Sea Bass". The name is a commercial moniker used primarily for retail purposes. More importantly, the toothfish is not a sustainable fishery and should be avoided (Parker *et al.*, 2002; Baldwin and Mounts, 2003).

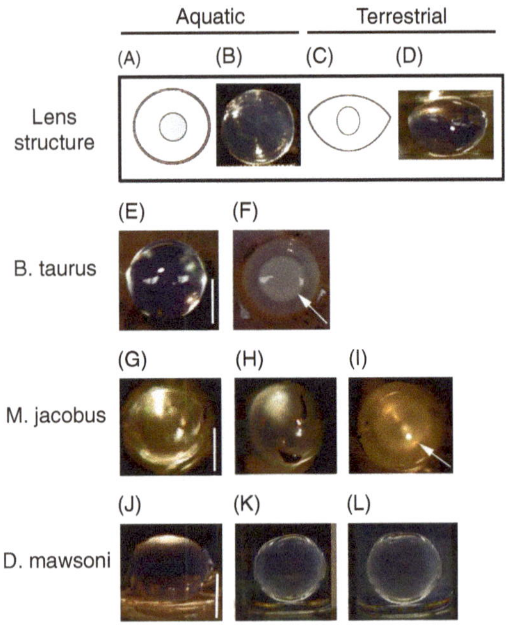

FIGURE 5.2 Lens shapes and cold cataract cooling experiment on the whole lenses of three species (*B. taurus*, *M. jacobus*, and *D. mawsoni*) from three different physiological temperatures (37°C, 25°C and −2°C, respectively). Schematics of the shapes and a picture of *D. mawsoni* (A & B) and *B. taurus* (C & D) lens. Cold cataract experiment results showing lenses from the cow, *B. taurus* (E) uncooled and (F) slightly warmed from ice bath with nucleus showing the cold cataract (arrow). The eye lens from the tropical marine blackbar soldierfish (*M. jacobus*) held at (G) 15°C for 6h and (H) 0°C for 6h. The third image (I) shows a lens that was held at 0°C for 48h with a definite inner nuclear region that is more opaque than the cortex region (arrow). Antarctic toothfish eye lens at the (J) endogenous clear −2°C lens contrasted (K) to a −12°C lens. Still clear toothfish lens after 48h at −12°C (L) to (K); minor opacity restricted to surface and not to the inner portions of the lens, as in the cow (F) and the soldierfish (I). Scale bars (E) 1.2cm, (G) 0.4cm, (J) 1.0cm. Adapted from Kiss *et al.* (2004)

biochemistry of the component proteins. Whole lens cooling experiments were performed on unfrozen lenses and were done in comparison with cow lens as well as a tropical marine blackbar soldierfish *Myripristis jacobus* lens. Mammalian lenses such as the cow lens demonstrate a cold sensitivity, known as a cold-cataract. This cold-cataract is a reversible phenomenon, manifesting itself at a few degrees below normal physiological temperatures and becoming progressively more distinct as the lens is cooled, easily seen with the unaided eye at ~10°C (Fig. 5.2F). Detailed analytical studies of the cold-cataract (Tanaka and Benedek, 1975; Clark and Benedek, 1980; Benedek, 1997) revealed that it was one (γ) of the three common lens crystallin proteins (α, β, γ) that was responsible for the cold-cataract (Thomson *et al.*, 1987; Broide *et al.*, 1991; Pande *et al.*, 1991; Berland *et al.*, 1992). In contrast, the ectothermic toothfish lens does not show any cold-cataract at its normal body temperature of −2°C. The clarity at −2°C by itself presents us with the possibility that the toothfish holds insights to crystallin stability that would extrapolate to crystallins from endothermic vertebrates. Cooling of the toothfish lens down to −12°C for a period of 48h did not induce a cold-cataract (Kiss *et al.*, 2004). However, fish lenses are extremely dense with some estimates placing the density at ~1000mg/mL (Kroger *et al.*, 1994; Pierscionek and Augusteyn, 1995). To address concerns that the lack of cold-cataract was due to lens density, cooling experiments were performed with a similarly dense tropical marine fish lens from a blackbar soldierfish. At 15°C the blackbar soldierfish lens began to show a faint cold-cataract approximately after 6h (Fig. 5.2G), which became progressively more pronounced by reducing the temperature slowly down to 0°C and incubation for a further 48h. The cold-cataract induced in the blackbar soldierfish was not as pronounced as those seen in cow lens, but occurred at a temperature above that (0°C) which was the normal body temperature of the toothfish (−2°C). Other previous reports of detailed cold-cataract studies in tropical and temperate fish lenses (Loewenstein and Bettelheim, 1979) noted that there was permanent cold-induced damage similar to what we observed for the blackbar soldierfish in the form of a halo seen at the interface of the lens nucleus and cortex (Fig. 5.2 I).

LENS BIOCHEMISTRY

Given the cold stability of the intact toothfish lens, the next necessary step was to evaluate whether or not the biochemical composition of the toothfish lens was comparable with previous reports of other vertebrates, including other fish lenses. An average adult toothfish lens is approximately 3g, which was one of the motivations for using this particular fish as a starting point for Antarctic fish lens biology. Standard biochemical methods were used to fractionate the toothfish lens, and its crystallins were separated by size-exclusion chromatography. Analysis of the separated fractions by SDS-PAGE and immunoblotting confirmed the presence three main isoform groups: α, β, and γ. The two main components of the soluble crystallin fraction were α- and γ-crystallins, with notably more γ-crystallins in the toothfish lens when compared with the amount of γ-crystallins found in comparable-sized mammalian (cow and human) lens (Bloemendal *et al.*, 2004; Kiss *et al.*, 2004). Temperature stability of the α- and γ-fraction of the toothfish were compared

TABLE 5.1 Empirically determined thermal stabilities (T_S) for α- and γ- crystallins from S200HR size fractionations from vertebrate species with three different organismal temperatures (OT). After Kiss (2004)

Species	OT °C	α-Crystallin (T_S) °C	γ-Crystallin (T_S) °C
D. mawsoni	−2	47	33
T. obesus	18	55	39
B. taurus	37	68	50

and contrasted to similar fractions from the bigeye tunafish *Thunnus obesus* and the cow. The thermal stabilities of the α- and γ-fractions were correlated with the organism temperature from which the crystallins were isolated (Table 5.1). These results of lens crystallin composition, increased γ-isoform abundance and thermal stability were expected with some parameters having been previously reported for other fishes, including species from the Antarctic (Smith, 1969, 1971; Ferguson *et al.*, 1971; Loewenstein and Bettelheim, 1979; McFall-Ngai and Horwitz, 1990; Wistow *et al.*, 2005).

Chaperone-like activity of the toothfish α-crystallin was evaluated using heat and chemically induced aggregation assays. These chaperone-like assays using substrate/labile proteins such as γ-crystallins and lysozyme were based on previous assays (Horwitz, 1992). Toothfish α-crystallin demonstrated a functional capacity to protect labile proteins from aggregation. This protective function of α-crystallin is a characteristic hallmark of the sHSP family of proteins in chaperone-like assays (Posner, 2003). One of the principal mechanisms of the chaperone-like activity of α-crystallin is believed to be the binding of partially, or improperly folded proteins *via* hydrophobic interactions (Narberhaus, 2002; Reddy *et al.*, 2006). As the labile proteins, usually assumed to be other crystallins (β and γ), undergo stress (age, UV, redox damage) and begin to unfold, they expose their inner hydrophobic residues. These exposed hydrophobic residues bind to receptive hydrophobic patches on α-crystallin, thus avoiding catastrophic crystallin aggregation and eventual cataracts. This sHSP function is thought to be non-specific and generalized in nature (Derham and Harding, 1999; Rajaraman *et al.*, 2001; Santhoshkumar and Sharma, 2001).

An unexpected result occurred when cow α-crystallin and toothfish γ-crystallin were used together in a chaperone-like assay. The cow (mammalian) α-crystallin offered no protection to the toothfish γ-crystallin (Fig. 5.3). Comparable assays using cow α-crystallin and bigeye tunafish γ-crystallin showed some protection, although not quite as great as with cow

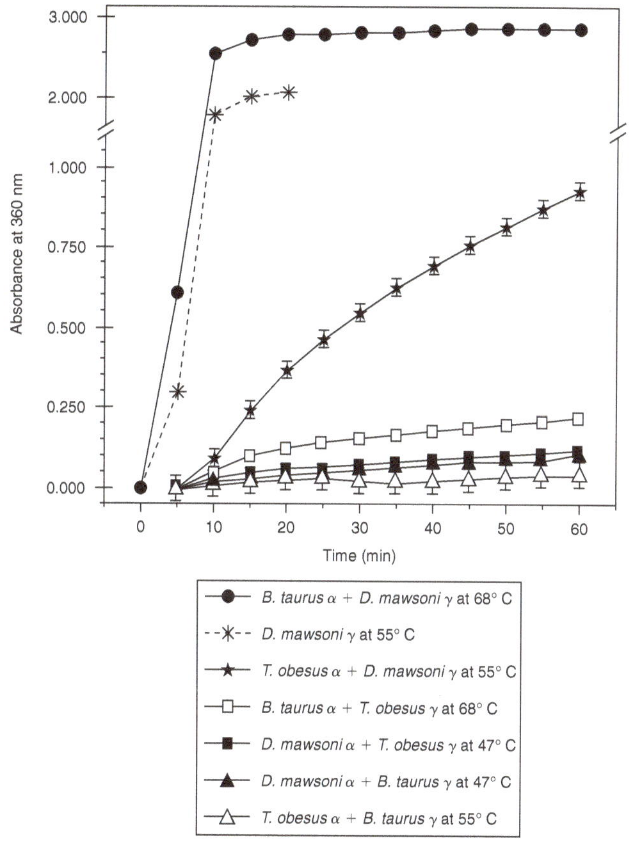

FIGURE 5.3 Cross species chaperone-like protection assay of γ-crystallin by α-crystallin from the three species: toothfish *D. mawsoni*, bigeye tuna *T. obesus*, and cow *B. taurus*. Chaperone-like assay temperature was at T_S (thermal stabilities) for the α-crystallin in the assay. Final concentration of both α- and γ-crystallin in the assay was 1 mg/mL. Combinations of α- and γ-crystallin are indicated in side panel. Standard error bars are obscured by symbols ($n = 3$), after Kiss *et al.* (2004).

α- + γ-crystallin in chaperone-like heat aggregation assays (Fig. 5.3) (Kiss *et al.*, 2004). The non-protection of toothfish γ-crystallin was important because it challenged the assumptions of the generalized nature of the α-crystallin/sHSP effect. For us, this result brought into sharp focus certain beliefs regarding the interactions between α-crystallins and the basis of stability in cold-adapted proteins (such as the toothfish γ-crystallins).

If the toothfish lens is stable at very cold temperatures, and mammalian γ-crystallins have been shown to be the cold sensitive components of eye lenses, then a reasonable conclusion would be that something about the γ-crystallins has changed to make the toothfish lens cold-stable. This conclusion is of course predicated on the assumption that mammalian and toothfish γ-crystallins are reasonably conserved proteins (they are – see below and Kiss *et al.*, 2008). Chilling experiments with the size-separated toothfish γ-crystallin fraction bear this hypothesis out, as the cold-cataract or liquid–liquid phase separation temperature was found to be −10°C for a solution of toothfish γ-crystallins at 58 mg/mL. This is approximately 14°C less than a comparable solution of cow γ-crystallins.

Thus, if the toothfish γ-crystallins are cold-stable and cannot interact with the cow α-crystallins, perhaps the basis of the cold-stability is precluding the chaperone-like activity. A long held belief is that cold-adapted, or cold-stable proteins have reduced number of hydrophobic residues. However, this has been difficult to demonstrate with non-enzymatic proteins in a convincing manner as seemingly contradictory evidence of increased hydrophobicity in cold-adapted proteins from Antarctic fishes has been detected (Detrich, 1997; Detrich *et al.*, 2000). Yet, in these cases where residues had increased hydrophobicity, the regions were protein–protein contact points and not residues readily exposed to the surface. In contrast, reduction of "hydrophobic" side-chains of surface accessible amino acids is thought to prevent cold denaturation of cold-stable proteins because it reduces the non-polar, or hydrophobic residues that are more easily solvated at low temperatures (≤4°C) (Privalov, 1990; Hochachka and Somero, 2002; Tsai *et al.*, 2002). As water cools, it becomes more ordered and is more receptive to being structured *via* a hydrogen bonding network. Non-polar residues can form better van der Waals interactions with water when it is in this state. This generates a favorable gain in enthalpy of the ice-like structure of water at low temperature which outweighs any unfavorable entropy loss from the increased order in the system (Tsai *et al.*, 2002). The end result is that at cold temperatures, non-polar residues actively participate in destabilizing proteins thus causing their so-called "cold-denaturation". A change in the 1° structure of the toothfish γ-crystallins to reduce their hydrophobic content would prevent cold denaturation at sub-zero temperatures. This structural change might also be the basis of the non-interaction with mammalian α-crystallin in the chaperone-like assays.

LENS CRYSTALLIN cDNA SEQUENCES

To ascertain how related at the molecular level the lens crystallins are to known vertebrates sequences, we isolated, cloned and sequenced 22 unique crystallin cDNAs from the Antarctic toothfish. They were two α (αA and αB), six β (βA1, βA2, βA4, βB1, βB2, and βB3) and fourteen γ (γN, γS1, γS2, γM1, γM3, γM4, γM5, γM7, γM8a, γM8b, γM8c, γM8d, γM8e, and γM9) (Kiss *et al.*, 2008). A detailed report and analysis of the sequences has just been published (Kiss *et al.*, 2008). However, in the interest of promoting the toothfish as a model systems, a summary of some of the salient points from this work is presented below.

The α-crystallin sequences are homologous with mammalian and other vertebrate sequences as was demonstrated by their chaperone-like function (see above and Kiss *et al.*, 2004). Detailed chaperone-like structure/function comparative studies of recombinant toothfish αA- and αB-crystallins are underway in collaboration with Prof. Mason Posner (Ashland University). Initial findings suggest functional optimization of α-crystallin reflective of the thermal habitat of the toothfish. More importantly in the context of establishing new model systems, the ability to express α-crystallins recombinantly underlines the feasibility of using the toothfish for lens biology.

Toothfish β-crystallins show a high degree of homology with vertebrate β-crystallins, with the notable exception of the toothfish βB1 does not have a long PAPA-like domain in the N-terminal region. The PAPA-domain of mammalian crystallins (mouse, rat, human) is believed to be a linker to cytoskeletal elements (Bloemendal *et al.*, 1984; Hejtmancik *et al.*, 1986; Coop *et al.*, 1998). Analysis of the toothfish β-crystallins revealed that it does have a long $(PAPA)_5$-domain at the C-terminal end of its βB3 crystallin (Fig. 5.4). The possible role of this hydrophobic PAPA-domain in the toothfish is unknown at this time, but it is worth noting that detailed studies of lens crystallins solubility, both in aqueous buffer and 6 M urea buffer demonstrated that a significant amount of β-crystallins were found in the insoluble, or albuminoid fraction of the lens homogenate (Kiss, 2005). At this point, this data suggests a correlation between the β-crystallins and the cytoskeletal components of the lens, which was also proposed as a function in the mid-1980s (Bloemendal *et al.*, 1984; Hejtmancik *et al.*, 1986;Coop *et al.*, 1998), however further work remains to be done to convincingly establish the function of the β-crystallin. Interestingly, the zebrafish *Danio rerio* has a long hydrophilic PNPN-domain at the C-terminal end of its βB3 crystallin whose function is also unknown (Fig. 5.4).

The γ-crystallins of the toothfish are not homologues of the known mammalian sequences, but several of them are homologous to zebrafish, and spotted green pufferfish *Tetraodon nigroviridis* γ-crystallins (Kiss, 2005). This is not to say that these toothfish γ-crystallins do not show characteristic γ-crystallin features, rather these toothfish isoforms do not have direct mammalian homologues. One notable biochemical characteristic common to γ-crystallins sequenced from fish is

```
                250       260       270       280       290       300
           ....|....|....|....|....|....|....|....|....|....|....|....|..
Dm_βB3     SVRRVRDMQWHKKGCFAAADPAPAPAPAPAPAPAPAPAPAPAPGPDPDPTP-----------APPAPPATAGAS
Dr_βB3     SVRRVRDMQWHKRGCFTVPTPDPAPKPNPNPNPNPNPNPNPNPNPNPAPNPAPAPAPPAPSATAASS
Bt_βB3     SVRRIRDQKWHKRGVFLSS---------------------------------------------------
Hs_βB3     SVRRIRDQKWHKRGRFPSS---------------------------------------------------
```

FIGURE 5.4 Amino acid alignment from the C-terminal regions of βB3 lens crystallins from toothfish (Dm_βB3), zebrafish (Dr_βB3), cow (Bt_βB3), human (Hs_βB3). Note the PAPA and PNPN domains of toothfish and zebrafish, but not in the mammalian species. After Kiss *et al.* (2008)

their high percentage of the amino acid methionine within the sequences. In fact, the isoform designation of the γ-crystallins is "γM" based on the first sequences that were analyzed (Chang *et al.*, 1988). Methionine is an interesting amino acid in that it has a unique structural plasticity that comes from the flexibility of the side-chain (R) group [$H_3C-S-CH_2=CH_2-$]. The flexibility that is imparted by methionine might make it uniquely suited to the extremely protein dense fish lens (>1000 mg/mL) as it would give the surface of the γM-crystallins a surface "squishiness" allowing optimal crystallin to crystallin contact. The toothfish lens is just such a high density protein system that would benefit from γ-crystallins with a methionine rich 1° structure.

OTHER ASPECTS OF TOOTHFISH EYE BIOLOGY

In addition to initial work on the biochemistry and isolation of crystallin sequences described above, detailed crystallin solubility and lipid composition studies were conducted on the toothfish lens (Kiss, 2005). From these studies it appears that the β-crystallins are correlated with insoluble, or cytoskeletal components of the lens upon homogenization. Also investigated was the lipid composition of the toothfish lens by extraction (Bligh and Dyer, 1959), and analysis using negative ion ESI mass spectroscopy (Han and Gross, 1994; Sweetman *et al.*, 1996; Terrados and Lopez-Jimenez, 1996). The identification of the fatty acyl species was done by comparison with molecular weights of known straight-chain fatty acids (FA). The overall finding was that the toothfish has a much greater proportion both in diversity and quantity of unsaturated fatty acyl chains relative to the cow (Rujoi *et al.*, 2004). This change was not unexpected and reflects yet another aspect of the toothfish lens adaptation to the extreme cold of the Southern Ocean. Mass spectroscopy analysis also detected a FA with a molecular mass of 267 mass units in the epithelial layer of the lens. This unidentified FA accounted for over 50% of the total FA content in the toothfish lens epithelia. The epithelia is the germinal tissue for newly differentiating lens fiber cells, and is thus among the tissue incorporating lipids into newly formed cells. The mass of the unknown fatty acid suggests either and odd-chain FA of 17 carbons or a furanoid. Furan fatty acids usually represent less than 1% of fish fatty acids, but can accumulate to levels as high as 50% (Sand *et al.*, 1984). However, fish are unable to synthesize furan FA, thus these are likely acquired via their diet, most likely via Antarctic krill species (Ju and Harvey, 2004). Thus, this unidentified FA may be a potential fatty acid trophic marker and could be used to trace the fishes food sources (Dalsgaard *et al.*, 2003). This raises the possibility that the lens lipids from ecologically sensitive fish could be utilized as a ready-made tag for food web studies. Furan fatty acid function is not entirely known, but they could be used as an antioxidant in these metabolically active lens epithelial cells.

The spectral sensitivity of the native and recombinant toothfish opsins and the spatial arrangement of photoreceptors has been recently reported (Pointer *et al.*, 2005). The toothfish has SWS1 UV, MWS, and Rh type opsins with γ_{max} around expected values for teleost fishes. The toothfish double and single cones are arranged in a row mosaic pattern, which differs from a shallow water Antarctic fishes, species *Trematomus hansoni*, which has a square mosaic arrangement. In this study by Pointer *et al.* (2005) there was little difference in spectral sensitivity for opsins across many species of related Antarctic notothenioid fishes which live at different depths experiencing differing light spectral and intensities. The implication drawn is that the retinal organization and not spectral tuning (as seen in Lake Baikal fishes, see Hunt *et al.*, 1996) are responsible for adaptation of different notothenioid fishes to different light quality regimes encountered at each species ecological niche (Pointer *et al.*, 2005).

STRENGTHS OF THE TOOTHFISH AS A MODEL SYSTEM

Although the toothfish has several strengths as a model system, one drawback is the accessibility of the animal to laboratory-based research. A remedy for whole organism physiology, or for whole lens studies would be through the auspices of national Antarctic research programs. The USAP (US Antarctic Program) is one agency which administers an active Antarctic research program, and investigators are funded *via* the NSF-OPP (National Science Foundation Office of Polar Programs:http://www.nsf.gov/dir/index.jsp?org=OPP) in the United States. Fully equipped laboratories, recirculating seawater aquaria, and comprehensive support staff and facilities exist at McMurdo Sound Research Station, Ross Isl, Antarctica. Thus, it is entirely feasible to do both laboratory and field research on these fishes. Periodically, the NSF-OPP will run workshops beginning in early January for 6 weeks that enable principle investigators, post-doctoral fellows, and graduate students to carry out proposed research projects under the supervision of experienced Antarctic researchers. In this way "new blood" can be brought into the Antarctic research program and new investigators can learn techniques and approaches for study in this extreme environment. Avenues other than direct travel are available to study the toothfish lens crystallins which include collaborative efforts with current Antarctic researchers, and in the case of the toothfish lens cDNA sequences, database retrieval will be publicly available *via* GenBank (kiss *et al.*, 2008).

As a well-studied polar fish, the toothfish has significant biochemistry, molecular evolutionary, anatomical, and now eye (lens and retina) studies published. The next logical step would be to propose this fish as a candidate for genome sequencing. In fact, a BAC (Bacterial Artificial Chromosomal) library was made at the Benaroya Research Institute (http://www.genome.gov/) and screening of the toothfish genome for evolutionary relationships among the crystallins from the toothfish is currently underway.

One of the strongest aspects of the toothfish as a model system is the ability to make meaningful comparisons. There is a "sister" species of the toothfish, the Patagonian toothfish *Dissostichus eleginoides* which lives in warmer waters, above freezing, around the South Georgia Islands and further up along the coast of South America to coastal waters of Peru and Uruguay. Additionally, there is a New Zealand Black Cod *Notothenia angustata* which is a non-Antarctic notothenioid fish believed to be an "escapee" from the Southern Oceans before the waters began to cool (Cheng *et al.*, 2003). This fish inhabits the waters about New Zealand (+12°C) and makes for an excellent temperate control. Phylogenetically further removed from the toothfish, the zebrafish (Wistow *et al.*, 2005), and the spotted green pufferfish (Genoscope) have been sequenced and provide an excellent source of comparative data with specimens easily accessible to most researchers.

Perhaps the most powerful tools we have to understand about evolutionary biology and adaptation are sequence-based computer bioinformatics software and analysis. Among the plethora of crystallin literature, there exist excellent studies on non-traditional model systems for lens biology (Wistow and Piatigorsky, 1987; Chiou, 1988; Piatigorsky, 1998a,b; Tomarev *et al.*, 1997; Werten *et al.*, 2000; Xu *et al.*, 2000; Swamynathan *et al.*, 2003; Kanungo *et al.*, 2004). These crystallins from a variety of animals demonstrate the evolutionary plasticity of the lens, and the adaptability of the eye. At the same time, the eye presents us with a modular organ system that seems to have more in common with "eyes" than "hosts". It is the similarity of the eye, and in particular the lens crystallins which will allow new methodologies to be applied with the proven approaches of comparative biology to further our understanding of protein stability and evolutionary biology. In the era of comparative genomics, we expect that crystallin cDNA sequence obtained from the Antarctic toothfish will enhance our rapidly growing understanding of adaptational biology and we look forward to elucidating the mechanisms that underlie the evolution of these genes.

REFERENCES

Baldwin CC, Mounts JH (2003). *One Fish, Two Fish, Crawfish, Bluefish: The Smithsonian Sustainable Seafood Cookbook*. Smithsonian, Washington, DC. Comment: 336

Benedek GB (1997). Cataract as a protein condensation disease: the proctor lecture. *Invest Ophthalmol Vis Sci* 38:1911–1921.

Berland CR, Thurston GM, Kondo M, Broide ML, Pande J, Ogun O, Benedek GB (1992). Solid–liquid phase boundaries of lens protein solutions. *P Natl Acad Sci USA* 89:1214–1218.

Bligh EG, Dyer WJ (1959). A rapid method of total lipid extraction and purification. *Can J Biochem Physiol* 37:911–917.

Bloemendal H, Berbers GA, De Jong WW, Ramaekers FC, Vermorken AJ, Dunia I, Benedetti EL (1984). Interaction of crystallins with the cytoskeletal-plasma membrane complex of the bovine lens. *Ciba F Symp* 106:177–190.

Bloemendal H, De Jong W, Jaenicke R, Lubsen NH, Slingsby C, Tardieu A (2004). Ageing and vision: structure, stability and function of lens crystallins. *Prog Biophys Mol Biol* 86:407–485.

Brierley AS, Thomas DN (2002). Ecology of southern ocean pack ice. *Adv Mar Biol* 43:171–276.

Broide ML, Berland CR, Pande J, Ogun OO, Benedek GB (1991). Binary-liquid phase separation of lens protein solutions. *P Natl Acad Sci USA* 88:5660–5664.

Chang T, Jiang YJ, Chiou SH, Chang WC (1988). Carp gamma-crystallins with high methionine content: cloning and sequencing of the complementary DNA. *Biochim Biophys Acta* 951:226–229.

Chen L, DeVries AL, Cheng CH (1997). Evolution of antifreeze glycoprotein gene from a trypsinogen gene in Antarctic notothenioid fish. *P Natl Acad Sci USA* 94:3811–3816.

Cheng CC, Chen L, Near TJ, Jin Y (2003). Functional antifreeze glycoprotein genes in temperate-water New Zealand nototheniid fish infer an Antarctic evolutionary origin. *Mol Biol Evol* 20:1897–1908.

Chiou SH (1988). A novel crystallin from octopus lens. *FEBS Lett* 241:261–264.

Clark JI, Benedek GB (1980). Phase diagram for cell cytoplasm from the calf lens. *Biochem Biophys Res Commun* 95:482–489.

Coop A, Goode D, Sumner I, Crabbe MJ (1998). Effects of controlled mutations on the N- and C-terminal extensions of chick lens beta B1 crystallin. *Graefe Arch Clin Exp Ophthalmol* 236:146–150.

Cossins AR, Murray PA, Gracey AY, Logue J, Polley S, Caddick M, Brooks S, Postle T, Maclean N (2002). The role of desaturases in cold-induced lipid restructuring. *Biochem Soc Trans* 30:1082–1086.

Dalsgaard J, St John M, Kattner G, Muller-Navarra D, Hagen W (2003). Fatty acid trophic markers in the pelagic marine environment. *Adv Mar Biol* 46:225–340.

Derham BK, Harding JJ (1999). Alpha-crystallin as a molecular chaperone. *Prog Retin Eye Res* 18:463–509.

Detrich HW (1997). Microtubule assembly in cold-adapted organisms: functional properties and structural adaptations of tubulins from Antarctic fishes. *Comp Biochem Physiol A Physiol* 118:501–513.

Detrich HW, Parker SK, Williams RC, Jr, Nogales E, Downing KH (2000). Cold adaptation of microtubule assembly and dynamics structural interpretation of primary sequence changes present in the alpha- and beta-tubulins of Antarctic fishes. *J Biol Chem* 275:37038–37047.

DeVries AL (1983). Antifreeze peptides and glycopeptides in cold-water fishes. *Annu Rev Physiol* 45:245–260.

Dissostichus mawsoni (Norman, 1937) [Internet]. Fishbase.org: Dewitt HH, Heemstra PandGon O; [updated Sep1990; cited Sep1990]. Available from: http://www.fishbase.org/Summary/speciesSummary.php?ID=7039&genusname=Dissostichus&speciesname=mawsoni

Eastman JT (1993). *Antarctic Fish Biology: Evolution in a Unique Environment*, 1st edn. Academic Press, Inc, New York. 322

Ferguson WE, Calhoun WB, III, Koenig VL (1971). Studies on the cold insoluble proteins from the lens of the striped mullet (Mugil cephalus). *Comp Biochem Physiol B* 40:959–972.

Fields PA, Somero GN (1998). Hot spots in cold adaptation: localized increases in conformational flexibility in lactate dehydrogenase A4 orthologs of Antarctic notothenioid fishes. *P Natl Acad Sci USA* 95:11476–11481.

Han X, Gross RW (1994). Electrospray ionization mass spectroscopic analysis of human erythrocyte plasma membrane phospholipids. *P Natl Acad Sci USA* 91:10635–10639.

Hejtmancik JF, Thompson MA, Wistow G, Piatigorsky J (1986). cDNA and deduced protein sequence for the beta B1-crystallin polypeptide of the chicken lens. Conservation of the PAPA sequence. *J Biol Chem* 261:982–987.

Hochachka PW, Somero G (2002). *Biochemical Adaptation: Mechanism and Process in Physiological Evolution*, 2nd edn. Oxford University Press, New York. 480

Horn P, Sutton C, DeVries AL (2003). Evidence to support the annual formation of growth zones in otoliths of Antarctic toothfish (*Dissostichus mawsoni*). *CCAMLR Sci* 10:125–138.

Horwitz J (1992). Alpha-crystallin can function as a molecular chaperone. *P Natl Acad Sci USA* 89:10449–10453.

Hunt BM, Hoefling K, Cheng CC (2003). Annual warming episodes in seawater temperatures in McMurdo Sound in relationship to endogenous ice in notothenioid fish. *Antarct Sci* 15:333–338.

Hunt DM, Fitzgibbon J, Slobodyanyuk SJ, Bowmaker JK (1996). Spectral tuning and molecular evolution of rod visual pigments in the species flock of cottoid fish in Lake Baikal. *Vision Res* 36:1217–1224.

Ju S, Harvey H (2004). Lipids as markers of nutritional condition and diet in the Antarctic krill Euphausia superba and Euphausia crystallorophias during austral winter. *Deep Sea Res II: Topical Stud Ocean* 51:2214.

Kanungo J, Swamynathan SK, Piatigorsky J (2004). Abundant corneal gelsolin in zebrafish and the 'four-eyed' fish, *Anableps anableps*: possible analogy with multifunctional lens crystallins. *Exp Eye Res* 79:949–956.

Kiss AJ (2005). *Functional, Biochemical and Molecular Analyses of the Cold Stable Eye Lens Crystallins from the Antarctic Toothfish Dissostichus mawsoni*. University of Illinois at Urbana-Champaign, Urbana-Champaign, IL.

Kiss AJ, Cheng CC (2008). Molecular diversity and genomic organization of the α, β and γ eye lens crystallins from the Antarctic toothfish *Dissostichus mawsoni*. *Comp Bioch Physiol D*: doi:10.1016/jcbd.2008.02.002.

Kiss AJ, Mirarefi AY, Ramakrishnan S, Zukoski CF, Devries AL, Cheng CH (2004). Cold-stable eye lens crystallins of the Antarctic nototheniid toothfish *Dissostichus mawsoni* Norman. *J Exp Biol* 207:4633–4649.

Knight CA, Cheng CC, DeVries AL (1991). Adsorption of alpha-helical antifreeze peptides on specific ice crystal surface planes. *Biophys J* 59:409–418.

Kroger RHH, Campbell MCW, Munger R, Fernald RD (1994). Refractive index distribution and spherical aberration in the crystalline lens of the African cichlid fish *Haplochromis burtoni*. *Vision Res* 34:1815–1822.

Loewenstein MA, Bettelheim FA (1979). Cold cataract formation in fish lenses. *Exp Eye Res* 28:651–663.

McFall-Ngai MJ, Horwitz JA (1990). Comparative study of the thermal stability of the vertebrate eye lens: Antarctic ice fish to the desert iguana. *Exp Eye Res* 50:703–709.

Metcalf VJ, Brennan SO, George PM (1999). The Antarctic toothfish (*Dissostichus mawsoni*) lacks plasma albumin and utilises high density lipoprotein as its major palmitate binding protein. *Comp Biochem Physiol B Biochem Mol Biol* 124:147–155.

Narberhaus F (2002). Alpha-crystallin-type heat shock proteins: socializing minichaperones in the context of a multichaperone network. *Microbiol Mol Biol Rev* 66:64–93.

Pande J, Berland C, Broide M, Ogun O, Melhuish J, Benedek G (1991). Suppression of phase separation in solutions of bovine gamma IV-crystallin by polar modification of the sulfur-containing amino acids. *P Natl Acad Sci USA* 88:4916–4920.

Parker RW, Paige KN, DeVries AL (2002). Genetic variation among populations of the Antarctic toothfish: evolutionary insights and implications for conservation. *Polar Biol* 25:256–261.

Piatigorsky J (1998a). Gene sharing in lens and cornea – facts and implications. *Prog Retin Eye Res* 17:145–174.

Piatigorsky J (1998b). Multifunctional lens crystallins and corneal enzymes. More than meets the eye. *Ann NY Acad Sci* 842:7–15.

Pierscionek BK, Augusteyn RC (1995). The refractive index and protein distribution in the blue eye trevally lens. *J Am Optom Assoc* 66:739–743.

Pointer MA, Cheng CC, Bowmaker JK, Parry JWL, Soto N, Jeffery G, Cowing JA, Hunt DM (2005). Adaptations to an extreme environment: retinal organisation and spectral properties of photoreceptors in Antarctic notothenioid fish. *J Exp Biol* 208:2363–2376.

Posner M (2003). A Comparative view of alpha crystallins: the contribution of comparative studies to understanding function. *Int Comp Biol* 43:481–491.

Privalov PL (1990). Cold denaturation of proteins. *Crit Rev Biochem Mol Biol* 25:281–305.

Rajaraman K, Raman B, Ramakrishna T, Rao CM (2001). Interaction of human recombinant alphaA- and alphaB-crystallins with early and late unfolding intermediates of citrate synthase on its thermal denaturation. *FEBS Lett* 497:118–123.

Reddy GB, Kumar PA, Kumar MS (2006). Chaperone-like activity and hydrophobicity of alpha-crystallin. *IUBMB Life* 58:632–641.

Romisch K, Collie N, Soto N, Logue J, Lindsay M, Scheper W, Cheng CH (2003). Protein translocation across the endoplasmic reticulum membrane in cold-adapted organisms. *J Cell Sci* 116:2875–2883.

Rujoi M, Estrada R, Yappert MC (2004). In situ MALDI-TOF MS regional analysis of neutral phospholipids in lens tissue. *Anal Chem* 76:1657–1663.

Sand D, Glass R, Olson D, Pike H, Schlenk H (1984). Metabolism of furan fatty acids in fish. *Biochim Biophys Acta* 793:429–434.

Santhoshkumar P, Sharma KK (2001). Analysis of alpha-crystallin chaperone function using restriction enzymes and citrate synthase. *Mol Vis* 7:172–177.

Smith AC (1969). Protein variation in the eye lens nucleus of the mackerel scad (*Decapterus pinnulatus*). *Comp Biochem Physiol* 28:1161–1168.

Smith AC (1971). The soluble proteins in eye lens nuclei of albacore, bluefin tuna and bonito. *Comp Biochem Physiol B* 39:719–724.

Swamynathan SK, Crawford MA, Robison WG, Jr, Kanungo J, Piatigorsky J (2003). Adaptive differences in the structure and macromolecular compositions of the air and water corneas of the "four-eyed" fish (*Anableps anableps*). *Faseb J* 17:1996–2005.

Sweetman G, Trinei M, Modha J, Kusel J, Freestone P, Fishov I, Joseleau-Petit D, Redman C, Farmer P, Norris V (1996). Electrospray ionization mass spectrometric analysis of phospholipids of *Escherichia coli*. *Mol Microbiol* 20:233–238.

Tanaka T, Benedek GB (1975). Observation of protein diffusivity in intact human and bovine lenses with application to cataract. *Invest Ophthalmol* 14:449–456.

Terrados J, Lopez-Jimenez JA (1996). Fatty acid composition and chilling resistance in the green alga *Cauterpa prolifera* (Forrskal) lamouroux (Chlorophyta, Caulerpales). *Biochem Mol Biol Int* 39:863–869.

Thomson JA, Schurtenberger P, Thurston GM, Benedek GB (1987). Binary liquid phase separation and critical phenomena in a protein/water solution. *P Natl Acad Sci USA* 84:7079–7083.

Tomarev SI, Callaerts P, Kos L, Zinovieva R, Halder G, Gehring W, Piatigorsky J (1997). Squid Pax-6 and eye development. *P Natl Acad Sci USA* 94:2421–2426.

Tsai CJ, Maizel JV, Nussinov R (2002). The hydrophobic effect: a new insight from cold denaturation and a two-state water structure. *Crit Rev Biochem Mol Biol* 37:55–69.

Werten PJ, Röll B, van Aalten DM, de Jong WW (2000). Gecko iota-crystallin: how cellular retinol-binding protein became an eye lens ultraviolet filter. *P Natl Acad Sci USA* 97:3282–3287.

Williams RC, Jr Correia JJ, DeVries AL (1985). Formation of microtubules at low temperature by tubulin from Antarctic fish. *Biochemistry* 24:2790–2798.

Wistow G, Wyatt K, David L, Gao C, Bateman O, Bernstein S, Tomarev S, Segovia L, Slingsby C, Vihtelic T (2005). γN-crystallin and the evolution of the $\beta\gamma$-crystallin superfamily in vertebrates. *FEBS J* 272:2276–2291.

Wistow G, Piatigorsky J (1987). Recruitment of enzymes as lens structural proteins. *Science* 236:1554–1556.

Xu YS, Kantorow M, Davis J, Piatigorsky J (2000). Evidence for gelsolin as a corneal crystallin in zebrafish. *J Biol Chem* 275:24645–24652.

CHAPTER

6

Xenopus, an Ideal Vertebrate System for Studies of Eye Development and Regeneration

Jonathan J. Henry[1], *Jason M. Wever*[1], *M. Natalia Vergara*[2], *Lisa Fukui*[1]
[1]Department of Cell and Developmental Biology, University of Illinois, Urbana, IL 61801, USA
[2]Department of Zoology, Miami University, Oxford, OH 45056, USA

OUTLINE

INTRODUCTION

History of *Xenopus* as a Model System for Cell, Developmental and Molecular Biology

Amphibians have served as model systems for the study of development and regeneration for well over 100 years (Callery, 2006). Some species exhibit remarkable abilities to replace complex body parts, including the tail, limbs, and various parts of the eye, such as the lens and retina, and this regeneration even occurs in the adults of certain species (Henry, 2003; Stocum, 2006). Representatives are readily available, easy to culture, and produce large numbers of embryos; however, the most widely used amphibians are anuran frogs in the genus *Xenopus*. Although the genus contains over 17 species (Kobel *et al.*, 1996), *Xenopus laevis* and, more recently, *Xenopus tropicalis* are most extensively used in research, though some work has also been carried out with other species, such as *X. borealis*.

Xenopus researchers are now found in nearly every major research institution, and studies using *Xenopus* have led to tremendous advances in our understanding of cell inductive interactions and signaling pathways underlying embryonic pattern formation, cell determination, and organogenesis. More specifically, experimental studies using *Xenopus* have made significant contributions toward understanding the process of eye development and embryonic lens induction. Though not as proficient as some other amphibians, *Xenopus* is capable of regenerating certain tissues, including parts of the eye, particularly during larval stages. Furthermore, studies conducted with *Xenopus* are shedding light on the mechanisms underlying these processes (Henry, 2003).

Members of the genus *Xenopus* represent archaeobatrachian frogs residing in a basal position compared to some other anurans, such as the ranid species in the Neobatrachia (Knöchel *et al.*, 1986; Graf, 1996; Kobel *et al.*, 1996). *Xenopus* species are found naturally in the waters of sub-Saharan Africa, and the most widely used member of this genus, *X. laevis* is commonly referred to as the "South African clawed frog" or "South African clawed toad" (Tinsley *et al.*, 1996, Fig. 6.1A). *Xenopus* emerged as a model experimental system after introduction as an *in vivo* assay for human pregnancy. Early studies in reproductive physiology demonstrated that female frogs could be stimulated to ovulate through subcutaneous injections of urine from pregnant women (due to the presence of chorionic gonadotropin; Bellerby, 1934; Shapiro and Zwarenstein, 1934; Ochsé, 1948). As colonies of *X. laevis* were subsequently maintained in laboratories throughout the world for such tests, the animals were readily accessible for investigators to adopt in other experimental settings (Gordon and Hopwood, 2000). *Xenopus* offered a tremendous advantage to cell and developmental biologists by providing a ready source of eggs and embryos for research. In captivity, *Xenopus* remains fertile throughout the year and mating and egg laying can be stimulated using pituitary extracts or human chorionic gonadotropin (hCG). By 1949, Pieter Nieuwkoop recognized the need for a thorough description of *Xenopus* development. Together with Job Faber he organized a team of scientists to describe the development of *Xenopus* from the fertilized egg through metamorphosis (Nieuwkoop and Faber, 1956). This "Normal Table of *Xenopus* (Daudin)" serves as an invaluable reference and has recently been reprinted by Garland Publishing, Inc. (Nieuwkoop and Faber, 1996; all *Xenopus* developmental stages used in this review refer to Niewkoop and Faber stages, unless otherwise noted). A number of other publications describe the biology, anatomy, histology, and specific uses of *Xenopus* as a model organism, including: Kay and Peng (1991), Hausen and Riebesell (1996), Seidman and Soreq (1996), Tinsley and Kobel (1996), Sive *et al.* (2000), and Wiechmann and Wirsig-Wiechmann (2003), which should also be consulted as authoritative references. Extensive on-line resources are available as well, including Xenbase (www.xenbase.org), the Joint Genome Institutes *X. tropicalis* genome database (http://genome.jgi-psf.org/Xentr4/Xentr4.home.html), the *X. tropicalis* web sites at the University of Virginia (http://faculty.virginia.edu/xtropicalis/), and the University of California, Berkeley (http://tropicalis.berkeley.edu/home/), as well as many others.

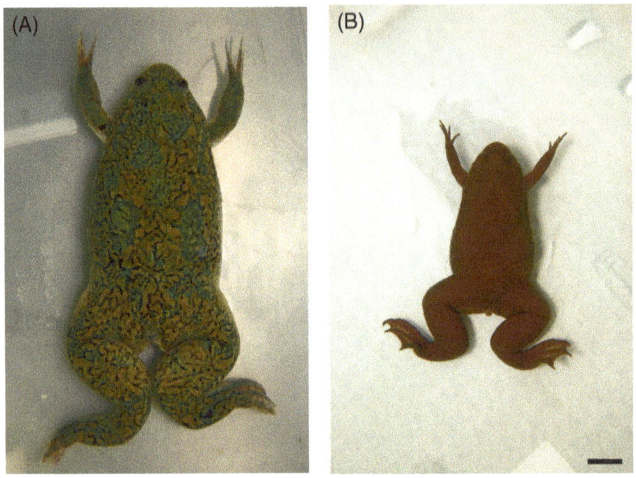

FIGURE 6.1 Dorsal views of adult female *X. laevis* (A) and *X. tropicalis* (B). Scale bar equals 2 cm.

Xenopus tropicalis: An Emerging Genetic System

Recently, *X. tropicalis* has been introduced as a more functional alternative to *X. laevis* (Fig. 6.1B, Amaya *et al.*, 1998; Amaya and Kroll, 1999; Offield *et al.*, 2000; Hirsch *et al.*, 2002a,b; Carruthers and Stemple, 2006), because it exhibits more rapid development with a shorter generation time, which is more amenable to genetic analyses. The generation time for *X. laevis* is approximately 1–2 years, while this may be as short as 4 months for *X. tropicalis*. In addition, *X. tropicalis* is diploid and has a smaller genome (10 chromosomes and a genome size of 1.7 gigabasepairs), unlike *X. laevis*, which is a pseudotetraploid (18 chromosomes with a genome size of 3.1 gigabasepairs). The *X. tropicalis* genome is small even compared to most amphibians, about the same size as that in the zebrafish, *Danio rerio*. These features favored *X. tropicalis* for genome sequencing, which has now been completed (at 8× coverage with approximately 22.5 million paired end sequencing reads) and fully annotated by the US Department of Energy's Joint Genome Institute (http://genome.jgi-psf.org/Xentr4/Xentr4.home.html). At the time of this printing, the sequences of over 600,000 ESTs (expressed sequence tags) are also available for *X. laevis*, and over 1.2 million ESTs have been sequenced for *X. tropicalis* (e.g. http://www.ncbi.nlm.nih.gov/dbEST/, http://www.sanger.ac.uk/Projects/X_tropicalis/, http://xgc.nci.nih.gov/, http://genome.jgi-psf.org/Xentr4/Xentr4.home.html, http://www.genoscope.cns.fr/externe/English/Projets/Projet_EC/EC.html, http://www.informatics.gurdon.cam.ac.uk/online/xt-fl-db.html, http://xenopus.nibb.ac.jp/, http://bibiserv.techfak.uni-bielefeld.de/xendb/). *Xenopus* microarrays are also available to screen for changes in gene expression (e.g. Affymetrix, Santa Clara, CA).

The availability of these EST and genome sequences enables one to identify and isolate practically any gene of interest for expression studies or functional analyses. Researchers can make predictions regarding the presence of specific protein domains and protein structure, which may be useful for functional studies, and can also examine the context in which particular genes may be regulated (e.g. identify cis-regulatory information, etc.). This also permits the investigator to identify specific promoters to drive embryonic expression of various genes in controlled spatiotemporal contexts for functional studies. The recent development of transgenic approaches in both *X. tropicalis* and *X. laevis* has also been a major breakthrough, as described below (e.g. Amaya *et al.*, 1998; Amaya and Kroll, 1999; Offield *et al.*, 2000; Hirsch *et al.*, 2002a,b).

TECHNICAL ADVANTAGES OF *XENOPUS* AS A MODEL SYSTEM

Basic Biology and Development

Xenopus offers a number of significant advantages as a model organism for cell developmental and molecular biology, which are not found in many other vertebrate systems (see Kay and Peng, 1991; Sive *et al.*, 2000; Callery, 2006).

Rearing of *Xenopus* is rather straightforward and does not require elaborate facilities or equipment. A number of publications describe in detail the care and handling (e.g. Nieuwkoop and Faber, 1956; Kay and Peng, 1991; Sive *et al.*, 2000). Embryos are obtained via either naturally induced matings stimulated by injections of hCG, as described above, or via *in vitro* fertilization (Heasman *et al.*, 1991; Sive *et al.*, 2000). The ability to carry out *in vitro* fertilization also ensures that embryos of desired stages can be obtained at anytime throughout the day. With each ovulation, one female can produce several hundred to a thousand eggs for *X. laevis*, or thousands of eggs for *X. tropicalis*. With a colony of 100 or more adult frogs, one can obtain embryos on a daily basis. Development is fairly synchronous, especially in the case of *in vitro* fertilized eggs. The ability to collect large quantities of gametes permits cellular, biochemical, as well as molecular analyses.

As they are poikliothermic, *Xenopus* will tolerate a fairly wide range of temperatures (e.g. 16–25°C for *X. laevis*), and developmental rates may be controlled by culturing embryos at different temperatures. For *X. laevis*, development at 23°C is approximately 1.5 times as rapid as that at 20°C, and twice as rapid as that at 16°C. The ability to adjust these developmental rates is convenient for many types of experiments (e.g. for heterochronic tissue transplantation). Development to the swimming larval stage is rapid, proceeding over the course of only a few days. For instance, in 3–4 days at 23°C, a stage 42 *X. laevis* tadpole larvae is developed, complete with well-differentiated eyes and other organ systems. This period is considerably shorter in *X. tropicalis*, which develops at a higher temperature of 25–27°C. Developmental analyses of complex organ systems can, therefore, be carried out rapidly. Further development of the juvenile metamorphosed frog takes longer, generally about 1–2 months for *X. laevis* and about half this time for *X. tropicalis*.

Xenopus embryos are easily manipulated, as development is external. Eggs are covered with only a thin transparent vitelline envelope and laid with an external coating of jelly. These extracellular investments may be easily removed by simple chemical and mechanical means, without any effect on normal development. As the eggs contain extensive internal yolk reserves, which are partitioned to each dividing cell, no external source of nutrition is required to support development; the embryos only need to be cultured in simple saline solutions. Unlike the case in many other chordates, eggs and embryos of *Xenopus* are very large (1–1.3 mm in diameter for *X. laevis* and 0.7–0.8 mm diameter for *X. tropicalis*). The large size of these embryos facilitates microinjection and other experimental manipulations such as tissue isolation and transplantation. Furthermore, the embryos exhibit a fairly regular cleavage pattern and possess a consistent cell lineage fate map, which enables one to target specific lineages for developmental analyses (see further discussion below).

As mentioned above, one key advantage afforded by amphibians such as *Xenopus,* is the ease with which one can isolate embryonic tissues and perform tissue transplantation. Wounds heal very rapidly and transplanted tissues are incorporated within a matter of minutes. The somewhat slower development and larger size of the embryos of *X. laevis* makes them better suited for these types of experiments, as compared to *X. tropicalis*. The fates of the transplanted tissues can be followed by microinjection of cell-autonomous lineage tracers (e.g. Henry and Grainger, 1987, 1990). Previously, other methods have also been employed to distinguish between host and donor tissues. For example, *X. borealis* possesses a convenient nuclear marker: their nuclei exhibit distinct bright fluorescent spots when stained with quinacrine (Thiébaud, 1983). In contrast, the nuclei of *X. laevis* stain uniformly with this dye. This difference has been used in cross-species (i.e. "xenoplastic") transplantation experiments to follow the development of the host and donor tissues (e.g. Thiébaud, 1983; Koga *et al.*, 1986; Sadaghiani and Thiebaud, 1987; Filoni *et al.*, 2006). Likewise, reciprocal transplants have also been performed using pigmented and albino strains of *X. laevis* to facilitate the tracing of tissue transplants (e.g. Conway and Hunt, 1987).

One may also carry out experiments using *Xenopus* cell lines. Unlike the case in many other vertebrates (such as mammals), it is rather easy to prepare amphibian tissue-specific cell lines (Smith and Tata, 1991; Peng *et al.*, 1991). Primary cultures can be established in defined media supplemented with serum, and immortal lines may be propagated from these cultures. The cells may be raised on the bench without the need of a CO_2 incubator or stringent temperature control, required of mammalian cells.

Tools for Molecular Level Analyses

A tremendous array of resources and technical approaches are available to facilitate molecular studies in *Xenopus* (Klein *et al.*, 2002, 2006). For instance, a variety of techniques have been devised to examine gene function. As mentioned above, many of these techniques are facilitated by the tremendous ease with which one can microinject the eggs and embryonic blastomeres. The ability to target specific cell lineages during development has substantial advantages, especially in cases where particular genes may affect the development of multiple tissues or global perturbations may lead to early embryonic lethality. Electroporation allows substances to be introduced into cells and tissues at later stages of development, including advanced larval or even adult stages (Swartz *et al.*, 2001; Haas *et al.*, 2002; Ogura *et al.*, 2002). Many useful experimental reagents such as lineage tracers, DNA constructs, synthetic RNAs, and morpholinos may be introduced using these various methods.

The sufficiency of different genes may be assayed via gain-of-function studies utilizing injection of synthetic RNAs or DNA expression constructs. In addition, overexpression or ectopic expression studies can be carried out using capped synthetic RNAs (e.g. Kreig and Melton, 1984; Moon and Christian, 1989; Sokol *et al.*, 1991) made from full-length clones or via DNA expression constructs. A variety of vectors have been designed to prepare synthetic RNA or serve as DNA expression constructs *in vivo* in *Xenopus* (e.g. pCSKA, Moon and Christian, 1989; pCS2+ , Turner and Weintraub, 1994; pT7TS and pXeX, Johnson and Krieg, 1994).

Loss-of-function studies may be carried out to assay the requirements of particular genes by injection of antisense RNA, morpholinos, or via the use of dominant negative approaches. Limited success has been reported using injected antisense RNAs (e.g. Lombardo and Slack, 1997) and using dsRNA interference ("RNAi", e.g. Zhou *et al.*, 2002; Anantharam *et al.*, 2003; Fruscoloni *et al.*, 2003); vector based RNAi expression may prove to be more efficient (Li and Rohrer, 2006). On the other hand, morpholinos have been shown to be highly effective at knocking down translation in both *X. laevis* and *X. tropicalis* (Ekker, 2000; Corey and Abrams, 2001; Nutt *et al.*, 2001; Heasman, 2002). Morpholinos are synthetic oligonucleotides designed to target specific sequences such as the 5′ UTR and/or translational start site to prevent translation of specific messenger RNAs, binding mRNAs irreversibly and serving as highly effective steric translational blocks to knockdown gene function

(Gene Tools, LLC, Philomath, OR; www.gene-tools.com). Morpholinos may also be designed to prevent proper splicing of the mRNA. Furthermore, the addition of covalent fluorescent tags (e.g. 3′ Lissamine red or Fluorescein green) makes them excellent cell lineage tracers to follow the progeny of the injected cells. One potential concern for the use of morpholinos in *X. laevis* is related to the pseudotetraploid condition of these organisms. It may be necessary to co-inject two different morpholinos, as the duplicated genes may have slightly different sequences but share conserved functions. Of course, this issue is not a concern for the diploid con-generic species, *X. tropicalis*. The availability of the fully sequenced *X. tropicalis* genome makes cloning-specific genes unnecessary, thus one can readily design morpholinos for loss-of-function analyses.

Trangenesis in *Xenopus*

In recent years, various techniques have been developed to prepare transgenic *Xenopus* frogs. In fact, the procedure is so effective that one can prepare hundreds of transgenic frogs in only a few hours from a single clutch of eggs. The earliest attempts to generate transgenic frogs involved injecting linearized DNA into fertilized eggs (Etkin *et al.*, 1984; Etkin and Pearman, 1987). Another attempt was to inject unfertilized eggs with nuclei isolated from stable cell lines transfected with the desired insert and it was also not met with great success (Kroll and Gerhart, 1994). More recently, an efficient method was developed by Kroll and Amaya (1996), which was adapted from a procedure originally developed for the slime mold, *Dictyostelium* (Kuspa and Loomis, 1992). In this method, sperm nuclei are first incubated in cell-free egg extracts (to decondense the sperm nuclei) and then the desired transgene is introduced with the addition of a restriction enzyme (i.e. restriction enzyme-mediated integration or "REMI"). The prepared sperm nuclei are then injected into unfertilized oocytes. An advantage of this method is that the resulting transgenic founder (F0) embryos express the inserted transgene, eliminating the need to wait for the F1 generation (Amaya and Kroll, 1999). In fact, for most experiments, one does not need to maintain transgenic animals for breeding purposes. Furthermore, in subsequent F1 and F2 generations, integrated transgenes have been shown to be stably incorporated into the germ line (Marsh-Armstrong *et al.*, 1999). Huang *et al.* (1999) subsequently improved this technique by adding an extra purification step to prepare the sperm and solubilize the sperm membranes using digitonin instead of lysolecithin. REMI has also been applied to produce transgenic *X. tropicalis* (Amaya *et al.*, 1998). Offield *et al.* (2000) made key modifications that increase the efficiency of producing transgenic embryos in *X. tropicalis* (which included the use of less cysteine for dejellying eggs and less egg extract when preparing sperm nuclei). Hirsch *et al.* (2002a) have further optimized transgenesis techniques for *X. tropicalis*.

Transgenesis using the REMI method has been used in a number of experiments, including some designed to study eye development. For example, Knox *et al.* (1998) and Moritz *et al.* (2001) produced transgenic *X. laevis* tadpoles to study retinal development. Knox *et al.* (1998) produced a line of transgenic *Xenopus* containing a rod opsin promoter driving GFP (green fluorescent protein) expression. Moritz *et al.* (2001) used a rhodopsin-GFP construct expressed in low levels within rod cells. They observed a position effect variegation pattern when measuring expression level differences between individual cells and temporally within single cells. In addition, they confirmed a localization defect from a C-terminal deletion in rhodopsin in another transgenic line. Mizuno *et al.* (2005) applied transgenesis to study the function of various promoter elements in the regulation of βB1-crystallin gene expression in *X. laevis*, which is discussed further below.

Transgenesis has also been combined with the gene trap approach to randomly produce GFP-fused proteins (Bronchain *et al.*, 1999). This was used as a high-throughput approach to generate and screen for insertions into the genome, where insertions were identified by GFP expression in embryos. Different embryos expressed GFP in a variety of tissue-specific locations including the lens, inner ear, intestine, and the brain. Furthermore, transgenic *Xenopus* have been used to verify the efficacy of using morpholinos to inhibit expression of specific genes (Nutt *et al.*, 2001).

Transgenic *X. tropicalis* lines carrying a γ1-crystallin promoter driving GFP expression have been generated, which are very useful for the study of lens development and regeneration (Offield *et al.*, 2000; Henry and Elkins, 2001, discussed further below). These transgenic lines enabled Offield *et al.* (2000) to accurately study the timing of lens differentiation, establishing that the onset of lens differentiation was significantly delayed when the presumptive lens ectoderm (PLE) was transplanted from a later (stage 19) embryo to an earlier one (stage 14), and that the delay was even longer in the case when the PLE was explanted in culture. Using transgenic larvae, Henry and Elkins (2001) demonstrated that lens regeneration can occur in *X. tropicalis* (discussed below).

Other methods have been combined with transgenesis to drive targeted, tissue-specific or inducible, stage-specific gene expression. For example, transgenesis has been successfully combined with the Gal4-UAS system to affect targeted expression. In this system, separate transgenic lines with activator and effector constructs, can be crossed to produce progeny expressing a particular gene of interest (Chae *et al.*, 2002; Hartley *et al.*, 2002). Similarly, the Cre/loxP system has been adapted for transgenic *Xenopus* (Ryffel *et al.*, 2003; Waldner *et al.*, 2006). Das and Brown (2004) implemented the mifepristone (RU 486) inducible system to create transgenic *Xenopus* lines; RU 486 is a ligand for a modified progesterone receptor domain fused to DNA binding and activation domains that upregulate expression of the desired transgene. The tetracycline inducible system, with addition of the ligand doxycycline, has also been used to induce transcription (Das and Brown, 2004). Using the *Xenopus* heat-shock protein promoter (hsp), Wheeler *et al.* (2000) adapted a temperature-inducible gene expression method to produce transgenic frogs that express a gene of interest following heat treatment. In a variant of this method, transgenic embryos incorporated a double promoter construct (called "Heat-shock Green Eye Monster") containing a γ1-*crystallin* promoter driving GFP expression and *hsp70* promoter driving expression of the gene of interest. The resulting embryos are sensitive to heat-shock, yet easy to identify, as the transgenic tadpoles exhibit green fluorescent lenses (Fu *et al.*, 2002; Beck *et al.*, 2003, 2006; Buchholz *et al.*, 2004; Slack *et al.*, 2004). The latter was used to study tail and limb regeneration, and transgenesis was further combined with grafting, where transgenic tissue expressing specific proteins was transferred to wild type embryos to discern the involvement of certain tissues under different genetic backgrounds (Beck and Slack, 2001; Beck *et al.*, 2003; Slack *et al.*, 2004). More specifically, Beck *et al.* (2003) and Slack *et al.* (2004) generated transgenic animals in which expression of specific genes were inducibly activated using the *hsp70* promoter (see Wheeler *et al.*, 2000). Expression of the inserted gene was induced by 30 min heat-shock (34°C) at particular stages of development, which can be confirmed via anti-myc antibodies since the HGEM cassette incorporates a 6x *myc* tag to verify transgene expression. Transgenic tadpoles and their tissues could be repeatedly heat-shocked on a daily basis to ensure prolonged activation of the transgene (Beck *et al.*, 2003). Using this method, Beck *et al.* (2003, 2006) and Slack *et al.* (2004) investigated the activity of the BMP and Notch signaling pathways during spinal cord and muscle regeneration in the tail. Knocking down either BMP or Notch inhibited tail regeneration while activation of either promoted tail regeneration. Furthermore, they found that BMP acts upstream of Notch in spinal cord regeneration. On the other hand, these pathways appear to act independently in the case of muscle regeneration.

Several simpler methods of transgenesis have recently been developed. Simplifying the method developed by Kroll and Amaya (1996), Sparrow *et al.* (2000) developed an approach that eliminates the REMI step and the need for cell-free egg extracts, which enables the use of smaller needles since the sperm nuclei do not undergo decondensation. In another approach plasmids containing the desired transgene are digested with *I-SceI* meganuclease and then the mixture is injected directly into fertilized embryos. *I-SceI* meganuclease has a long 18 bp recognition site that is estimated to occur once every 7×10^{10} bp in random genomic DNA. This is a simpler method than handling sperm nuclei, and has been shown to work for both *X. laevis* and *X. tropicalis* (Ogino *et al.*, 2006; Pan *et al.*, 2006). Likewise, in a method using the *Sleeping Beauty (SB)* transposon system, a plasmid containing the *SB* transposon with the gene of interest is directly injected into fertilized eggs along with *SB* transposase mRNA. This results in the creation of mosaic and hemi-transgenic F0 embryos and ubiquitous expression in later generations, observed up to the F2 generation (Sinzelle *et al.*, 2006). The integrase from the bacteriophage ϕC31 can also be used to mediate recombination between bacterial and phage attachment sites. In this method, ϕC31 mRNA is injected into fertilized eggs along with a plasmid containing the bacterial attachment site and the gene of interest. The *Xenopus* genome is hypothesized to have 100–1000 sites that can effectively be cleaved by the ϕC31 integrase as phage attachment sites, thus enabling integration of the transgene into the genome (Allen and Weeks, 2005).

OVERVIEW OF EYE DEVELOPMENT, ANATOMY AND MORPHOLOGY

Embryonic Origins of Eye Tissues in *Xenopus* (Cell Lineage Analyses)

Xenopus exhibits a fairly regular, holoblastic cleavage pattern. A large number of studies have been undertaken to examine early embryonic axial relationships and to establish a fate map for specific blastomeres (including: Nakamura and Kishiyama, 1971; Keller, 1975, 1976; Nakamura *et al.*, 1978; Jacobson and

Hirose, 1978, 1981; Hirose and Jacobson, 1979; Gimlich and Cook, 1983; Jacobson, 1983; Gimlich and Gerhart, 1984; Heasman *et al.*, 1984; Cooke and Weber, 1985; Gimlich, 1986; Masho and Kubota, 1986; Dale and Slack, 1987; Klein, 1987; Moody, 1987a,b; Masho, 1988, 1990; Wetts *et al.*, 1989; Huang and Moody, 1993, 1995, 1997; Li *et al.*, 1997). These studies indicate that there is a close, though not perfect, relationship between the plane of the first cleavage division and the dorsal midline (i.e. plane of bilateral symmetry, see Nakamura and Kishiyama, 1971; Klein, 1987; Danilchik and Black, 1988; Masho, 1990). Different systems of nomenclature have been devised to identify specific cells within the embryos (see Nakamura *et al.*, 1978; Hirose and Jacobson, 1979; Jacobson and Hirose, 1981; Gimlich and Cooke, 1983; Dale and Slack, 1987; Wetts and Fraser, 1989). Here we refer to the nomenclature developed by Hirose and Jacobson (1979), Jacobson and Hirose (1981), and subsequently used by Moody (1987a,b). The cleavage pattern and nomenclature are illustrated in Fig 6.2. The precision of these fate mapping studies has been ensured through the use of cell autonomous lineage tracers including horseradish peroxidase (Weisblat *et al.*, 1978; Hirose and Jacobson, 1979) and fluorescent lineage tracers such as fluorescent dextrans (e.g. Wiesblat *et al.*, 1980; Gimlich and Braun, 1985). Regularity of the fate map was further enhanced by pre-selecting embryos based on visible axial relationships of the early cleavage planes relative to the pigmented animal cap and dorsal "grey crescent." Wetts and Fraser (1989) demonstrated that a very slow intermixing of cells helps contribute to the consistency of the fate map in *Xenopus* and to the well-defined boundaries of each clonal domain (see also Hirose and Jacobson, 1979).

Fate mapping studies (see Moody, 1987a,b) reveal that the retinal rudiments are primarily derived from the D1.1 and D1.2 cells, with some variable and minor contribution from V1.2 in the 16-cell embryo. The lens is primarily derived from the D1.2 and V1.2 blastomeres with some variable and minor contribution from the D1.1 and V1.1 cells. This fate map was further refined for blastomeres at the 32-cell stage (Moody, 1987b): the retinal rudiments are derived mainly from D1.1.2 and V1.2.1 cells (Fig. 6.2A,C,D); the lens is derived mainly from the D1.2.1, and V1.2.1, with minor contributions from D1.1.1, D1.1.2, D1.2.2 and V1.1.1 cells (Fig. 6.2B–D). Other analyses have been conducted to examine the development of the central nervous system, and more specifically, the contributions of various cells at later stages including the 512-cell stages (Jacobson and Hirose, 1978, 1981; Hirose and Jacobson, 1979; Jacobson, 1983). Studies have also been undertaken to examine the origins of specific subsets of retinal neurons (Huang and Moody, 1993, 1995, 1997). Lineage tracing at even later stages of gastrulation and neurulation relate various landmarks and specific topographical regions to the formation of various tissues, including those of the eye (Nieuwkoop and Florschutz, 1950; Keller, 1975; Holt, 1980; Brun, 1981). These latter studies indicate that the retinal rudiments are located in the anterior neural plate at early neural plate stages (e.g. stage 13/14), but extend into the region of the anterolateral neural folds at later stages of development prior to optic vesicle formation in both *Xenopus* and the axolotl (stages 15–19, Brun, 1981). The close proximity of the eye primorida to the presumptive lens ectoderm may be significant, as planar cell signaling plays a critical role in lens induction, and such signals could emanate specifically from the retinal rudiments of the neural plate (Henry and Grainger, 1987). Holt's (1980) analyses indicated that ventral components of the *Xenopus* eye (the neuroepithelium) arise from ventral region of the forebrain that undergo migration via the optic stalk later during the process of optic cup formation. Ventral cells of the early optic vesicle are therefore displaced dorsally by these late-arriving cells. Holt's (1980) findings challenged the interpretation of earlier studies examining eye polarization and retinotectal mapping. Hirose and Jacobson (1979) claimed that some ventral parts of each eye appear to be derived from the contralateral side of the embryo. These observations were, however, derived from injections of lineage tracers performed during early cleavage stages and are probably related to the somewhat variable relationship between the first cleavage plane and the plane of bilateral symmetry in *Xenopus* (described above). Actually, the results of Li *et al.* (1997) in *Xenopus* and the chicken indicate that the paired eyes are derived from one initially broad eye field located in the anterior neural plate. This field is split via signals from the underlying prechordal mesoderm suppressing the expression of retinal progenitors along the midline. In fact, further study has shown that midline cells in the neural plate of neural plate stage embryos remain in these locations and do not migrate laterally to contribute to retinal development at later stages of development (Li *et al.*, 1997).

Early Stages of Eye Development

The eyes of *Xenopus* are similar to those of other vertebrates. Many details regarding the development of the eyes are described in Nieuwkoop and Faber (1956). As in other vertebrates, the central nervous system forms

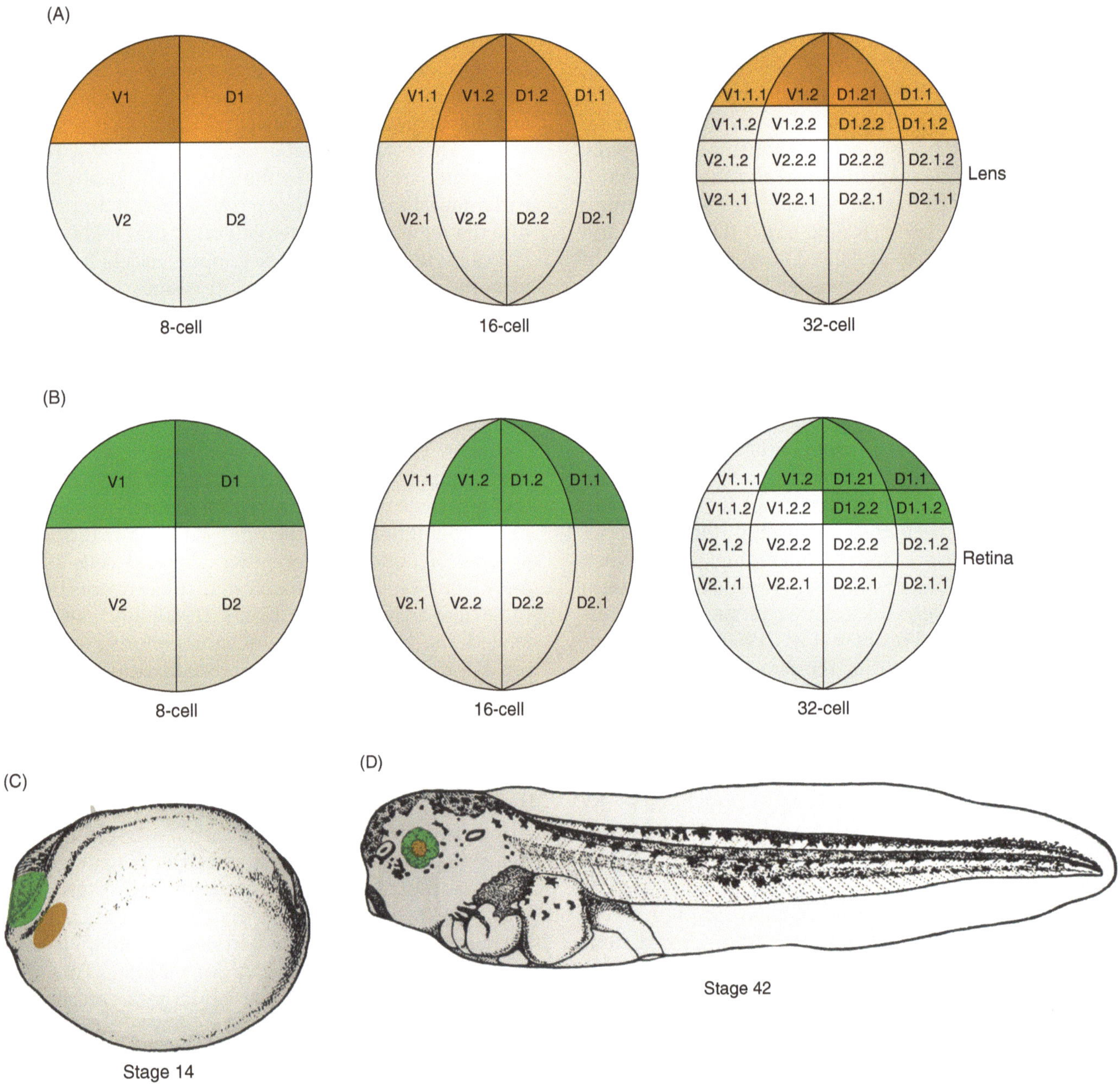

FIGURE 6.2 Diagrams depicting contributions of early cleavage blastomeres (8–32 cell stages) to the formation of the lens and retina (A, B). The ultimate locations of these tissues are also depicted for the neural plate (stage 14) and the young larval stage (stage 42) in (C, D). Stages are as labeled and follow those of Nieuwkoop and Faber (1956). Lens contributions are depicted in shades of orange. Retinal contributions are depicted in shades of green. Darker shading identifies those cells with the greatest contributions to those structures. Lighter shading identifies cells that make minor and variable contributions to the formation of specified structures. The cornea epithelium is derived from the same cells as the lens. Note that the progenitors of these tissues become more highly localized within specific sub-lineages, as development progresses to later stages. Left lateral views are shown of all stages depicted. In (A) and (B) the animal pole of the embryo is located toward the top of the figure. Dorsal is located to the right. In (C) and (D), the anterior end of the embryo or larvae is located to the left of the figure and the dorsal side is located toward the top of the figure. Lineage contributions and nomenclature follows that of Moody (1985, 1987a,b). Diagrams in (C) and (D) are after Nieuwkoop and Faber (1956).

from the neural plate, located on the dorsal side of the developing embryo (prominent at stage 14 in *Xenopus*). This flattened plate subsequently rolls up to form the neural tube, as the lateral neural folds fuse along the dorsal midline (stages 15–20). The paired eyes are derived from a unified eye field located in the developing forebrain (diencephalon) near the anterior end of the embryo. During neurulation the eye field becomes separated by midline signaling events, defects of which cause cyclopia (Li *et al.*, 1997; Patten and Placzek, 2000; Roessler and Muenke, 2001). At stage 18/19 the optic vesicles begin to protrude from the sides of the developing neural tube, just prior to fusion of the anterior neuropore (stage 19/20). Between stages 19–21, the optic vesicles, which give rise to the retinal tissues, come in contact with the overlying head ectoderm, which will form the future lens and cornea epithelium. The optic vesicles are not apparent externally until stage 21 when they begin to protrude from beneath the overlying head ectoderm (presumptive lens ectoderm). At this stage the central cavity of the brain (continuous with the ventricles) begins to extend into the optic vesicles. Reciprocal inductive interactions take place that control the development of the lens and eyecup (Henry and Grainger, 1987, 1990; Grainger *et al.*, 1992; Grainger, 1992, 1996; discussed further below). In *Xenopus*, the embryonic ectoderm is comprised of an outer pigmented layer and an inner un-pigmented sensorial layer. The lens is derived from the inner sensorial layer as a placode that enlarges to form the rounded, solid lens rudiment which ultimately separates from the sensorial layer as the eyecup is formed. The ultimate fate of the outer pigmented epithelium is unclear. It should be noted that the lens in *Xenopus* does not form via a process of invagination of the surface ectoderm as in some other vertebrates, such as the chick, mouse, and human. In *Xenopus* the thickened lens placode initially forms at stage 26/27. The lens rudiment enlarges and normally separates from the sensorial ectoderm by stage 33/34. A cavity subsequently appears within this rudiment to form a lens vesicle no earlier than stage 35/36. Centrally located cells on the proximal side of the lens vesicle, facing the eyecup, give rise to the elongated primary fibers that synthesize lens crystallins beginning at stage 35/36; nuclei of some fiber cells begin to degenerate at stage 40. Distal cells facing away from the eyecup form the mitotically active lens epithelium. By stage 41 the lens cavity disappears as the lens epithelium contacts the growing fiber cell mass. After the formation of the primary lens fibers, secondary lens fiber cells are added at the periphery (equatorial region), which continues through the juvenile stages of life.

Development of the Lens

McDevitt and Brahma (1973) created a normal table of lens development defined by seven stages. During stage 1 (Nieuwkoop and Faber stages 23–24), the optic vesicle is in contact with the presumptive lens ectoderm, which consists of the pigmented and sensorial layers, but there is no discernable morphological change in the ectoderm compared to the surrounding tissue. The initial thickening of the lens placode within the inner sensorial layer occurs at stage 2 (Nieuwkoop and Faber stages 26–27). By stage 3 (Nieuwkoop and Faber stages 29–30), a thickened lens rudiment has formed with a flattened, irregular shape. The lens rudiment fills the cavity of the invaginating eyecup, and at the center of the rudiment lies a slightly condensed mass of cells, which will subsequently give rise to the initial primary fiber cells. McDevitt and Brahma (1973) describe the connection between the lens rudiment and the surface ectoderm as being variable at this stage. At stage 4 (Nieuwkoop and Faber stage 31), the lens rudiment assumes a more regular appearance, but never attains a spherical shape. A central core of compact cells is surrounded by more loosely arranged cells. Generally, the lens fully separates from the sensorial ectoderm at this stage of development. At stage 5 (Nieuwkoop and Faber stages 35–36) the central cell mass begins to form elongated fiber cells. McDevitt and Brahma (1973) describe evidence of cell degeneration in the lens rudiment at this stage. As mentioned above, stage 5 represents the earliest stage of lens vesicle formation in which a cavity forms between the developing lens epithelium and lens fiber cell mass (though the definitive lens vesicle may not be apparent until stage 6, in some cases). At stage 6 (Nieuwkoop and Faber stages 37–41) the lens epithelium forms a single cell layer that overlies the core of differentiating primary fiber cells. The transitional (equatorial) zone located between the lens epithelium and lens fiber cell mass is substantial at this stage. Finally, at stage 7 (Nieuwkoop and Faber stages 45+) the lens epithelium is tightly apposed to the fiber cell mass, and many secondary lens fiber cells are apparent. The study of McDevitt and Brahma (1973) reveals key differences in the development of the lens in *Xenopus*. Unlike the case in other vertebrates, a definitive lens vesicle (containing a central lumen) represents a transient stage formed relatively late during development. Furthermore, in *Xenopus* fiber cells begin to differentiate prior to lens vesicle formation and the elaboration of the lens epithelium.

As lens fiber cells begin to move from their birthplace at the marginal zone of the lens epithelium toward the center of the lens fiber mass, they undergo distinct morphological changes including loss of various organelles including the nucleus, elongation and attenuation, and dense packing with filamentous, regularly spaced crystallin proteins (Balinsky, 1965; Chow and Lang, 2001). Brakenhoff *et al.* (1992) have developed a device that may be used to examine the transparency of the lens in *Xenopus*.

Analyses of Crystallin Expression During Lens Development

Various studies have characterized lens crystallin expression during development in *X. laevis*. Early studies examined the timing of expression of various crystallins using polyclonal antibodies. α-, β-, and γ-crystallins are expressed in larval and adult frog lenses with some stage-specific differences (Brahma and van Doorenmaalen, 1968; Brahma and Bours, 1972; Nöthinger *et al.*, 1971; McDevitt and Brahma, 1973, 1979, 1982; Brahma, 1980; Campbell *et al.*, 1986; Shastry, 1989; Henry and Grainger, 1990; Shastry and Reddy, 1990; Brunekreef *et al.*, 1997). γ-crystallins make up the vast majority of lens crystallin proteins, while α-crystallins make up the smallest fraction. The expression of lens crystallins (e.g. γ-crystallins) is first detected at lens developmental stage 3 (stages of McDevitt and Brahma, 1973, described above). Crystallins initially appear in the central mass of prospective fiber cells, prior to their elongation. Expression increases as fiber cells undergo differentiation and additional fibers are added. Some expression (presumably α- and/or β-crystallins) can also be detected in the lens epithelium, but not until later stages (lens developmental stage 7, McDevitt and Brahma, 1973).

The transcription of crystallin genes has also been examined during development in *Xenopus* (Smolich *et al.*, 1993, 1994; Brunekreef *et al.*, 1997; Mizuno *et al.*, 1999a, 2005; Offield *et al.*, 2000; Hirsch *et al.*, 2002a,b). Differences in the timing of expression of various crystallins have been noted when compared to other vertebrates (Brunekreef *et al.*, 1997). For instance, the observations of Brunekreef *et al.* (1997) substantiate the findings of McDevitt and Brahma (1973) in showing that presumptive fiber cells begin to differentiate before cells of the lens epithelium in amphibians, including *Xenopus*. As in other systems, the expression of some crystallins is not restricted to lens cells (Smolich *et al.*, 1994; Brunekreef *et al.*, 1997). Analyses have also led to the characterization of some crystallin promoters (Smolich *et al.*, 1993; Offield *et al.*, 2000; Hirsch *et al.*, 2002a,b; Mizuno *et al.*, 2005). Such cis-regulatory elements have been used to drive lens-specific expression of other genes such as GFP, and in the generation of transgenic frogs, as described above (Offield *et al.*, 2000; Hirsch *et al.*, 2002a,b; Mizuno *et al.*, 2005). The tremendous level of conservation of certain regulatory elements of these crystallin genes was demonstrated in an early study by Brakenhoff *et al.* (1991), which showed that a rodent γ-crystallin promoter could drive lens-specific expression of CAT reporter in developing *Xenopus* lens cells.

In a very nice set of experiments, Mizuno *et al.* (2005) generated transgenic F0 larvae to examine the requirements of specific promoter elements in the regulation of βB1-crystallin in both development and regeneration of the lens in *X. laevis*. They found significant sequence similarity between the promoters of *Xenopus* and chicken βB1-crystallin genes, which included the PL1 and PL2 binding sites for MAF, Pax6, and Prox1 (transcription factors described below), indicating a conserved mechanism of gene regulation in both chicken and frog. Stepwise deletion of the promoter sequence also uncovered additional regulatory element(s) in *Xenopus*.

Development of the Retina

As lens vesicle formation takes place, the optic vesicle is transformed into the eyecup through the process of invagination, beginning at stage 27. The distal part of the optic vesicle, in contact with the lens placode, ultimately forms the neural retina, while the more proximal region gives rise to the retinal pigmented epithelium (RPE). Invagination is initiated distally at the anterior-dorsal margin. The choroid fissure is formed as a ventral groove in the optic vesicle at stage 29. The margins of the choroid fissure come into contact at stage 32. The more proximal portion (located close to the optic stalk) closes at stage 37/38. The more distal region closes by stage 46. The hyaloid artery reaches the interior of the eye through a small persistent opening of the choroid fissure located in the ciliary body near the pupillary opening. Most of the original cavity located within the optic vesicle (between the inner neural retina and outer RPE of the eyecup) is obliterated by stage 32. A portion of this cavity persists within the optic stalk (optic nerve), which remains in communication with the ventricle of the forebrain until stage 35/36. As mentioned above, lineage tracing experiments of Holt (1980) showed that ventral components of the eye (neuroepithelial cells) arise from ventral regions of the forebrain, that undergo migration later during the process of optic cup formation via the optic stalk.

As further differentiation takes place, the RPE forms a thin layer of flattened cells on the outer surface of the eyecup, while the neuroepithelium becomes thickened. Pigmentation first appears in the outer RPE at stage 32. By stage 35/36 the outer surface of the developing eyes appear to be entirely black. These pigment cells, called xanthophores, differ from the pigmented melanophores associated with the body ectoderm. In addition, iridiophores, which have a metallic silver appearance, are also formed on the outer surface of the eye. Unlike melanophores, pigmented cells of the eye are formed in albino strains of *Xenopus*.

Within the neural retina, the pars optic retinae begins to undergo differentiation by stages 35/36 and 37/38, the nuclei are arranged in three discrete layers consisting of the outer nuclear layer, the inner nuclear layer, and the ganglion cell layer. These layers become separated by the inner and outer plexiform layers, which are visible at stage 37/38. The visual photoreceptors (rods and cones) are readily distinguishable by stage 42 (reached in just 3–4 days at 22–24°C). By stage 47/48 rods and cones reach a fully differentiated state. These cells will undergo further growth beginning at stage 49 through stage 66. Prior to metamorphosis, retinal cells proliferate symmetrically within the ciliary marginal zone (CMZ). At metamorphosis considerable asymmetric growth occurs at the CMZ, which continues in the juvenile frog. Much of the adult retina is apparently derived from ventral CMZ cells (Beach and Jacobson, 1979). By stage 39 a defined layer of mesenchyme surrounds the eye which gives rise to the inner choroid and outer sclera. This tissue is continuous with that which forms the inner cornea endothelium (see below). The sclera and choroid layers begin to segregate by stage 44. The mesenchymal rudiments of the eye muscles are also apparent at stage 39.

Development of the Cornea and Other Eye Tissues

After the lens vesicle separates from the sensorial ectoderm (generally around stages 33–34), the remaining ectoderm gives rise to the inner and outer layers of the cornea epithelium. The larval cornea epithelium exhibits a high degree of transparency and has no specialized secretory cells or pigmented melanophores, which are found in the adjacent head epidermis. The underlying cornea endothelium forms by stage 40 and is derived from migrating neural crest cells that populate the area between the lens and overlying ectoderm. Throughout most of the early period of development the inner cornea endothelium and the outer cornea epithelium remain separated except for a small central connection that overlies the pupillary opening. This "stalk-like" connection is established at stage 42 and may include the remains of a stalk that initially connected the developing lens to the sensorial ectoderm. On rare occasions a tiny ball of cells (reactive to anti-lens crystallin antibodies) may be present within this stalk (Henry, personal observation). Presumably these "secondary lenses" represent induced lens cells that remained after separation of the lens rudiment. The ultimate fate of these secondary lenses is uncertain. Further fusion of the inner and outer cornea takes place at stages 55 and by stage 66 the cornea epithelium and endothelium are fully fused.

As it is the case in other amphibians, further remodeling of the eye occurs during the process of metamorphosis to form the juvenile frog. Readers are referred to the review of Hoskins (1990), for further details.

INDUCTIVE INTERACTIONS IN EYE DEVELOPMENT

Embryonic Lens Induction

Over 100 years of experimental studies have led to a considerable understanding of the inductive interactions that control eye development and, in particular, development of the lens. To a large extent, these studies were carried out using various amphibian species, due to the greater ease with which one can obtain and manipulate their embryos in transplantation and explant culture experiments.

Previously, it had been thought that the optic vesicle was both necessary and sufficient for vertebrate lens development (Grainger *et al.*, 1988, 1992; Saha *et al.*, 1989, 1992; Grainger, 1992, 1996). This notion was supported by the results of many classical embryology experiments (reviewed by Reyer, 1958a,b). For instance, Spemann (1901) observed that ablation of eyecup rudiments prevented lens formation in the frog *Rana temporari*. Lewis (1904, 1907a,b) claimed that a lens could be induced in non-lens ectoderm by juxtaposition with transplanted optic vesicles in the frogs *Rana palustris* and *Rana sylvatica*. Unfortunately, many of these early experiments were fraught with problems related to tissue contamination, which complicates the interpretation of these results. Either the presumptive lens cells were not completely removed from the host embryos, or they were inadvertently transplanted with the donor tissues. More recent experiments with *Xenopus* revealed the need for careful host and donor marking to track the fates of transplanted tissues, which had not been carried out in most of the earlier studies (Henry and Grainger, 1987; Grainger *et al.*, 1988). Furthermore, the use of definitive molecular markers (e.g. antibodies against lens crystallin proteins) is particularly important, especially in cases where the differentiated response may be weak and lacks characteristic lens morphology (Henry and Grainger, 1990). In fact, more careful experiments reveal that the optic vesicle is actually a rather weak lens inducer in *Xenopus* (Henry and Grainger, 1987; Grainger *et al.*, 1988; Saha *et al.*, 1989, 1992; Grainger *et al.*, 1997). Lens induction is actually a multi-step process, involving a critical "early" phase of induction that precedes that of the optic vesicles (the so-called "late" phase, Fig. 6.3A). A similar conclusion was subsequently reached for other vertebrates (e.g. the frog *Rana palustris*, Grainger *et al.*, 1988, 1997; the axolotl, *Ambystoma mexicanum*,Servetnick *et al.*, 1996; and the chicken, Sullivan *et al.*, 2004); although, the extent to which the early and late phases of lens induction play a role in lens development may vary in some vertebrates (Grainger *et al.*, 1997; Mizuno *et al.*, 1998).

Experimental studies, and in particular, those carried out with *Xenopus,* have led to the formulation of a conceptual model that defines certain properties inherent to the lens inducing and responding tissues (Fig. 6.3). This model serves as a paradigm for understanding the general nature of inductive interactions in other systems (Henry and Grainger, 1987, 1990; Saha *et al.*, 1989, 1992; Grainger, 1992, 1996; Grainger *et al.*, 1992, 1997; Hirsch and Grainger, 2000; Henry *et al.*, 2002). As mentioned above, embryonic lens development is a multi-step process directed by a series of inductive tissue interactions. These interactions take place over two principal phases (Fig. 6.3A). The existence of an early period of lens induction, which precedes induction by the optic vesicle, was established by a number of early investigators using different vertebrates (Liedke, 1942, 1951, 1955; Reyer, 1950, 1954, 1958a,b; Jacobson, 1966; reviewed by Saha *et al.*, 1989; Grainger, 1992, 1996). In the frog *Xenopus* (Henry and Grainger, 1987; Servetnick and Grainger, 1991), the "early phase" begins during gastrulation (approximately stage 11.5). This phase of

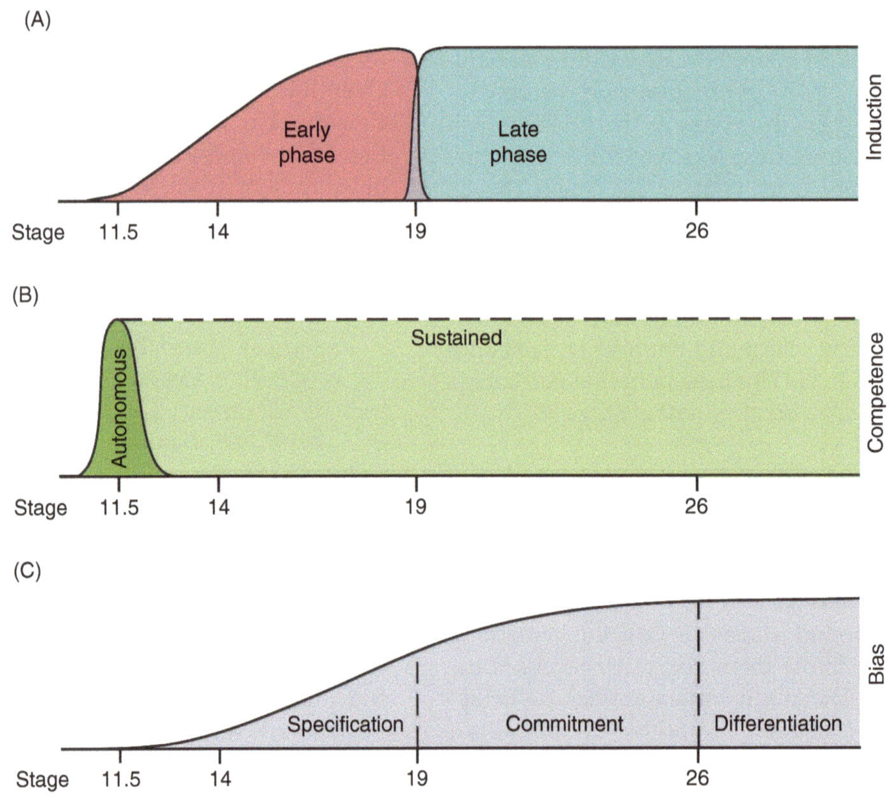

FIGURE 6.3 Schematic diagrams depicting a contemporary model of lens induction. (A) Multi-step nature of lens induction involving "early" and "late" phases of induction. (B) Establishment of "competence" in presumptive lens ectoderm to respond to lens inductive interactions. (C) Establishment of "lens-forming bias" in presumptive lens ectoderm following exposure to lens inductive interactions. Presumptive lens ectoderm is "specified" by stage 19, and is presumably "committed" (or "determined") by stage 26 when the lens placode is formed. Following this time, the lens ectoderm undergoes a prolonged period of "differentiation" in which crystallin proteins are synthesized, fiber cells are formed, and the lens exhibits normal polarity properties. Refer to the text for further explanation. Diagram after Henry *et al.* (2002).

induction involves "planar" signals emanating from the anterior neural plate, which begin to specify the surrounding contiguous crescent of placodal ectoderm (Henry and Grainger, 1990). If anterior neural tissue is removed early during development (prior to stage 19), lens formation will not take place. Transverse inductive signals from tissues underlying the placodal ectoderm (endoderm and mesoderm) also play a role in this early phase of lens induction (Jacobson, 1966; Henry and Grainger, 1990). The "late phase" of lens induction begins when the optic vesicles contact the overlying placodal head ectoderm (beginning at stage 19) and continues through later stages of development. In *Xenopus*, a large field of head ectoderm is initially induced to form a lens (Grainger *et al.*, 1997). The optic vesicles serve to pinpoint the exact site of lens formation in a larger field of presumptive lens-forming ectoderm to ensure coordinated development of the lens and retinal tissues. The continued presence of the eyecup is also required to support normal differentiation of the lens.

For lens induction to proceed, the embryonic ectoderm must be competent to respond to the appropriate inductive signals. Embryonic ectoderm first develops an autonomous window of "competence" to respond to lens inductive interactions early during development (Fig. 6.3B), which corresponds with the initiation of lens induction (between stages 11–12, Henry and Grainger, 1987; Servetnick and Grainger, 1991; Grainger *et al.*, 1997). Younger embryonic ectoderm responds to these inductive signals, in a different context, to form neural tissues. Competence is maintained once ectodermal tissues receive lens inductive signals. In non-head ectoderm, this property is quickly lost at later stages of development. Competence is assayed in an experimental context by subjecting ectodermal tissues isolated from different regions of the embryo and different stages of embryonic development to defined periods of lens induction. The response is measured in terms of the percentage of cases that form lenses, as well as the overall extent of lens epithelial and fiber cell differentiation that ultimately takes place in the transplanted tissues.

As further induction takes place, the responding ectoderm develops an increasing "lens-forming bias" or propensity to form a lens (Fig. 6.3C). Like competence, bias is assayed experimentally and is correlated with the quality and duration of the inductive interactions. Growing bias represents the process of "specification" and "commitment" that ultimately leads to lens differentiation. In *Xenopus* the presumptive lens ectoderm is "specified" by stage 19, as this tissue will differentiate into lens cells when cultured in isolation (Fig. 6.3C, Henry and Grainger, 1990). Specified tissues, however, may be re-programmed along different developmental pathways if subjected to other inductive interactions. In *Xenopus*, specification is accomplished via the early phase of lens induction. During the process of neurulation, a larger region of head ectoderm is biased to form lenses than actually will ultimately participate in lens formation (Grainger *et al.*, 1997), and subsequent interactions with the optic vesicle serve to pinpoint the exact site of lens formation. The optic vesicle (first formed at stage 19) can only serve as a weak inducer of lens formation in maximally competent ectoderm (i.e. ectoderm isolated during early gastrula stages, before a lens-forming bias has been established, which can only take place in an experimental context). On the other hand, the optic vesicle plays an important role as a late lens inductor in ectoderm (e.g. stage 19) that has been previously biased by virtue of the early phase of lens induction (Henry and Grainger, 1987; Grainger *et al.*, 1997). Later during development the lens is irreversibly determined or "committed," which presumably takes place by stage 26/27 when the lens placode forms and crystallin expression is initiated (Fig. 6.3C, general definitions according to Slack, 1991; Grainger, 1992, 1996).

The significance of the early phase of lens induction is highlighted by the fact that lenses will form in the absence of the differentiating optic vesicle and eyecup (so-called "free-lenses," reviewed by Reyer, 1958a,b). These observations indicate that the early phase of lens induction is sufficient to induce the formation of lens cells. This process of free-lens formation has been carefully described for *Xenopus* (Balinsky, 1951, 1957; de Graff, 1960; Babcock, 1961; Brahma and Grunz, 1988; Henry and Grainger, 1990). Babcock (1961) defined four stages (1–4) of free-lens formation in *Xenopus*. During stage 1, an epidermal thickening is observed in the embryonic ectoderm. During stage 2, a spheroidal cellular aggregate forms within the thickened ectoderm, which begins to separate from the ectoderm. At stage 3, the free-lens is fully separated from the surface ectoderm. Stage 4 is defined when the spherical free-lens has a central core of elongated fiber cells surrounded by lens epithelium. This free-lens lacks normal polarity properties, which are not established in the absence of the eyecup. During the process of free-lens formation crystallins are first detected at Babcock stage 4 (Brahma and Grunz, 1988; see also Henry and Grainger, 1990), which corresponds roughly to lens developmental stages 4–5 of McDevitt and Braham (1973, described above). This represents a later stage compared to that normally seen during embryonic lens development (as described above), and the lack of normal inductive interactions from the optic vesicle may delay lens differentiation.

Induction of the Retina

Reciprocal interactions from presumptive lens ectoderm also influence the development of the optic cup and retina (Hyer *et al.*, 2003). Using the chicken as an experimental system, Hyer *et al.* (2003) demonstrated that these important interactions occur early during development between the presumptive lens ectoderm and the developing optic vesicle, prior to lens and optic cup formation. A similar situation likely takes place in *Xenopus,* as the eyecup and retina do not differentiate normally when the presumptive lens ectoderm is removed during neural plate stages (e.g. Henry and Grainger, 1987).

CONTRIBUTIONS TO OUR UNDERSTANDING OF THE MOLECULAR BASIS OF EYE DEVELOPMENT

Molecular Level Control of Retinal Development

Although the first morphological evidence of eye development is observed at stage 19/20 with the evagination of the optic vesicles, neural induction and eye field specification begins at a much earlier stage of development. Work performed using *Xenopus* has contributed greatly to our understanding of CNS and eye development. For instance, the establishment of anterior neural fates involves antagonizing bone morphogenetic protein (BMP) signaling through various molecules including: chordin, noggin, follistatin, cereberus, and xnr3 (Harland, 2000), and anterior–posterior polarity is established by gradients of secreted factors including: FGFs, retinoids, and Wnts. Some data suggests that steps involved in retinal fate commitment may take place even before neural induction begins, possibly involving maternal asymmetries of certain patterning factors (Moore and Moody, 1999; Yan and Moody, 2007). A full discussion of neural induction and patterning is beyond the scope of this paper, and readers are referred to the reviews of Weinstein and Hemmati-Brivanlou (1999), Harland (2000), and Wilson and Edlund (2001); De Robertis and Kuroda (2004), Stern (2005), for further information.

After neural induction has begun (around stage 11), the transcription factor *Otx2* is strongly expressed in the presumptive anterior neural plate of *Xenopus* late gastrula stage embryos. *Otx2*, an *orthodenticle*-related gene containing a bicoid class homeodomain, is required for development of forebrain and midbrain derived structures (reviewed by Chow and Lang, 2001 and see Martinez–Morales et al., 2001). At stage 12.5, *Otx2* is significantly down-regulated in the medial region of the presumptive anterior neurectoderm, demarcating an area where expression of a suite of genes commonly called eye field transcription factors (EFTFs) begins to specify the eye field (Zuber *et al.*, 2003). This group of EFTFs includes ET, Rx1, Pax6, Six3, Lhx2, tll, and Optx2 (also known as Six6), which play key roles in eye development. While expression of the EFTFs is restricted to the *Otx2*-negative anterior neurectoderm, their expression patterns differ somewhat in both a spatial and temporal context.

Work in *Xenopus* using double *in situ* hybridization revealed the unique, overlapping expression patterns of EFTFs during eye field specification and later at mid-neurula stages (Zuber *et al.*, 2003). Through a series of overexpression experiments, Zuber *et al.* (2003) constructed a model for the epigenetic relationships of certain EFTFs in *Xenopus* eye field specification (see Fig. 6.4A–D). The proposed model is certainly neither definitive nor inclusive but begins to show that eye field specification occurs through complex interactions of the EFTFs. *ET* and *Rx1* are believed to be early initiators of eye field specification because of their regulatory relationships with both the neural inducer *noggin* and the forebrain specification gene *Otx2*. ET is a member of the *Tbx2* T-box subfamily and is believed to be upstream of *Rx1*, a paired-like homeobox gene. *Rx1* expression induces *Pax6*, a highly conserved *paired*-box homeodomain gene related to *Drosophila eyeless* that is crucial for normal eye development. As is the case in other vertebrates, misexpression of *Pax6* causes ectopic eye formation in *Xenopus,* and knockdown experiments create reduced eye or eyeless phenotypes (Altmann *et al.*, 1997). Both *Six3*, of the SIX-homeodomain family related to *sine oculis*, and *Lhx2*, a LIM-homeodomain gene, are believed to be downstream of *Pax6* and are also required for eye development. Interestingly, *Six3*, *Lhx2*, and *Pax6* appear to cross-regulate each other's expression (Fig. 6.4D). Compared to other EFTFs, *tll* and *Optx2* have less affect in the overall specification of the eye field and are believed to act later in the patterning of eye tissue. *Optx2*, another member of the SIX-homeodomain family, is directly inducible by *Pax6* expression but seems to lie outside the *Six3/Lhx2/Pax6* feedback pathway. Conversely, *tll*, a *Drosophila tailless* homolog, can provide positive feedback to this pathway (Fig. 6.4D, Zuber *et al.*, 2003).

Many other transcription factors such as *Pax2, Vax1*, and *Vax2* are expressed in the eye field with the

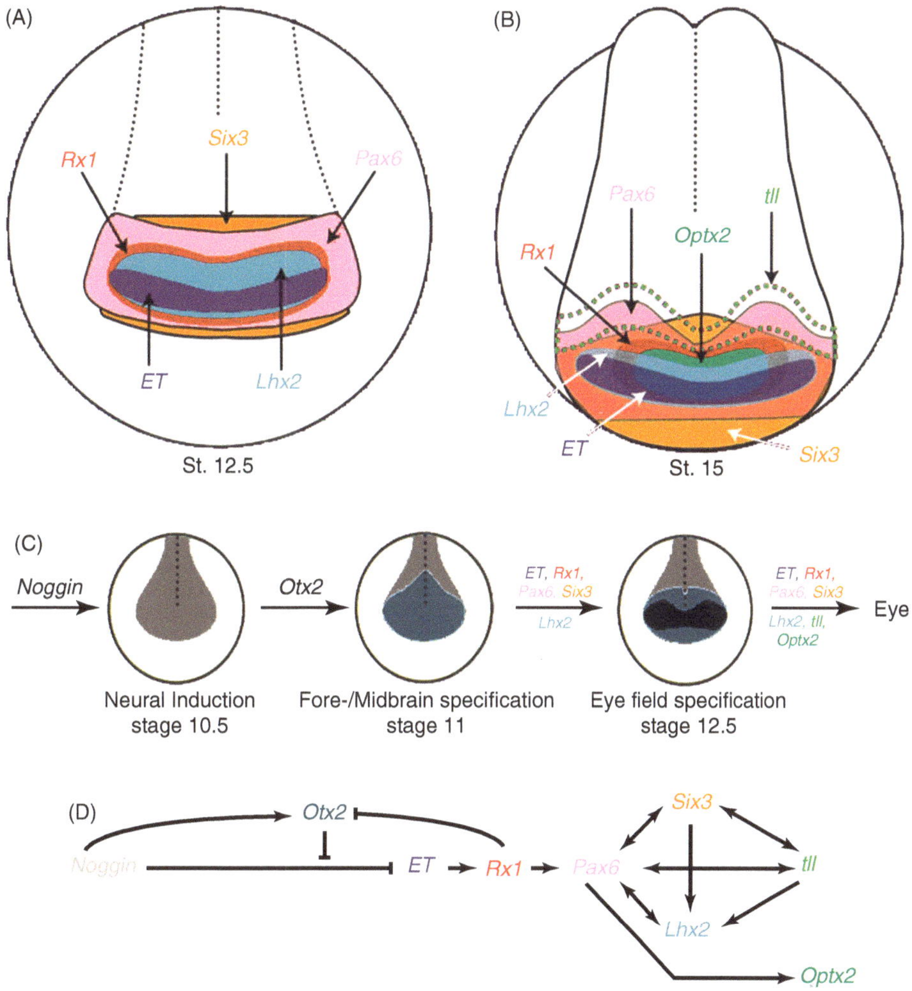

FIGURE 6.4 Neural induction and eye field specification in *Xenopus*. All representations of embryos are viewed from a dorsal-anterior perspective with the anterior most ends located toward the bottom of each diagram. The dorsal-midline is indicated by a straight, vertical dotted line. The model is not inclusive of all factors involved in neural induction and eye field specification but represents a summary of some major factors involved in early eye development. (A, B) A schematic summary of the overlapping gene expression of eye field transcription factors at stage 12.5 (A) and stage 15 (B). (C) Relative timing of gene expression in the patterning of the eye field in the anterior neural plate. Gray signifies *noggin* expression and the field of neural induction. Blue-gray represents the specification of the forebrain and midbrain by *Otx2* expression. Black is the specified eye field where eye field transcription factor expression occurs. (D) Proposed model of a gene expression cascade in eye field specification and further eye development. See text for further details. Figure adapted from Zuber *et al.* (2003).

EFTFs, but their involvement in eye field specification has not been extensively studied (Barbieri *et al.*, 1999; Lupo *et al.*, 2000). On the other hand, their function has been shown to be important in later eye development by encouraging differentiation or proliferation of certain tissues. For example, *Pax2* is a *paired* homeobox gene known to be involved in the establishment of a distinct boundary between the optic stalk and the ventral retina (Dressler *et al.*, 1990; Nornes *et al.*, 1990; Beebe, 1994; Schwarz *et al.*, 2000). *Vax1* and *Vax2* are homeobox genes involved in optic nerve and ventral retina development, respectively (Fig. 6.5, Hallonet *et al.*, 1998; Barbieri *et al.*, 1999).

Early during neural development, the eye field exists as a single region spanning the width of the anterior neural plate (Li *et al.*, 1997). Normally cells at the midline of the anterior neural plate do not contribute to eye formation (Li *et al.*, 1997); however, experimental embryological studies demonstrated that transplants derived from the midline can form eye structures (Adelmann, 1936). On the other hand, as neurulation begins, expression of EFTFs and other essential eye development genes begin to resolve into two distinct regions lateral to the midline. These observations raise the question of how two distinct, lateral eyes are ultimately derived from a single eye field. Early

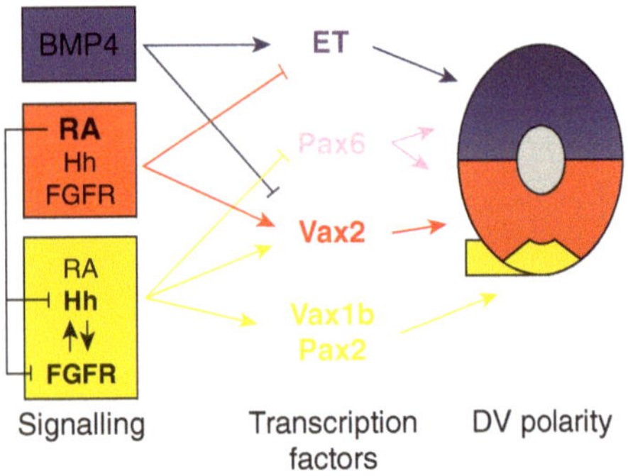

FIGURE 6.5 Proposed model for signaling events involved in the dorsoventral patterning of the *Xenopus* eye (optic vesicle/eyecup). Blue signifies BMP4 signaling and the gradient biasing dorsal retina cell fates. Red represents high level of retinoic acid (RA) signaling with low hedgehog (Hh) and low fibroblast growth factor receptor (FGFR) signaling as well as a gradient biasing ventral retina cell fates. Yellow indicates high levels of Hh and FGFR signaling (which are co-activating) with low levels of RA signaling and bias toward optic stalk cell fates. Figure adapted from Lupo *et al.* (2005).

experiments that address this issue include the work of Mangold (1931) which showed that removal of the prechordal mesoderm in *Triton* led to cyclopic tadpoles. Repetition of these experiments in *Xenopus* resulted in the same cyclopic condition as well as persistent midline expression of key eye development genes, such as *Pax6* and *ET* (Li *et al.*, 1997; Pera and Kessel, 1997). The molecular basis for eye field separation involves hedgehog (Hh) signaling, an EGF-CFC family receptor called *one-eyed pinhead*, and the secreted nodal-related transforming growth factor-β molecule *Cyclops* (reviewed in Chow and Lang, 2001). Work, primarily in zebrafish, has shown the function of these genes to be necessary for down-regulating eye gene expression at the midline, for eye field separation, and for normal eye development (refer to Chow and Lang, 2001).

After eye field separation, eye development genes continue to resolve into more discrete expression patterns, regionalizing the optic vesicle into three dorsoventral (DV) compartments. Data collected by Lupo *et al.* (2005) is summarized in Fig. 6.5. The ventral most compartment represents the future optic stalk and expresses *Pax2*, *Vax1*, and *Vax2*. The middle compartment represents the ventral retinal region and contains *Vax2* and *Pax6* expressing cells. Finally, the dorsal retinal compartment later forms the functional dorsal retina and expresses *Pax6* and *ET* (Torres *et al.*, 1996; Bertuzzi *et al.*, 1999; Hallonet *et al.*, 1999; Koshiba-Takeuchi *et al.*, 2000; Schwarz *et al.*, 2000; Liu *et al.*, 2001; Barbieri *et al.*, 2002; Mui *et al.*, 2002). Work performed in *Xenopus* by Lupo *et al.* (2005) showed that the expression of these DV patterning genes and thus the fate of cells in the developing eye are under control of BMP4, retinoic acid (RA), Hh, and fibroblast growth factor (FGF) signaling. Furthermore, DV patterning in the developing eye depends on the dosage and combination of these different signaling molecules. High levels of BMP4 signaling activate *ET* expression and promote dorsal retina fates while also repressing *Vax2* and the differentiation of ventral retina fates. High RA levels repress Hh and FGF signaling. Furthermore, high levels of RA with concomitant low levels of Hh and FGF leads to the repression of *ET*, and to expression of *Vax2*, favoring ventral retina fates. Low levels of RA signaling combined with high Hh and FGF repress *Pax6*, which normally acts to promote both dorsal and ventral retina differentiation. In addition, this combination of signaling activates *Vax2* and *Vax1/Pax2* expression, which encourage ventral retina and optic stalk fates, respectively.

Xenopus has also contributed greatly to understanding later stages of retinal development, shedding light on the signaling pathways and transcription factors involved in the differentiation of neural cell types in various layers of the retina as well as on axon guidance from the eye to proper locations within the brain (Perron *et al.*, 1998; Mann *et al.*, 2004). Analyses reveal a complex timing of combinations of Notch–Delta signaling and proneural genes and these are summarized in Fig. 6.6A–C. Studies have shown that alterations in the expression of genes in the Notch–Delta lateral inhibition pathway can change the fate of neurons in the retina (Dorsky *et al.*, 1997). Levels of Delta expression in retinal progenitors vary over time, specifying certain neuronal cell types and inhibiting Delta expression and differentiation of the same cell type in neighboring cells. Notch–Delta signaling has also been implicated in controlling the expression of a number of *Drosophila achaete-* and *scute*-related proneural factors that induce differentiation of different neuronal cell types in the retina. These proneural factors include basic helix-loop-helix (bHLH) transcription factors such as Ash1, Ash3, Ath3, and Ath5 to name a few (Ferreiro *et al.*, 1992; Zimmerman *et al.*, 1993; Turner and Weintraub, 1994; Kanekar *et al.*, 1997; Perron *et al.*, 1999). Specific neuronal cell types in the different layers of the retina are generated in a highly conserved temporal order: first being the retinal ganglion, horizontal, and cone cells; and ending with amacrine, rod, bipolar, and Müller glial cells. To better understand the molecular basis for the differentiation of these cell types, many researchers have turned to examination of the peripheral area of the *Xenopus* retina, known as the ciliary marginal zone (CMZ), because an active stem cell population located therein continues to add new cells to the retina through the life of the organism (Straznicky and Gaze, 1971; Johns, 1977; Reh, 1989; Wetts *et al.*,

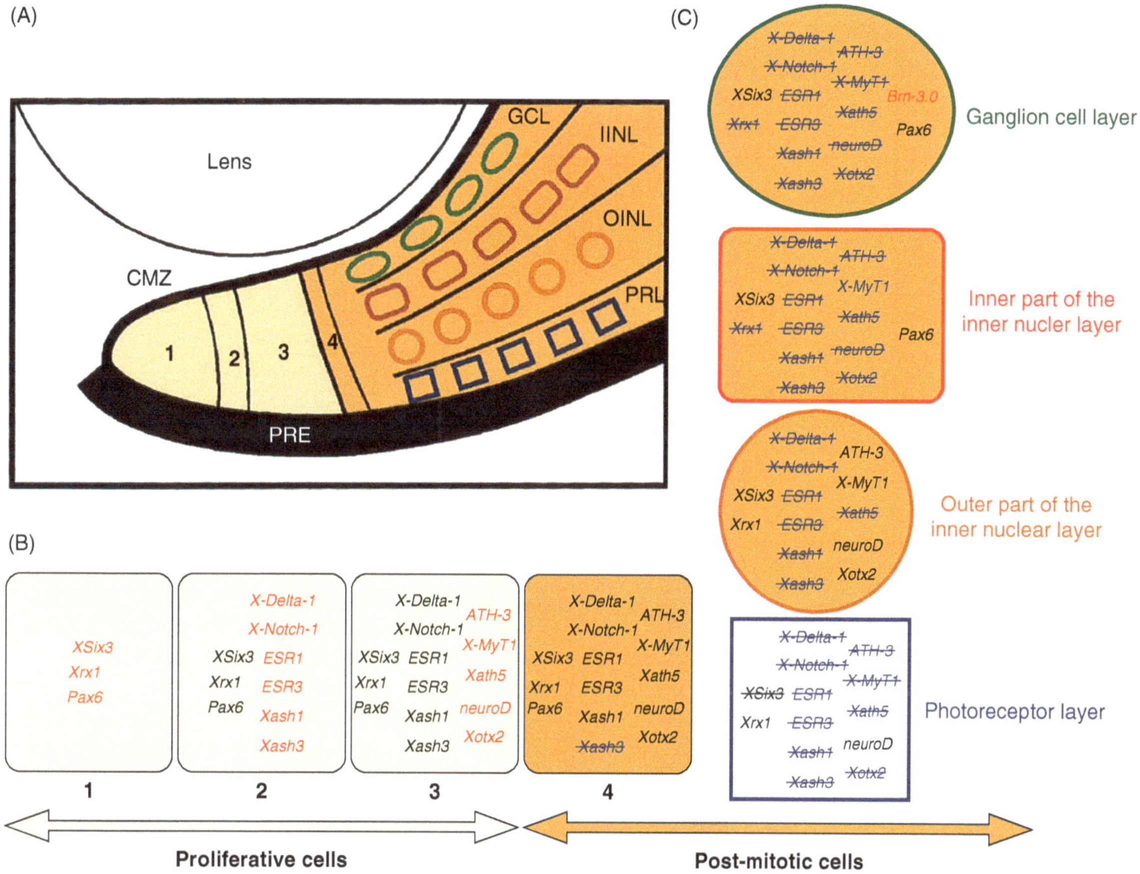

FIGURE 6.6 Gene expression during retinogenesis in the *Xenopus* eye. (A) Diagram of the *Xenopus* eye showing both the CMZ split into four parts, as well as the different cell layers of the retina. Throughout the figure, yellow represents zones where cells are dividing while light orange shows areas containing post-mitotic cells. GCL, ganglion cell layer; IINL, inner part of the inner nuclear layer; OINL, outer part of the inner nuclear layer; PRL, photoreceptor layer. RPE, retinal pigmented epithelium. (B) Summary of gene expression in the four regions of the CMZ. (C) Summary of gene expression in the differentiating layers of the retina. Red text represents the first onset of gene expression; black text corresponds to gene expression that persisted from a previous stage; gray text with a strikethrough line symbolizes cessation of gene expression that was active in the previous stage. For more information refer to the text and see Perron *et al.* (1998), from which this figure was adapted.

1989). Furthermore the spatial ordering of the cells in the CMZ mirrors the events in cellular development and differentiation that occur during the establishment of different neuronal cell types in the retina. Work by Perron *et al.* (1998) determined the expression patterns of the cells undergoing retinogenesis from stem cell to proliferating retinoblast and finally through the differentiation of specific neuronal cell types.

BMP signaling has also been implicated in late retinal development, but its exact role has not been fully elucidated (Hocking and McFarlane, 2007). During the period of retinal cell proliferation, *Bmp4* transcripts are found in the dorsal retina, and *Bmp7* is expressed in the distal periphery. Furthermore, BMP type-1a and type-2b receptor transcripts localize to the ventral retina and the retinal ganglion cell layer, respectively (Hocking and McFarlane, 2007). Knockout studies done in mice suggest the BMP signaling pathway is required for eye development and patterning of many retinal tissues (Dudley *et al.*, 1995; Furuta and Hogan, 1998; Sasagawa *et al.*, 2002; Murali *et al.*, 2005).

The retinal ganglion cells (RGCs) are the only neurons in the *Xenopus* retina to extend axons out of the eye into the brain. These axons exit through the optic disc region at the center of the retina, bundle together into the optic nerves, cross the midline at the optic chiasm in the hypothalamus, and finally form synaptic connections with their targets in the contralateral tectum. However, some of these axons do not cross the midline and instead connect with the ipsilateral tectum, a process required for normal binocular vision. Work in *Xenopus* has been highly instructive in explaining the molecular mechanisms behind the various aspects of RGC axon guidance. Changes in attractive/repulsive cues from laminin and netrin molecules and their receptors in the retina have been implicated in guidance to the optic disc, while gradients of the excreted ephrin-B and of ephrin-B receptor expression have been shown to control axon guidance at the optic chiasm and in retinotectal targeting (reviewed by Mann *et al.*, 2004).

Molecular Level Control of Lens Development

Studies have also begun to illuminate the molecular level controls of lens development and differentiation. Figure 6.7 is a proposed model of the gene expression pathways involved in these processes. The information presented in Fig. 6.7 is not complete, but focuses primarily on principal regulatory factors, especially those discerned from work done with *Xenopus*.

As discussed above, *Pax6* represents a key regulatory gene involved in eye development. Ectopic expression of *Pax6* in *Xenopus* not only leads to ectopic eye formation (Chow *et al.*, 1999) but also results in the formation of supernumerary lenses (Altmann *et al.*, 1997). Furthermore, Pax6 can induce expression of many downstream lens developmental genes (Chow *et al.*, 1999, discussed below). Pax6 also appears to play roles in lens differentiation, including lens epithelial maintenance, proliferation and fiber cell differentiation.

The factors responsible for the initial regulation/activation of *Pax6* expression are unclear. As previously discussed, Otx2 is important for the formation of forebrain and midbrain structures, preceding both retinal and lens development. Expression analyses reveal that *Xotx2* transcripts are present in the presumptive lens ectoderm at the time *Pax6* is first expressed in those tissues (Pannese *et al.*, 1995; Kablar *et al.*, 1996; Zyger *et al.*, 1998; Schaefer *et al.*, 1999). The experiments of Zyger *et al.* (1998) in *Xenopus* suggest that *Xotx2* may represent an important factor acting upstream of *Pax6* in lens development. In addition, Notch mis-expression experiments performed in *Xenopus* result in the activation of *Pax6* expression, as well as other downstream lens markers (Onuma *et al.*, 2002), suggesting that Notch signaling may also be involved to initiate lens development. Knockout (KO) studies in zebrafish have shown that Bmp7 and members of the FGF family are required later during development to maintain *Pax6* expression in the developing lens placode (reviewed by Chow and Lang, 2001).

Data from various systems indicate that Pax6 is responsible either directly or indirectly for regulating a number of downstream targets (Fig. 6.7, Kamachi *et al.*, 1998; Blixt *et al.*, 2000; Chauhan *et al.*, 2002a,b,c; Goudreau *et al.*, 2002; Yamada *et al.*, 2003), including various *crystallin* genes (Cvekl and Piatigorsky, 1996). For instance, Pax6 induces expression of the *Xenopus* forkhead transcription factor, *Xlens1* (Kenyon *et al.*, 1999). Ectopic expression studies indicate that *Xlens1* may be important for placode formation and the maintenance of the lens epithelium.

Likewise, Pax6 activates expression of the homeodomain-containing gene *Prox1* (the prospero-like transcription factor), which is first expressed in the developing lens placode, and later becomes restricted to the differentiating lens (Oliver *et al.*, 1993; Tomarev *et al.*, 1996, 1998; Schaefer *et al.*, 1999; Chow and Lang, 2001; Reza *et al.*, 2002). When *Prox1* function is disrupted in the mouse the lens does not differentiate, but remains as a hollow vesicle (Wigle *et al.*, 1999). Prox1 appears to control fiber cell differentiation and serves as an activator of crystallin expression (see Cvekl and Piatigorsky, 1996; Ring *et al.*, 2000). Pax6 also activates the *sine oculis-like* transcription factor gene, *Six3*. *Six3* is implicated in lens epithelial maintenance, and *Pax6* and *Six3* interact in a co-inductive fashion. *Xenopus* Six3 contains a defining "six" domain, as well as a homeodomain, and ectopic expression of *Six3* results in lens formation within the area of the otic vesicles of *Medaka* (Oliver *et al.*, 1996) and mice (Lagutin *et al.*, 2001). Pax6 directly induces another gene, *Mab21/2*, which is also expressed in the PLE just prior to lens placode formation (Lau *et al.*, 2001; Yamada *et al.*, 2003). *Mab21* knockdown experiments in *Xenopus* result in eye defects related to reduced proliferation of lens placode cells, although lens placode invagination does occur (Lau *et al.*, 2001; Wong and Chow, 2002). The winged helix transcription factor *FoxE3* is a downstream target of *Mab21/1* expression (Yamada *et al.*, 2003). Mouse loss-of-function studies demonstrate that *FoxE3* is required for both lens epithelium proliferation and lens fiber differentiation (Blixt *et al.*, 2000; Brownell *et al.*, 2000). In addition, *FoxE3* negatively regulates *Prox1*. These findings suggest that *FoxE3* and *Prox1* are involved in the transition from lens epithelium to fiber cell differentiation.

Pax6 also regulates expression of members of the SOX family, a subset of the SRY testis determining factor family of HMG box transcription factors, known to be important in eye development (Penzel *et al.*, 1997; Furuta and Hogan, 1998; Nishiguchi *et al.*, 1998; Zygar *et al.*, 1998; Ashery-Padan *et al.*, 2000). Three family members, originally characterized in the chick, called *Sox1*, *Sox2*, and *Sox3*, are expressed in the lens, and play important roles in initiating transcription of δ-crystallins in the avian lens (Uwanogho *et al.*, 1995; Kamachi *et al.*, 1995, 1998). Upstream regulators of *Sox1* have not been identified, but *Sox1* expression in the mouse has been shown

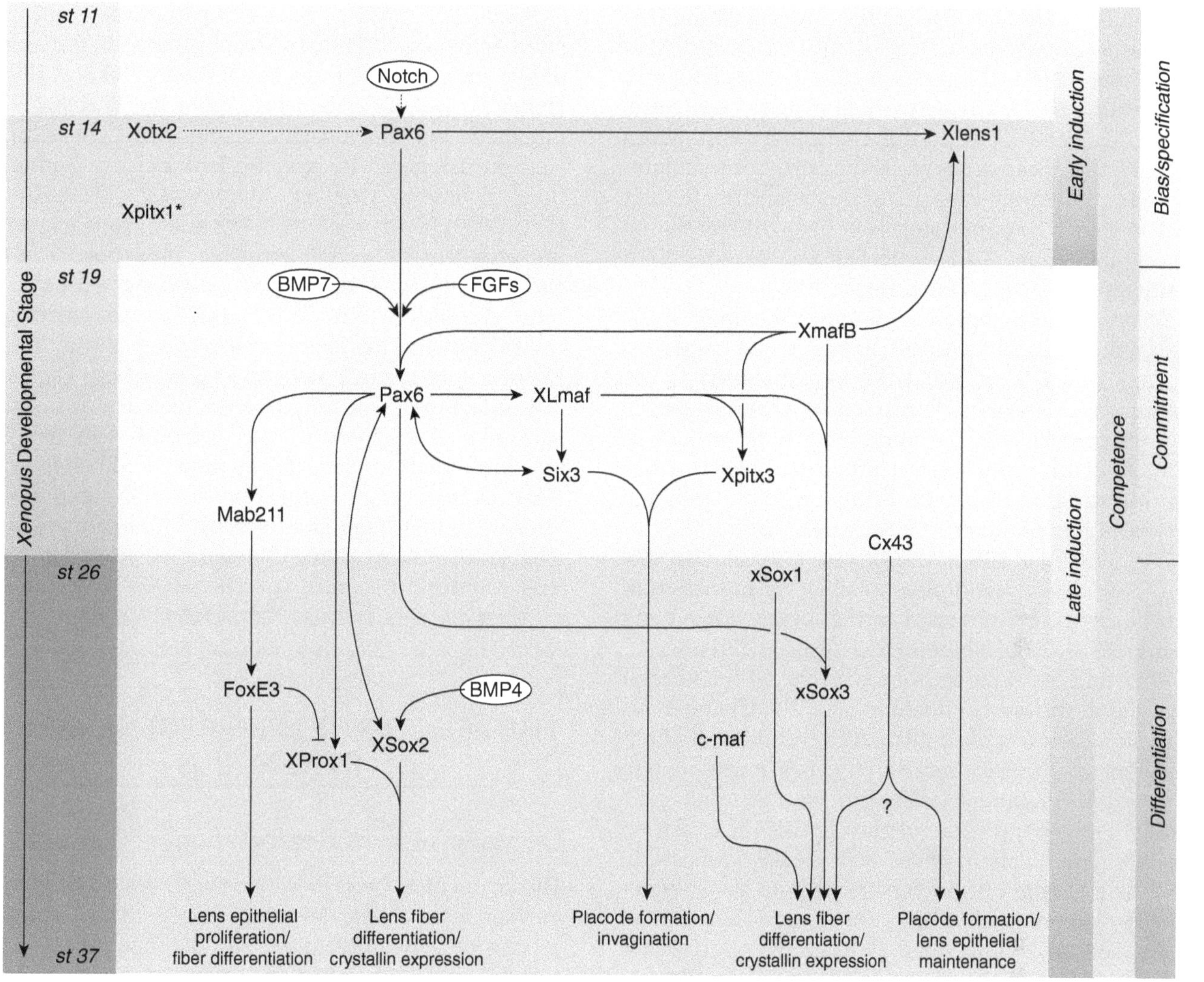

FIGURE 6.7 Elements of emerging networks and the putative functions of specific genes involved in lens development. These diagrams are not inclusive of all factors involved in lens development, but rather focus on certain major elements and those established particularly from research using *Xenopus*. Some established extrinsic growth factor interactions are encircled within ovals. A timeline of *Xenopus* development is provided on the left and phases of lens induction and development are shown on the right. Some liberty has been taken to align periods of gene expression of specific genes with particular periods of development/induction. Arrows do not necessarily imply direct activating or inhibitory interactions. See text for further details. *Xpitx1** is expressed early in the presumptive lens ectoderm.

to directly activate *crystallin* expression (Nishiguchi *et al.*, 1998; Chow and Lang, 2001; Ishibashi and Yasuda, 2001). Both *Sox2* and *Sox3* are downstream of *Pax6* but can reciprocally activate *Pax6* expression. In chick lenses, *Sox2* expression through *Pax6* points toward a co-activation role in δ-crystallin expression (Kamachi *et al.*, 2001). *Sox3* is also activated by *Xenopus* XmafB, suggesting it may also be involved in lens epithelium maintenance (see below). Later during development, BMP-4 signaling from the optic vesicle plays a critical role in lens development and has been shown to positively regulate *Sox2* expression (Furuta and Hogan, 1998; Mizuseki, *et al.*, 1998; see also Belecky-Adams *et al.*, 2002).

Other key components of lens development include members of the Maf oncogene family of basic leucine zipper transcription factors (e.g. bZIP, Ishibashi and Yasuda, 2001, see Fig. 6.7). A number of studies demonstrated the necessity of Maf family activity for lens fiber differentiation. For instance, *maf* mutant mice are deficient in fiber elongation and crystallin expression (Kim *et al.*, 1999; Ring *et al.*, 2000). *Xenopus XmafB* is expressed in the inner sensorial layer of the presumptive lens ectoderm beginning during neural tube stages, and expression persists through lens placode formation, finally being confined to the anterior lens epithelium. *XmafB* expression appears to be Pax6 independent and has no identified inducers, but overexpression studies show it can induce *Pitx3*, *Pax6*, and *Xlens1* expression (discussed further below; Ishibashi and Yasuda, 2001). The network of expression of *Pax6*, *XmafB*, *Six3*, and *Xlens1* in the anterior lens epithelium suggests that these genes are responsible for maintaining the proliferative capacity of the lens epithelium at least in part by suppressing lens fiber differentiation (Chow and Lang, 2001; Ishibashi and Yasuda, 2001; Goudreau *et al.*, 2002; Onuma *et al.*, 2002).

Xenopus XLmaf represents another member of the Maf family, but, unlike *XmafB*, its expression is induced by Pax6 (Reza *et al.*, 2002). *XLmaf* expression in the presumptive lens ectoderm, unlike that of *XmafB*, requires contact with the underlying optic vesicle, as ablation of the optic vesicle prevents *XLmaf* expression but not that of *XmafB*. Ectopic *XLmaf* expression can induce expression of *xSox3*, *Six3*, and *Pitx3*, as well as lens fiber differentiation and crystallin expression, both directly and also possibly via induction of *xSox3* expression (discussed further below; Ogino and Yasuda, 1998, 2000; Ishibashi and Yasuda, 2001).

Pitx1 and *Pitx3* are expressed in the PLE and later continue in the lens epithelium (Fig. 6.7, Chang *et al.*, 2001; Pommereit *et al.*, 2001). *Pitx1* and *Pitx3* are both members of the *Pitx* family of genes implicated in pituitary development (related to the *Drosophila* gene *bicoid*). *Pitx1* expression begins in the PLE during the lens specification phase and weakly persists in the lens placode, but no functional data is yet available (Hollemann and Pieler, 1999). Expression of *Pitx3* begins later at around stage 24 when the optic vesicle and the PLE make contact. The dorsal lens placode shows the strongest *Pitx3* expression, but expression is later restricted to the lens epithelium and excluded from lens fiber cells (Pommereit *et al.*, 2001). Functional analyses reveal that Pitx3 can induce *Pax6* and *Six3* expression and establish that *Pitx3* expression is required for lens development and proper retinal differentiation (Khosrowshahian *et al.*, 2005). Furthermore, the promoter region of *Pitx3* contains binding sites for Maf transcription factors. A loss of *Pitx3* in the lens placode in the mouse resulted in an *aphakia* phenotype in which affected mice developed abnormally small eyes that lacked lenses (Semina *et al.*, 2000).

Cx43, a connexin gene, which has been cloned from *X. tropicalis*, is also expressed in the lens placode just prior to lens fiber differentiation and becomes limited to the lens epithelium as well (Fig. 6.7, van der Heyden *et al.*, 2001). Localization of *Cx43* to the lens epithelium suggests that it may be involved in proliferation and maintenance of these cells, but further functional analyses are required to establish this role.

Specific changes in gene expression may be correlated with some of the properties defined in the model of lens induction (Figs 6.3 and 6.7, Schaefer *et al.*, 1999; Henry *et al.*, 2002; Walter *et al.*, 2004). Zyger *et al.* (1998) demonstrated that specific changes in gene expression are triggered by specific lens inductive interactions in *Xenopus* (see also Kenyon *et al.*, 1999; Köster *et al.*, 2000). Expression of *Xotx2* and *Pax6* is triggered in competent embryonic ectoderm by exposure to the early inductors. Their expression (along with that of other genes shown in Fig. 6.7) may serve as early indicators of increasing lens-forming bias in those tissues. Experiments suggest that expression of both *Pax6* and *Otx2* may confer or maintain competence of the presumptive lens ectoderm and establish lens-forming bias in different systems (Fujiwara *et al.*, 1994; Li *et al.*, 1994; Zyger *et al.*, 1998). Likewise, *xSOX3* expression appears to be triggered as a result of the late phase of lens induction involving the optic vesicles, after the lens ectoderm becomes specified (Figs. 6.3 and 6.7, Zyger *et al.*, 1998; see also Kamachi *et al.*, 1998).

THE PROCESS OF LENS REGENERATION IN *XENOPUS*

Overview of Lens Regeneration

The larvae of *X. laevis* are able to regenerate lenses, once they are removed (reviewed by Henry, 2003), through a process of transdifferentiation of the inner layer of the corneal epithelium after the original lens is removed from the eye of pre-metamorphic tadpoles (Fig. 6.8, Freeman, 1963). This process differs from that of Wolffian lens regeneration in which the new lens regenerates from the dorsal iris pigmented epithelium in some urodeles (e.g. newts and salamanders, reviewed by Del Rio-Tsonis and Tsonis, 2003; Henry, 2003; and see Tsonis, in this volume). Freeman (1963) defined five stages of lens regeneration or cornea-lens transdifferentiation, "CLT" (Fig. 6.8). These stages are reached at different times after lens removal depending on the age of the larva; older larvae pass through these stages at a slower rate. During Freeman stage 1, the cells of the inner layer of the cornea epithelium assume a cuboidal appearance, occurring 24h after lens removal. Thickening of this inner layer of the cornea epithelium occurs during stage 2, similar to the formation of the lens placode within the sensorial layer during embryonic development. At this stage, the nuclei in many of these presumptive lens cells typically have only one nucleolus (characteristic of lens epithelial cells), rather than two (characteristic of

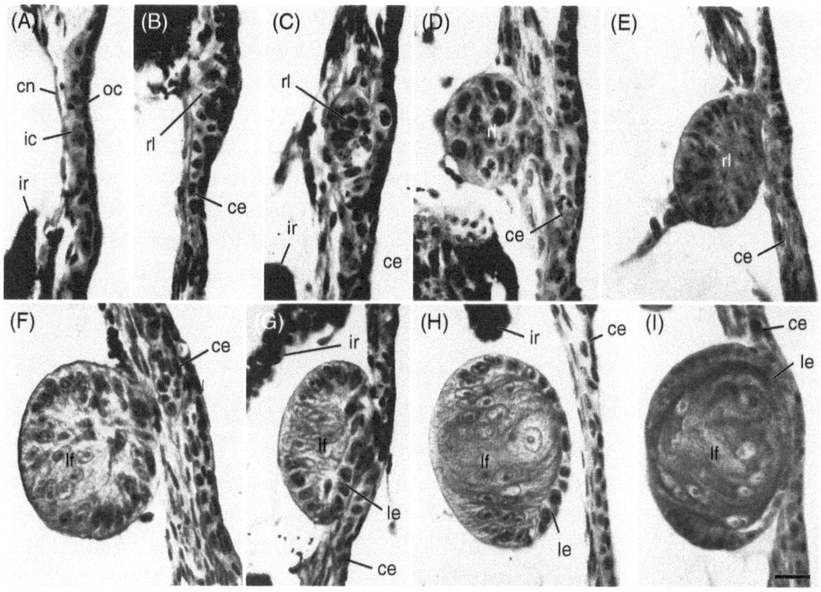

FIGURE 6.8 Cornea-lens transdifferentiation in *X. laevis*. Sections show different stages of lens regeneration, following the convention of Freeman (1963). (A) Stage 1. (B) Stage 2. (C) Early stage 3. (D) Middle stage 3. (E) Late stage 3. (F) Early stage 4. (G) Middle stage 4. (H) Late stage 4. (I) Early stage 5. ce, cornea epithelium; cn, cornea endothelium; ic, inner layer of the cornea epithelium; ir, iris; oc, outer layer of the cornea epithelium; le, lens epithelium; lf, lens fibers; rl, regenerating lens vesicle. Refer to text for further details. Figure from Henry (2003), after Freeman (1963). Scale bar equals 25 μm.

the larval cornea epithelium). During stage 3, a loosely organized aggregate of cells begins to separate from the cornea epithelium. These cells begin to orient with differential apical and basal polarity in preparation for the formation of a lens vesicle. A definitive lens vesicle is established at stage 4 when the cells located closest to the vitreous chamber begin to elongate to form primary fiber cells. The lens vesicle is typically separated from the cornea epithelium by this stage of regeneration. At stage 5, secondary fiber cells are added from the equatorial zone and the nuclei of the primary fiber cells begin to disappear. The lens then continues to grow and add additional secondary fiber cells.

The process of CLT is triggered by factors present in the eyecup, which appear to be synthesized by the neural retina (Freeman, 1963; Henry and Mittleman, 1995; Bosco *et al.*, 1997a). Normally, diffusion of these factors is physically blocked from reaching the outer cornea due to the presence of the lens and the inner corneal endothelium (Filoni *et al.*, 1997; Henry and Elkins, 2001). The identity of these signals has not yet been determined; however, there is evidence that FGFs may play a key role in this process (see further discussion below).

The process of CLT is strikingly similar to that of embryonic lens development. The morphological events associated with these processes are nearly identical (Freeman, 1963; McDevitt and Brahma, 1973; Brahma and McDevitt, 1974). In fact, the corneal epithelium and, hence, the regenerated lens are derived from the same tissue that originally gave rise to the lens during embryogenesis. Furthermore, both processes rely on the presence of the neural retina (e.g. optic vesicle/eye cup). In fact, in a series of heterochronic tissue transplantation experiments, it was demonstrated that the signals involved in triggering CLT appear to be related to those involved in embryonic lens induction (Henry and Mittleman, 1995). Furthermore, many genes expressed during embryonic lens development are re-expressed during CLT (as described below). These findings suggest that that lenses derived from cornea tissue arise via a cellular and molecular pathway similar to that taken during embryogenesis. On the other hand, a few differences have also been noted which are described further below. Unlike the case in Wolffian lens regeneration, it is unclear to what extent the process of cellular de-differentiation may play in larval cornea-lens transdifferentiation in *Xenopus* (Henry, 2003).

In *X. laevis*, the capacity to undergo transdifferentiation is restricted to the cornea epithelium and the surrounding pericorneal ectoderm. Normally, the capacity to regenerate a lens declines in older larvae and is eventually lost by later stages just prior to metamorphosis (by stage 66, Freeman, 1963). On the other hand, Freeman and Overton (1962) and Filoni *et al.* (1997) demonstrated that even the cornea of post-metamorphic frogs is able to undergo transdifferentiation to form lens cells when implanted directly into the vitreous chamber of either pre- or post-metamorphic frogs (see also Bosco *et al.*, 1992; Bosco and Willems, 1992). The normal decline in lens regeneration capacity is due to the increasing rate with which the inner cornea endothelium heals to cover the pupillary opening, cutting off key substances required to support lens regeneration. Cannata *et al.* (2003) have demonstrated that the capacity of the ectoderm to undergo transdifferentiation to form lens cells is imparted by virtue of embryonic lens inductive interactions and the presence of the larval eye. The capacity of embryonic ectoderm to form lens cells is normally lost by stage 30/31 for flank ectoderm and by stage 48 for head epidermis (Arresta *et al.*, 2005b). After these stages, only the cornea and pericorneal epidermis retain the capacity to undergo CLT. The main source of the signals that promote lens regeneration competence in the ectoderm appears to be derived from the lens and the retina. Interestingly, lens regeneration capacity in larval cornea and pericorneal ectoderm persists even after the underlying retinal tissues have been removed surgically (Bosco and Filoni, 1992; Arresta *et al.*, 2005b). For example, if implanted into the vitreous chamber, cornea and pericorneal ectoderm are still able to undergo transdifferentiation weeks after removal of the eye. Likewise, implantation of an eyecup beneath the flank ectoderm at later stages of embryonic development will impart this foreign ectoderm with the capacity to undergo CLT during later larval stages.

The process of CLT was actually first described for the urodele *Hynobius japonicus* (Ikeda, 1936, 1939). Within the anura, this process appears to be restricted to members of the genus *Xenopus* (Bosco, 1988a; Henry and Elkins, 2001; Henry, 2003; Filoni *et al.*, 2006). Aside from *X. laevis*, Henry and Elkins (2001) demonstrated that CLT also takes place in *X. tropicalis*, but at a lower frequency, using transgenic frogs carrying a GFP transgene driven by a 2.2-kb γ-crystallin enhancer/promoter, which served as a cell autonomous marker of lens differentiation *in vivo* (Fig. 6.9). The lower frequency of regeneration occurs due to the more rapid rate at which the cornea endothelium heals back following surgery, cutting off critical retinal factors required to support regeneration. A similar situation appears to exist in *X. borealis*, which is more closely related to *X. laevis* (Filoni *et al.*, 2006). The larvae of *X. borealis* are normally unable to regenerate a lens due to the rapidity with which the cornea endothelium heals to cover the pupillary opening, coupled with a somewhat slower response of the cornea epithelium to critical retinal factors that support lens regeneration. In an experimental context, however, the larval cornea of *X. borealis* is able to undergo transdifferentiation and form a lens when implanted directly into the vitreous chamber.

Analyses of Crystallin Expression During Lens Regeneration

Crystallin expression has been characterized during lens regeneration in *X. laevis* (Campbell, 1965; Brahma and McDevitt, 1974; Campbell and Truman, 1977; Reeve and Wild, 1978; Brahma, 1980; Henry and Mittleman, 1995; Schaefer *et al.*, 1999; Mizuno *et al.*, 1999a). Some differences were detected compared to the patterns observed during embryonic lens development (McDevitt and Brahma, 1973; Campbell and Truman, 1977; Mizuno *et al.*, 1999a). Brahma and McDevitt (1974) and Henry and Mittleman (1995) found that the initial expression of crystallins is seen during lens regeneration at late stage 3 to early stage 4 (stages of Freeman, 1963), which corresponds roughly to stages 3/4 of embryonic lens development (stages of McDevitt and Brahma, 1973; note also that Campbell (1965) reported lens crystallin expression at the earliest stages of lens regeneration, even within the cornea epithelium, one day after lens removal, using antibodies against total soluble lens proteins). As is the case in lens development, γ-crystallin expression was restricted to fiber cells within the lens. Unlike embryonic lens development, however, other crystallins (presumably α- and β-crystallins) were also detected in the lens epithelium at mid stage 4 of regeneration (roughly stage 4/5 of embryonic

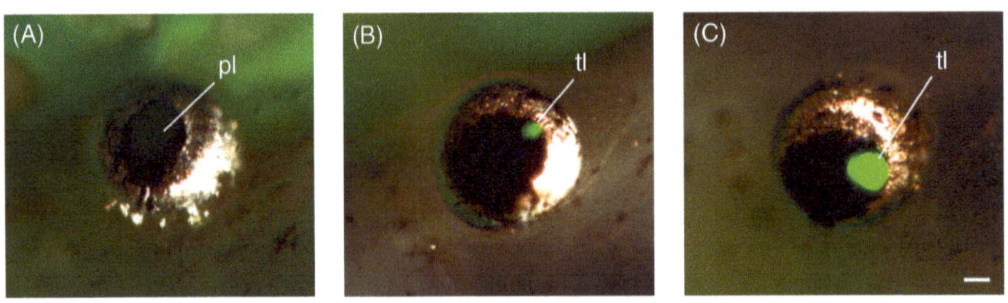

FIGURE 6.9 Cornea-lens transdifferentiation in transgenic *X. tropicalis* larvae carrying a transgene encoding the jellyfish green fluorescent protein (GFP) coupled to a γ-crystallin enhancer/promoter. Combined light and epifluorescence micrographs show the expression of GFP in the regenerating lenses of these examples. (A) Eye following removal of the lens (lentectomy). Note absence of differentiated lens cells and GFP in the eye at 1 week following lens removal. (B) Eye 10 days following lens removal. Note the presence of a small regenerating lens with GFP expressing cells. (C) Eye 21 days following lens removal. Note the presence of a larger regenerating lens with GFP expressing cells. pl, pupil; tl, transdifferentiating lens. Scale bar equals 200 μm.

lens development), which represents an earlier stage compared to that found during the process of embryonic lens development (normally first appearing in the lens epithelium at lens developmental stage 7). Mizuno *et al.* (1999) found that there are other differences in the timing of crystallin gene expression in lens development versus lens regeneration. During embryonic lens development, *αA-*, *βB1-*, and *γ-crystallin* mRNA transcripts are all detected simultaneously in the lens placode (beginning at stage 26/27 or McDevitt and Brahma lens developmental stage 2), while at later stages these same transcripts are expressed only in lens fiber cells. During lens regeneration, however, *αA-* and *βB1-crystallin* are first expressed in the presumptive lens fiber cells of the regenerated lens vesicle (middle to late Freeman stage 3). *γ-crystallin* is not detected until early Freeman stage 4, and only in the differentiating lens fiber cells. Ultimately, by late Freeman stage 4, the expression of all three of these genes was restricted to lens fiber cells, similar to the pattern seen later during embryogenesis. These findings indicate that there are some differences between the process of embryonic lens development and lens regeneration.

Contributions to Our Understanding of the Molecular Basis of Lens Regeneration in *Xenopus*

There has been some debate over the relationships between lens development and regeneration (Del Rio-Tsonis and Tsonis, 2003; Henry, 2003), but most studies suggest that CLT appears to be closely related to embryonic lens development at the molecular level (Mizuno *et al.*, 1999b; Schaefer *et al.*, 1999; Henry *et al.*, 2002). Analyses of gene expression indicate that a number of transcription factors (including *Pax6*, *xSox3*, *Xotx2*, and *Xprox1*) are re-expressed during lens regeneration, in a similar spatial and temporal context to that observed during lens development (Schaefer *et al.*, 1999; Mizuno *et al.*, 1999b). Henry *et al.* (2002) showed that *Pax6* mRNA transcripts are present in the larval cornea prior to lens regeneration. As mentioned above, *Pax6* expression has been correlated with the competence of embryonic ectoderm to respond to lens inductive interactions (Fujiwara *et al.*, 1994; Li *et al.*, 1994; Zygar *et al.*, 1998). The presence of *Pax6* in cornea epithelium may also be important for conferring this tissue with the competence to regenerate a lens.

The study by Mizuno *et al.* (2005), described above, demonstrated that expression of *βB1-crystallin* during lens regeneration in *Xenopus* requires the same promoter elements as that required during embryonic lens development. These observations suggest that the mechanisms regulating crystallin gene expression are shared between these two lens-forming processes.

A large suite of genes upregulated during the process of CLT in *X. laevis* have been identified (Henry *et al.*, 2002; Henry, 2003; Walter and Henry, 2004; Walter *et al.*, 2004; Wolfe *et al.*, 2004; Elkins and Henry, 2006; Wolfe and Henry, 2006). The sequences of over 700 unique genes have been deposited in the NCBI database, and information regarding each of these is available online through a publicly accessible database (http://www.life.uiuc.edu/henry, Henry *et al.*, 2002). This information is continually updated and lists all cDNA sequences and their accession numbers (NCBI), gene similarities/identities, links to related EST and genomic sequences (e.g. in *X. tropicalis*), and available images of *in situ* mRNA expression patterns for various stages of development. These genes encode a wide variety of proteins including: lens crystallins (e.g. β-, and γ-crystallins); proteins involved in DNA replication, transcription and translation (including a number of transcription factors); matrix metalloproteinases (e.g. MMP-9, -13, -14, and -18); and extracellular matrix, transmembrane and various cell signaling proteins (including Wnt7b and the retinoic acid receptor, RXRγ). Obviously, these genes may play critical roles in the regeneration process, perhaps involved in the de-differentiation of cornea tissue, cornea wound-healing, and lens cell determination and differentiation. Remarkably, most of these genes also appear to be expressed during embryonic lens development, and this has proven to be a valuable resource for identifying genes that play important roles in eye/lens development in general (Henry *et al.*, 2002; Walter *et al.*, 2004; Elkins and Henry, 2006; Wolfe and Henry, 2006). As mentioned above, there may be some differences between lens regeneration and embryonic lens development (e.g. patterns of crystallin expression, Mizuno *et al.*, 1999b). Nearly one-third of the genes examined that are expressed during CLT do not appear to be expressed in the embryonic eye/lens (Henry *et al.*, 2002).

For example, different matrix metalloproteinases are expressed during CLT, including *MMP-9*, *MMP-13*, *MMP-14*, and *MMP-18* (Carinato *et al.*, 2000; Henry *et al.*, 2002; Walter *et al.*, 2004; Henry, unpublished observations). MMPs, such as MMP-9, have been implicated in the breakdown of the basal lamina associated with the corneal epithelium (Bowman's membrane) following injury, and may play a role in controlling the reassembly of the damaged epithelial basement membrane (Berman, 1989; Matsubara *et al.*, 1991a,b; Fini *et al.*, 1996, 1998; Barro *et al.*, 1998; Ye and Azar, 1998; Ye *et al.*, 2000; Li de *et al.*, 2003). On the other hand, MMPs have been implicated in the process of amphibian limb regeneration (Yang and Bryant, 1994; Miyazaki *et al.*, 1996; Kherif *et al.*, 1999; Yang *et al.*, 1999). In *Xenopus* larvae, *Xmmp-9* is expressed in the wounded corneal epithelium within 5h following lens removal, and persists for 2 days following injury (Carinato *et al.*, 2000). Expression is concentrated at the site of the corneal wound, and not in the central region where CLT occurs (Carinato *et al.*, 2000). Furthermore, expression is also seen in sham-operated corneas which do not undergo CLT. During embryogenesis, *Xmmp-9* mRNA is first expressed at stage 26 by a relatively small number of highly dispersed mesenchymal cells, which may represent fibroblasts (Carinato *et al.*, 2000). It is also expressed in a few other tissues, but no embryonic expression was detected in the developing eye or lens. Therefore, it appears that MMP-9 may not play an important role in lens development or regeneration; however, other MMPs (i.e. MMP-13, -14, and -18) may play more significant roles in these processes. For example, in *Xenopus* embryos, *MMP-18* displays a similar expression pattern as *MMP-9*, but *MMP-13* appears to be expressed in developing placodal lens cells and the retina beginning at stage 28 (Walter *et al.*, 2004; Henry, unpublished observations).

Although the identity of the signals involved in triggering CLT are not known, one study has demonstrated that FGF-1 is able to trigger CLT in primary cultures of *X. laevis* corneas (Bosco *et al.*, 1994, 1997b). Furthermore, Arresta *et al.* (2005b) observed the presence of FGFR-2 (*bek* variant) protein in cornea epidermis, but not in normal head epidermis, which is not capable of undergoing CLT. Interestingly, this protein is also expressed in head epidermis exposed to transplanted eyes, which has gained competence to undergo CLT (as discussed above, Bosco and Filoni, 1992; Arresta *et al.*, 2005a,b). These findings suggest that activated FGF receptor signaling is important in the process of CLT. It is also interesting to note that rare cases of transdifferentiation of the cornea have also been reported for the newt *Notophthalmus viridescens* following experimental treatment with RAR antagonists (Tsonis *et al.*, 2000). These findings might suggest some involvement of retinoic acid signaling in these transdifferentiation processes, supported by the fact that retinoic acid receptor RxRγ is expressed in *Xenopus* cornea epithelium (as mentioned above, Henry, unpublished observations).

Functional Studies with cDNA Library Clones

A number of the genes identified by Henry *et al.* (2002) as being upregulated during the process of CLT appear to be important in embryonic lens and retinal development. One gene, *Psf2* represents a member of the GINS heterotetramer involved in DNA replication. GINS serves as a "sliding clamp" for DNA polymerase-ε to promote the initiation and elongation of DNA replication (Kubota *et al.*, 2003). However, our own analysis suggests that *Psf2* may play a DNA replication-independent role in retinal differentiation and subsequent lens induction, as morpholino knockdown experiments resulted in missing or small eyes and lenses. This is in contrast to studies designed to specifically block DNA replication during key developmental periods, which do not result in this same pattern of defects (Walter *et al.*, 2008).

Another gene, *xMADML* encodes a novel protein containing a shared kinase domain and a SH2 binding domain with human nuclear binding receptor protein (NBRP in humans or MADM in other animals). *xMADML* is expressed very specifically in the developing lens, and morpholino knockdown experiments show a wide range of developmental defects, including loss of eye/lens, coloboma (the failure of the RPE to fuse in the ventral portion of the eye), and depigmentation of the retina (Elkins and Henry, 2006).

A third gene, *XlNLRR-6* is a novel member of the neuronal leucine-rich repeat (NLRR) family of transcription factors, and may possibly be involved in potentiating EGF signaling by mediating endocytosis of EGF ligand-receptor complexes. *XlNLRR-6* morpholino knockdown and reciprocal tissue transplantation experiments point toward autonomous roles of *XlNLRR-6* in both lens and retinal development (Wolfe and Henry, 2006).

Another gene, identified by blast homology as *mind bomb homolog 1 (mib1)*, codes for an E3 ubiquitin protein ligase that is known to positively regulate Delta-mediated Notch signaling by tagging Delta receptors for endocytosis through ubiquitination of their intracellular domain (Itoh *et al.*, 2003; Koo *et al.*, 2005). Morpholino knockdown is highly dose dependent, causing lethality at high doses and reduced eye/head size and possible loss of lenses at lower doses (Henry and Perry, unpublished results).

CHD4, or chromodomain helicase 4, is a gene belonging to the SNF/RAD54 helicase family responsible for epigenetic transcription repression through nucleosome remodeling and histone deacetylation (Linder *et al.*, 2007). Morpholino knockdown in *Xenopus* embryos causes an overall reduction in brain size, enlargement of the notochord, duplication of CNS structures, varying degrees of cyclopia, fused eyes or micropthalmia, and loss of lenses; however, further experimentation is required to fully understand the implications of CHD4 loss of function (Henry and Krebs, unpublished data).

A final gene shows homology to G-protein-coupled receptor 84 identified in *X. tropicalis*, similar to trehalose receptor 1. The protein contains a seven transmembrane receptor of the rhodospin superfamily, which includes both G-protein-coupled receptors and opsins. Opsins have been implicated in both light absorption and G-protein activation. Morpholino knockdown of this clone results in coloboma, abnormal angular-shaped eyes and defects in the RPE, and the shape of the lens (Henry and Perry, unpublished data).

REGENERATION OF THE NEURAL RETINA IN *XENOPUS*

Overview of Retinal Regeneration

The vertebrate retina is a complex neural structure and its proper organization and integrity are essential for normal visual function. Only certain vertebrates possess the ability to regenerate their retina upon damage, and the extent to which they are able to do so varies among species, developmental stage, and type of injury (reviewed by Hitchcock and Raymond, 1992; Del Rio-Tsonis and Tsonis, 2003). Retina regeneration is an intricate and multifactorial process, since not only neurogenesis of different cell types has to be achieved but also the correct specification of cell positional identities and connectivity patterns has to occur for functionality to be restored.

It is well established that anuran amphibians, including *Xenopus*, are able to regenerate the retina. The retinal regenerative capacity in these animals is generally not as extensive as that of urodeles; however, in many cases the eye can re-gain functionality. Moreover, this process occurs at stages in which the damaged eye is already fully differentiated, which represents an important advantage over embryonic animal models.

Various strategies have been developed to study the regenerative potential of the *Xenopus* retina. Depending on the type of lesion inflicted, regenerated tissues appear to arise from different cellular sources, which implies that different mechanisms may be involved in these processes. Much of the work done in this area, however, has focused on the phenomenology, rather than on the mechanisms regulating the production of new retinal cells or on the correct patterning of these tissues. Early *in vivo* studies involving retinal ablations were carried out primarily to analyze the patterns of retino-tectal connectivity (Ide *et al.*, 1984, 1986; O'Gorman *et al.*, 1987; Wunsh and Ide, 1990; Underwood and Ide, 1992; Underwood *et al.*, 1992, 1993; Ide, 1998). These studies involved the ablation of approximately one- to two-thirds of the tadpole eye, and regeneration of missing tissues was observed to occur from cells derived from the remaining retinal tissue. Axotomy, the excision of the optic nerve, has also been widely used to study optic nerve regeneration in *Xenopus* (Gaze, 1959; Beazley, 1981; Jenkins and Straznicky, 1986; Taylor *et al.*, 1989; Beaver *et al.*, 2001). A third approach has been to ablate the whole neural retina leaving the rest of the eye intact. Many of these studies were performed decades ago before the availability of modern molecular tools. Since then, there have been only a handful of studies on this topic (e.g. Mitashov, 1997; Mitashov and Maliovanova, 1982; Yoshii *et al.*, 2007). Finally, *in vitro* studies have also been carried out to identify the secreted growth factors that might induce retinal regeneration (Lopashov, 1991; Sakaguchi *et al.*, 1997; Yoshii *et al.*, 2007).

In Vivo Studies: Ablation of Eye Fragments in *Xenopus* Tadpoles. Healing Modes and Their Correlation to the Patterning of Retino-tectal Projections

Within a few days following removal of up to two-thirds of the eye in *Xenopus* tadpoles (beginning at stage 32) regeneration takes place to restore the retina and within a few weeks, this tissue grows to the normal size in 70% of the cases (reviewed by Ide, 1998). During the growth phase, the new ganglion cells project their axons to the tectum, and the pattern of connections that are established during that process depends on the mode of healing. Two classes of growth can be distinguished within the first 24 h. In some cases, a "tongue" shaped structure is formed by division and migration of cells, mainly from the ventral side of the remaining retina. The cells in the dorsal edge proliferate as well, but do not contribute to the formation of the "tongue" (Wunsh and Ide, 1990; Underwood and Ide, 1992). This tongue of cells elongates to close the wound and reforms a complete eyeball over the initial days of healing that later differentiates to form a histologically normal retina. The source of these cells is from the remaining retinal tissue, with possible participation of the RPE (Underwood and Ide, 1992). In the second class of growth, the remaining retinal fragment simply "rounds up" to close the wound, with no associated cell movements into the region of the ablation. Some cases show an "intermediate" healing mode, combining the two classes of growth described above.

If the animals are allowed to undergo metamorphosis and electrophysiological measurements are taken to determine retino-tectal projection patterns, it was observed that eyes that healed via formation of the "tongue" tended to produce pattern duplications. If the remaining eye fragment was nasal, the regenerated temporal part of the eye projected to the same location in the cortex as the nasal ganglion cells. The opposite occurred if the remaining fragment was derived from the temporal side. Interestingly, if an eye was healing in this mode and the tongue was cut, the later regeneration of the retina presented less pattern duplication (Ide *et al.*, 1986). On the other hand, the animals that healed through the "round up" mode tended to produce normal, unduplicated projections. The eyes that healed via intermediate modes produced pattern duplications only in some cases (Underwood *et al.*, 1993). Interestingly, after regeneration was completed, and the animals were raised to adulthood, if the optic nerve was cut and then allowed to re-connect, these projection patterns remained the same. This suggests that information intrinsic to the regenerated tissues was responsible for the pattern of connectivity to the tectum (Underwood *et al.*, 1992).

In studies where the temporal two-thirds of the eye were removed, the remaining nasal fragments tended to heal by forming a "tongue" in the majority of the cases and produced duplicated projections. On the other hand, if the remaining eye fragment was temporal, healing mainly occurred through the "round up" mode, producing few duplications (Ide *et al.*, 1984). It is important to mention that about one-third of the operated animals did not regenerate the retina at all, and tended to undergo eye degeneration.

O' Gorman *et al.* (1987) developed a model to analyze the establishment of retino-tectal conections in which xenoplastic, compound eyes were created. The nasal, temporal, dorsal, or ventral half of one eye was substituted for half of an operated eye between *X. borealis* and *X. laevis* tadpoles between stages 31 and 36. In this way individual donor cells could be distinguished from host cells by their differential staining with quinacrine, as mentioned earlier. Like the others mentioned above, however, this study did not seek to explain how the new retina was formed, or what mechanisms underlay these regenerative processes, but this clearly represents an interesting technical approach to further investigate the regenerative phenomenon.

Axotomy in *Xenopus* Tadpoles: Optic Nerve Regeneration and Ganglion Cell Number

When the optic nerve is severed (axotomy) in *X. laevis* tadpoles or adults without interrupting the blood supply to the eye, the optic nerve connection can be re-established (Gaze, 1959). Cell counts using autoradiographic methods demonstrated that after axotomy there was a substantial loss in the retinal ganglion cell population, which decreased about 43% (Jenkins and Straznicky, 1986). The surviving ganglion cells, however, were able to regenerate their axons and re-establish connections with the tectum. If the optic nerve did not reconnect, permanent isolation of the retina from visual centers caused a loss of about 80% of the ganglion cell population. Since the ganglion cell layer contains approximately 20% of displaced amacrine cells, it is possible that those are the cells that remain in the ganglion cell layer in this situation (Beazley, 1981; Jenkins and Straznicky, 1986). BrdU studies have been performed to test if these re-connections are linked to the continued neurogenesis of retinal ganglion cells that occurs at the ciliary margin of young *Xenopus*, but no relationship was found between the two (Taylor *et al.*, 1989; Beaver *et al.*, 2001). This method of optic nerve transection has been widely used in the study of axonal pathfinding mechanisms and the specification of retino-tectal connections, but it is also a useful system to examine the general mechanisms involved in the regeneration of axons in the central nervous system.

Retinal Ablation and Eye Restoration in Post-metamorphic Frogs: Sources of New Retinal Cells

As mentioned above, the intact eyes of anuran amphibians continue to grow after metamorphosis, even after they are fully differentiated. H3-thymidine experiments have shown that the source of proliferating cells that contribute to this retinal growth in adult intact eyes is the CMZ of the eye (Beach and Jacobson, 1979; Svistunov and Mitashov, 1985; Amato *et al.*, 2004). One interesting difference between *Xenopus* and urodeles is that, in the former, more extensive proliferation is observed in the ventral region of the eye compared to the dorsal region.

When small lesions were inflicted in the retina and adjacent RPE of post-metamorphic *X. laevis*, regeneration was able to restore the morphology and function of the eye (Levine, 1981), but the extent of retina repair depended on the size of the ablation. Cell proliferation could be observed approximately 1 week after surgery, closing the wounded region between days 11 and 14. Differentiation of the newly formed neuroepithelium began around 13 days post-surgery, and in some cases it continued for up to 30 days. The regenerated retina possessed a higher cell density than the intact one, but it returned to normal size within 60 days post-surgery (Levine, 1981). Cell proliferation seemed to take place at the ciliary region of the eye, and also in intraretinal nests of cells that exist in the inner nuclear layer and outer nuclear layer, but the contribution of these cells to the newly formed retina was not definitively assessed, and participation of the RPE could not be ruled out (Levine, 1981).

On the other hand, when the retina was completely removed from the eye of post-metamorphic *Xenopus*, leaving the RPE behind, partial regeneration took place, mainly due to proliferation of cells in the ciliary region (Mitashov and Maliovanova, 1982). In this case regeneration was incomplete and the eye did not reach the dimensions of control, unoperated eyes, even after 70 days post-retinectomy. As opposed to regeneration in urodeles, transdifferentiation of the RPE does not seem to play a major role in *Xenopus* under these conditions *in vivo*.

Recently, another approach has been explored in which the retina was completely removed from the eyes of post-metamorphic *X. laevis*, while leaving the intact vascular membrane in the eye (Yoshii *et al.*, 2007). That study revealed that pigmented cells migrate from the RPE layer and attach to the vascular membrane to re-form a neuroepithelium that later differentiates into a complete retina by day 30 post-surgery (Yoshii *et al.*, 2007). Yoshii *et al.* (2007) argue that the vascular membrane provides the factors necessary for retinal pigment cells to undergo transdifferentiation. This is the only work in which a full regeneration of the retina was obtained in post-metamorphic *Xenopus* after complete retina removal, and it opens interesting possibilities for the exploration of this phenomenon in greater detail.

Potential of the Pigmented Eye Tissues to Transdifferentiate into Neural Retina: Experiences from *In Vitro* Culture and Transplantation Experiments

The capacity of urodele amphibians to regenerate their retina through transdifferentiation of the RPE cells has been widely documented. However, the RPE of frogs seems to possess different regenerative capacities compared to that of urodeles. An early approach to evaluate the capacity of *Xenopus* RPE to transdifferentiate into neural retina, and to identify associated inducing molecules that promote this process, involved the transplantation of RPE explants from differentiated eyes of tadpoles or adult *X. laevis* into the eyes of tadpoles whose lenses had been removed (Sologub, 1975). In such explants, depigmentation started at 4–5 days post-transplantation followed by proliferation within 7–10 days post-transplantation. At 20 days, some differentiation of retinal cells was observed. From these experiments, it can be concluded that transdifferentiation of the RPE into retina in *X. laevis* requires the influence of certain factors provided by the neural retina, since explants transplanted into the enucleated orbit, as well as those transplanted into the anterior chamber of host eyes failed to transdifferentiate. In addition, a critical mass of RPE and a certain integrity of the explant were required for transdifferentiation to occur. Furthermore, adhesion of the RPE explant to different parts of the eye inhibited its ability to transdifferentiate. It is interesting to note that the lens regeneration process that takes place at the same time did not seem to affect the transdifferentiation of the RPE.

Similar results were obtained in a more recent study in which RPE fragments were transplanted into the vitreous chamber of a host tadpole and the lens was put back in place (Arresta *et al.*, 2005a). In this work, the peak of BrdU incorporation in the depigmenting RPE fragments occurred at 10 days post-transplantation and then gradually decreased to be observed only at the peripheral edge of the newly formed retinas at 30 days post-transplantation. *Pax6* was expressed in the depigmenting explants starting at 10–15 days post-transplantation, and expression became more restricted with time as the differentiation process progressed. After 30 days, it was only observed in the ganglion cell layer and inner nuclear layer, a pattern similar to that of the normal retina (Arresta *et al.*, 2005a). Additional studies support the idea that inducing factors from surrounding eye tissues direct transdifferentiation; one such study revealed that early gastrula ectoderm of *X. laevis*, when exposed to mature retina or to lens epithelium *in vitro*, can transdifferentiate into retina or lens (reviewed by Lopashov, 1991; Henry and Mittleman, 1995; Lopashov *et al.*, 1997).

What are the factors produced by the mature retina that induce such fate changes in RPE cells? Studies performed in culture using RPE explants from stages 47–53 *Xenopus* tadpoles suggest that a good candidate for such a molecule is FGF2. Incubation of the explants in the presence of FGF2 for up to 30 days induced their transdifferentiation *in vitro* into different types of retinal neurons and glia (Sakaguchi *et al.*, 1997). Interestingly, in the cases in which the explants were allowed to attach to the surface of the culture dish, transdifferentiation did not occur. In a different study, cultures of RPE from post-metamorphic frogs that were attached to a collagen coated membrane, transdifferentiated into neurons after 30 days if the choroid membrane was still present or if the isolated RPE was exposed to FGF2 or IGF1. The latter suggests that the choroid might provide a source of these crucial factors that supports trasdifferentiation of the RPE (Yoshii *et al.*, 2007).

It is important to point out that the pigmented iris epithelium of *X. laevis* tadpoles can also transdifferentiate into retina when transplanted into the vitreous chamber of a host tadpole (Cioni *et al.*, 1986, reviewed by Bosco, 1988b). Moreover, this type of transdifferentiation was also observed when the dorsal iris was implanted into the tail fin together with the pituitary, or when iris fragments were implanted into the stump of an amputated hindlimb of stage 54/55 tadpoles (Cioni *et al.*, 1987, 1990).

FUTURE DIRECTIONS

Xenopus leaps out as one of the most promising systems for understanding the mechanisms underlying the processes of vertebrate eye development and regeneration, providing significant technical advantages to study these processes. The recent availability of the *X. tropicalis* genome sequence and the powerful array of functional tools (e.g. transgenesis) will ultimately permit us to decipher the molecular pathways underlying lens development and regeneration. Other key questions remain to be answered, such as why is it that certain animals are able to regenerate various body parts, while others cannot? Clearly, an understanding of the molecular relationships between development and regeneration of eye tissues will lead to the development of new therapeutic approaches to treat injured and diseased eyes.

ACKNOWLEDGMENTS

Jonathan J. Henry, Jason M. Wever, and Lisa Fukui have been supported by NIH/NEI research grant EY09844.

REFERENCES

Adelmann HB (1936). The problem of cyclopia. Part 1. *Q Rev Biol* 11:161–182.

Allen BG, Weeks DL (2005). Trasngenic *Xenopus laevis* embryos can be generated using C31 integrase. *Nat Meth* 2:975–979.

Altmann CR, Chow RL, Lang RA, Hemmati-Brivanlou A (1997). Lens induction by Pax-6 in *Xenopus laevis*. *Dev Biol* 185:119–123.

Amato MA, Arnault E, Perron M (2004). Retinal stem cells in vertebrates: parallels and divergences. *Int J Dev Biol* 48:993–1001.

Amaya E, Kroll KL (1999). A method for generating transgenic frog embryos. *Molecular Embryology: Methods and Protocols* 97:393–414.

Amaya E, Offield MF, Grainger RM (1998). Frog genetics, *Xenopus tropicalis* jumps into the future. *Trends Genet* 14:253–255.

Anantharam A, Lewis A, Panaghie G, Gordon E, McCrossan ZA, Lerner DJ, Abbott GW (2003). RNA interference reveals that endogenous *Xenopus* MinK-related peptides govern mammalian K^+ channel function in oocyte expression studies. *J Biol Chem* 278:11739–11745.

Arresta E, Bernardini S, Bernardini E, Filoni S, Cannata SM (2005a). Pigmented epithelium to retinal transdifferentiation and *Pax6* expression in larval *Xenopus laevis*. *J Exp Zool* 303A:958–967.

Arresta E, Bernardini S, Gargioli C, Filoni S, Cannata SM (2005b). Lens-forming competence in the epidermis of *Xenopus laevis* during development. *J Exp Zool A Comp Exp Biol* 303:1–12.

Ashery-Padan R, Marquardt T, Zhou X, Gruss P (2000). Pax6 activity in the lens primordium is required for lens formation and for correct placement of a single retina in the eye. *Genes Dev* 14:2701–2711.

Babcock, RG (1961). Ribonucleic acid and lens induction in *Xenopus laevis* (Daudin). Doctoral Dissertation, University of Virginia, VA

Balinsky BI (1951). On the eyecup-lens correlation in some South African amphibians. *Experientia* 7:180–181.

Balinsky BI (1957). On the factors determining the size of the lens rudiment in amphibian embryos. *J Exp Zool* 135:255–300.

Balinsky BI (1965). *An introduction to embryology*, 2nd edn. W.B.: Saunders Company, Philadelphia, PA. 673 pp.

Barbieri AM, Lupo G, Bulfone A, Anreazzoli M, Mariani M, Fougerousse F, Consalez GG, Borsani G, Beckman JS, Barasacchi G, Ballabio A, Banfi S (1999). A homeobox gene, *vax2*, controls the patterning of the eye dorsoventral axis. *P Natl Acad Sci USA* 96:10729–10734.

Barbieri AM, Broccoli V, Bovolenta P, Alfano G, Marchitiello A, Mocchetti C, Crippa L, Bulfone A, Marigo V, Ballabio A (2002). *Vax2* inactivation in mouse determines alteration of the eye dorsal–ventral axis, misrouting of the optic fibres and eye coloboma. *Dev Dis* 129:805–813.

Barro CD, Romanet JP, Fdili A, Guillot M, Morel F (1998). Gelatinase concentration in tears of corneal-grafted patients. *Curr Eye Res* 17:174–182.

Beach DH, Jacobson M (1979). Patterns of cell proliferation in the retina of the clawed frog during development. *J Comp Neurol* 183:603–613.

Beaver RS, Dunlop SA, Harman AM, Stirling RV, Easter SS, Roberts JD, Beazley LD (2001). Continued neurogenesis is not a pre-requisite for regeneration of a topographic retino-tectal projection. *Vision Res* 41:1765–1770.

Beazley LD (1981). Retinal ganglion cell death and regeneration of abnormal retinotectal projections after removal of a segment of optic nerve in *Xenopus* tadpoles. *Dev Biol* 85:164–170.

Beck CW, Christen B, Barker D, Slack JM (2006). Temporal requirement for bone morphogenetic proteins in regeneration of the tail and limb of *Xenopus* tadpoles. *Mech Dev* 123:674–688.

Beck CW, Christen B, Slack JM (2003). Molecular pathways needed for regeneration of spinal cord and muscle in a vertebrate. *Dev Cell* 5:429–439.

Beck CW, Slack JMW (2001). An amphibian with ambition: a new role for *Xenopus* in the 21st century. *Genome Biol* 2:10291–10295.

Beebe DC (1994). Homeobox genes and vertebrate eye development. *Invest Ophthalmol Vis Sci* 35:2897–2900.

Bellerby CW (1934). A rapid test for the diagnosis of pregnancy. *Nature* 133:494–495.
Belecky-Adams TL, Adler R, Beebe DC (2002). Bone morphogenetic protein signaling and the initiation of lens fiber cell differentiation. *Dev Cell* 129:3795–3802.
Berman M (1989). The pathogenesis of corneal epithelial defects. *Acta Ophthalmol Suppl* 192:55–64.
Bertuzzi S, Hindges R, Mui SH, O'Leary DD, Lemke G (1999). The homeodomain protein vax1 is required for axon guidance and major tract formation in the developing forebrain. *Genes Dev* 13:3092–3105.
Blixt A, Mahlapuu M, Aitola M, Pelto-Huikko M, Enerback S, Carlesson P (2000). A forhead gene, *FoxE3*, is essential for lens epithelial proliferation and closure of the lens vesicle. *Genes Dev* 14:245–254.
Bosco L (1988a). The problem of lens regeneration in anuran amphibian tadpoles. *Acta Embryol Morphol Exp* 9:25–38.
Bosco L (1988b). Transdifferentiation of ocular tissues in larval *Xenopus laevis*. *Differentiation* 39:4–15.
Bosco L, Filoni S (1992). Relationship between presence of the eye cup and maintenance of lens-forming capacity in larval *Xenopus laevis*. *Dev Growth Differ* 34:619–625.
Bosco L, Willems D (1992). Persistence of the lens-inducing capacity of the neural retina in adult Anura, *Xenopus laevis* and *Rana esculenta*. *Rend Acc Naz Lincei* 3:345–351.
Bosco L, Willems D, Stefanelli SA (1992). Persistence if the lens-inducing capacity of the neural retina in adult Anura, *Xenopus laevis* and *Rana esculenta Rend Fis Acc Lincei ser* 3:345–351.
Bosco L, Venturini G, Willems D (1994). First evidence of lens-transdifferentiation of larval *Xenopus laevis* induced by brain-derived acidic FGF *Rend Fis Acc Lincei* 5:261–268.
Bosco L, Testa O, Venturini G, Willems D (1997a). Lens fibre transdifferentiation in cultured larval *Xenopus laevis* outer cornea under the influence of neural retina-conditioned medium. *Cell Mol Life Sci* 53:921–928.
Bosco L, Venturini G, Willems D (1997b). *In vitro* lens transdifferentiation of *Xenopus laevis* outer cornea induced by fibroblast growth factor (FGF). *Development* 124:421–428.
Brahma SA (1980). Isofocusing and immunoelectrophoretic studies of soluble eye lens proteins from regenerated and normally developed *Xenopus laevis*. *Exp Eye Res* 30:269–275.
Brahma SK, Bours J (1972). Thin layer isoelectric focusing of the soluble lens extracts from larval stages and adult *Xenopus laevis*. *Exp Eye Res* 13:308–314.
Brahma SA, Grunz H (1988). Cytallins during *Xenopus laevis* free-lens formation. *Roux Arch Dev Biol* 197:190–192.
Brahma SA, McDevitt DS (1974). Ontogeny and localization of the lens crystallins in *Xenopus laevis* lens regeneration. *J Embryol Exp Morphol* 32:783–794.
Brahma SA, van Doorenmaalen WJ (1968). Studies on lens regeneration in *Xenopus laevis*. *Experientia* 24:519–521.
Brakenhoff RH, Elliott GS, Schut FH, Gelsing RR, Thijssen HM, Lubsen NH, Schoenmakers JG (1992). A device for quantitative measurement of the opacity of small animal lenses: application to *Xenopus laevis*. *Exp Eye Res* 54:631–634.
Brakenhoff RH, Ruuls RC, Jacobs EH, Schoenmakers JG, Lubsen NH (1991). Transgenic *Xenopus laevis* tadpoles: a transient in vivo model system for the manipulation of lens function and lens development. *Nucleic Acids Res* 19:1279–1284.
Bronchain OJ, Hartley KO, Amaya E (1999). A gene trap approach in *Xenopus*. *Curr Biol* 9:1195–1198.
Brownell I, Dirksen M, Jamrich M (2000). Forkhead *FoxE3* maps to the dysgenetic lens locus and is critical in lens development and differentiation. *Genesis* 27:81–93.
Brun RB (1981). The movement of the prospective eye vesicles from the neural plate into the neural fold in *Ambystoma mexicanum* and *Xenopus laevis*. *Dev Biol* 88:192–199.
Brunekreef GA, van Genesen ST, Destree OH, Lubsen NH (1997). Extralenticular expression of *Xenopus laevis alpha-, beta-,* and *gamma-crystallin* genes. *Invest Ophthalmol Vis Sci* 38:2764–2771.
Buchholz DR, Tomita A, Fu L, Paul BD, Shi YB (2004). Transgenic analysis reveals that thyroid hormone receptor is sufficient to mediate the thyroid hormone signal in frog metamorphosis. *Mol Cell Biol* 24:9026–9037.
Callery EM (2006). There's more than one frog in the pond: a survey of the Amphibia and their contributions to developmental biology. *Semin Cell Dev Biol* 17:80–92.
Campbell JC (1965). An immunofluorescent study of lens regeneration in larval *Xenopus laevis*. *J Embryol Exp Morphol* 13:171–179.
Campbell JC, Truman DE (1977). Variations in differentiation in the regenerating lens of *Xenopus laevis*. *Exp Eye Res* 25:99–100.
Campbell JC, Clayton RM, Truman DE (1986). Antigens of the lens of *Xenopus laevis*. *Exp Eye Res* 7:4–10.
Cannata SM, Arresta E, Bernardini S, Gargioli C, Filoni S (2003). Tissue interactions and lens-forming competence in the outer cornea of larval *Xenopus laevis*. *J Exp Zool A Comp Exp Biol* 299:161–171.
Carinato ME, Walter BE, Henry JJ (2000). *Xenopus laevis* gelatinase B (*Xmmp-9*), development, regeneration, and wound healing. *Dev Dynam* 217:377–387.
Carruthers S, Stemple DL (2006). Genetic and genomic prospects for *Xenopus tropicalis* research. *Semin Cell Dev Biol* 17:146–153.
Chae J, Zimmerman LB, Grainger RM (2002). Inducible control of tissue-specific transgene expression in Xenopus tropicalis transgenic lines. *Mech Dev* 117:235–241.
Chang W, KhosrowShahian F, Chang R, Crawford MJ (2001). *xPitx1* plays a role in specifying cement gland and head during early *Xenopus* development. *Genesis* 29:78–90.
Chauhan BK, Reed NA, Yang Y, Cermak L, Reneker L, Duncan MK, Cvekl A (2002a). A comparative cDNA microarray analysis reveals a spectrum of genes regulated by Pax6 in mouse lens. *Genes Cells* 7:1267–1283.
Chauhan BK, Zhang W, Cveklova K, Kantorow M, Cvekl A (2002b). Identification of differentially expressed genes in mouse *Pax6* heterozygous lenses. *Invest Ophthalmol Vis Sci* 43:1884–1890.
Chauhan BK, Reed NA, Zhang W, Duncan MK, Kilimann MW, Cvekl A (2002c). Identification of genes downstream of Pax6 in the mouse lens using cDNA microarrays. *J Biol Chem* 277:11539–11548.
Chow RL, Altmann CR, Lang RA, Hemmati-Brivanlou A (1999). *Pax6* induces ectopic eyes in a vertebrate. *Development* 126:4213–4222.
Chow RL, Lang RA (2001). Early eye development in vertebrates. *Annu Rev Cell Dev Biol* 17:255–296.
Cioni C, Filoni S, Aquila C, Bernardini S, Bosco L (1986). Transdifferentiation of eye tissues in anuran amphibians: analysis of the transdifferentiation capacity of the iris of *Xenopus laevis* larvae. *Differentiation* 32:215–220.
Cioni C, Filoni S, Bosco L, Aquila C, Bernardini S (1987). Transdifferentiation of larval *Xenopus laevis* iris implanted into the amputated hindlimb. *Experientia* 43:443–444.
Cioni C, Bosco L, Filoni S (1990). Transdifferentiation of larval *Xenopus laevis* iris under the influence of the pituitary. *Experientia* 46:1078–1080.
Cooke J, Weber JA (1985). Dynamics of the control of body pattern in the development of *Xenopus laevis*: I. Timing and pattern in the

development of dorsoanterior and posterior blastomeres pairs, isolated at the 4-cell stage. *J Embryol Exp Morphol* 88:85–112.

Conway KM, Hunt RK (1987). Whole eyes reconstituted from embryonic half anlagen: alterations in donor-derived territories in *Xenopus* pigment chimerae. *J Exp Zool* 244:231–241.

Corey DR, Abrams JM (2001). Morpholino antisense oligonucleotides: tools for investigating vertebrate development. *Genome Biol* 2:1015.

Cvekl A, Piatigorsky J (1996). Lens development and crystallin gene expression: many roles for Pax-6. *Bioessays* 18:621–630.

Dale L, Slack JMW (1987). Fate map for the 32-cell stage of *Xenopus laevis*. *Development* 99:527–551.

Danilchik MV, Black SD (1988). The first cleavage plane and the embryonic axis are determined by separate mechanisms in *Xenopus laevis*. I. Independence in undisturbed embryos. *Dev Biol* 128:58–64.

Das B, Brown DD (2004). Controlling transgene expression to study *Xenopus laevis* metamorphosis. *P Natl Acad Sci USA* 101:4839–4842.

De Graff AB (1960). A study on the factors controlling lens competence in amphibian epidermis. *Acta Embryol Morphol Exp* 3:40–52.

Del Rio-Tsonis K, Tsonis PA (2003). Eye regeneration at the molecular age. *Dev Dynam* 226:211–224.

De Robertis EM, Kuroda H (2004). Dorsal–ventral patterning and neural induction in *Xenopus* embryos. *Annu Rev Cell Dev Biol* 20:285–308.

Dorsky RI, Chang WS, Rapaport DH, Harris WA (1997). Regulation of neuronal diversity in the *Xenopus* retina by delta signaling. *Nature* 385:67–70.

Dressler GR, Deutsch U, Chowdhury K, Nornes HO, Gruss P (1990). *Pax2*, a new murine paired-box-containing gene and its expression in the developing excretory system. *Development* 109:787–795.

Dudley AT, Lyons KM, Robertson EJ (1995). A requirement for bone morphogenetic protein-7 during development of the mammalian kidney and eye. *Genes Dev* 9:2795–2819.

Ekker SC (2000). Morphants: a new systematic vertebrate functional genomics approach. *Yeast* 17:302–306.

Elkins MB, Henry JJ (2006). Isolation and characterization of a novel gene, *xMADML*, involved in *Xenopus laevis* eye development. *Dev Dynam* 235:1845–1857.

Etkin LD, Pearman B (1987). Distribution, expression and germ line transmission of exogenous DNA sequences following microinjection into *Xenopus laevis* eggs. *Development* 99:15–23.

Etkin LD, Pearman B, Roberts M, Bektesh SL (1984). Replication, integration and expression of exogenous DNA injected into fertilized eggs of *Xenopus laevis*. *Differentiation* 26:194–202.

Ferreiro B, Skoglund P, Bailey A, Dorsky R, Harris WA (1992). *XASH1*, a *Xenopus* homolog of *achaete-scute*: a proneural gene in anterior regions of the vertebrate CNS. *Mech Dev* 40:25–36.

Filoni S, Bernardini S, Cannata SM (2006). Experimental analysis of lens-forming capacity in *Xenopus borealis* larvae. *J Exp Zoolog A Comp Exp Biol* 305:538–550.

Filoni S, Bernardini S, Cannata SM, D'Alessio A (1997). Lens regeneration in larval *Xenopus laevis*, experimental analysis of the decline in regeneration capacity during development. *Dev Biol* 187:13–24.

Fini ME, Cook JR, Mohan R (1998). Proteolytic mechanisms in corneal ulceration and repair. *Arch Dermatol Res* 290:S12–S23.

Fini ME, Parks WC, Rinehart WB, Girard MT, Matsubara M, Cook JR, West-Mays JA, Sadow PM, Burgeson RE, Jeffrey JJ, Raizman MB, Krueger RR, Zieske JD (1996). Role of matrix metalloproteinases in failure to re-epithelialize after corneal injury. *Am J Pathol* 149:1287–1302.

Freeman G (1963). Lens regeneration from the cornea in *Xenopus laevis*. *J Exp Zool* 154:39–66.

Freeman G, Overton J (1962). The effects of thyroxin on the competence of lens regeneration in *Xenopus*. *Anat Rec* 142:305.

Fruscoloni P, Zamboni M, Baldi MI, Tocchini-Valentini GP (2003). Exonucleolytic degradation of double-stranded RNA by an activity in *Xenopus laevis* germinal vesicles. *P Natl Acad Sci USA* 100:1639–1644.

Fu L, Buchholz D, Shi Y (2002). Novel double promoter approach for identification of transgenic animals: a tool for in vivo analysis of gene function and development of gene-based therapies. *Mol Reprod Dev* 62:470–476.

Fujiwara M, Uchida T, Osumi-Yamashita N, Eto K (1994). Uchida rat (*rSey*): a new mutant rat with craniofacial abnormalities resembling those of the mouse *Sey* mutant. *Differentiation* 57:31–38.

Furuta Y, Hogan BLM (1998). *BMP4* is essential for lens induction in the mouse embryo. *Genes Dev* 12:3764–3775.

Gaze RM (1959). Regeneration of the optic nerve in *Xenopus laevis*. *Q J Exp Physiol CMS* 44:290–308.

Gimlich J (1986). Acquisition of developmental autonomy in the equatorial region of the *Xenopus* embryo. *Dev Biol* 115:340–352.

Gimlich RL, Braun J (1985). Improved fluorescent compounds for tracing cell lineage. *Dev Biol* 109:509–514.

Gimlich RL, Cooke J (1983). Cell lineage and the induction of second nervous systems in amphibian development. *Nature* 306:471–473.

Gimlich RL, Gerhart JC (1984). Early cell interactions promote embryonic axis formation in *Xenopus laevis*. *Dev Biol* 104:117–130.

Gordon JB, Hopwood N (2000). The introduction of *Xenopus laevis* into developmental biology: of empire, pregnancy testing and ribosomal genes. *Int J Dev Biol* 44:43–50.

Goudreau G, Petrou P, Reneker LW, Graw J, Loster J, Gruss P (2002). Mutually regulated expression of *Pax6* and *Six3* and its implications for the *Pax6* haploinsufficient lens phenotype. *P Natl Acad Sci USA* 99:8719–8724.

Graf JD (1996). Molecular approaches to the phylogeny of *Xenopus*. In: Tinsley RC, Kobel HR (eds), *The Biology of Xenopus*. Clarendon Press, Oxford, pp. 379–389.

Grainger RM (1992). Embryonic lens induction: shedding light on vertebrate tissue determination. *Trends Genet* 8:349–355.

Grainger RM (1996). New perspectives on embryonic lens induction. *Semin Cell Dev Biol* 7:149–155.

Grainger RM, Henry JJ, Henderson R (1988). Reinvestigation of the role of the optic vesicle in embryonic lens induction. *Development* 102:517–526.

Grainger RM, Henry JJ, Saha MS, Servetnick M (1992). Recent progress on the mechanisms of embryonic lens formation. *Eye* 6:117–122.

Grainger RM, Mannion JE, Cook TL, Zygar CA (1997). Defining intermediate stages in cell determination: acquisition of a lens-forming bias in head ectoderm during lens determination. *Dev Genet* 20:246–257.

Haas K, Jensen K, Sin WC, Foa L, Cline HT (2002). Targeted electroporation in *Xenopus* tadpoles *in vivo* from single cells to the entire brain. *Differentiation* 70:148–154.

Hallonet M, Hollemann T, Wehr R, Jenkins NA, Copeland NG, Pieler T, Gruss P (1998). *Vax1* is a novel homeobox-containing gene expressed in the developing anterior ventral forebrain. *Development* 125:2599–2610.

Hallonet M, Hollemann T, Pieler T, Gruss P (1999). *Vax1*, a novel homeobox-containing gene, directs development of the basal forebrain and visual system. *Genes Dev* 13:3106–3114.

Harland R (2000). Neural induction. *Curr Opin Genet Dev* 10:357–362.

Hartley KO, Nutt SL, Amaya E (2002). Targeted gene expression in transgenic *Xenopus* using the binary Gal4-UAS system. *P Natl Acad Sci USA* 99:1377–1382.

Hausen P, Riebesell M (1996). *The Early Development of Xenopus Laevis: An Atlas of the Histology.* Springer Verlag Telos, Max Planck Institute, Tubingen, Germany.

Heasman J (2002). Morpholino oligos: making sense of antisense? *Dev Biol* 243:209–214.

Heasman J, Holwill S, Wylie CC (1991). *Fertilization of cultured Xenopus oocytes and use in studies of maternally inherited molecules.* In: Kay BK, Peng HB (eds), *Xenopus laevis*: Practical Uses in Cell and Molecular Biology. In: Methods in Cell Biology, Vol. 36, Academic Press, Inc., San Diego, CA. pp. 214–230.

Heasman J, Wylie CC, Hausen P, Smith JC (1984). Fates and states of determination of single vegetal pole blastomeres of *X. laevis*. *Cell* 37:185–194.

Henry JJ (2003). Cell and molecular biology of lens regeneration. *Int Rev Cytol* 228:195–264.

Henry JJ, Elkins MB (2001). Cornea-lens transdifferentiation in the anuran, *Xenopus tropicalis*. *Dev Genes Evol* 211:377–387.

Henry JJ, Grainger RM (1987). Inductive interactions in the spatial and temporal restriction of lens-forming potential in embryonic ectoderm of *Xenopus laevis*. *Dev Biol* 124:200–214.

Henry JJ, Grainger RM (1990). Early tissue interactions leading to embryonic lens formation in *Xenopus laevis*. *Dev Biol* 141:149–163.

Henry JJ, Mittleman JM (1995). The matured eye of *Xenopus laevis* tadpoles produces factors that elicit a lens-forming response in embryonic ectoderm. *Dev Biol* 171:39–50.

Henry JJ, Carinato ME, Schaefer JJ, Wolfe AD, Walter BE, Perry KJ, Elbl TN (2002). Characterizing gene expression during lens formation in *Xenopus laevis*, evaluating the model for embryonic lens induction. *Dev Dynam* 224:168–185.

Hirose G, Jacobson M (1979). Clonal organization of the central nervous system of the frog. I. Clones stemming from individual blastomeres of the 16-cell and earlier stages. *Dev Biol* 71:191–202.

Hirsch N, Grainger RM (2000). Induction of the lens. *Res Probl Cell Differ* 31:51–68.

Hirsch N, Zimmerman LB, Gray J, Chae J, Curran KL, Fisher M, Ogino H, Grainger RM (2002a). *Xenopus tropicalis* transgenic lines and their use in the study of embryonic induction. *Dev Dynam* 225:522–535.

Hirsch N, Zimmerman LB, Grainger RM (2002b). *Xenopus*, the next generation, *X. tropicalis* genetics and genomics. *Dev Dynam* 225:422–433.

Hitchcock PF, Raymond PA (1992). Retinal regeneration. *Trend Neurosci* 15:103–108.

Hocking JC, McFarlane S (2007). Expression of *Bmp* ligands and receptors in the developing *Xenopus* retina. *Int J Dev Biol* 51:161–165.

Hollemann T, Pieler T (1999). *Xpitx*-1: a homeobox gene expressed during pituitary and cement gland formation of *Xenopus* embryos. *Mech Dev* 88:249–255.

Holt C (1980). Cell movements in *Xenopus* eye development. *Nature* 287:850–852.

Hoskins SG (1990). Metamorphosis of the amphibian eye. *J Neurobiol* 21:970–989.

Huang H, Marsh-Armstrong N, Brown D (1999). Metamorphosis is inhibited in transgenic *Xenopus laevis* tadpoles that overexpress type III deiodinase. *P Natl Acad Sci USA* 96:962–967.

Huang S, Moody SA (1993). The retinal fate of *Xenopus* cleavage stage progenitors is dependent upon blastomeres position and competence: studies of normal and regulated clones. *J Neurosci* 13:3183–3210.

Huang S, Moody SA (1995). Asymmetrical blastomeres origin and spatial domains of dopamine and neuropeptide Y amicrine subtypes in *Xenopus* tadpole retinas. *J Comp Nuerol* 360:2–13.

Huang S, Moody SA (1997). Three types of serotonin-containing amacrine cells in tadpole retina have distinct clonal origins. *J Comp Neurol* 387:42–52.

Hyer J, Kuhlman J, Afif E, Mikawa T (2003). Optic cup morphogenesis requires pre-lens ectoderm but not lens differentiation. *Dev Biol* 259:351–363.

Ide CF (1998). Role of cell displacement, cell division, and fragment size in pattern formation during embryonic retinal regeneration in *Xenopus*. *Acta Biol Hung* 39:179–189.

Ide CF, Reynolds P, Tompkins R (1984). Two healing patterns correlate with different adult neural connectivity patterns in regenerating embryonic *Xenopus* retina. *J Exp Zool* 230:71–80.

Ide CF, Blankenau A, Morrow J, Tompkins R (1986). Cell movements and novel growth patterns during early healing in regenerating embryonic *Xenopus* retina. *Prog Clin Biol Res* 217B:133–136.

Ikeda Y (1936). Beiträge zur Frage der Fähigkeit zur Linsenregeneration bei einer Art von *Hynobius* (*Hynobius unnangso* Tago). Arb Anat Inst Kais Jpn Univ Sendai 18:17–50.

Ikeda Y (1939). Zur Frage der Lensenpotenz der Hornhaut in spätembryonalen und larvvalen Stadien bei ener Art von Hynobius (*Hynobius unnangso* Tago). Arb Anat Inst Univ Sendai 22:27–52.

Ishibashi S, Yasuda K (2001). Distinct roles of *maf* genes during Xenopus lens development. *Mech Dev* 101:155–166.

Itoh M, Kim CH, Palardy G, Oda T, Jiang YJ, Maust D, Yeo SY, Lorick K, Wright GJ, Ariza-McNaughton L, Weissman AM, Lewis J, Chandrasekharappa SC, Chitnis AB (2003). Mind bomb is a ubiquitin ligase that is essential for efficient activation of Notch signaling by Delta. *Dev Cell* 4:67–82.

Jacobson AG (1966). Inductive processes in embryonic development. *Science* 152:25–34.

Jacobson M (1983). Clonal organization of the central nervous system of the frog. III. Clones stemming form individual blastomeres of the 128-, 256-, and 512-cell stages. *J Neurosci* 3:1019–1038.

Jacobson M, Hirose G (1978). Origin of the retina from both sides of the embryonic brain: a contribution to the problem of crossing at the optic chiasma. *Science* 202:637–639.

Jacobson M, Hirose G (1981). Clonal organization of the central nervous system of the frog. II. Clones stemming from individual blastomeres of the 32- and 64-cell stages. *J Neurosci* 1:271–284.

Jenkins S, Straznicky C (1986). Naturally occurring and induced ganglion cell death. A retinal whole-mount autoradiographic study in *Xenopus*. *Anat Embryol* 174:59–66.

Johns PR (1977). Growth of the adult goldfish eye. III. Source of the new retinal cells. *J Comp Neurol* 176:343–357.

Johnson AD, Krieg PA (1994). pXeX, a vector for efficient expression of cloned sequences in *Xenopus* embryos. *Gene* 147:223–226.

Kablar B, Vignali R, Menotti L, Pannese M, Andreazzoli M, Polo C, Giribaldi MG, Boncinelli E, Barsacchi G (1996). *Xotx* genes in the developing brain of *Xenopus laevis*. *Mech Dev* 55:145–158.

Kamachi Y, Sockanathan S, Liu Q, Breitman M, Lovell-Badge R, Kondoh H (1995). Involvement of SOX proteins in lens-specific activation of crystallin genes. *Embo J* 14:3510–3519.

Kamachi Y, Uchikawa M, Collignon J, Lovell-Badge R, Kondoh H (1998). Involvement of Sox1, 2 and 3 in the early and subsequent molecular events of lens induction. *Development* 125:2521–2532.

Kamachi Y, Uchikawa M, Tanouchi A, Sekido R, Kondoh H (2001). Pax6 and SOX2 form a co-DNA-binding partner complex that regulates initiation of lens development. *Genes Dev* 205: 79–97.

Kanekar S, Perron M, Dorsky R, Harris WA, Jan LY, Jan YN, Vetter ML (1997). *Xath5* participates in a network of bHLH genes in the developing *Xenopus* retina. *Neuron* 19:981–994. Erratum in: Neuron 1998 21:following 1221.

Kay BK, Peng HB (1991). *Xenopus laevis*: practical uses in cell and molecular biology. In: Kay BK, Peng HB (eds), *Methods in Cell Biology*, Vol. 36. Academic Press, Inc, San Diego, CA.

Keller RE (1976). Vital dye mapping of the gastrula and neurula of *Xenopus laevis*. II. Prospective areas and morphogenetic movements of the deep layer. *Dev Biol* 51:118–137.

Keller RE (1975). Vital dye mapping of the gastrula and neurula of *Xenopus laevis*. I. Prospective areas and morphogenetic movements of the superficial layer. *Dev Biol* 42:222–241.

Kenyon KL, Moody SA, Jamrich M (1999). A novel fork head gene mediates early steps during *Xenopus* lens formation. *Development* 126:5107–5116.

Kherif S, Lafuma C, Dehaupas M, Lachkar S, Fournier JG, Verdiere-Sahuque M, Fardeau M, Alameddine HS (1999). Expression of matrix metalloproteinases 2 and 9 in regenerating skeletal muscle, a study in experimentally injured and *mdx* muscles. *Dev Biol* 205:158–170.

Khosrowshahian F, Wolanski M, Chang WY, Fujiki K, Jacobs L, Crawford MJ (2005). Lens and retina formation require expression of *Pitx3* in *Xenopus* pre-lens ectoderm. *Dev Dynam* 234: 577–589.

Kim JI, Li T, Ho IC, Grusby MJ, Glimcher LH (1999). Requirement for the c-Maf transcription factor in crystallin gene regulation and lens development. *P Natl Acad Sci USA* 96:3781–3785.

Klein SL (1987). The first cleavage furrow demarcates the dorsal – ventral axis in *Xenopus* embryos. *Dev Biol* 120:299–304.

Klein SL, Gerhard DS, Wagner L, Richardson P, Schriml LM, Sater AK, Warren WC, McPherson JD (2006). Resources for genetic and genomic studies of *Xenopus*. *Meth Mol Biol* 322:1–16.

Klein SL, Strausberg RL, Wagner L, Pontius J, Clifton SW, Richardson P (2002). Genetic and genomic tools for *Xenopus* research: the NIH *Xenopus* initiative. *Dev Dynam* 225:384–391.

Knöchel W, Korge E, Basner A, Meyerhof W (1986). Globin evolution in the genus *Xenopus*, comparative analysis of cDNAs coding for adult globin polypeptides of *Xenopus borealis* and *Xenopus tropicalis*. *J Mol Evol* 23:211–223.

Knox BE, Schlueter C, Sanger BM, Green CB, Besharse JC (1998). Transgene expression in *Xenopus* rods. *FEBS Lett* 423:117–121.

Kobel HR, Loumont C, Tinsley RC (1996). The extant species. In: Tinsley RC, Kobel HR (eds), *The Biology of Xenopus*. Clarendon Press, Oxford, pp. 9–33.

Koga M, Kageura H, Yamana K (1986). Use of hybrids between *Xenopus laevis* and *Xenopus borealis* in chimera formation: dorsalization of ventral cells. *Dev Growth Differ* 28:177–183.

Koo BK, Lim HS, Song R, Yoon MJ, Yoon KJ, Moon JS, Kim YW, Kwon MC, Yoo KW, Kong MP, Lee J, Chitnis AB, Kim CH, Kong YY (2005). Mind bomb 1 is essential for generating functional Notch ligands to activate Notch. *Development* 132:3459–3470.

Koshiba-Takeuchi K, Takeuchi JK, Matsumoto K, Momose T, Uno K, Hoepker V, Ogura K, Takahashi N, Nakamura H, Yasuda K et al. (2000). *Tbx5* and the retinotectum projection. *Science* 287:134–137.

Köster RW, Kühnlein RP, Wittbrodt J (2000). Ectopic *Sox3* activity elicits sensory placode formation. *Mech Dev* 95:175–187.

Kreig RA, Melton DA (1984). Functional messenger RNA's are produced by SP6 in vitro transcription of cloned cDNA's. *Nucleic Acid Res* 12:7057–7070.

Kroll KL, Amaya E (1996). Transgenic *Xenopus* embryos from sperm nuclear transplantations reveal FGF signaling requirements during gastrulation. *Development* 122:3173–3183.

Kroll KL, Gerhart JC (1994). Transgenic *X. laevis* embryos from eggs transplanted with nuclei of transfected cultured cells. *Science* 266:650–653.

Kubota Y, Takase Y, Komori Y, Hashimoto Y, Arata T, Kamimura Y, Araki H, Takisawa H (2003). A novel ring-like complex of *Xenopus* proteins essential for the initiation of DNA replication. *Genes Dev* 17:1141–1152.

Kuspa A, Loomis WF (1992). Tagging developmental genes in *Dictyostelium* by restriction enzyme-mediated integration of plasmid DNA. *P Natl Acad Sci USA* 89:8803–8807.

Lagutin O, Zhu CC, Furuta Y, Rowitch DH, McMahon AP, Oliver G (2001). Six3 promotes the formation of ectopic optic vesicle-like structures in mouse embryos. *Dev Dynam* 221:342–349.

Lau GT, Wong OG, Chan PM, Kok KHG, Wong RL, Chin KT, Lin MC, Kung HF, How KL (2001). Embryonic XMab21/2 expression is required for gastrulation and subsequent neural development. *Biochem Ciophys Res Commun* 280:1378–1384.

Levine RL (1981). La regenerescence de la retine chez *Xenopus laevis*. *Revue Canadienne de Biologie* 40:19–27.

Lewis WH (1904). Experimental studies on the development of the eye in Amphibia. I. On the origin of the lens, *Rana paulustris*. *Am J Anat* 3:505–536.

Lewis WH (1907a). Experimental studies on the development of the eye in Amphibia. III. On the origin and differentiation of the lens. *Am J Anat* 6:473–509.

Lewis WH (1907b). Lens formation form strange ectoderm in *Rana sylvatica*. *Am J Anat* 7:145–169.

Li M, Rohrer B (2006). Gene silencing in *Xenopus laevis* by DNA vector-based RNA interference and transgenesis. *Cell Res* 16:99–105.

Li H, Tierney C, Wen L, Wu JY, Rao Y (1997). A single morphogenetic field gives rise to two retina primordia under the influence of the prechordal plate. *Development* 124:603–615.

Li H-S, Yang J-M, Jacobson RD, Pasko D, Sundin O (1994). Pax-6 is first expressed in a region of ectoderm anterior to the early neural plate: implications for stepwise determination of the lens. *Dev Biol* 162:181–194.

Li de Q, Shang TY, Kim HS, Solomon A, Lokeshwar BL, Pflugfelder SC (2003). Regulated expression of collagenases MMP-1, -8, and -13 and stromelysins MMP-3, -10, and -11 by human corneal epithelial cells. *Invest Ophthalmol Vis Sci* 44:2928–2936.

Liedke KB (1942). Lens competence in *Rana pipiens*. *J Exp Zool* 90:331–351.

Liedke KB (1951). Lens competence in *Ambystoma punctatum*. *J Exp Zool* 130:353–379.

Liedke KB (1955). Studies on lens induction in *Amblystoma punctatum*. *J Exp Zool* 130:353–379.

Linder B, Mentele E, Mansperger K, Straub T, Kremmer E, Rupp RA (2007). CHD4/Mi-2 activity is required for the positioning of the mesoderm/neuroectoderm boundary in *Xenopus*. *Genes Dev* 21:973–983.

Liu Y, Lupo G, Marchitiello A, Gestri G, He RQ, Banfi S, Barsacchi G (2001). Expression of the *Xvax2* gene demarcates presumptive ventral telencephalon and specific visual structures in *Xenopus laevis*. *Mech Dev* 100:115–118.

Lombardo A, Slack JM (1997). Inhibition of *eFGF* expression in *Xenopus* embryos by antisense mRNA. *Dev Dynam* 108:162–169.

Lopashov GV (1991). Regenerative capacity of retinal cells and the maintenance of their differentiation. Regeneration of vertebrate sensory receptor cells. Chiba Foundation Symposium, Vol. 160. Wiley, Chichester, England, pp. 209–218.

Lopashov GV, Golubeva ON, Zviadadze KG (1997). Inductive capacity of living eye tissues from adult frogs. *Differentiation* 61:237–242.

Lupo G, Andreazzoli M, Gestri G, Liu Y, He R, Barsacchi G (2000). Homeobox genes in the genetic control of eye development. *Int J Dev Biol* 44:627–636.

Lupo G, Liu Y, Chandraratna RAS, Barsacchi G, He RQ, Harris WA (2005). Dorsoventral patterning of the *Xenopus* eye: a collaboration of retinoid, hedgehog, and FGF receptor signaling. *Development* 132:1737–1748.

Mangold O (1931). Das Determinationproblem. III. Das Wirbeltierauge in der Entwicklung und Regeneration. *Erg Biol* 7:193–403.

Mann F, Harris WA, Holt CE (2004). New views on retinal axon development: a navigation guide. *Int J Dev Biol* 48:957–964.

Marsh-Armstrong N, Huang H, Berry DL, Brown DD (1999). Germline transmission of transgenes in *Xenopus laevis*. *P Natl Acad Sci* 96:14389–14393.

Martinez-Morales JR, Signore M, Acampora D, Simeone A, Bovolenta P (2001). Otx genes are required for tissue specification in the developing eye. *Development* 128:2019–2030.

Masho R (1988). Fates of the animal-dorsal blastomeres of eight-cell stage *Xenopus* embryos vary according to the specific patterns of the third cleavage plane. *Dev Growth Differ* 30:347–359.

Masho R (1990). Close correlation between the first cleavage plane and the body axis in early *Xenopus* embryos. *Dev Growth Differ* 32: 57–64.

Masho R, Kubota HY (1986). Developmental fates of blastomeres of eight-cell stage *Xenopus laevis* embryos. *Dev Growth Differ* 28:113–123.

Matsubara M, Girard MT, Kublin CL, Cintron C, Fini ME (1991a). Differential roles for two gelatinolytic enzymes of the matrix metalloproteinase family in the remodeling cornea. *Dev Biol* 147:425–439.

Matsubara M, Zieske JD, Fini ME (1991b). Mechanism of basement membrane dissolution preceding corneal ulceration. *Invest Ophthalmol Vis Sci* 32:3221–3237.

McDevitt DS, Brahma SK (1973). Ontogeny and localization of the crystallins during *Xenopus laevis* embryonic lens development. *J Exp Zool* 186:127–140.

McDevitt DS, Brahma SK (1979). Embryonic appearance of alpha, beta, and gamma crystallins in the periodic albinism (ap) mutant of *Xenopus laevis*. *Differentiation* 14:107–112.

McDevitt DS, Brahma SK (1982). Ontogeny and localization of the crystallins in eye development and regeneration. In: McDevitt DS (ed.), *Cell Biology of the Eye*, pp. 143–191.

Mitashov VI (1997). Retinal regeneration in amphibians. *Int J Dev Biol* 41:893–905.

Mitashov VI, Maliovanova SD (1982). Proliferative potencies of cells of pigment and cilliary epithelia of the eye of *Xenopus laevis* under normal conditions and on regeneration. *Ontogenez* 13:228–234.

Miyazaki K, Uchiyama K, Imokawa Y, Yoshizato K (1996). Cloning and characterization of cDNAs for matrix metalloproteinases of regenerating newt limbs. *P Natl Acad Sci USA* 93:6819–6824.

Mizuno N, Mochii M, Takagi C, Takahashi TC, Eguchi G, Okada TS (1998). A critical role for the optic vesicle in lens development; a reinvestigation of free lens formation in *Cynops pyrrhogaster*. *Differentiation* 63:247–252.

Mizuno N, Mochii M, Takahashi TC, Eguchi G, Okada TS (1999a). Lens regeneration in *Xenopus* is not a mere repeat of lens development, with respect to crystallin gene expression. *Differentiation* 64: 143–149.

Mizuno N, Mochii M, Yamamoto TS, Takahashi TC, Eguchi G, Okada TS (1999b). *Pax-6* and *Prox 1* expression during lens regeneration from *Cynops* iris and *Xenopus* cornea: evidence for a genetic program common to embryonic lens development. *Differentiation* 65:141–149.

Mizuno N, Ueda Y, Kondoh H (2005). Requirement for *βB1-crystallin* promoter of *Xenopus laevis* in embryonic lens development and lens regeneration. *Dev Growth Differ* 47:131–140.

Mizuseki K, Kishi M, Matsui M, Nakanishi S, Sasai Y (1998). *Xenopus* Zic-related-1 and Sox-2, two factors induced by chordin, have distinct activities in the initiation of neural induction. *Development* 125:579–587.

Moody SA (1987a). Fates of the blastomeres of the 16-cell stage *Xenopus* embryo. *Dev Biol* 119:560–578.

Moody SA (1987b). Fates of the blastomeres of the 32-cell-stage *Xenopus* embryo. *Dev Biol* 122:300–319.

Moon RT, Christian JL (1989). Microinjection and expression of synthetic mRNAs in *Xenopus*. *Technique* 1:76–89.

Moore KB, Moody SA (1999). Animal-vegetal asymmetries influence the earliest steps in retina fate commitment in *Xenopus*. *Dev Biol* 212:25–41.

Moritz OL, Tam BM, Papermaster DS, Nakayama T (2001). A functional rhodopsin-green fluorescent protein fusion protein localizes correctly in transgenic *Xenopus laevis* retinal rods and is expressed in a time-dependent pattern. *J Biol Chem* 276:28242–28251.

Mui SH, Hindges R, O'Leary DD, Lemke G, Bertuzzi S (2002). The homeodomain protein Vax2 patterns the dorsoventral and nasotemporal axes of the eye. *Development* 129:797–804.

Murali D, Yoshikawa S, Corrigan RR, Plas DJ, Crair MC, Oliver G, Lyons KM, Mishina Y, Futura Y (2005). Distinct developmental programs require different levels of Bmp signaling during mouse retinal development. *Development* 132:913–923.

Nakamura O, Kishiyama K (1971). Prospective fates of blastomeres at the 32-cell stage of *Xenopus laevis* embryos. *P Japan Acad Sci* 47:407–412.

Nakamura O, Takasaki H, Nagata A (1978). Further studies of the prospective fates of blastomeres at the 32-cell stage of *Xenopus laevis* embryos. *Med Biol* 56:355–360.

Nieuwkoop PD, Faber J (1956). *Normal Table of Xenopus laevis (Daudin)*. North Holland Publishing, Amsterdam.

Nieuwkoop PD, Faber J (1996). *Normal Table of Xenopus laevis (Daudin)*. Garland Publishing, Inc, New York.

Nieuwkoop PD, Florschutz PA (1950). Quelques caracteres speciaux de la gastrulation et de la neurulation de l'oeuf de *Xenopus laevis* Daud. et de quelques autres anoures. *Arch Biol* 61:113–150.

Nishiguchi S, Wood H, Kondoh H, Lovell-Badge R, Episkopou V (1998). *Sox1* directly regulates the gamma-crystallin genes and is essential for lens development in mice. *Genes Dev* 12:776–781.

Nornes HO, Dressler GR, Knapik EW, Deutsch U, Gruss P (1990). Spatially and temporally restricted expression of *Pax2* during murine neurogenesis. *Development* 109:797–809.

Nöthinger R, McDevitt DS, Yamada T (1971). Comparison of γ-crystallins from frog and newt. *Exp Eye Res* 12:94–98.

Nutt SL, Bronchain OJ, Hartley KO, Amaya E (2001). Comparison of morpholino based translational inhibition during the development of *Xenopus laevis* and *Xenopus tropicalis*. *Genesis* 30: 110–113.

Ochsé W (1948). Die Zucht des südafrikanischen Krallenfrosches *Xenopus laevis* (Daudin). *Gynaecologia* 126:57–77.

Offield MF, Hirsch N, Grainger RM (2000). The development of *Xenopus tropicalis* transgenic lines and their use in studying developmental timing in living embryos. *Development* 127:1789–1797.

Ogino H, McConnell WB, Grainger RM (2006). Highly efficient transgenesis in *Xenopus tropicalis* using I-SceI meganuclease. *Mech Dev* 123:103–113.

Ogino H, Yasuda K (1998). Induction of lens differentiation by activation of a bZIP transcription factor, L-Maf. *Science* 280:115–118.

Ogino H, Yasuda K (2000). Sequential activation of transcription factors in lens induction. *Dev Growth Differ* 42:437–448.

O'Gorman S, Kilty J, Hunt RK (1987). Healing and growth of half-eye "compound eyes" in *Xenopus*: application of an interspecific cell marker. *J Neurosci* 7:3764–3782.

Ogura T (2002). In vivo electroporation: a new frontier for gene delivery and embryology. *Differentiation* 70:163–171.

Oliver G, Sosa-Pineda B, Geisendorf S, Spana EP, Doe CQ, Gruss P (1993). *Prox 1*, a *prospero*-related homeobox gene expressed during mouse development. *Mech Dev* 44:3–16.

Onuma Y, Takahashi S, Asashima M, Kurata S, Gehring WJ (2002). Conservation of Pax6 function and upstream activation by Notch signaling in eye development of frogs and flies. *P Natl Acad Sci USA* 99:2020–2025.

Pan FC, Chen Y, Loeber J, Henningfeld K, Pieler T (2006). I-SceI meganuclease-mediated transgenesis in *Xenopus*. *Dev Dynam* 235:247–252.

Pannese M, Polo C, Andreazzoli M, Vignali R, Kablar B, Barsacchi G, Boncinelli E (1995). The *Xenopus* homologue of *Otx2* is a maternal homeobox gene that demarcates and specifies anterior body regions. *Development* 121:707–720.

Patten I, Placzek M (2000). The role of Sonic hedgehog in neural tube patterning. *Cell Mol Life Sci* 57:1695–1708.

Peng HB, Baker LP, Chen Q (1991). Tissue culture of *Xenopus* neurons and muscle cells as a model for studying synaptic induction. In: Kay BK, Peng HB (eds), *Methods in Cell Biology, Xenopus laevis: Practical Uses in Cell and Molecular Biology*, Vol. 36. Academic Press, Inc, San Diego, CA, pp. 511–526.

Penzel R, Oschwald R, Chen Y, Tacke L, Grunz H (1997). Characterization and early embryonic expression of a neural specific transcription factor *xSOX3* in *Xenopus laevis*. *Int J Dev Biol* 41:667–677.

Pera EM, Kessel M (1997). Patterning of the chick forebrain anlage by the prechordal plate. *Development* 124:4153–4162.

Perron M, Kanekar S, Vetter ML, Harris WA (1998). The genetic sequence of retinal development in the ciliary margin of the *Xenopus* eye. *Dev Biol* 199:185–200.

Perron M, Opdecamp K, Butler K, Harris WA, Bellefroid EJ (1999). *X-ngnr-1* and *Xath3* promote ectopic expression of sensory neuron markers in the neurula ectoderm and have distinct inducing properties in the retina. *P Natl Acad Sci* 96:14886–15001.

Pommereit D, Pieler T, Hollemann T (2001). *Xpitx3*: a member of the Rieg/Pitx gene family expressed during pituitary and lens formation in *Xenopus laevis*. *Mech Dev* 102:255–257.

Reeve JG, Wild AE (1978). Lens regeneration from cornea of larval *Xenopus laevis* in the presence of the lens. *J Embryol Exp Morphol* 48:205–214.

Reh TA (1989). The regulation of neuronal production during retinal neurogenesis. In: Finlay BL, Sengelaub DR (eds), *Development of the Vertebrate Retina*. Plenum Press, New York, pp. 43–67.

Reyer RW (1950). An experimental study of lens regeneration in *Trituris viridescens viridescens*: lens development from the dorsal iris in the absence of the embryonic lens. *J Exp Zool* 113:317–353.

Reyer RW (1954). Regeneration of the lens in the amphibian eye. *Quart Rev Biol* 29:1–46.

Reyer RW (1958a). Studies on lens induction in *Ambystoma punctatum* and *Triturus viridecens*. I. Transplants of prospective belly ectoderm. *J Exp Zool* 138:505–555.

Reyer RW (1958b). Studies on lens induction in *Ambystoma punctatum* and *Triturus viridecens*. II. Transplants of prospective lateral head ectoderm. *J Exp Zool* 139:137–179.

Reza HM, Ogino H, Yasuda K (2002). L-Maf, a downstream target of Pax6, is essential for chick lens development. *Mech Dev* 116:61–73.

Ring BZ, Cordes SP, Overbeek PA, Barsh GS (2000). Regulation of mouse lens fiber cell development and differentiation by the *Maf* gene. *Development* 127:307–317.

Roessler E, Muenke M (2001). Midline and laterality defects: left and right meet in the middle. *Bioessays* 23:888–900.

Ryffel GU, Werdien D, Turan G, Gerhards A, Goobes S, Senkel S (2003). Tagging muscle cell lineages in development and tail regeneration using Cre recombinase in transgenic *Xenopus*. *Nucleic Acids Res* 31:e44.

Sadaghiani B, Thiebaud CH (1987). Neural crest development in the *Xenopus laevis* embryo, studied by interspecific transplantation and scanning electron microscopy. *Dev Biol* 124:91–110.

Saha MS, Servetnick M, Grainger RM (1992). Vertebrate eye development. *Curr Opin Genet Dev* 2:582–588.

Saha MS, Spann CL, Grainger RM (1989). Embryonic lens induction: more than meets the optic vesicle. *Cell Differ Dev* 28:153–171.

Sakaguchi DS, Janick LM, Reh TA (1997). Basic fibroblast growth factor (FGF-2) induced transdifferentiation of retinal pigment epithelium: generation of retinal neurons and glia. *Dev Dynam* 209:387–398.

Sasagawa S, Takabatake T, Takabatake Y, Muramatsu T, Takeshima K (2002). Axes establishment during eye morphogenesis in *Xenopus* by coordinate and antagonistic actions of BMP4, Shh and RA. *Genesis* 33:86–96.

Schaefer JJ, Oliver G, Henry JJ (1999). Conservation of gene expression during embryonic lens formation and cornea-lens transdifferentiation in *Xenopus laevis*. *Dev Dynam* 215:308–318.

Schwarz M, Cecconi F, Bernier G, Andrejewski N, Kammandel B, Wagner M, Gruss P (2000). Spatial specification of mammalian eye territories by reciprocal transcriptional repression of Pax2 and Pax6. *Development* 127:4325–4334.

Seidman S, Soreq H (1996). *Transgenic Xenopus: microinjection methods and developmental neurobiology*. In: Neuromethods, Vol. 38. Humana Press, Totowa, NJ.

Semina EV, Murray JC, Reiter R, Hrstka RF, Graw J (2000). Deletion in the promoter region and altered expression of *Pitx3* homeobox gene in *aphakia* mice. *Hum Mol Genet* 9:1575–1585.

Servetnick MD, Cook TL, Jr, Grainger RM (1996). Lens induction in axolotls: comparison with inductive signaling mechanisms in *Xenopus laevis*. *Int J Dev Biol* 40:755–761.

Servetnick M, Grainger RM (1991). Changes in neural and lens competence in *Xenopus* ectoderm: evidence for an autonomous developmental timer. *Development* 112:177–188.

Shapiro HA, Zwarenstein H (1934). A rapid test for the diagnosis of pregnancy in *Xenopus*. *Nature* 133:762.

Shastry BS (1989). Immunological studies on gamma crystallins from *Xenopus*: localization, tissue specificity and developmental expression of proteins. *Exp Eye Res* 49:361–369.

Shastry BS, Reddy VN (1990). Isolation and characterization of a distantly related member of the Beta-crystallin super family from *Xenopus*. *Biochem Biophys Res Commun* 171:1333–1337.

Sinzelle L, Vallin J, Coen L, Chesneau A, Du Pasquier D, Pollet N, Demeneix B, Mazabraud A (2006). Generation of transgenic *Xenopus laevis* using the sleeping beauty transposon system. *Transgenic Res* 15:751–760.

Sive HL, Grainger RM, Harland RM (2000). *Early Development of Xenopus laevis. A Laboratory Manual*. Cold Spring Harbor Laboratory Press, Cold Spring Harbor, NY.

Slack JMW (1991). *From Egg to Embryo*. Cambridge University Press, Cambridge.

Slack JM, Beck CW, Gargioli C, Christen B (2004). Cellular and molecular mechanisms of regeneration in *Xenopus*. *Philos T R Soc B* 29(359):745–751.

Smith JC, Tata JR (1991). Xenopus cell lines. In: Kay BK, Peng HB (eds), *Methods in Cell Biology, Xenopus laevis: Practical Uses in Cell and Molecular Biology*, Vol.36. Academic Press, Inc, San Diego, CA.

Smolich BD, Tarkington SK, Saha MS, Stathakis DG, Grainger RM (1993). Characterization of *Xenopus laevis*γ-crystallin encoding genes. *Gene* 128:189–195.

Smolich BD, Tarkington SK, Saha MS, Grainger RM (1994). Xenopus gamma-crystallin gene expression: evidence that the gamma-crystallin gene family is transcribed in lens and non-lens tissues. *Mol Cell Biol* 14:1355–1363.

Sokol S, Christian JL, Moon RT, Melton DA (1991). Injected Wnt RNA induces a complete body axis in *Xenopus* embryos. *Cell* 67:741–752.

Sologub AA (1975). Metaplasia of eye tissues in *Xenopus laevis* in the tadpole and adult forms. *Ontogenez* 6:563–571.

Sparrow DB, Latinkic B, Mohun TJ (2000). A simplified method of generating transgenic *Xenopus*. *Nucleic Acids Res* 28:e12.

Spemann H (1901). Über Korrelationen in der Entwickelung des Auges. *Verh Anat Ges* 15:61–79.

Straznicky K, Gaze RM (1971). The growth of the retina in *Xenopus laevis*, an autoradiographic study. *J Embryol Exp Morphol* 26:67–79.

Stern CD (2005). Neural induction: old problem, new findings, yet more questions. *Development* 132:2007–2021.

Stocum DL (2006). *Regenerative Biology and Medicine*. Academic Press/Elsevier, New York and San Diego.

Sullivan CH, Braunstein L, Hazard-Leonards RM, Holen AL, Samaha F, Stephens L, Grainger RM (2004). A re-examination of lens induction in chicken embryos: in vitro studies of early tissue interactions. *Int J Dev Biol* 48:771–782.

Svistunov SA, Mitashov VI (1985). Radioautographic investigation of retinal growth in mature amphibians. *Ontogenez* 16:474–482.

Swartz M, Eberhart J, Mastick GS, Krull CE (2001). Sparking new frontiers: using in vivo electroporation for genetic manipulations. *Dev Biol* 233:13–21.

Taylor JS, Jack JL, Easter SS (1989). Is the capacity for optic nerve regeneration related to continued retinal ganglion cell production in the frog? *Eur J Neurosci* 1:626–638.

Thiébaud CH (1983). A reliable new marker in *Xenopus*. *Dev Biol* 198:245–249.

Tinsley RC, Kobel HR (1996). In: Tinsley RC, Kobel HR (eds), *The Biology of Xenopus*. Clarendon Press, Oxford, Great Britain.

Tinsley RC, Loumont C, Kobel HR (1996). Geographical distribution and ecology. In: Tinsley RC, Kobel HR (eds), *The Biology of Xenopus*. Clarendon Press, Oxford, Great Britain, pp. 379–389.

Tomarev SI, Sundin O, Banerjee-Basu S, Duncan MK, Yang JM, Piatigorsky J (1996). Chicken homeobox gene *Prox 1* related to *Drosophila prospero* is expressed in the developing lens and retina. *Dev Dyn*. 206:354–367.

Tomarev SI, Zinovieva RD, Chang B, Hawes NL (1998). Characterization of the mouse *Prox1* gene. *Biochem Biophys Res Commun*. 248:684–689.

Torres M, Gomez-Pardo E, Gruss P (1996). *Pax2* contributes to inner ear patterning and optic nerve trajectory. *Development* 122:3381–3391.

Tsonis PA, Trombley MT, Rowland T, Chandraratna RAS, Del-Rio Tsonis K (2000). Role of retinoic acid in lens regeneration. *Dev Dynam* 219:588–593.

Turner DL, Weintraub H (1994). Expression of achaete-scute homolog 3 in *Xenopus* embryos converts ectodermal cells to a neural fate. *Genes Dev* 8:1434–1447.

Underwood LW, Ide CF (1992). An autoradiographic time study during regeneration in fully differentiated *Xenopus* eyes. *J Exp Zool* 262:193–201.

Underwood LW, Nelson P, Noelke E, Ide CF (1992). Embryonic retinal ablation and post-metamorphic optic-nerve crush-effects upon the pattern of regenerated retinotectal connections. *J Exp Zool* 261:18–26.

Underwood LW, Carruth MR, Vandecaride A, Ide CF (1993). Relationship between local cell-division and cell displacement during regeneration of embryonic *Xenopus* eye fragments. *J Exp Zool* 265:165–177.

Uwanogho D, Rex M, Cartwright EJ, Pearl G, Healy C, Scotting PJ, Sharpe PT (1995). Embryonic expression of the chicken *Sox2*, *Sox3* and *Sox11* genes suggests an interactive role in neuronal development. *Mech Dev* 49:23–36.

van der Heyden MAG, Roeleveld L, Peterson J, Destrée OHJ (2001). *Connexin43* expression during Xenopus development. *Mech Dev* 108:217–220.

Waldner C, Sakamaki K, Ueno N, Turan G, Ryffel GU (2006). Transgenic *Xenopus laevis* strain expressing Cre recombinase in muscle cells. *Dev Dynam* 235:2220–2228.

Walter BE, Henry JJ (2004). Embryonic expression of pre-initiation DNA replication factors in *Xenopus laevis*. *Gene Expr Patterns* 5:81–89.

Walter BE, Tian Y, Garlisch AK, Carinato ME, Elkins MB, Wolfe AD, Schaefer JJ, Perry KJ, Henry JJ (2004). Molecular profiling: gene expression reveals discrete phases of lens induction and development in *Xenopus laevis*. *Mol Vis* 10:186–198.

Walter BE, Perry KJ, Fukui L, Malloch EL, Wever J, Henry JJ (2008). *Psf2* plays important roles in normal eye development in *Xenopus laevis*. *Mol. Vis*. (In Press).

Wetts R, Fraser SE (1989). Slow intermixing of cells during *Xenopus* embryogenesis contributes to the consistency of the blastomere fate map. *Development* 105:9–15.

Wetts R, Serbedzija GN, Fraser SE (1989). Cell lineage analysis reveals multipotent precursors in the ciliary margin of the frog retina. *Dev Biol* 136:254–263.

Weinstein DC, Hemmati-Brivanlou A (1999). Neural induction. *Annu Rev Cell Dev Biol* 15:411–433.

Weisblat DA, Sawyer RT, Stent GS (1978). Cell lineage analysis by intracellular injection of a tracer enzyme. *Science* 202:1295–1298.

Weisblat DA, Zackson SL, Blair SS, Young JD (1980). Cell lineage analysis by intracellular injection of fluorescent tracers. *Science* 209:1538–1541.

Wheeler GN, Hamilton FS, Hoppler S (2000). Inducible gene expression in transgenic *Xenopus* embryos. *Curr Biol* 10:849–852.

Wiechmann AF, Wirsig-Wiechmann CR (2003). *Color Atlas of Xenopus laevis Histology"*. Kluwer Academic Publishers, Norwell, MA.

Wigle JT, Chowdhury K, Gruss P, Oliver G (1999). *Prox1* function is crucial for mouse lens-fibre elongation. *Nat Genet* 21:318–322.

Wilson SI, Edlund T (2001). Neural induction: toward a unifying mechanism. *Nat Neurosci* 4:1161–1168.

Wolfe AD, Crimmins G, Cameron JA, Henry JJ (2004). Early regeneration genes: building a molecular profile for shared expression in cornea-lens transdifferentiation and hind-limb regeneration. *Dev Dynam* 230:615–629.

Wolfe AD, Henry JJ (2006). *Neuronal leucine-rich repeat 6 (XlNLRR-6)* is required for late lens and retina development in *Xenopus laevis*. *Dev Dynam* 235:1027–1041.

Wong YM, Chow KL (2002). Expression of zebrafish *mab21* genes marks the differentiating eye, midbrain and neural tube. *Mech Dev* 113:149–152.

Wunsh LM, Ide CF (1990). Fully differentiated *Xenopus* eye fragments regenerate to form pattern-duplicated visuo-tectal projections. *J Exp Zool* 254:192–201.

Yamada R, Mizutani-Koseki Y, Hasegawa T, Osumi N, Koseki H, Takahashi N (2003). Cell-autonomous involvement of *Mab21/1* is essential for lens placode development. *Development* 130:1759–1770.

Yan B, Moody SA (2007). The competence of *Xenopus* blastomeres to produce neural and retinal progeny is repressed by two endo-mesoderm promoting pathways. *Dev Biol* 305:103–119.

Yang EV, Bryant SV (1994). Developmental regulation of a matrix metalloproteinase during regeneration of axolotl appendages. *Dev Biol* 166:696–703.

Yang EV, Gardiner DM, Carlson MRJ, Nugas CA, Bryant SV (1999). Expression of *Mmp-9* and related matrix metalloproteinase genes during axololt limb regeneration. *Dev Dynam* 216:2–9.

Ye HQ, Azar DT (1998). Expression of gelatinases A and B, and TIMPs 1 and 2 during corneal wound healing. *Invest Ophthalmol Vis Sci* 39:913–921.

Ye HQ, Maeda M, Yu FS, Azar DT (2000). Differential expression of MT1-MMP (MMP-14) and collagenase III (MMP-13) genes in normal and wounded rat corneas. *Invest Ophthalmol Vis Sci* 41:2894–2899.

Yoshii C, Ueda Y, Okamoto M, Araki M (2007). Neural retinal regeneration in the anuran amphibian *Xenopus laevis* post-metamorphosis: transdifferentiation of retinal pigmented epithelium regenerates the neural retina. *Dev Biol* 303:45–56.

Zhou Y, Ching YP, Kok KH, Kung HF, Jin DY (2002). Post-transcriptional suppression of gene expression in *Xenopus* embryos by small interfering RNA. *Nucleic Acids Res* 30:1664–1669.

Zimmerman K, Shih J, Bars J, Collazo A, Anderson DJ (1993). *XASH-3*, a novel *Xenopus achaete-scute* homolog, provides an early marker of planar neural induction and position along the mediolateral axis of the neural plate. *Development* 119:221–232.

Zuber ME, Gestri G, Viczian AS, Barsacchi G, Harris WA (2003). Specification of the vertebrate eye by a network of eye field transcription factors. *Development* 130:5155–5167.

Zyger CA, Cook LT, Grainger RM (1998). Gene activation during early stages of lens induction in *Xenopus*. *Development* 125:3509–3519.

CHAPTER

7

The Newt as a Model for Eye Regeneration

Meagan Roddy, Panagiotis A. Tsonis
Department of Biology and Center for Tissue Regeneration and Engineering, University of Dayton, Dayton, OH 45469-2320, USA

OUTLINE

BACKGROUND

The human eye is afflicted by many diseases which can alter the perception of an image and lead to blindness. Some diseases are congenital, but many people develop diseases later in life, and must learn to cope until better treatments, or even a cure, is found for their condition. People develop these diseases for many reasons. Some major contributing factors to disease include age, obesity, and cancer.

In the age-related category, macular degeneration is a disease resulting in vision loss and blindness through two avenues. The macula is located in the middle area of the retina. Slow damage to the macula occurs over time, as the light sensitive cells of the macula begin to breakdown, causing sharpness of image perception to decline. More rapid damage develops as blood vessels begin to grow and spread under the macula, lose fluids in the form of blood as they go. The increase in fluidity in this region of the eye causes a shift in the retina, and the movement of these vital parts within the eye. This movement also causes the loss of sharp visual acuity. Luckily, there are routine treatments available in the form of laser surgery, as well as pharmaceutical treatments of drugs which either destroy the developing blood vessels in the eye or cause a block in the action of native growth factors involved in the production of these blood vessels. Currently, there is no successful treatment for the slow form of this condition, which is caused by the breakdown of the macular cells (Pauleikhoff and Koch, 1995).

Also in the age-related category is the eye condition known as cataract. A cataract is clouding of the lens, which makes the retina unable to process a clear image. Problems with several proteins within the lens lead to a cloud, resulting in blurred vision, as well as a loss of color distinction. Cataracts can occur in a single eye, or affect both, and there are many different kinds of cataracts, some of which or due to injury and diseases, however, mainly the cause is increased age. Cataracts can be treated through surgeries where after capsulorrhexis the lens is removed, but the lens capsule stays in place to accept an artificial lens. A major issue resulting from these surgeries is the formation of secondary cataracts. Even in the case of artificial lens implantation, secondary cataracts result from the presence of lens epithelial cells which remain on the capsule. These cells differentiate to mesenchymal cells and cause the re-clouding of the eye, or secondary cataracts. This then results in more surgery and reoccurrence of surgical complications (Livingston *et al.*, 1995; Schein *et al.*, 1994; Liu *et al.*, 1996).

These examples of common diseases of the eye are routinely treated with the methods described above. However, due to the importance of vision and the complicated nature of the eye, new treatments are constantly being explored. Gene replacement therapy holds promise in early prevention and treatment, particularly in cases of children born with congenital birth defects of the eye. Candidate genes are identified through many resources, particularly through the genetic studies on animal models. Once a population of candidate genes has been identified, gene therapy, like the delivery of healthy genes where variations on a genetic level have caused eye defect, may change the course of defect in a healthy direction. Animal models hold much promise for clinical trials in gene therapy to commence, leading to novel treatments for vision loss and blindness. An example of this type of research is the identification of RPE65, a gene in the retina, which when mutated, causes a congenital disease in children, which leads to severe vision loss and blindness. Clinical trials are currently exploring the delivery of healthy RPE65 to the retinas in these patients, with the hope of restoring vision (Bainbridge *et al.*, 2006).

Stem cell therapy is also a promising method for disease treatment currently being explored. It has been shown that transplantation of healthy retinal cells to the diseased retina slows the onset of retinal diseases. Moreover, current studies in animal models are showing that the transplantation of neural stem cells and stem cells collected from the umbilical cord after birth, can in fact rescue vision loss and blindness in models exhibiting diseases like macular degenerations. This most recent set of studies not only provides means of treatment, it may hold the cure for vision loss and blindness caused by diseases of the retina and lens all together (MacLaren *et al.*, 2006; Bernardos *et al.*, 2007).

The models described above certainly hold much promise in the race for treatments and cures for eye diseases. They are not, however, without complications. Stem cells are subject to much ethical debate, and from a practical standpoint, transplanted cells can also be rejected by the host. A third, and perhaps most exiting model for treatment of disease, is the idea of regeneration of one's own tissue to repair damages leading to vision loss and blindness. Instead of going to an outside source and possible delivery to the injured area, regeneration offers treatment at the source and using one's own resources. In some circles, regeneration can be defined as a "new beginning" and for humans who lose a limb or suffer vision loss regeneration may be the key to the door of healing. A wide variety of organisms can regenerate, however, most can only undergo this process during a short part of the developmental stage of life. Mammals, like humans, can regenerate a variety of parts; hair, skin, and a portion of the liver. One animal model can regenerate virtually anything, and more importantly throughout adulthood, is the newt. The newt is an indispensable model for the study of the eye for this reason. Two main components of the eye are readily studied in the newt, those being the lens and the retina.

Retina and lens regeneration occurs in some species of frog, but here again, only through early development (Freeman, 1963; Sanchez Alvarado and Tsonis, 2006). As stated above, (some species of) the newt offers a unique and interesting opportunity to study regeneration of the eye from larvae into adulthood stages (Fig. 7.1). The complex mechanisms of regeneration in the newt involve processes of transdifferentiation, which will be described in more detail later in the chapter. It is necessary to also mention that although the newt is a very valuable model for regeneration research, it is also a challenging model due to slow progress in genetic and molecular biology research and the scarce availability of resources (Del Rio-Tsonis and Tsonis, 2003). As these resources become available more and more information about regeneration will be gathered, which might result in breakthroughs in the regeneration fields.

RETINA REGENERATION

Retina regeneration can occur in a wide variety of organisms, from fish through mammals (Sanchez Alvarado and Tsonis, 2006). The process of regeneration in most of these species seems to occur via

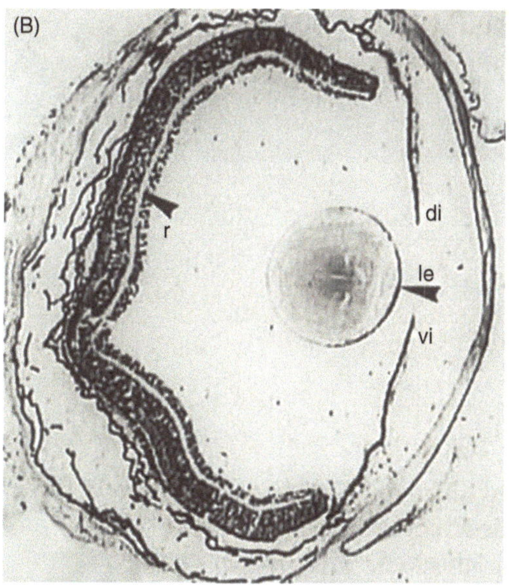

FIGURE 7.1 The research model. (A) *Notophthalmus viridescens*, the American adult newt, used as a model for eye regeneration. The adult newt can regenerate among many other tissues, both the lens and retina. (B) A histological section through the newt eye, showing the dorsal iris (di), the ventral iris (vi), the lens (l), the lens epithelium (le), and the retina (r).

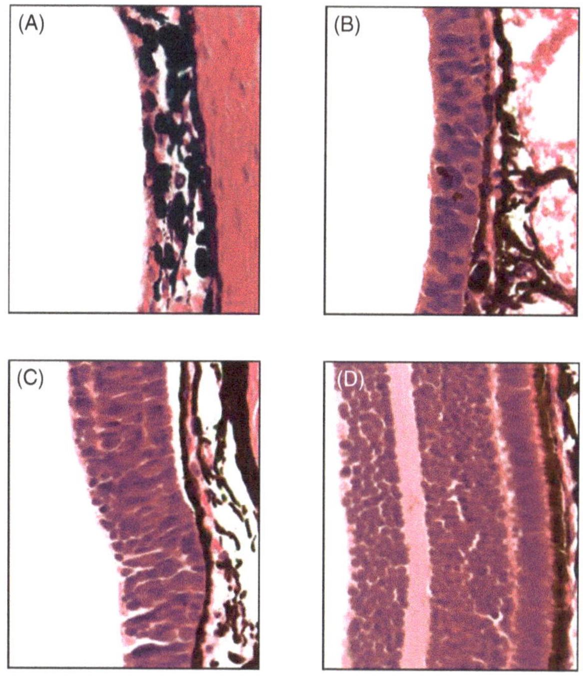

FIGURE 7.2 Retina regeneration in the adult newt. (A) Five days after retinectomy, the retinal PECs begin to dedifferentiated and proliferate. (B) Around 14 days, the neuroepithelial layer has formed and those cells will differentiate into all the cells of the retina. (C) Thirty days into regeneration, the retina has reformed. (D) At 45 days, the fully differentiated layered retina has formed. Courtesy of Dr. K. Del Rio-Tsonis.

transdifferentiation of the pigmented epithelial cells (PECs) of the retina in the early embryo or larvae. Transdifferentiation can be defined as a process where a cell changes its distinctiveness and becomes a divergent cell type. Retinal regeneration can also occur in some organisms by the differentiation of precursor cells located in specific areas of the retina, like the ciliary margin zone (CMZ). A precursor cell is a stem cell with a distinct lineage, meaning a cell can only become one type of a small assortment of cells in a lineage. For example, the retinal precursor cells can only differentiate into the cells of the retina.

The newt regenerates retina by means of transdifferentiation of retinal pigmented epithelial cells (also known as rPECs or RPEs), which lose their pigmentation and detach themselves from the basement membrane, both characteristics of their terminally differentiated state, and proceed to re-enter the cell cycle. They then go on to form a neuroepithelial cell layer, a precursor cell-like state, and further differentiate into all cell types of the retina. Retinal neurons and glial cells are produced and these go on to create the functional neural retina. As well as undergoing this process, the rPECs also divide their own population, replenishing themselves.

Figure 7.2 serves to guide the readers through the regeneration process. Within 5 days after the removal of the newt retina, dedifferentiation and proliferation of the rPECs commences. At around 14 days after removal of the retina, the neuroepithelial layer has formed and that layer gives rise to the cells which make up the regenerating retina. Thirty days into the regeneration event, the cells of the neuroepithelial layer have differentiated into the layers of the retina, which include the outer nuclear layer, the inner nuclear layer, the ganglion cell layer, as well as the renewed pigmented epithelium (Tsonis, 2000; Tsonis, 2002; Tsonis and Del Rio-Tsonis, 2004).

Gene Regulation

Genes and signaling pathways have been investigated through various molecular tools available for the newt. Retinal regeneration is being dissected in order to examine what specific genes play important roles in regeneration, inhibition of regeneration and alteration of the process. One such protein is RPE65, a membrane-bound protein found in high amounts in the RPE cells. RPE65 is not detected in retina development during embryogenesis, but it can be found in terminally differentiated RPE cells. This protein seems to be downregulated during the early regeneration process (Fig. 7.3). Pax-6, however, a gene involved in early eye development during embryogenesis, seems to be upregulated during some of the stages of retinal regeneration. Pax-6 expression level seems to increase from day 10 after retinectomy. Starting a little later in the regeneration event is Cx43, a gene which encodes for connexin, a protein subunit of gap junctions. Also expressed during the latter stages of retinal regeneration is Notch-1. Studies have shown that Notch signaling plays an important role in neurogenesis during retinal regeneration as well as later neuronal regulation during this process (Chiba *et al.*, 2006; Nakamura and Chiba, 2007).

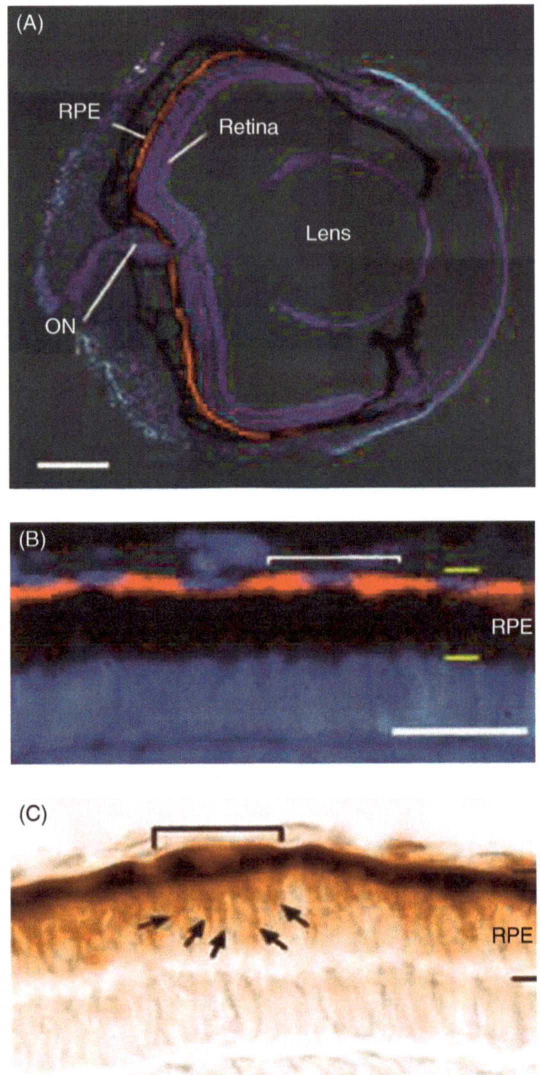

FIGURE 7.3 Regulation during retina regeneration. (A) Immunofluorescence of the eye acknowledging different components of the eye. Retinal epithelial cells, in red, are the cells responsible for retina regeneration. (B) RPE layer in higher magnification. Yellow brackets represent the width of one cell and the white brackets represent the length of one cell. (C) Location of RPE65, a protein found heavily expressed in the retina during regeneration. Courtesy of Dr. C. Chiba.

LENS REGENERATION

Lens regeneration also involves transdifferentiation, of the pigmented epithelial cells of the iris, or PECs. The PECs re-enter the cell cycle, dedifferentiate, and lose their characteristic pigmentation. *In vivo*, regeneration occurs from the dorsal iris population of PECs. Despite the apparent similarity between the dorsal and ventral PECs, the ability to regenerate through transdifferentiation under normal conditions, belongs exclusively to the dorsal PECs. Interestingly, however, culturing of both dorsal and ventral PECs results in lentoid body formation (Tsonis and Del Rio-Tsonis, 2004; Tsonis *et al.*, 2004a). This means that the ventral PECs have the potential for transdifferentiation, but such a potential is not allowed *in vivo*.

Early events of dedifferentiation occur through day 8 of the regeneration process. By day 10 of the process, a lens vesicle is formed, which is a kind of precursor to what will become a full lens later in the process. After the lens vesicle has formed, the posterior cells start to elongate, express crystallins and differentiate to lens fibers. The anterior cells become lens epithelium (Fig. 7.4) (Eguchi, 1963; Eguchi, 1964).

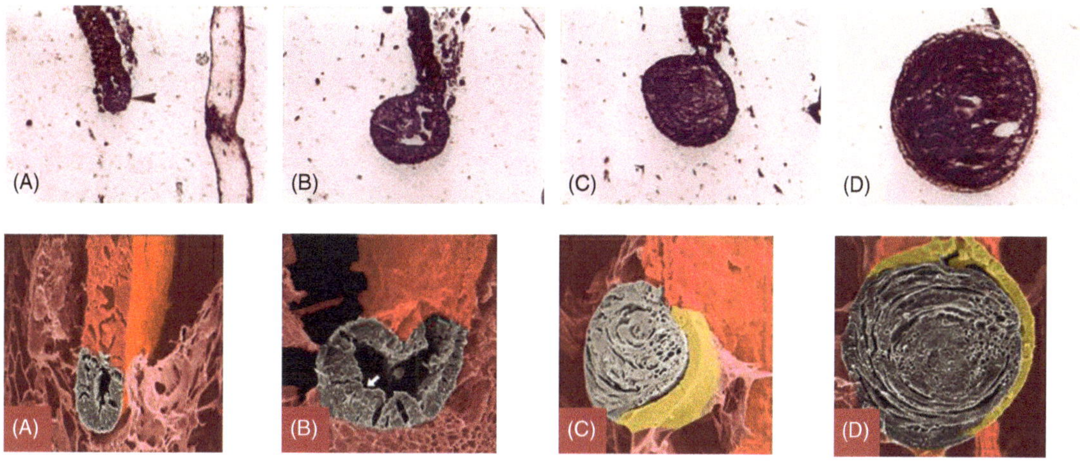

FIGURE 7.4 Lens formation from the dorsal iris during regeneration. Top, events of lens regeneration shown through immunohistochemisty. Bottom, same events shown through scanning electron microscopy (SEM). (A) Day 10 after lentectomy, a lens vesicle forms at dorsal margin (arrowhead). (B) Day 14, elongation and further lens development as differentiation of lens fibers occur at the posterior part (arrow) of the vesicle. (C) Day 20 of lens reformation. (D) Day 30, regeneration has concluded. (A–C) top is showing FGF1 expression, (D) top shows the presence of γ-crystallin.

Gene Regulation

Focusing first on the genetic regulation, it would only make sense to hypothesize that different genes are at play in the dorsal and ventral iris creating two very different results, a side which regenerates and a side which does not. There are of course, many genes and proteins to consider in the regeneration process. Widely considered the master regulator gene in eye development, *pax-6* is found in both the dorsal and ventral iris during lens regeneration (Del Rio-Tsonis *et al.*, 1995). Development of knockdown technology in the adult newt was quite informative on the role of *pax-6*. When *pax-6* expression is decreased in the eye, lens regeneration suffers dramatically due to the decreases in proliferation of the PECs and when *pax-6* is knocked down during later regeneration events, crystallins are not made and lens fibers production is decreased (Madhavan *et al.*, 2006). Crystallin identification is important in determining the stages of regeneration after the formation of the lens vesicle (Madhavan *et al.*, 2006) (Fig. 7.5). *Six-3*, as well as *pax-6*, are genes known to be involved in eye development. *Six-3* is shown to increase drastically in the dorsal iris during normal regeneration. Members of the hedgehog signaling pathway are also expressed through the regeneration process; Sonic hedgehog (Shh) and Indian hedgehog (Ihh) only being expressed in the regenerating and developing lens, and no longer expressed once the lens is intact again. When one interferes with this signaling pathway, the overall regeneration process is inhibited. Also, cell proliferation rates are decreased and differentiation in the regenerating lens suffers (Ekker *et al.*, 1995; Tsonis *et al.*, 2004b). Retinoic acid receptors are also expressed in the lens during regeneration and their inhibition might account for aberrant lens formation (Tsonis *et al.*, 2000; Tsonis *et al.*, 2002).

Other signaling pathways have also provided interesting results. Exogenous fibroblast growth factors (FGFs) have been shown to elicit regeneration of a second lens from the dorsal iris (Del Rio-Tsonis *et al.*, 1997). It has been suggested that this action of FGFs is mediated via an induction of cell proliferation. The FGF pathway seems to collaborate with the Wnt pathway. When explants of dorsal iris were treated with FGF2 in the presence of Wnt inhibitors, the action of FGF2 was inhibited. Combined addition of FGF2 and Wnt3a was in fact able to induce lens transdifferentiation of ventral explant as well (Hayashi *et al.*, 2004; Hayashi *et al.*, 2006).

Interestingly, research on another signaling pathway, the bone morphogenetic protein (BMP) pathway, revealed that its inhibition allows the ventral iris to transdifferentiate into lens. Importantly, such induction was also observed when ventral PECs were transfected with *six-3* and treated with retinoic acid (Grogg *et al.*, 2005). These manipulation of signaling pathways and regulatory genes clearly indicates that induction of the ventral iris is possible and thus such observation might open new avenues in experimenting with higher animals that are unable to regenerate eye tissues. We should mention here that in addition to induction these important players were found to be

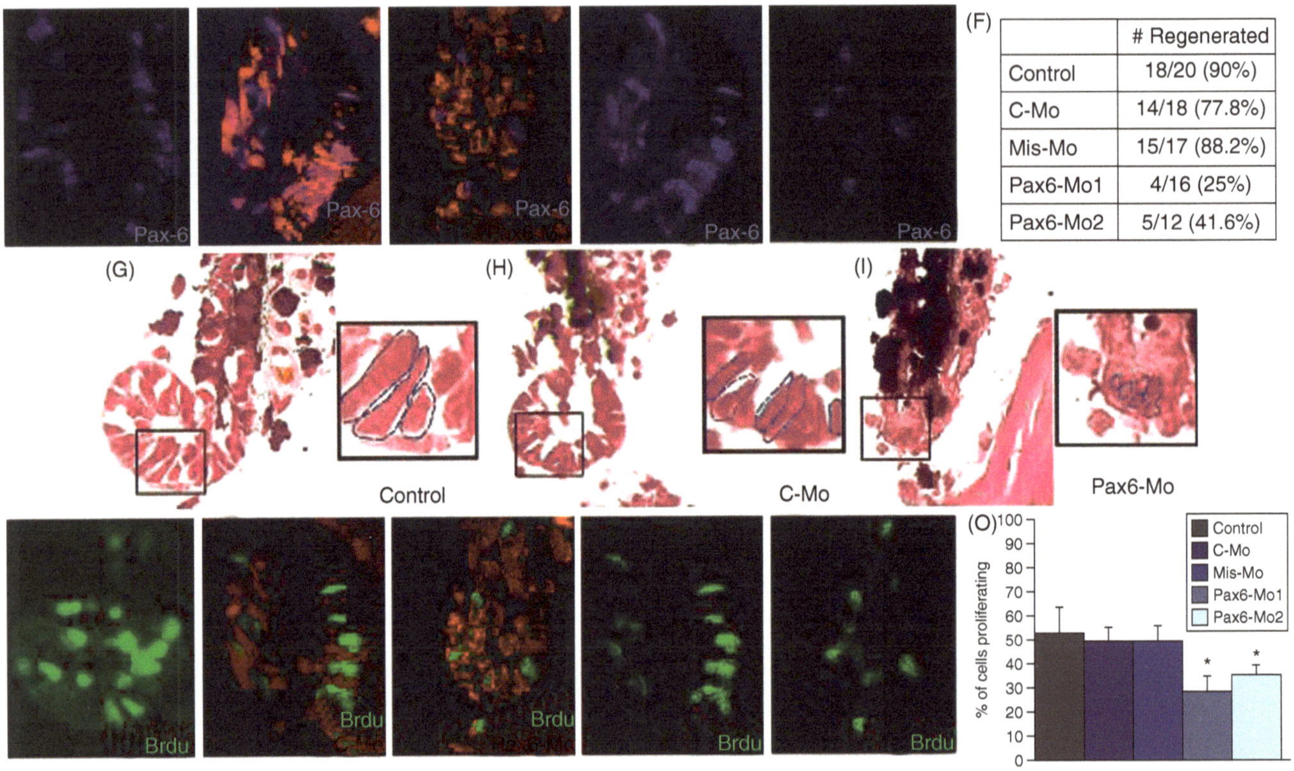

FIGURE 7.5 Pax-6 regulation of lens regeneration. Pax-6 is shown in blue and the Pax-6 knockdown probe, a morpholino, is shown in red. Animals were treated with morpholinos at 10 days after lentectomy and collected at day 13 after lentectomy. Untreated control eyes (a) and eyes treated with control morpholino, C-Mo, (B) express similar levels of Pax-6 (blue) in the regenerating lens vesicle. (C) Treatment with *Pax6*-Mo1 reduces expression of Pax-6. (D and E) Pax-6 staining only from (B) and (C), respectively. (F–I) Animals were injected with morpholinos at day 4 and day 10 after lentectomy and collected at 15 days after lentectomy. (F) Animals treated with Pax-6 morpholinos showed a marked reduction in lens vesicle formation compared with untreated animals or animals injected with the C-Mo or Mis-Mo. (G and H). The lens vesicle from untreated animals and animals treated with C-Mo have elongating cells (Inserts). The lens vesicle fails to form in animals treated with *Pax6*-Mo1; however, dedifferentiation at the dorsal tip does occur (I). (J–N) Animals were injected with morpholinos at 10 day after lentectomy and collected at 13 days after lentectomy. Lens vesicles of untreated control eyes (J) and of eyes treated with C-Mo (K) have a higher number of BrdU-positive cells compared with eyes treated with *Pax6*-Mo1 (L). (M and N) BrdU staining of (K) and (L), respectively. (O) Morpholino knockdown of Pax-6 results in a statistically significant reduction in proliferation. The number of proliferating cells is shown as a percentage of the total number of cells in each group. All error bars are SEM. Asterisks indicate statistical significance of $p < 0.01$ when compared with animals treated with C-Mo or Mis-Mo. Sections shown in (B) and (K) and in (C) and (L) are identical and were stained for Pax-6 and BrdU, but the results are presented separately.

present in both dorsal and ventral iris. This was kind of unexpected and it means that the ventral iris does initially undergo similar events as the dorsal iris, but maybe there is a general inhibitory effect later.

To receive more insights about gene expression patterns, a microarray analysis with newt cDNA was utilized and it revealed that gene signatures in the dorsal and ventral iris are in fact very similar. Even genes which are responsible for tissue remodeling, like collagenase and cathepsin are present and upregulated in both dorsal and ventral irises during regeneration. Surprisingly expression levels for some of these genes are even higher in the ventral iris. It is not unreasonable to suggest given these unexpected results on gene expression that we might encounter novel regulatory events during newt regenerative processes (Makarev *et al.*, 2007). Expressed sequence tag (EST) analysis is underway and it will provide more valuable information about patterns of gene expression (Maki *et al.*, unpublished).

MicroRNAs

MicroRNAs, or miRNA are short RNAs, about 22 nucleotides long, which can bind to complementary sequences of RNA and subsequently block mRNA translation. miRNAs have multiple binding sequences

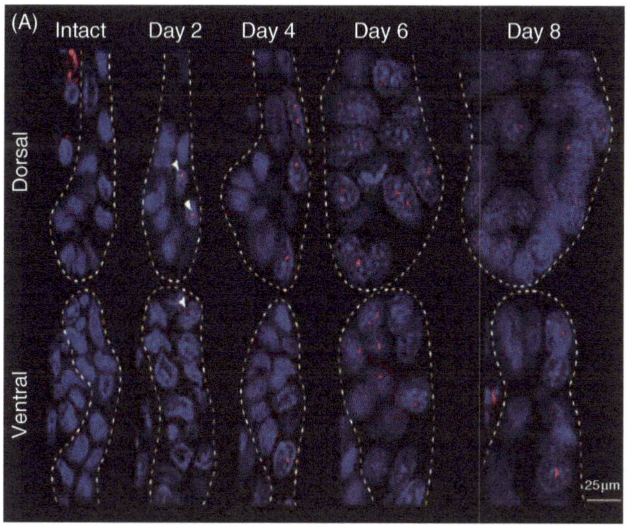

FIGURE 7.6 Nuclear regulation of regeneration (A) Nucleostemin accumulation in the nucleus of the dorsal and ventral PECs (shown in pink). The entire nuclei have been counterstained with DAPI (blue). (From Maki *et al.*, 2007; courtesy of Dr. N. Maki).

due to the short size and can block translation on a wide scale. As such they might be involved in the transition from one cell type to another as we see during regeneration. Cloning of newt miRNA from the eye has pinpointed differential regulation in both dorsal and ventral iris. Some of the targets of the cloned miRNAs were predicted to be *FGFR2*, and *SOX9* (for miR-124a), *PAX3*, *chordin*, and *TGFβR1* for *let7b* (Makarev *et al.*, 2006). Based on this, further study of the role of miRNA regulation in regeneration showed that members of the *let7* family were found to be downregulated in the regenerating dorsal iris. Examination of the ventral iris revealed that miR-148 is upregulated in both intact and regenerating ventral iris when compared with dorsal counterparts (Tsonis *et al.*, 2007). Thus, miRNAs might be very useful regulators of the regenerative processes.

Transdifferentiation in Newts: A Model for Stem Cell Differentiation?

One could argue that during the process of lens regeneration the dedifferentiated cells become stem cell-like, going from a terminally differentiated state to an undifferentiated state and can then further differentiate into the cells needed to create a new lens. Such a hypothesis was first proposed by us in 2000, using the example of mesenchymal stem cell similarity to limb blastema cells (Tsonis, 2000; Tsonis, 2004). Research in this area is now pursued but some initial studies support this hypothesis. For example, the stem cell nuclear protein nucleostemin is found highly expressed in the nucleus of undifferentiated cells, like pluripotent embryonic stem cells, as well as multipotent stem cells of the nervous system and primitive cells of the bone marrow (Maki *et al.*, 2007) (Fig. 7.6). As the onset of differentiation of these cells occurs, accumulation of nucleostemin within the nucleus is shown to decrease. When dedifferentiated PECs were studied during regeneration as compared to non-regenerating PECs, the stem cell nuclear protein was found to be highly expressed in both the dorsal and ventral regenerating PECs, although the amount of protein present decreased faster in the ventral PECs, and after differentiation the amount of nucleostemin decreased (Maki *et al.*, 2007). This type of regulation goes a step further in defining dedifferentiated PECs as molecularly similar to stem cells, although more research must be done in order to obtain definitive answers.

Immunity and Regulation

Proteins of the cell's membrane, serum and of a regulatory nature, as well as receptors and different cell types which fall into the immunology category surprisingly seems to play an important role in regeneration. The complement system is a major part of the host's innate immune system. It employs a wide range of purpose; from lysis of foreign invaders to assisting in the recruitment of immune-specific cells and triggering cell functions. The complement system is made up of a cascade of proteins, where one protein is activated, and triggers the activation of another protein and so on. It is made up of two intertwining cascades referred to as the classical pathway and the alternative pathway. For immunity purposes, the pathway results in the classical antibody–antigen binding, this then creates the onset of a full blown immune attack. Aside from their respective duties involved with innate immunity, C3 and C5 proteins have been shown to be involved in certain cell proliferation. The fragments which C3 and C5 further divide into during the complement cascade, C3→C3a and C3b, C5→C5a and C5b, also provide assistance in processes like differentiation, apoptosis, and cell activation. However, it has been shown that these two members of the complement cascade are present in the lens of the eye during the regeneration process. Protein studies have shown that in an intact lens, both proteins are absent. However, during regeneration of the lens, both proteins turn up in the regenerating lens, C3 found earlier in the dedifferentiated PECs and C5 found in the lens vesicle a little later in the regeneration event. It was also shown that these cells synthesize the mRNA for these proteins (Kimura *et al.*, 2003).

Also interesting is the recruitment of other cell types to the inner eye before regeneration can commence. Macrophages, in particular, play a specific role in the breakdown and removal of a damaged lens in order for the onset of regeneration to begin. Dendritic cells also move into the lens area from surrounding areas of the eye to play an important role in the engulfment of the remaining parts of a digested lens for further removal. One study goes so far as to show that collecting these dendritic cells after lens engulfment, and subsequent transplantation into the normal eye of the newt, induced the onset of lens regeneration and resulted in a double lens formation from the dorsal iris (Kanao and Miyachi, 2006).

The Promise of the Newt

Being the only vertebrate that can regenerate body parts, the newt is naturally quite promising in the field of regenerative biology. We must answer the fundamental questions pertaining to such amazing capabilities. We should utilize the newt to learn and also to compare with other strategies, such as stem cells. Only then we can hope to achieve the goals set by regenerative biology and medicine.

REFERENCES

Bainbridge JW, Tan MH, Ali RR (2006). Gene therapy progress and prospects: the eye. *Gene Ther* 16:1191–1197.

Bernardos RL, Barthel LK, Meyers JR, Raymond PA (2007). Late-stage neuronal progenitors in the retina are radial Muller glia that function as retinal stem cells. *J Neurosci* 27(26):7028–7040.

Chiba C, Hoshino A, Nakamura K, Susaki K, Yamano Y, Kaneko Y, Kuwata O, Maruo F, Saito T (2006). Visual cycle protein RPE65 persists in new retinal cells during retinal regeneration of adult newt. *J Comp Neurol*(495):391–407.

Del Rio-Tsonis K, Tsonis PA (2003). Eye regeneration at the molecular age. *Dev Dyn* 226(2):211–224.

Del Rio-Tsonis K, Washabaugh CH, Tsonis PA (1995). Expression of pax-6 during urodele eye development and lens regeneration. *Proc Natl Acad Sci USA* 92:5092–5096.

Eguchi G (1963). Electron microscopic studies on lens regeneration. I. Mechanisms of depigmentation of the iris. *Embryologia* 8:45–62.

Eguchi G (1964). Electron microscopic studies on lens regeneration. I. Formation and growth of lens vesicle and differentiation of lens fibers. *Embryologia* 8:247–287.

Ekker SC, Ungar AR, Greenstein P, von Kessler DP, Porter JA, Moon RT, Beachy PA (1995). Patterning activities of vertebrate hedgehog proteins in the developing eye and brain. *Curr Biol* 5(8):944–955.

Freeman GJ (1963). Lens regeneration from the cornea in Xenopus laevis. *Exp Zool* (154):39–66.

Grogg MW, Call MK, Okamoto M, Vergara MN, Del Rio-Tsonis K, Tsonis PA (2005). BMP inhibition-driven regulation of six-3 underlies induction of newt lens regeneration. *Nature* 438(7069):858–862.

Hayashi T, Mizuno N, Owaribe K, Kuroiwa A, Okamoto M, Kondoh H (2004). FGF2 triggers iris-derived lens regeneration in newt eye. *Mech Dev* (121):519–526.

Hayashi T, Mizuno N, Takada R, Takada S, Kondoh H (2006). Determinative role of Wnt signals in dorsal iris-derived lens regeneration in newt eye. *Mech Dev* (123):793–800.

Kanao T, Miyachi Y (2006). Lymphangiogenesis promotes lens destruction and subsequent lens regeneration in the newt eyeball, and both processes can be accelerated by transplantation of dendritic cells. *Dev Biol* 290(1):118–124.

Kimura Y, Madhavan M, Call MK, Santiago W, Tsonis PA, Lambris JD, Del Rio-Tsonis K (2003). Expression of complement 3 and complement 5 in newt limb and lens regeneration. *J Immunol* 170(5):2331–2339.

Liu CS, Wormstone IM, Duncan G, Marcantonio JM, Webb SF, Davies PD (1996). A study of human lens cells *in vitro*. A model for posterior capsule opacification. *Invest Ophthalmol Vis Sci* 37:906–914.

Livingston PM, Carson CA, Taylor HR (1995). The epidemiology of cataract: a review of the literature. *Ophthalmic Epidemiol* 3:151–164.

MacLaren RE, Pearson RA, MacNeil A, Douglas RH, Salt TE, Akimoto M, Swaroop A, Sowden JC, Ali RR (2006). Retinal repair by transplantation of photoreceptor precursors. *Nature* 444(7116):203–207.

Madhavan M, Haynes TL, Frisch NC, Call MK, Minich CM, Tsonis PA, Del Rio-Tsonis K (2006). The role of Pax-6 in lens regeneration. *Proc Natl Acad Sci USA* 103(40):14848–14853.

Makarev E, Spence JR, Del Rio-Tsonis K, Tsonis PA (2006). Identification of microRNAs and other small RNAs from the adult newt eye. *Mol Vis* 12:1386–1391.

Makarev E, Call MK, Grogg MW, Atkinson DL, Milash B, Odelberg SJ, Tsonis PA (2007). Gene expression signatures in the newt irises during lens regeneration. *FEBS Lett* 581(9):1865–1870.

Maki N, Takechi K, Sano S, Tarui H, Sasai Y, Agata K (2007). Rapid accumulation of nucleostemin in nucleolus during newt regeneration. *Dev Dyn* 236:941–950.

Nakamura K, Chiba C (2007). Evidence for Notch signaling involvement in retinal regeneration of adult newt. *Brain Res* 113:628–642.

Pauleikhoff D, Koch JM (1995). Prevalence of age-related macular degeneration. *Curr Opin Ophthalmol* 6(3):51–56.

Sanchez Alvarado A, Tsonis PA (2006). Bridging the regeneration gap: genetic insights from diverse animal models. *Nat Rev Genet* 7(11):873–884.

Schein OD, Steinberg EP, Javitt JC, Cassard SD, Tielsch JM, Steinnwachs DM, Legro MW, Diener-West M, Sommer A (1994). Variation in cataract surgery practice and clinical outcomes. *Ophthalmology* (6):1142–1152.

Tsonis PA (2000). Regeneration in vertebrates. *Dev Biol* 221(2): 273–284.

Tsonis PA (2002). Regenerative biology: the emerging field of tissue repair and restoration. *Differentiation* 70(8):397–409.

Tsonis PA (2004). Stem cells from differentiated cells. *Mol Interv* 4(2):81–83.

Tsonis PA, Del Rio-Tsonis K (2004). Lens and retina regeneration: transdifferentiation, stem cells and clinical applications. *Exp Eye Res* 78(2):161–172.

Tsonis PA, Trombley MT, Rowland T, Chandraratna RA, Del Rio-Tsonis K (2000). Role of retinoic acid in lens regeneration. *Dev Dyn* 219(4):588–593.

Tsonis PA, Tsavaris M, Call MK, Chandraratna RA, Del Rio-Tsonis K (2002). Expression and role of retinoic acid receptor alpha in lens regeneration. *Dev Growth Differ* 44(5):391–394.

Tsonis PA, Madhavan M, Tancous EE, Del Rio-Tsonis K (2004a). A newt's eye view of lens regeneration. *Int J Dev Biol* 48(8-9):975–980.

Tsonis PA, Vergara MN, Spence JR, Madhavan M, Kramer EL, Call MK, Santiago WG, Vallance JE, Robbins DJ, Del Rio-Tsonis K (2004b). A novel role of the hedgehog pathway in lens regeneration. *Dev Biol* 267(2):450–461.

Tsonis PA, Call MK, Grogg MW, Sartor MA, Forge A, Fyffe R, Goldenberg R, Cowper-Sallari R, Tomlinson CR (2007). MicroRNAs and regenerations: let-7 as potential regulators of dedifferentiation in lens and inner ear hair cell regeneration of the adult newt. *Biochem Biophys Res Commun* 362(4):940–945.

CHAPTER

8

The Chick as a Model for Retina Development and Regeneration

Teri L. Belecky-Adams[1], *Tracy Haynes*[2], *Jonathan M. Wilson*[1], *Katia Del Rio-Tsonis*[2]

[1]Department of Biology and Center for Regenerative Biology and Medicine, Indiana University-Purdue University Indianapolis, Indianapolis, IN 46202, USA

[2]Department of Zoology, Miami University, Oxford, OH, USA

OUTLINE

"Our real teacher has been and still is the embryo, who is, incidentally, the only teacher who is always right." Viktor Hamburger

THE CHICK EMBRYO AS A MODEL SYSTEM

Introduction

The ancient Egyptians and Greeks were history's first recorded embryologists, both of which used the chick as a model system to understand how human development occurred. In 343 BC (historians' best estimation), Aristotle studied the chick embryo as a means to discover secrets of the formation of life. In Book II of *Generation of Animals*, Aristotle highlights the chick as the greatest model with which to study eye development (Aristotle, 343 BC), and later, in Book VI, he gives a gross anatomical description of the entire developmental process of the chick embryo (Aristotle, 343; 350 BC). Understandably, Aristotle takes note of one of the most noticeable and prominent features of the chick embryo – the eyes, telling the readers that the eyes are "swollen out to a great extent" and that "this condition of the eyes lasts on for a good while" (Aristotle, 343 BC). On the 10th day of development, he said, "the head is still larger than the rest of its body, and the eyes larger than the head, but still devoid of vision. The eyes, if removed about this time, are found to be larger than beans, and black; if the cuticle be peeled off there is a white and cold liquid inside, quite glittering in the sunlight" (Aristotle, 350 BC).

As vague as Aristotle's descriptions were, no observations of the chick eye surpassed those of the great teacher and philosopher until almost 2,000 years later. All of the gross anatomical features of eye development were first discovered in the chick, including two hallmarks of eye development; the choroid fissure which was first described by Marcello Malpighi in 1672, and much later, the evagination of the optic vesicle from the neural tube which was described by Christian Pander in 1817 (Adelmann, 1966). Descriptions of the embryonic retina were first recorded by Antoine Maitre-Jan in 1722, who said at the 9th day of development, it "is white and has the consistency of a coagulum" (Adelmann, 1966).

These and many other investigators using the chick as a model system paved the way for the late 19th and 20th century research to bring basic research to where it is today. Within this chapter, the reader will find a discussion of the advantages of the chick embryo as a model system for eye research, both in development and regeneration research. There is also a discussion of the techniques that have been used extensively with the chick embryo in the past, as well as new advances that will propel the use of the chick in eye research far into the future.

The Advantages of the Chick Embryo

Chick embryos are wonderful to work with in a variety of aspects. The following are general points that make chick embryos such a useful model system. (1) The eggs are a cheap and readily available source of material that is available year-round from a local or regional supplier. In the day and age of the transgenic mouse, it has become an issue to find systems that can be used, either as alternative vertebrate models or models to be used in conjunction with more expensive model systems, to define the functions of various genes. In comparison to the mouse, the chick is very inexpensive and has very little cost associated with housing. This has led to studies using the chick embryo as a high-throughput tool in which genes and reporter constructs driven by untranslated genomic sequences are introduced into the embryo as an initial determination of gene function, necessary cis-acting regions, etc. (Timmer *et al.*, 2001; Nakamura *et al.*, 2004; Uchikawa *et al.*, 2004). (2) Chicks undergo a series of successive and reproducible changes during development that have been well documented by several embryologists, most notably Malpighi, Lillie, Huxley, and Hamburger and Hamilton (Malpighi, 1672; 1675; Lillie, 1908; Huxley, 1934; Hamburger and Hamilton, 1951). This is a critical issue primarily because investigators would like to be able to manipulate embryos at specific stages, so a time scale of approximately when embryos incubated at a specific temperature will become a particular stage is necessary. In addition, a large number of eggs can be incubated at one time in order to obtain embryos that are at the desired stage. (3) *In ovo* embryonic studies are more easily accomplished than *in vivo* mammalian embryonic studies. For instance, experiments in which dividing cells are labeled in the chick embryo do not have to deal with the label, tritiated thymidine or bromodeoxyurdine (BrdU), being diluted by the maternal vascular system. (4) For many tissues, including the eye, the tissue is easily accessible for various manipulations. Using some very cheap and readily available instruments, windows in the eggs can be opened, revealing the embryo and creating space for the insertion of instruments for surgical manipulations, etc. (5) Many experimental methods have been well established to study the chick eye, including retinal, lens and retinal pigmented epithelial (RPE) cultures, retinal wholemount *in situ* hybridization and immunohistology, *in ovo* electroporation, and expression of genes via retroviral infection (Belecky-Adams *et al.*, 1996; 1997; 1999; 2001; 2002; Weng *et al.*, 1998; Adler *et al.*, 2001; 2002; Sehgal *et al.*, 2006; Wilson *et al.*, 2007). There will be a discussion of some of these techniques later in the "Toolbox" section of this chapter. (6) The period over which the eye develops is relatively short and occurs entirely within the embryonic period of development. The short period over which the retina develops is a significant advantage when considering functional studies with genes of interest. In addition, it is also an advantage that the majority of differentiation within the retina occurs embryonically (Fujita and Horii, 1963; Prada *et al.*, 1991), hence investigators do not have the added stress that birth places on the developing systems to complicate analysis. (7) Chick embryonic eyes are enormous! (Fig. 8.1(A) and (B)). This can be a substantial advantage when considering techniques such as single cell or explant cultures, due to the availability of large amounts of tissue. (8) The chicken genome is available (Wallis *et al.*, 2004) and methods for making the chick embryo more accessible to genetic manipulations are being quickly developed. This may be of interest to investigators for a variety of reasons, including comparative analyses of various homologs or orthologs in other species, the study of gene organization and regulation, and the study of the evolution of genes, gene families, and signaling pathways. (9) The retina can regenerate during early development (Coulombre and Coulombre, 1965; Park and Hollenberg, 1989; 1991; Spence *et al.*, 2004; 2007a,b) (Fig. 8.2). This is a substantial advantage if one wishes to study how the nervous

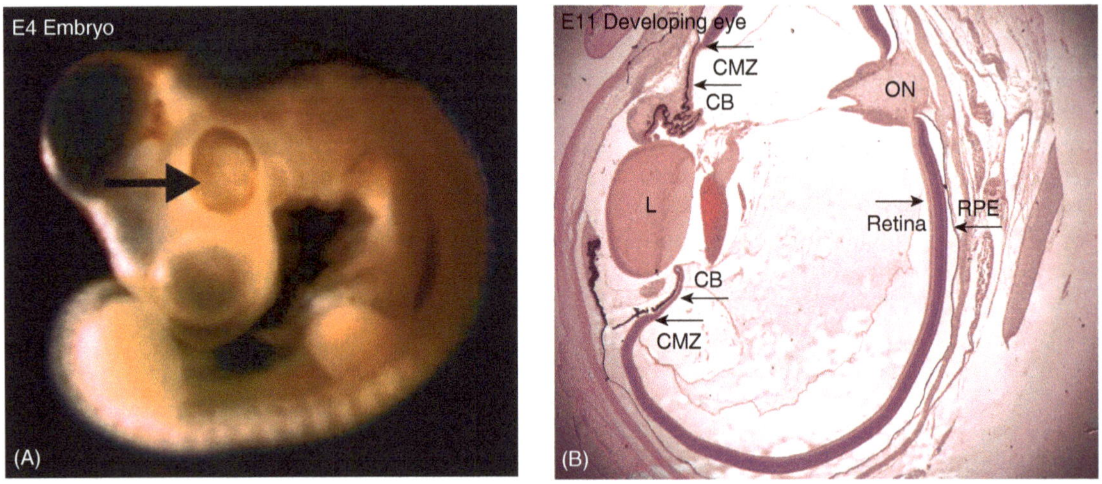

FIGURE 8.1 (A) A photograph of a chick embryo taken at E4 showing the location of the large developing eye. (B) A cross section of a developing eye at E11 showing the location of the retina, retinal pigmented epithelium (RPE), ciliary marginal zone (CMZ), ciliary body (CB), lens, (L), and the optic nerve (ON).

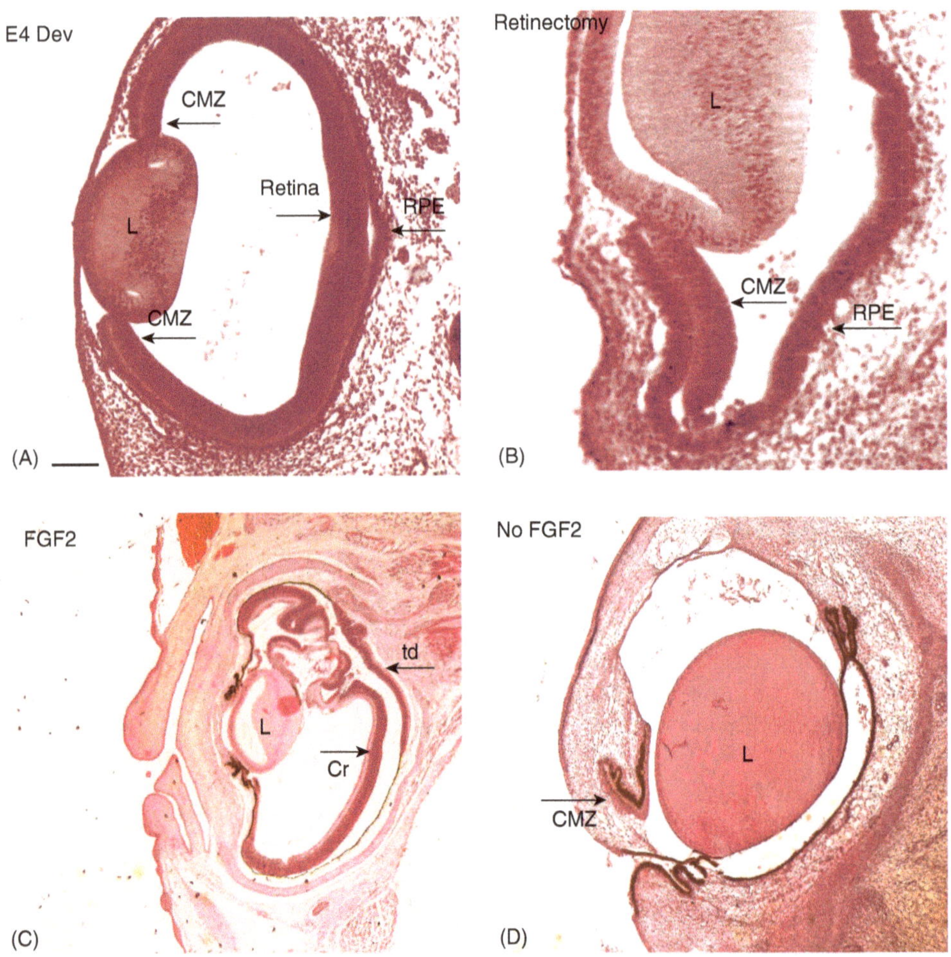

FIGURE 8.2 (A) A cross section of a developing eye at E4 showing the location of the ciliary marginal zone (CMZ), lens (L), retina, and retinal pigmented epithelium (RPE). (B) Cross section of the chick eye after retinectomy at E4. The CMZ is not removed and the RPE is thickened but not yet pigmented. (C) Cross section of retina regenerated in the presence of FGF2 at 7 days post-retinectomy (E11). Retina is regenerated from the retinal stem/progenitor cells present in the ciliary margin (cr) and transdifferentiation of the RPE (td). (D) Cross section of the chick eye 3 days post-retinectomy showing the lack of regeneration in the absence of FGF2.

system regenerates and/or compare early timepoints, when regeneration is possible with timepoints when it is not possible. Regeneration of the chick retina will be discussed in more detail in the second half of this chapter.

The Embryonic Chick Toolbox

Surgical Manipulations

The chick system has a long and venerable history using ablations, rotations, and auto-, allo- and xeno-transplantations. These surgical manipulations in the chick embryo have led to some of developmental biology's most important findings concerning induction of various tissues, fate mapping, patterning, axonal pathfinding, cell lineage, and differentiation. There have been several recent articles concerning the use of the chick in developmental biology that hit on many of these manipulations, so we will not repeat what was discussed in these articles (Stern, 2005). Rather, we will focus primarily on examples of procedures used in the visual system of the chick.

Naturally, the accessibility of the chick embryo has led to elegant analyses using chimeras of labeled chick cells transplanted back into host chickens or chimeras composed of quail and chick. Several of these studies have been directed at determining the fate maps of cells that give rise to the eye or parts of the eye (Hyer *et al.*, 1998; 2003). A large body of work using surgical manipulations has centered on the role of various tissues in patterning of the eye. For instance, removal of the lens ectoderm has shown the importance of the ectoderm in retinal differentiation and showed that the presence of the lens ectoderm is necessary for the morphological development of the optic cup (Fernandez-Garre *et al.*, 2002). Importantly, this same study has established that the lens *ectoderm* is necessary at a certain stage for development of the optic cup, however, the presence of the lens following its invagination into the optic cup appears not to be necessary for the survival and development of the optic cup (Fernandez-Garre *et al.*, 2002). To determine when dorso-ventral eye polarity is established, Araki and colleagues utilized rotations of optic cup explants, using the choroid fissure as a marker of polarity (Uemonsa *et al.*, 2002). Ablations and rotations and quail-chick chimeras of various portions of the optic vesicle have helped to determine when the naso-temporal development of the retina is specified (Dutting *et al.*, 1995a,b; Thanos *et al.*, 1996; Mueller *et al.*, 1998). Transplantation and rotation of the lens, has been used to show that the size and polarity of the lens can be changed *in vivo* (Coulombre and Coulombre, 1969). Using similar techniques, polarity of the chick tectum and the role of the tectum in retinal differentiation and apoptosis has been defined (Cohen *et al.*, 1989; Ichijo *et al.*, 1990; Itasaki *et al.*, 1991; de Curtis *et al.*, 1993; Le Douarin, 1993; Nakamura *et al.*, 1994; Yamagata *et al.*, 1995; Cook *et al.*, 1998; Borsello *et al.*, 2002). Finally, there have also been several studies in which the interaction between the developing cornea and lens has been documented (Zinn, 1970; Lwigale *et al.*, 2007).

Bead Implantation

As investigators began overexpression/misexpression studies, the use of the chick as a model system was stymied for a short period of time because the cells of the chick were too small to inject DNA or mRNA (Stern, 2005). This led to the use of either grafts of transfected cells or insertion of inert beads to deliver factors to a given tissue. Acrylic, ethylene/vinyl acetate copolymer or agarose beads were used that had high affinity for many different molecules, and could slowly release the bound factors. Beads have been used to deliver a variety of growth factors to the developing eyefield and/or eye. One of the best known studies of this type is one in which the phenomenon of RPE transdifferentiation into retina was described by Park and Hollenberg (1989). Following removal of the retina, RPE treated with beads soaked in fibroblast growth factor (FGF) can generate a new retina (Park and Hollenberg, 1989; 1991; Spence *et al.*, 2004; 2007b). Beads have also been used to deliver growth factors in a number of studies to the developing forebrain and eye cup to affect eyefield and/or optic cup development (Ohkubo *et al.*, 2002). Further, beads can be used to deliver other substances to the developing eye, such as function-blocking antibodies or inhibitors of signaling pathways (Martinez-Morales *et al.*, 2005; Spence *et al.*, 2007a). Also explant cultures have been developed to allow optic vesicles to be exposed to growth factors (Trousse *et al.*, 2001).

Chemical Genetics

Chemical genetics is defined as the use of small molecules to affect biological events (Yeh *et al.*, 2003). This section includes examples of chemicals that have been used *in ovo* to specifically stimulate or inhibit various signaling pathways. The strength of the chick system here is that various reagents can be added dropwise to the egg, injected intravenously for systemic uptake, or injected into the optic cup at various stages. Further, multiple additions or injections over time can be easily done. While the list of chemicals that may be added is endless, we would like to consider molecules that interfere with two signaling pathways: the sonic hedgehog (Shh) and FGF pathways.

Cyclopamine, a chemical originally identified as a teratogen, inhibits Shh signaling. Exposure of embryos to this chemical results in cyclopia stemming from the improper patterning in the ventral forebrain (Coventry *et al.*, 1998). Recent studies have compared the effects of cyclopamine to a cholesterol synthesis inhibitor *in ovo* to show that the mechanism of action within the forebrain was due to the direct antagonism of the Shh pathway rather than effects on cholesterol linkage of the Shh molecule (Incardona *et al.*, 1998). In a separate study, cyclopamine was injected directly into the developing eye cup to show the effects of decreased Shh signaling on axonal pathfinding within the retina (Kolpak *et al.*, 2005). On the other hand, SU5402 is a member of a family of FGF signaling inhibitors that bind specifically to the active sites of FGFR kinase domains (Mohammadi *et al.*, 1997). SU5402 has been used *in ovo* to block ganglion cell differentiation and lens fiber elongation (McCabe *et al.*, 1999; Huang *et al.*, 2003). Another more potent FGFR inhibitor PD173074 has been used to dissect the effects of FGF signaling during chick retina regeneration (Spence *et al.*, 2004; 2007a,b). These small molecules and many others have an enormous range of possibilities and combinations that can be tested.

Embryonic Cultures, Explants, Single Cell, Recombined Tissue

Culture systems are widely used to determine the effect of growth factors, toxins, inhibitors, and any other substance in different types of cells or tissues, when a certain amount of precision is required to ensure that all the cells are treated with a specific concentration of the given factor. Several types of chick culture systems have been used to tease out mechanisms of differentiation in the retina, namely eye cup cultures, explants, dispersed cell culture (low and high density), reaggregation of dispersed cells, and immortalized cell lines. We would like to consider three of the most common types of questions that have been addressed using chick retinal, RPE, lens, and corneal cultures and give a few examples of each from the literature. (1) How does one cell type or tissue affect the differentiation/development of another? In the first example of this type of study, Fuhrmann *et al.* (2000) (showed that a signal from extraocular mesenchyme upregulated RPE markers and downregulated retinal markers in optic cup cultures. In a second example, the innervation of the developing cornea by the trigeminal nerve was shown to be dependent on the expression of semaphorin A in the adjacent lens epithelium (Lwigale *et al.*, 2007). (2) How do cells from the same tissue influence one another during development? This question has been addressed using the various culture techniques listed above in a variety of permutations. For instance, low density cultures have been used to show that the stage at which retinal progenitor cells are cultured, determines the type of retinal cell they will form *in vitro,* demonstrating the importance of the *in vivo* environment in dictating cell fate (Adler *et al.*, 1989; Repka *et al.*, 1992a,b; Belecky-Adams *et al.*, 1996). Other studies have used heterochronic cultures to investigate the effects of earlier born cells on progenitor cell differentiation (Waid *et al.*, 1998), and the importance of cell–cell communication in retinal differentiation (Austin *et al.*, 1995). Finally, work from Layer and colleagues has explored the possibility of reconstituting the laminar formation and differentiation of the retina using reaggregation cultures of dispersed retinal cells (Rothermel *et al.*, 1997; 2006). (3) How does treatment of cells at various stages of development with growth factors affect development of retinal progenitors? There are an enormous range of growth factors that have been used in cultures of chick retina, including FGFs, BMPs, activins, CNTF, Shh, and NGF to mention only a few (Pittack *et al.*, 1991; Fuhrmann *et al.*, 1995; Matsuo *et al.*, 1997; Belecky-Adams *et al.*, 1999; Frade, 2000; Cirillo *et al.*, 2001; Le *et al.*, 2001; Zhang *et al.*, 2001; Belecky-Adams *et al.*, 2002; Nakagawa *et al.*, 2003; Kolpak *et al.*, 2005; Sehgal *et al.*, 2006).

DNA Transfer

In this section, we will consider several techniques that enable the investigator to introduce DNA, in the form of expression vectors, RNA interfering molecules (including morpholinos, siRNA, dsRNA, and shRNA) and reporter constructs to test cis-acting sequences in non-coding regions of the genome. Several techniques will be included in this section, including retroviral transfer, electroporation, and transfection using lipid-based reagents.

The chicken-specific replication competent retrovirus (RCAS) has been used for misexpression of genes in the chick since the late 1980s, and was the first technical revolution that allowed the introduction of exogenous genes into chick cells *in ovo* (Hughes *et al.*, 1984a,b; Morgan *et al.*, 1992; Riddle *et al.*, 1993). This retrovirus is derived from the SR-A strain of the Rous sarcoma (src) virus, and was made by deleting sequences that encode the src gene. Deletion of this portion of the viral genome allows insertions of genes of interest at this site (Hughes *et al.*, 1984a). It has become so commonly used in the chick system that it has its own website (http://www.retrovirus.info/RCAS), run by one of the originators of the RCAS retrovirus, Stephen Hughes (Hughes *et al.*, 1984a,b). Since its arrival on the scene, there have been various modifications to the virus that allow it to be used in different ways. For instance, adaptor plasmids were made to aid in the insertion of exogenous genes into the RCAS retrovirus, mutations have been made in the genes that encode envelope proteins that allow investigators to target host range, other deletions have been made in viral genes to allow larger insertions, mutations to the long terminal repeat (LTR) enhancer allow the inserted gene to be expressed at different levels, and tetracycline inducible elements have been added to the RCAS A retrovirus so that expression of genes inserted into the retrovirus can be induced (Hughes *et al.*, 1987; Greenhouse *et al.*, 1988; Sato *et al.*, 2002). The RCAS system has been used by many to advance our understanding of the visual system, for example, the retrovirus has been used to study patterning (Nakamoto *et al.*, 1996; Schulte *et al.*, 1999; Sakuta *et al.*, 2001; Adler *et al.*, 2002; Kim *et al.*, 2006), mitosis (Crisanti *et al.*, 2001), axonal pathfinding (Kolpak *et al.*, 2005), differentiation (Blancher *et al.*, 1996; Jiang *et al.*, 1998; Ogino *et al.*, 1998; Yan *et al.*, 2000a; Li *et al.*, 2001; Liu *et al.*, 2001; Yan *et al.*, 2001; Esteve *et al.*, 2003; Canger *et al.*, 2004; Cho *et al.*, 2006; Moreira *et al.*, 2006), survival (Pimentel *et al.*, 2000), and regeneration (Spence *et al.*, 2004; Spence *et al.*, 2007a,b; Haynes *et al.*, 2007).

The RCAS retrovirus also has several drawbacks associated with it, including (1) the upper restriction on the size of insertions to the viral coding sequence is about 2 Kb, so that it is unlikely that one could introduce more than one gene into the retrovirus, (2) it cannot be used to target post-mitotic cells, (3) there is an increase in the cost and time associated with making a retroviral stock, (4) it takes between 16 and 24 h to get expression of the viral proteins *in ovo*, and (5) the investigator must use the substantially more expensive virus-free eggs.

Few technical advances have made the chick system more amenable to the types of studies performed today than electroporation. The basic idea behind electroporation is that an electrical pulse delivered by electrodes placed in the tissue disrupts the cell membranes, allowing DNA to enter cells. The negatively charged DNA will move toward the anode side of the electrode, resulting in transfection of tissue on the side of the anode. There have been a raft of articles discussing *in ovo* electroporation and the best parameters to use to enhance survival and increase transfection efficiency (Muramatsu *et al.*, 1997; Itasaki *et al.*, 1999; Nakamura *et al.*, 2000; Yasuda *et al.*, 2000; Yasugi *et al.*, 2000; Nakamura *et al.*, 2001; Swartz *et al.*, 2001; Katahira *et al.*, 2003; Chen *et al.*, 2004; Krull, 2004; Nakamura *et al.*, 2004; Uchikawa *et al.*, 2004; Sato *et al.*, 2007). Several investigators have also determined how to introduce various types of interfering molecules into the developing chick using electroporation or viruses, making knock-down experiments feasible (Hu *et al.*, 2002; Katahira *et al.*, 2003; Kos *et al.*, 2003; Pekarik *et al.*, 2003; Chesnutt *et al.*, 2004; Rao *et al.*, 2004; Hernandez *et al.*, 2005; Canto-Soler and Adler, 2006; Harpavat and Cepko, 2006; Watanabe *et al.*, 2007). The use of electroporation has several advantages over the use of viruses to introduce DNA, such as there is no longer a need to clone sequences into the retroviral plasmid, no size restriction on insertions to the expression vector, no need to expend the effort and funds in making a viral stock with which to infect tissues, and no need to purchase the more expensive virus-free eggs. Further, because directionality of the transfection can be controlled somewhat by placement of the electrodes, the electroporation method has more precision over where DNA can be targeted. Introducing DNA via electroporation is not limited to dividing cells, as is the retrovirus, and the expression of plasmids introduced by electroporation is generally detectable within a few hours post-electroporation. One limitation that electroporation does have is that the DNA is not incorporated into the genome; hence its expression is lost over time. A recent advancement in this area is the stable incorporation of genes into the genome through the co-electroporation of a transposon-containing expression vector with a separate expression vector containing a transposase (Sato *et al.*, 2007). This combination led to the persistence of the electroporated green fluorescent protein (GFP) marker. This has also been combined with the tetracycline inducible elements, such that transgenes could be introduced fairly early in development, when accessibility of the embryo is at its highest, and turned on later in development by addition of tetracycline (Sato *et al.*, 2007). Until recently, electroporation had been used in very early embryos, primarily because later embryos turn inside such that the head is no longer visible and the embryo becomes covered with a dense vasculature. Two changes have been made recently to address introduction of genes into older embryos via electroporation. The first is the *ex ovo* electroporation of embryos grown in petri dishes and the second advancement is that of electroporation in hatchlings (Luo *et al.*, 2005; Yamaguchi *et al.*, 2007).

Last, there have been a variety of methods used to transfect cells with lipid-based technology (Iwakiri *et al.*, 2005; Muramatsu *et al.*, 1997; Yasugi *et al.*, 2000; Decastro *et al.*, 2006). The basis of this technique is the ability of liposomes loaded with DNA to fuse with cellular membranes and deliver their cargo to the cytosol.

This method has been used to generate chimeric chick embryos (Fraser *et al.*, 1993), and to transfect a number of tissues (Brazolot *et al.*, 1991; Demeneix *et al.*, 1994; Rosenblum *et al.*, 1995; Decastro *et al.*, 2006). While most early lipid delivery systems were not as efficient as electroporation in the developing embryo (Decastro *et al.*, 2006), enhanced Lipofectamine delivery through the addition of disulfide linked pegylated lipid has lead to a substantial increase in the transient transfection of a variety of tissues, including the neural tube and optic cup.

It is unlikely that these are the last of the advances for delivery of genes and other molecules into the developing and regenerating chick system (Kawakami *et al.*, 2008). One promising technology being developed is that of sonication (Ohta *et al.*, 2003; Fischer *et al.*, 2006). There have been some recent advances in "sonoporation" which make transfections *in vivo* more efficient and more likely to be used in the future (Gvili *et al.*, 2007; Saito *et al.*, 2007).

Disadvantages of the Chick Embryo

There are also some disadvantages of the chick system. Until a few years ago, the biggest disadvantage of the system was the inability to genetically modify chickens (Stern, 2005). However, this difficulty has been overcome by several groups, and the practice of making transgenic chickens will soon become more standardized and catch up with the powerful techniques currently available for other models such mice, zebrafish, and *Xenopus* tropicalis (Mozdziak *et al.*, 2003; Koo *et al.*, 2004; 2006; Mozdziak and Petitte, 2004; 2006; Kwon *et al.*, 2004; Chapman *et al.*, 2005). Another weakness that the chick model has been associated with is the lack of natural mutants and/or a long-term storage facility for such mutants. There are some mutants available, as has been reviewed recently, however, even some of those few are in danger of being lost (Delany, 2004). Coupled with this is the challenge of chemical mutagenesis in chickens. It is unlikely that the chick will ever be able to take advantage of mutagenesis screening that is common in models such as *Drosophila* and zebrafish.

CHICK RETINA REGENERATION

Introduction

As mentioned earlier, one of the great advantages of working with the embryonic chick eye is that the retina

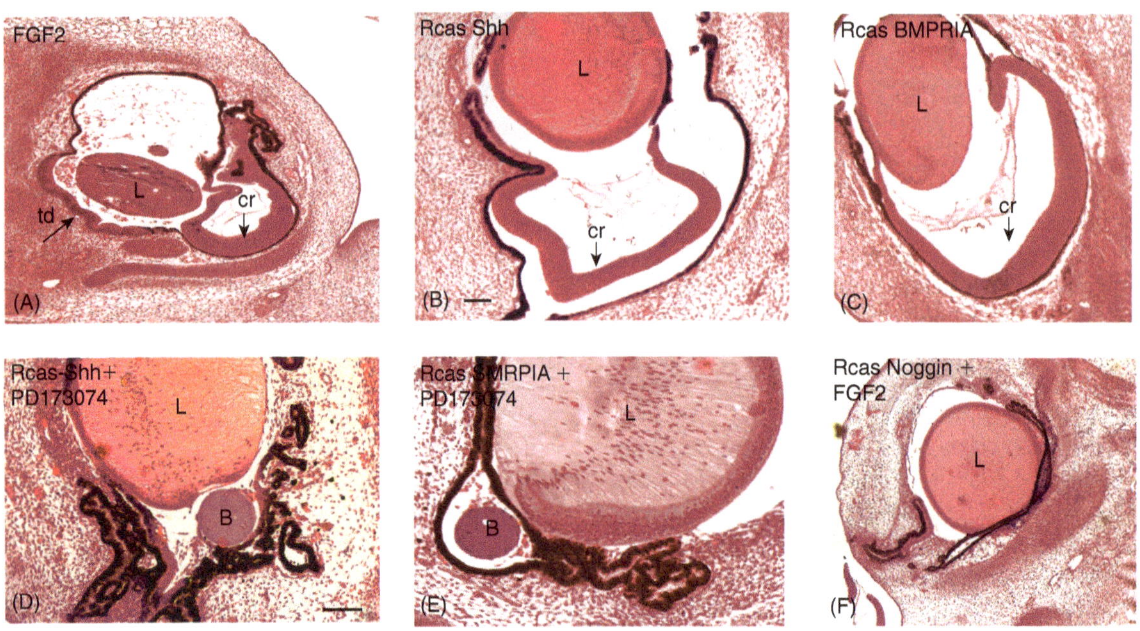

FIGURE 8.3 (A) A cross section from a regenerating eye 3 days post-retinectomy showing regeneration from both the retinal stem/progenitor cells (cr) and transdifferentiation of the RPE (td) in the presence of FGF2. (B and C) A cross section from a regenerating eye 3 days post-retinectomy showing regeneration from the retina stem/progenitor cells (cr) when the Shh (RCAS Shh) (B) or the BMP (RCAS BMPRIA) (C) pathway is constitutively activated. (D–F) A cross section from a regenerating eye at 3 days post-retinectomy showing the lack of regeneration in the presence of RCAS Shh and PD173074, an inhibitor of the FGF pathway (D), RCAS BMPRIA + PD173074 (E), and RCAS noggin, an inhibitor of the BMP pathway, and FGF2 (F).

can be repaired or replaced if damaged or removed. The accessibility to the embryo for microsurgery combined with the availability of molecular tools in the chick has made this a great system to study and dissect the early molecular events that take place during retina regeneration. The chick genome was also recently sequenced (Wallis *et al.*, 2004) and this provides a vast range of possibilities to study the early stages of retina regeneration, including the use of gene array technology to identify critical genes regulated during chick retina regeneration.

The embryonic chick can regenerate its retina via two modes. One requires the activation of stem/progenitor cells present in the ciliary margin, while the other involves the use of the classic process of transdifferentiation (Fig. 8.2). The phenomenon of retina regeneration in the embryonic chick has been observed since the early 1900s, however, it was not until Coulombre and Coulombre (1965) that the process of retina regeneration was described in more detail. Park and Hollenberg (1989, 1991) discovered that in order for any retina regeneration to take place a source of FGF had to be present. Recently, we have shown that other signaling pathways including the hedgehog (Hh) and bone morphogenetic protein (BMP) pathways regulate the process of retina regeneration (Spence *et al.*, 2004, 2007a, b; Haynes *et al.*, 2007). We will discuss the mechanisms by which each mode of retina regeneration is regulated.

Regeneration by Stem/progenitor Cell Activation

Regeneration from the stem/progenitor cells in the ciliary margin requires an induction process whereby the stem/progenitor cells are activated to proliferate and differentiate into the retinal cell types. The stem/progenitor cells in this region are used by the embryo to provide for the continuous growth for the retina, however, there is always a population of cells that remain undifferentiated and will not spontaneously respond to injury. However, after removal of the retina, the stem/progenitor cells can be activated with exogenous growth factors to proliferate and differentiate into each of the retinal cell types reforming a complete retina in about 1 week (Spence *et al.*, 2004; Fig. 8.2(C)). Activation of the retinal stem/progenitor cells is most robust if the retina is removed on embryonic day 4, although some activation does occur at later stages but at a reduced level.

Role of FGF/MAPK Signaling Pathway

As mentioned, FGF was the first exogenous growth factor to be identified as an inducer of retina regeneration in the embryonic chick. Park and Hollenberg (1991) used FGF1 to induce chick retina regeneration from the stem/progenitor cells of the ciliary margin. More recently, we have used FGF2 (which was originally used by Park and Hollenberg (1989) to induce transdifferentiation in the chick retina) and studied its ability to activate the retinal stem/progenitor cells (Spence *et al.*, 2004, 2007a) (Fig. 8.2(C), 3(A). FGF2 can activate several signaling pathways within the cell, but the activation of the mitogen-activated kinase (MAPK) signaling cascade by FGF2 is critical for retina regeneration since the addition of an inhibitor for this pathway in the presence of FGF2 results in a significant reduction in regeneration (Spence *et al.*, 2007a). The activation of MAPK by FGF2 induces proliferation of the retinal stem/progenitor cells and is required for cell survival (Spence *et al.*, 2007a).

Role of Shh Signaling Pathway

Other signaling pathways are also involved in the regulation of retina regeneration from the ciliary margin. One of these pathways is the Shh pathway. Like, FGF2, overexpression of Shh has been shown to induce retina regeneration from the stem/progenitor cells (Fig. 8.3(B)). However, induction of regeneration by either of these molecules is dependent on the other pathway being functional, since reduced regeneration from the ciliary margin occurs in eyes treated with either FGF2 and an inhibitor of the Shh pathway or a virus overexpressing Shh and an inhibitor of the FGF pathway (Fig. 8.3(D) and 8.4(A); Spence *et al.*, 2004).

Detailed studies have been done to dissect the role of FGF2 and Shh in retina regeneration. It has been found that Shh can induce regeneration from the ciliary margin by activating transcription of FGF ligands and FGF receptors thereby inducing proliferation through the FGF/MAPK pathway described above (Spence *et al.*, 2007a). A functional Shh pathway is also necessary because Shh works with FGF2 to promote cell survival and Shh alone is required for the maintenance of progenitor cell identity (Spence *et al.*, 2007a). In addition to stem/progenitor cell induction, overexpression of Shh has been shown to reduce the number of regenerating ganglion cells, demonstrating a role for Shh in retina differentiation (Spence *et al.*, 2004).

Role of BMP Signaling Pathway

In addition to FGF2 and Shh, BMP has also been shown to induce retina regeneration from the ciliary margin (Fig. 8.3(C)) (Haynes *et al.*, 2007). BMP can also activate the FGF/MAPK pathway by increasing the transcription of FGF receptors. The BMP pathway and the FGF/MAPK pathway are both necessary for proliferation of retinal progenitor cells during the initial induction period of regeneration because if one pathway is blocked, proliferation and therefore regeneration does not occur (Fig. 8.3(E) and (F)). During this initial induction period of regeneration BMP activates the canonical BMP pathway (via SMADs). However, during the later stages of regeneration, BMP switches and activates a non-canonical BMP pathway (via TAK1) which leads to p38 activation and apoptosis. Inhibition of p38 is necessary to maintain BMP-induced regeneration otherwise the regenerated retina will undergo massive cell death. Even the addition of ectopic FGF2 does not prevent the high level of cell death because BMP decreases the transcription of FGF receptors at this stage. There is some evidence that BMP also regulates the differentiation of ganglion cells and the cells of the inner nuclear layer because these cells do not form in the absence of BMP (Haynes *et al.*, 2007).

While we are still deciphering how FGF2, Shh, and BMP pathways work together as well as in cooperation with other pathways yet to be studied, it is clear that functional FGF2, Shh, and BMP pathways are necessary for induction of regeneration from the stem/progenitor cells present in the anterior region of the eye. Further studies will help delineate whether the pathways work in concert or parallel to regulate proliferation, cell survival, and differentiation.

Regeneration by Transdifferentiation

An In Vivo Model

The second mode of regeneration that takes place in the chick retina is via the process of transdifferentiation. When a complete retinectomy is performed in embryonic day 4 chick eyes, and an exogenous source of FGF is introduced in the eye, the retinal pigmented epithelium (RPE) undergoes a reprogramming where the cells dedifferentiate, losing their pigment and become "embryonic-like." These cells enter the cell cycle and build a neuroepithelium which will eventually differentiate to give rise to the newly regenerated retina. This process of transdifferentiation has been described histologically (Coulombre and Coulombre, 1965; Park and Hollenberg, 1989; Spence *et al.*, 2004) as well as with cell and molecular markers (Spence *et al.*, 2004; 2007b).

Other species can also regenerate their retina via transdifferentiation during early stages of their development (review in Lopashov and Stroeva, 1964; Mitashov, 1996, 1997), however, studying this process *in vivo* can be challenging in animals such as mice or even fish as these animal models are either not easily accessible or are too small to manipulate during early stages of their development. Some anurans such as newts have unsurpassed regeneration abilities and can regenerate their retina via transdifferentiation throughout their lifetime. A month and a half after retina removal, a complete functional retina is restored (Mitashov, 1996, 1997; Del Rio-Tsonis and Tsonis, 2003; Tsonis and Del Rio-Tsonis, 2004; Chapter 7). These virtues qualifies the newt as one of the best animal models to study transdifferentiation; however, the lack of molecular tools for newt studies has greatly limited the use of this model for dissecting the molecular regulation of transdifferentiation. The availability of molecular tools in the embryonic chick as well as the fast rate of retina regeneration (it only takes 7 days after retina removal to obtain a complete laminated retina with all the mayor retinal cell types present), qualifies this animal as the preferred model for the dissection of molecular mechanisms during RPE to retina transdifferentiation.

The embryonic chick eye only provides a good model to study early events of retina regeneration and transdifferentiation but will not address the restoration of vision since the transdifferentiated retina does eventually degenerate due to the lack of RPE, which fails to restore itself during the process of transdiffrentiation (Coulombre and Coulombre, 1964; Park and Hollengberg, 1989). The lack of RPE in the transdifferentiated retina accounts for its reverse orientation when compared to the original retina or even to the one that regenerates via stem/progenitor cell activation (see Figure 8.2(C)).

A window of transdifferentiation

It is interesting to note that there is a small window during chick eye development where the RPE is competent to transdifferentiate (Coulombre and Coulombre, 1964; Park and Hollengberg, 1989). It is feasible to remove the retina as early as E3.5 and if a source FGF is added then, the RPE will transdifferentiate into retina. This competence is present until about E4.5. During this time, the RPE expresses micropthalmia (Mitf) and has stopped expressing Pax-6 (Spence *et al.*, 2007b). In the absence of neural retina (NR), RPE transdifferentiation in chick eyes has not been reported after E5 *in vivo* with any known treatment. However, RPE to retina transdifferentiation has been reported in developing eyes when Pax-6 is overexpressed in the RPE of chick eyes up to stage 35 (Azuma *et al.*, 2005), or in *in vitro* E5-6 (HH stages 28-29) explant cultures where activin/ TGF-beta/

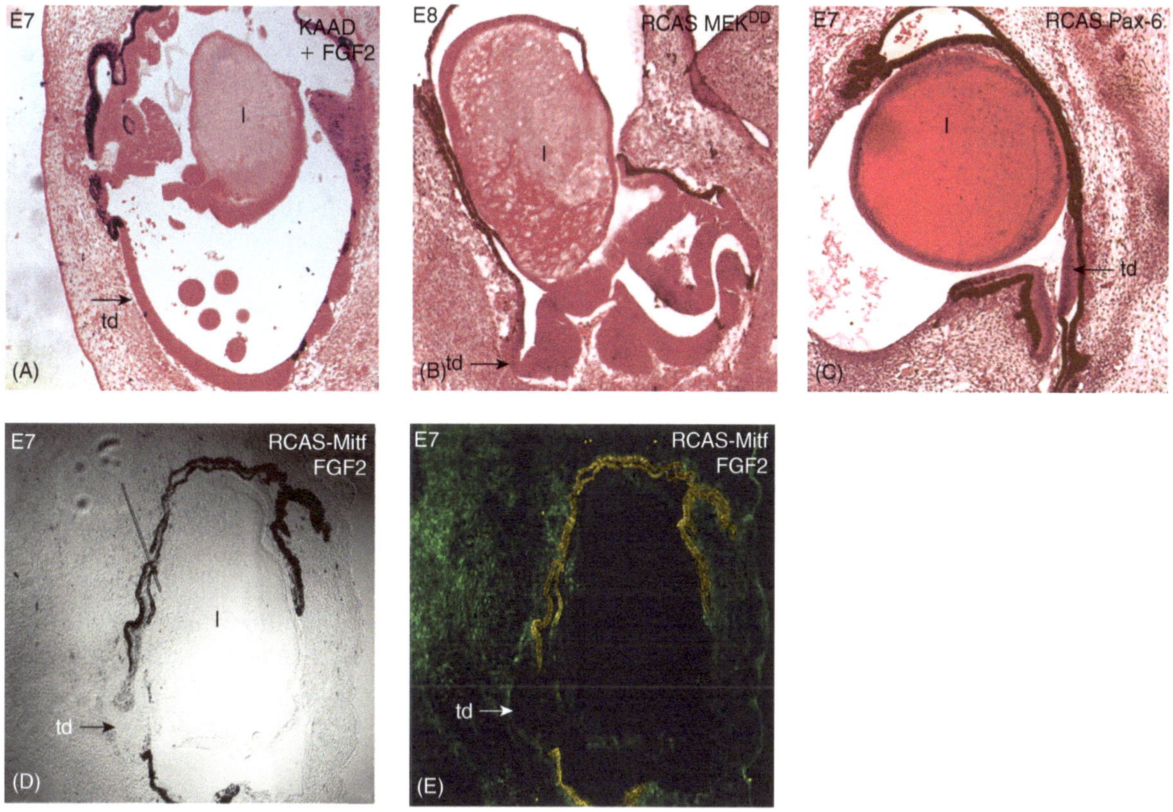

FIGURE 8.4 (A–C) A cross section of a regenerating eye at 3 days post-retinectomy (A, C) or 4 days post-retinectomy (B) showing transdifferentiation of RPE induced by KAAD, an inhibitor of the Shh pathway, and FGF2 (A), activation of MAPK pathway (RCAS MekDD (B), and overexpression of Pax-6 (RCAS Pax-6) (C). (D and E) Inhibition of transdifferentiation by overexpression of Mitf (RCAS Mitf) in the presence of FGF2 is shown by immunohistochemistry on a cross section of a regenerating eye at 3 days post-retinectomy using an antibody for Mitf (red) and an antibody for a protein from the viral coat (green). Yellow cells show the location of infected RPE (E). Transdifferentiation only occurs in area of the RPE that are not infected (td). DIC is shown in D.

nodal receptors are inhibited in the presence of FGF (Sakami *et al.*, 2008), or even in RPE explants of post-hatched chicks transfected with Optx2 (Toy *et al.*, 1998).

Dissecting the molecular pathway of RPE transdifferentiation

There are several molecular players involved in the process of transdifferentiation that have been unraveled by a disruption on their pathway or function during either retina development or regeneration. Two different groups of molecules have been identified in the saga of transdifferentiation. On one side, are the genes that protect the RPE phenotype and on the other, the ones that define the retina phenotype. Mitf (Mochii *et al.*, 1998a; b; Planque *et al.*, 1999; 2001; 2004; Bumsted and Barnstable, 2000; Nguyen and Arnheiter, 2000), Otx (Martinez-Morales *et al.*, 2001; 2003; 2004; Sakami *et al.*, 2005), Wnt13 (Fuhrmann *et al.*, 2000), BMPs (Muller *et al.*, 2007;), Shh (Zhang and Yang, 2001; Perron *et al.*, 2003; Spence *et al.*, 2004) and activin (Fuhrmann *et al.*, 2000; Sakami *et al.*, 2008) are associated with the induction and maintenance of the RPE, whereas Pax-6 (Belecky-Adams *et al.*, 1997; reviewed in Levine and Green, 2004; Chx10 (Rowan *et al.*, 2004; Horsford *et al.*, 2005), Msx-2 (Holme *et al.*, 2000), Optx2 (Toy *et al.*, 1998), Neuro D (reviewed in Yan *et al.*, 2005) and FGF/MAPK (Vogel-Höpker *et al.*, 2000; Galy *et al.*, 2002; and reviewed in Yang *et al.*, 2004) are associated with retina.

In chicks, Pax-6 overexpression in the RPE is sufficient for the induction of transdifferentiation during retina regeneration (Spence *et al.*, 2007b) (Fig. 8.4(C)) and even during development (Azuma *et al.*, 2005) while Mitf overexpression is sufficient to protect the RPE from transdifferentiating during FGF-induced retina regeneration (Spence *et al.*, 2007b) (Fig. 8.4(D) and (E)).

An In Vitro Model

The embryonic chick has been used for the study of transdifferentiatiation by several researchers using *in vitro* systems including isolated RPE cells or explants.

Transdifferentiation of RPE to NR

RPE explants

RPE cells have been cultured from chick embryos and tested for their ability to transdifferentiate into NR cells (Pittack *et al.*, 1991; Guillemot and Cepko, 1992). If the RPE is removed from the chick at E4.5-E5.5 (HH stages 24-28), dissociated and treated with FGF, the cultured RPE cells lose their pigment but do not express markers of neural cells (Pittack *et al.*, 1991). However, if the RPE cells are not dissociated, but instead left as an intact sheet of cells and treated with FGF, the RPE cells will lose their pigment and express markers indicative of retinal progenitor cells and even express markers of NR cells (Pittack *et al.*, 1991; Guillemot and Cepko, 1992). Recently, Sakami *et al.* (2008), had used this explant system to test the potential of activin to block FGF-induced RPE transdifferentiation using E4 explants, and have shown that when inhibiting the activin/TGF-beta/nodal pathway, E5 incompetent RPE can transdifferentiate. Interestingly, according to Zhou and Opas (1994), FGF does not act on the fully differentiated RPE, but only on those cells that have been stimulated to change their identity, probably via changes in their adhesive status. In addition, once FGF is able to direct RPE explants to transdifferentiate, the substratum where the cells are grown dictates their differentiation (Opas and Dziak, 1994). Transfecting RPE explants with key genes is another way to induce transdifferentiation effects such as the ones incurred by transfecting Optx2 unto E7-E8 as well as post-hatched chick RPE explants (Toy *et al.*, 1998).

RPE isolated cultures

While the dissociated RPE cells did not express neuronal markers when cultured from chick embryos at E4.5-E5.5, they did begin the transdifferentiation process by losing their pigment if treated with FGF2. Additional studies of these cultured RPE cells revealed that an overexpression of Mitf, a transcription factor involved in defining RPE identity, inhibited FGF from triggering transdifferentiation of the RPE (Mochii *et al.*, 1998b). Furthermore, addition of Msx-2, a gene only expressed in NR, to the cultured RPE cells caused a decrease in Mitf and an increase in the neuronal marker, class III beta-tubulin (Holme *et al.*, 2000). Therefore, transdifferentiation of cultured RPE cells from E4.5-E5.5 into neuronal cells requires the downregulation of RPE genes, such as Mitf, and/or the upregulation of neuronal specific genes, such as Msx-2.

RPE cells cultured at a slightly later day in development, at E6, have also been used to study the ability of RPE to transdifferentiate into NR. Addition of FGF to cultured E6 RPE cells did result in an increase in cells expressing an early ganglion cell marker, RA4, but there was not a transdifferentiation to neuronal morphology. However, addition of NeuroD did result in a transdifferentiation of E6 RPE cells to photoreceptors (Yan and Wang, 1998; Yan and Wang, 2000a; b) while the addition of neurogenin 2 (ngn2) resulted in transdifferentiation of the E6 RPE cells to photoreceptors and retinal ganglion cells (Yan *et al.*, 2001). Cath5 and NSCL1 were also able to induce transdifferentiation of E6 RPE cells into retinal ganglion cells (Ma *et al.*, 2004; Xie *et al.*, 2004). Studying the induction potential of RPE cells *in vitro* will be beneficial in deciphering the molecules needed to induce transdifferentiation of the RPE *in vivo* at both E4 and at later stages.

Transdifferentiation of NR to RPE

NR from early chick embryos also possesses the plasticity to transdifferentiate into RPE *in vitro*. Studies performed by Opas *et al.* (2001) have shown that dissociated 6-day-old embryonic NR can transdifferentiate into RPE spontaneously. These pigmented transdifferentiating cells express RPE-specific protein, eRPEAG and lack of expression of the neural cell adhesion molecule, NCAM (Opas *et al.*, 2001).

In Vitro–In Vivo

RPE cells cultured *in vitro* have also been transplanted into the embryonic chick eyes and shown to integrate into the developing eye. Cells cultured from the developing RPE of an E5.5 chick embryo and grown until they develop the morphology of RPE cells will integrate into the developing RPE when transplanted into the embryonic chick at E11-E18 (Liang *et al.*, 2006). However, if, before transplantation, the cultured RPE cells are treated with an RCAS virus expressing NeuroD, which has been shown to be important for photoreceptor development (Yan and Wang, 1998), the infected RPE cells will begin to express visinin, an early marker for cone photoreceptors, and integrate into the outer nuclear layer of the retina indicative of transdifferentiation of the transplanted RPE cells into photoreceptor cells (Liang *et al.*, 2006). These transplanted cells continue the differentiation process expressing advanced photoreceptor markers such as opsin and extend axons into the inner nuclear layer or ganglion cell layer. Although these transdifferentiated photoreceptor cells do integrate into the correct location and express the appropriate markers for photoreceptors, the photoreceptors are not all organized perpendicular to the RPE and some advanced markers are expressed in the cell body instead of the axon (Liang *et al.*, 2006). This is believed to occur because there is not an intimate association of the transplanted cells with the developing RPE that is needed for proper organizational cues. Despite the organizational problems that need to be solved, these studies involving the chick embryo provide hope that transplanted RPE cells can someday be directed to differentiate *in vivo* to replace lost or damaged photoreceptors.

Using the Embryonic Chick Eye to Probe for Retina Repair Potential of Mammalian Cells

Embryonic stem cells isolated from the mammalian blastocyst and retinal stem cells isolated from rodents and post-mortem humans have been cultured and directed to differentiate into ocular structures including lens (Oota, *et al.*, 2003; Takahashi, *et al.*, 2006), retina (Zhao *et al.*, 2002; Hirano *et al.*, 2003; Haruta, 2005; Banin, *et al.*, 2006; Lamba, *et al.*, 2006; Limb, *et al.*, 2006; Zhao, *et al.*, 2006, and Vugler, *et al.*, 2007) and RPE (Haruta, *et al.*, 2004; Klimanskaya *et al.*, 2004; Aoki, *et al.*, 2006; and Takahashi *et al.*, 2006). The embryonic chick has proven to be an excellent model to determine the ability of these stem cells to integrate and differentiate *in vivo* (Coles *et al.*, 2004; Aoki *et al.*, 2006).

Embryonic Stem Cells

Mouse embryonic stem cells incubated with basic FGF, cholera toxin, dexamethasone and Wnt2b resulted in these stem cells expressing retinal precursor markers and differentiating into eye-like structures resembling lens, RPE, and retina with a high frequency within 10–12 days *in vitro* (Hirano *et al.*, 2003 and Aoki *et al.*, 2006). When these eye-like structures were developed for 11 days *in vitro* and then transplanted into the developing chick eye, they most often migrated to the developing RPE layer and differentiated into mature RPE cells expressing the RPE marker, RPE65 (Aoki *et al.*, 2006). A few of these transplanted eye-like structures also expressed markers indicative of a ganglion cell lineage (Aoki *et al.*, 2006; 2007). Embryonic stem cell transplanted after only 6 days in culture also integrated into the retina of the chick and were induced to form lens tissue or express markers of a ganglion cell lineage. Based on these studies, we can speculate that human embryonic stem cells have the potential to integrate into different tissues of the eye and differentiate into functional cells of the lens, retina, and RPE if manipulated correctly.

Adult Stem cells

Retinal stem cells isolated from the ciliary margin of post-mortem human eyes were also tested for their potential to differentiate *in vivo* using the embryonic chick eye. These retinal stem cells were able to proliferate and differentiate spontaneously into all retinal cell types when cultured *in vitro* although the addition of FGF, epidermal growth factor (EGF), and heparin increased the rate at which this occurred (Coles *et al.*, 2004). These retinal stem cells were able to respond to environmental cues in the developing chick eye and express markers of ganglion and horizontal cells when transplanted at the time these cells would normally be developing in the chick eye (Coles *et al.*, 2004). These studies show great promise for the future use of either embryonic or adult stem cells in the treatment of retina degenerative diseases. They also demonstrate the conservation between environmental cues in human and chicks making the chick a reliable model in which to study the potential of these cells.

The Post-hatch Chick and Its Potential Sources of Retina Repair

The Ciliary Margin

Although retinal stem/progenitor cells continue to proliferate for up to 3 weeks after hatching, they are unable to regenerate a complete retina even in the presence of exogenous growth factors (Fischer and Reh, 2000). The ciliary margin which houses the retinal stem/progenitor cells is composed of two distinct regions in the fully developed chicken eye. The more anterior structure is the ciliary body (CB) which is composed of two cellular layers, the pigmented epithelial layer (PE) and the non-pigmented epithelial layer (NPE). Posterior to the CB at the tip of the NR is the ciliary marginal zone (CMZ) (Fischer and Reh, 2003a). EGF, insulin, and IGF-1 increase proliferation and induce differentiation of the cells in the CMZ (Fischer and Reh, 2000; 2003a), whereas FGF2, insulin and EGF stimulate the cells of the NPE to proliferate and differentiate (Fischer and Reh, 2003a). While differentiation can be induced, it is limited in the post-hatch chick. Cells in the CMZ will differentiate into amacrine and bipolar cells and cells from the NPE differentiate to form amacrine and ganglion cells but other cell types including photoreceptors are not formed by either group of cells. Regardless, retinal injury will not stimulate the cells of the CMZ to regenerate or repair the retina (Fischer and Reh, 2000).

Müller Glia

Müller Glia are another possible source of regeneration in the post-hatch chick. Injection of toxins that cause cell death in certain retinal neurons or the addition of FGF2 or insulin causes the Müller Glia to proliferate, lose their characteristic Müller Glia markers and begin to express markers indicative of retinal progenitors (Fischer and Reh, 2001; Fischer *et al.*, 2002; Fischer and Reh, 2003b). Many of the activated Müller Glia remain undifferentiated but a small percentage of them do differentiate into ganglion, amacrine, or bipolar cells (Fischer and Reh, 2001; Fisher *et al.*, 2002) under certain treatments. It has been shown that the Notch pathway is necessary for the dedifferentiation and proliferation of Müller Glia but if the Notch pathway remains active, it will inhibit the differentiation of the newly formed progenitors into neural cells (Hayes *et al.*, 2007). In addition, NeuroD has been shown to induce dedifferentiation of Müller Glia cultured from toxin-damaged retina and promote the differentiation of immature photoreceptors (Fischer *et al.*, 2004).

CONCLUSION

The chick provides an excellent system to explore cell and molecular events during retina development and regeneration, including cell fate determination, stem and progenitor cell biology, cell differentiation, cell division, cell death, cell signaling, axon path finding, retinotectal projections and neural circuitry to name a few. It is an inexpensive, molecularly friendly system with many tools currently available.

ACKNOWLEDGMENTS

We would like to thank Dr. Natalia Vergara for helping with the editing of this chapter and grant support NEI EY017319-02, NIA grant AG 24397-01 and Prevent Blindness America grant PBA 0720 to KDRT.

REFERENCES

Adelmann HB (1966). *Marcello Malpighi and the Evolution of Embryology*. Cornell University Press, Ithica, NY.

Adler R, Belecky-Adams TL (2002). The role of bone morphogenetic proteins in the differentiation of the ventral optic cup. *Development* 129(13):3161–3171.

Adler R, Hatlee M (1989). Plasticity and differentiation of embryonic retinal cells after terminal mitosis. *Science* 243(4889):391–393.

Adler R, Tamres A, Bradford RL, Belecky-Adams TL (2001). Microenvironmental regulation of visual pigment expression in the chick retina. *Dev Biol* 236(2):454–464.

Aoki H, Hara A, Nakagawa S, Motohashi T, Hirano M, Takahashi Y, Kunisada T (2006). Embryonic stem cells that differentiate into RPE cell precursors in vitro develop into RPE cell monolayers in vivo. *Exp Eye Res.* 82(2):265–274.

Aoki H, Hara A, Niwa M, Motohashi T, Suzuki T, Kunisada T (2007). Transplantation of cells from eye-like structures differentiated from embryonic stem cells in vitro and in vivo regeneration of retinal ganglion-like cells. *Graefes Arch Clin Exp Ophthalmol.* Nov 15 [Epub ahead of print].

Aristotle. *On the Generation of Animals, Book II. Adelaide.* eBooks@Adelaide, South Australia, 343 BC.

Aristotle. *The History of Animals, Book VI. Adelaide.* eBooks@Adelaide, South Australia, 350 BC.

Austin CP, Feldman DE, Ida JA, Jr., Cepko CL (1995). Vertebrate retinal ganglion cells are selected from competent progenitors by the action of Notch. *Development* 121(11):3637–3650.

Azuma N, Tadokoro K, Asaka A, Yamada M, Yamaguchi Y, Handa H, Matsushima S, Watanabe T, Kida Y, Ogura T, Torii M, Shimamura K, Nakafuku M (2005). Transdifferentiation of the retinal pigment epithelia to the neural retina by transfer of the Pax6 transcriptional factor. *Hum Mol Genet* 14(8):1059–1068.

Banin E, Obolensky A, Idelson M, Hemo I, Reinhardtz E, Pikarsky E, Ben-Hur T, Reubinoff B (2006). Retinal incorporation and differentiation of neural precursors derived from human embryonic stem cells. *Stem Cells* 24(2):246–257.

Belecky-Adams T, Adler R (2001). Developmental expression patterns of bone morphogenetic proteins, receptors, and binding proteins in the chick retina. *J Comp Neurol* 430(4):562–572.

Belecky-Adams T, Cook B, Adler R (1996). Correlations between terminal mitosis and differentiated fate of retinal precursor cells in vivo and in vitro: analysis with the "window-labeling" technique. *Dev Biol* 178(2):304–315.

Belecky-Adams T, Tomarev S, Li HS, Ploder L, McInnes RR, Sundin O, Adler R (1997). Pax-6, Prox 1, and Chx10 homeobox gene expression correlates with phenotypic fate of retinal precursor cells. *Invest Ophthalmol Vis Sci* 38(7):1293–1303.

Belecky-Adams TL, Adler R, Beebe DC (2002). Bone morphogenetic protein signaling and the initiation of lens fiber cell differentiation. *Development* 129(16):3795–3802.

Belecky-Adams TL, Scheurer D, Adler R (1999). Activin family members in the developing chick retina: expression patterns, protein distribution, and in vitro effects. *Dev Biol* 210(1):107–123.

Blancher C, Omri B, Bidou L, Pessac B, Crisanti P (1996). Nectinepsin: a new extracellular matrix protein of the pexin family. Characterization of a novel cDNA encoding a protein with an RGD cell binding motif. *J Biol Chem* 271(42):26220–26226.

Borsello T, Mottier V, Castagne V, Clarke PG (2002). Ultrastructure of retinal ganglion cell death after axotomy in chick embryos. *J Comp Neurol* 453(4):361–371.

Brazolot CL, Petitte JN, Etches RJ, Verrinder Gibbins AM (1991). Efficient transfection of chicken cells by lipofection, and introduction of transfected blastodermal cells into the embryo. *Mol Reprod Dev* 30(4):304–312.

Bumsted KM, Barnstable CJ (2000). Dorsal retinal pigment epithelium differentiates as neural retina in the microphthalmia (mi/mi) mouse. *Invest Ophthalmol Vis Sci* 41(3):903–908.

Canger AK, Rutishauser U (2004). Alteration of neural tissue structure by expression of polysialic acid induced by viral delivery of PST polysialyltransferase. *Glycobiology* 14(1):83–93.
Canto-Soler MV, Adler R (2006). Optic cup and lens development requires Pax6 expression in the early optic vesicle during a narrow time window. *Dev Biol* 294(1):119–132.
Chapman SC, Lawson A, Macarthur WC, Wiese RJ, Loechel RH, Burgos-Trinidad M, Wakefield JK, Ramabhadran R, Mauch TJ, Schoenwolf GC (2005). Ubiquitous GFP expression in transgenic chickens using a lentiviral vector. *Development* 132(5):935–940.
Chen YX, Krull CE, Reneker LW (2004). Targeted gene expression in the chicken eye by in ovo electroporation. *Mol Vis* 10:874–883.
Chesnutt C, Niswander L (2004). Plasmid-based short-hairpin RNA interference in the chicken embryo. *Genesis* 39(2):73–78.
Cho SH, Cepko CL (2006). Wnt2b/beta-catenin-mediated canonical Wnt signaling determines the peripheral fates of the chick eye. *Development* 133(16):3167–3177.
Cirillo A, Chifflet S, Villar B (2001). Neural retina of chick embryo in organ culture: effects of blockade of growth factors by suramin. *Cell Tissue Res* 304(3):323–331.
Cohen J, Nurcombe V, Jeffrey P, Edgar D (1989). Developmental loss of functional laminin receptors on retinal ganglion cells is regulated by their target tissue, the optic tectum. *Development* 107(2):381–387.
Coles BL, Angenieux B, Inoue T, Del Rio-Tsonis K, Spence JR, McInnes RR, Arsenijevic Y, van der Kooy D (2004). Facile isolation and the characterization of human retinal stem cells. *Proceedings of the National Academy of Sciences of the United States of America* 101(44):15772–15777.
Cook B, Portera-Cailliau C, Adler R (1998). Developmental neuronal death is not a universal phenomenon among cell types in the chick embryo retina. *J Comp Neurol* 396(1):12–19.
Coulombre JL, Coulombre AJ (1965). Regeneration of neural retina from the pigmented epithelium in the chick embryo. *Dev Biol* 12(1):79–92.
Coulombre JL, Coulombre AJ (1969). Lens development. IV. Size, shape, and orientation. *Invest ophth* 8(3):251–257.
Coventry S, Kapur RP, Siebert JR (1998). Cyclopamine-induced holoprosencephaly and associated craniofacial malformations in the golden hamster: anatomic and molecular events. *Pediatr Dev Pathol* 1(1):29–41.
Crisanti P, Raguenez G, Blancher C, Neron B, Mamoune A, Omri B (2001). Cloning and characterization of a novel transcription factor involved in cellular proliferation arrest: PATF. *Oncogene* 20(39):5475–5483.
Decastro M, Saijoh Y, Schoenwolf GC (2006). Optimized cationic lipid-based gene delivery reagents for use in developing vertebrate embryos. *Dev Dyn* 235(8):2210–2219.
de Curtis I, Reichardt LF (1993). Function and spatial distribution in developing chick retina of the laminin receptor alpha 6 beta 1 and its isoforms. *Development* 118(2):377–388.
Delany ME (2004). Genetic variants for chick biology research: from breeds to mutants. *Mech Dev* 121(9):1169–1177.
Del Rio-Tsonis K, Tsonis PA (2003). Eye regeneration at the molecular age. *Dev Dyn* 6(2):211–224.
Demeneix BA, Abdel-Taweb H, Benoist C, Seugnet I, Behr JP (1994). Temporal and spatial expression of lipospermine-compacted genes transferred into chick embryos in vivo. *Biotechniques* 16(3):496–501.
Dutting D, Meyer SU (1995a). Transplantations of the chick eye anlage reveal an early determination of nasotemporal polarity. *Int J Dev Biol* 39(6):921–931.
Dutting D, Thanos S (1995b). Early determination of nasal-temporal retinotopic specificity in the eye anlage of the chick embryo. *Dev Biol* 167(1):263–281.
Esteve P, Trousse F, Rodriguez J, Bovolenta P (2003). SFRP1 modulates retina cell differentiation through a beta-catenin-independent mechanism. *J Cell Sci* 116(Pt 12):2471–2481.
Fernandez-Garre P, Rodriguez-Gallardo L, Gallego-Diaz V, Alvarez IS, Puelles L (2002). Fate map of the chicken neural plate at stage 4. *Development* 129(12):2807–2822.
Fischer AJ, Reh TA (2000). Identification of a proliferating marginal zone of retinal progenitors in postnatal chickens. *Dev Biol* 220(2):197–210.
Fischer AJ, Reh TA (2001). Muller glia are a potential source of neural regeneration in the postnatal chicken retina. *Nat Neurosci* 4(3):247–252.
Fischer AJ, Reh TA (2003a). Growth factors induce neurogenesis in the ciliary body. *Dev Biol* 259(2):225–240.
Fischer AJ, Reh TA (2003b). Potential of Muller glia to become neurogenic retinal progenitor cells. *Glia* 43(1):70–76.
Fischer AJ, McGuire CR, Dierks BD, Reh TA (2002). Insulin and fibroblast growth factor 2 activate a neurogenic program in Muller glia of the chicken retina. *J Neurosci* 22(21):9387–9398.
Fischer AJ, Stanke JJ, Omar G, Askwith CC, Burry RW (2006). Ultrasound-mediated gene transfer into neuronal cells. *J Biotechnol* 122(4):393–411.
Fischer AJ, Wang SZ, Reh TA (2004). NeuroD induces the expression of visinin and calretinin by proliferating cells derived from toxin-damaged chicken retina. *Dev Dyn* 229(3):555–563.
Frade JM (2000). Unscheduled re-entry into the cell cycle induced by NGF precedes cell death in nascent retinal neurones. *J Cell Sci* 113(Pt 7):1139–1148.
Fraser RA, Carsience RS, Clark ME, Etches RJ, Gibbins AM (1993). Efficient incorporation of transfected blastodermal cells into chimeric chicken embryos. *Int J Dev Biol* 37(3):381–385.
Fuhrmann S, Kirsch M, Hofmann HD (1995). Ciliary neurotrophic factor promotes chick photoreceptor development in vitro. *Development* 121(8):2695–2706.
Fuhrmann S, Levine EM, Reh TA (2000). Extraocular mesenchyme patterns the optic vesicle during early eye development in the embryonic chick. *Development* 127(21):4599–4609.
Fujita S, Horii M (1963). Analysis of Cytogenesis in Chick Retina by Tritiated Thymidine Autoradiography. *Arch Histol Jpn* 23:359–366.
Galy A, Néron B, Planque N, Saule S, Eychène A (2002). Activated MAPK/ERK kinase (MEK-1) induces transdifferentiation of pigmented epithelium into neural retina. *Dev Biol* 248(2):251–264.
Greenhouse JJ, Petropoulos CJ, Crittenden LB, Hughes SH (1988). Helper-independent retrovirus vectors with Rous-associated virus type O long terminal repeats. *J Virol* 62(12):4809–4812.
Guillemot F, Cepko CL (1992). Retinal fate and ganglion cell differentiation are potentiated by acidic FGF in an in vitro assay of early retinal development. *Development* 114:743–754.
Gvili K, Benny O, Danino D, Machluf M (2007). Poly(D,L-lactide-co-glycolide acid) nanoparticles for DNA delivery: waiving preparation complexity and increasing efficiency. *Biopolymers* 85(5–6):379–391.
Hamburger V, Hamilton HL (1951). A series of normal stages in the development of the chick embryo. *Journal of Morphology* 88:49–92.
Harpavat S, Cepko CL (2006). RCAS-RNAi: a loss-of-function method for the developing chick retina. *Dev Biol* 22, 6:2.
Haruta M (2005). Embryonic stem cells: potential source for ocular repair. *Semin. Ophthalmol* 20(1):17–23.
Haruta M, Sasai Y, Kawasaki H, Amemiya K, Ooto S, Kitada M, Suemori H, Nakatsuji N, Ide C, Honda Y, Takahashi M (2004). In vitro and in vivo characterization of pigment epithelial cells differentiated from primate embryonic stem cells. *Invest Ophthalmol Vis Sci* 45(3):1020–1025.

Hayes S, Nelson BR, Buckingham B, Reh TA (2007). Notch signaling regulates regeneration in the avian retina. *Dev. Biol* 312(1):300–311.

Haynes T, Gutierrez C, Aycinena JC, Tsonis PA, Del Rio-Tsonis K (2007). BMP signaling mediates stem/progenitor cell-induced retina regeneration. *Proc Natl Acad Sci USA* 104(51):20380–20385.

Hernandez VH, Bueno D (2005). RNA interference is ineffective as a routine method for gene silencing in chick embryos as monitored by fgf8 silencing. *Int J Biol Sci* 1(1):1–12.

Hirano M, Yamamoto A, Yoshimura N, Tokunaga T, Motohashi T, Ishizaki K, Yoshida H, Okazaki K, Yamazaki H, Hayashi S, Kunisada T (2003). Generation of structures formed by lens and retinal cells differentiating from embryonic stem cells. *Dev Dyn.* 228(4):664–671.

Holme RH, Thomson SJ, Davidson DR (2000). Ectopic expression of Msx2 in chick retinal pigmented epithelium cultures suggests a role in patterning the optic vesicle. *Mech Dev* 91(1-2):175–187.

Horsford DJ, Nguyen MT, Sellar GC, Kothary R, Arnheiter H, McInnes RR (2005). Chx10 repression of Mitf is required for the maintenance of mammalian neuroretinal identity. *Development* 132(1):177–187.

Hu WY, Myers CP, Kilzer JM, Pfaff SL, Bushman FD (2002). Inhibition of retroviral pathogenesis by RNA interference. *Curr Biol* 12(15):1301–1311.

Huang JX, Feldmeier M, Shui YB, Beebe DC (2003). Evaluation of fibroblast growth factor signaling during lens fiber cell differentiation. *Invest Ophthalmol Vis Sci* 44(2):680–690.

Hughes SH, Greenhouse JJ, Petropoulos CJ, Sutrave P (1987). Adaptor plasmids simplify the insertion of foreign DNA into helper-independent retroviral vectors. *J Virol* 61(10):3004–3012.

Hughes S, Kosik E (1984). Mutagenesis of the region between env and src of the SR-A strain of Rous sarcoma virus for the purpose of constructing helper-independent vectors. *Virology* 136(1):89–99.

Hughes S, Mellstrom K, Kosik E, Tamanoi F, Brugge J (1984). Mutation of a termination codon affects src initiation. *Mol Cell Biol* 4(9):1738–1746.

Huxley J, De Beer GR (ed.) (1934). *The Elements of Experimental Embryology*. Cambridge, Cambridge University Press.

Hyer J, Kuhlman J, Afif E, Mikawa T (2003). Optic cup morphogenesis requires pre-lens ectoderm but not lens differentiation. *Dev Biol* 259(2):351–363.

Hyer J, Mima T, Mikawa T (1998). FGF1 patterns the optic vesicle by directing the placement of the neural retina domain. *Development* 125(5):869–877.

Ichijo H, Fujita S, Matsuno T, Nakamura H (1990). Rotation of the tectal primordium reveals plasticity of target recognition in retinotectal projection. *Development* 110(2):331–342.

Incardona JP, Gaffield W, Kapur RP, Roelink H (1998). The teratogenic Veratrum alkaloid cyclopamine inhibits sonic hedgehog signal transduction. *Development* 125(18):3553–3562.

Itasaki N, Bel-Vialar S, Krumlauf R (1999). 'Shocking' developments in chick embryology: electroporation and in ovo gene expression. *Nat Cell Biol* 1(8):E203–E207.

Itasaki N, Ichijo H, Hama C, Matsuno T, Nakamura H (1991). Establishment of rostrocaudal polarity in tectal primordium: engrailed expression and subsequent tectal polarity. *Development* 113(4):1133–1144.

Iwakiri R, Kobayashi K, Okinami S, Kobayashi H (2005). Suppression of Mitf by small interfering RNA induces dedifferentiation of chick embryonic retinal pigment epithelium. *Exp Eye Res* 81(1):15–21.

Jiang JX, Goodenough DA (1998). Retroviral expression of connexins in embryonic chick lens. *Invest Ophthalmol Vis Sci* 39(3):537–543.

Katahira T, Nakamura H (2003). Gene silencing in chick embryos with a vector-based small interfering RNA system. *Dev Growth Differ* 45(4):361–367.

Kawakami S, Higuchi Y, Hashida M (2007). Nonviral approaches for targeted delivery of plasmid DNA and oligonucleotide. *J Pharm Sci* Sep 6.

Kim JW, Lemke G (2006). Hedgehog-regulated localization of Vax2 controls eye development. *Genes Dev* 20(20):2833–2847.

Klimanskaya I, Hipp J, Rezai KA, West M, Atala A, Lanza R (2004). Derivation and comparative assessment of retinal pigment epithelium from human embryonic stem cells using transcriptomics. *Cloning Stem Cells* 6(3):217–245.

Kolpak A, Zhang J, Bao ZZ (2005). Sonic hedgehog has a dual effect on the growth of retinal ganglion axons depending on its concentration. *J Neurosci* 25(13):3432–3441.

Koo BC, Kwon MS, Choi BR, Kim JH, Cho SK, Sohn SH, Cho EJ, Lee HT, Chang W, Jeon I, Park JK, Park JB, Kim T (2006). Production of germline transgenic chickens expressing enhanced green fluorescent protein using a MoMLV-based retrovirus vector. *Faseb J* 20(13):2251–2260.

Koo BC, Kwon MS, Choi BR, Lee HT, Choi HJ, Kim JH, Kim NH, Jeon I, Chang W, Kim T (2004). Retrovirus-mediated gene transfer and expression of EGFP in chicken. *Mol Reprod Dev* 68(4):429–434.

Kos R, Tucker RP, Hall R, Duong TD, Erickson CA (2003). Methods for introducing morpholinos into the chicken embryo. *Dev Dyn* 226(3):470–477.

Krull CE (2004). A primer on using in ovo electroporation to analyze gene function. *Dev Dyn* 229(3):433–439.

Kwon MS, Koo BC, Choi BR, Lee HT, Kim YH, Ryu WS, Shim H, Kim JH, Kim NH, Kim T (2004). Development of transgenic chickens expressing enhanced green fluorescent protein. *Biochem Biophys Res Commun* 320(2):442–448.

Lamba DA, Karl MO, Ware CB, Reh TA (2006). Efficient generation of retinal progenitor cells from human embryonic stem cells. *Proc Natl Acad Sci* 103:12769–12774.

Le AC, Musil LS (2001). FGF signaling in chick lens development. *Dev Biol* 233(2):394–411.

Levine EM, Green ES (2004). Cell-intrinsic regulators of proliferation in vertebrate retinal progenitors. *Seminars in Cell & Developmental Biology* 15(1):63–74.

Le Douarin NM (1993). Embryonic neural chimaeras in the study of brain development. *Trends Neurosci* 16(2):64–72.

Li CM, Yan RT, Wang SZ (2001). Atrophy of Muller glia and photoreceptor cells in chick retina misexpressing cNSCL2. *Invest Ophthalmol Vis Sci* 42(13):3103–3109.

Liang L, Yan RT, Ma W, Zhang H, Wang SZ (2006). Exploring RPE as a source of photoreceptors: differentiation and integration of transdifferentiating cells grafted into embryonic chick eyes. *Invest Ophthalmol Vis Sci* 47(11):5066–5074.

Lillie FR (ed.) (1908). *The Development of the Chick; An Introduction to Embryology*. Henry Holt, New York.

Liu W, Mo Z, Xiang M (2001). The Ath5 proneural genes function upstream of Brn3 POU domain transcription factor genes to promote retinal ganglion cell development. *Proc Natl Acad Sci USA* 98(4):1649–1654.

Limb GA, Daniels JT, Cambrey AD, Secker GA, Shortt AJ, Lawrence JM, Khaw PT (2006). Current prospects for adult stem cell-based therapies in ocular repair and regeneration. *Curr Eye Res* 31(5):381–390.

Lopashov GV, Stroeva OG (1964). *Development of the Eye. Experimental Studies*. Israel Program for Scientific Transations, Jerusalem.

Luo J, Redies C (2005). Ex ovo electroporation for gene transfer into older chicken embryos. *Dev Dyn* 233(4):1470–1477.

Lwigale PY, Bronner-Fraser M (2007). Lens-derived Semaphorin3A regulates sensory innervation of the cornea. *Dev Biol* 306(2):750–759.

Ma W, Yan RT, Xie W, Wang SZ (2004). bHLH genes cath5 and cNSCL1 promote bFGFstimulated RPE cells to transdifferentiate toward retinal ganglion cells. *Dev Biol* 265(2):320–328.

Malpighi M (ed.) (1672). *De formactione pulli in ovo.* Royal Society of London, London.

Malpighi M (1675). *Repetitas auctasque de ovo incubato observationes continens.* Johannis Martyn, London.

Martinez-Morales JR, Del Bene F, Nica G, Hammerschmidt M, Bovolenta P, Wittbrodt J (2005). Differentiation of the vertebrate retina is coordinated by an FGF signaling center. *Dev Cell* 8(4):565–574.

Martinez-Morales JR, Dolez V, Rodrigo I, Zaccarini R, Leconte L, Bovolenta P, Saule S (2003). OTX2 activates the molecular network underlying retina pigment epithelium differentiation. *J Biol Chem* 278(24):21721–21731.

Martinez-Morales JR, Rodrigo I, Bovolenta P (2004). Eye development: a view from the retina pigmented epithelium. *Bioessays.* 26(7):766–777.

Martinez-Morales JR, Signore M, Acampora D, Simeone A, Bovolenta P (2001). Otx genes are required for tissue specification in the developing eye. *Development* 128(11):2019–2030.

Matsuo T, Takabatake M, Matsuo N (1997). The effects of growth factors on multicellular spheroids formed by chick embryonic retinal cells. *Acta Med Okayama* 51(5):251–260.

McCabe KL, Gunther EC, Reh TA (1999). The development of the pattern of retinal ganglion cells in the chick retina: mechanisms that control differentiation. *Development* 126(24):5713–5724.

Mitashov VI (1996). Mechanisms of retina regeneration in urodeles. *Int. J. Dev. Biol* 40(4):833–844.

Mitashov VI (1997). Retinal regeneration in amphibians. *Int J Dev Biol* 41(6):893–905.

Mochii M, Ono T, Matsubara Y, Eguchi G (1998a). Spontaneous transdifferentiation of quail pigmented epithelial cell is accompanied by a mutation in the Mitf gene. *Dev Biol* 196(2):145–159.

Mochii M, Mazaki Y, Mizuno N, Hayashi H, Eguchi G (1998b). Role of Mitf in differentiation and transdifferentiation of chicken pigmented epithelial cell. *Dev Biol* 193(1):47–62.

Mozdziak PE, Borwornpinyo S, McCoy DW, Petitte JN (2003). Development of transgenic chickens expressing bacterial beta-galactosidase. *Dev Dyn* 226(3):439–445.

Mozdziak PE, Petitte JN (2004). Status of transgenic chicken models for developmental biology. *Dev Dyn* 229(3):414–421.

Mozdziak PE, Wysocki R, Angerman-Stewart J, Pardue SL, Petitte JN (2006 Oct). Production of chick germline chimeras from fluorescence-activated cell-sorted gonocytes. *Poult Sci.* 85(10):1764–1768.

Mohammadi M, McMahon G, Sun L, Tang C, Hirth P, Yeh BK, Hubbard SR, Schlessinger J (1997). Structures of the tyrosine kinase domain of fibroblast growth factor receptor in complex with inhibitors. *Science* 276(5314):955–960.

Moreira EF, Adler R (2006). Effects of follistatin overexpression on cell differentiation in the chick embryo retina. *Dev Biol* 298(1):272–284.

Morgan BA, Izpisua-Belmonte JC, Duboule D, Tabin CJ (1992). Targeted misexpression of Hox-4.6 in the avian limb bud causes apparent homeotic transformations. *Nature* 358(6383):236–239.

Mueller BK, Dutting D, Haase A, Feucht A, Macchi P (1998). Partial respecification of nasotemporal polarity in double-temporal chick and chimeric chick-quail eyes. *Mech Dev* 74(1–2):15–28.

Müller F, Rohrer H, Vogel-Höpker A (2007). Bone morphogenetic proteins specify the retinal pigment epithelium in the chick embryo. *Development (Cambridge, England)* 134(19):3483–3493.

Muramatsu T, Mizutani Y, Ohmori Y, Okumura J (1997). Comparison of three nonviral transfection methods for foreign gene expression in early chicken embryos in ovo. *Biochem Biophys Res Commun* 230(2):376–380.

Nakagawa S, Takada S, Takada R, Takeichi M (2003). Identification of the laminar-inducing factor: Wnt-signal from the anterior rim induces correct laminar formation of the neural retina in vitro. *Dev Biol* 260(2):414–425.

Nakamoto M, Cheng HJ, Friedman GC, McLaughlin T, Hansen MJ, Yoon CH, O'Leary DD, Flanagan JG (1996). Topographically specific effects of ELF-1 on retinal axon guidance in vitro and retinal axon mapping in vivo. *Cell* 86(5):755–766.

Nakamura H, Funahashi J (2001). Introduction of DNA into chick embryos by in ovo electroporation. *Methods* 24(1):43–48.

Nakamura H, Itasaki N, Matsuno T (1994). Rostrocaudal polarity formation of chick optic tectum. *Int J Dev Biol* 38(2):281–286.

Nakamura H, Katahira T, Sato T, Watanabe Y, Funahashi J (2004). Gain- and loss-of-function in chick embryos by electroporation. *Mech Dev* 121(9):1137–1143.

Nakamura H, Watanabe Y, Funahashi J (2000). Misexpression of genes in brain vesicles by in ovo electroporation. *Dev Growth Differ* 42(3):199–201.

Nguyen M, Arnheiter H (2000). Signaling and transcriptional regulation in early mammalian eye development: a link between FGF and MITF. *Development* 127(16):3581–3591.

Ogino H, Yasuda K (1998). Induction of lens differentiation by activation of a bZIP transcription factor, L-Maf. *Science* 280(5360):115–118.

Ohkubo Y, Chiang C, Rubenstein JL (2002). Coordinate regulation and synergistic actions of BMP4, SHH and FGF8 in the rostral prosencephalon regulate morphogenesis of the telencephalic and optic vesicles. *Neuroscience* 111(1):1–17.

Ohta S, Suzuki K, Tachibana K, Yamada G (2003). Microbubble-enhanced sonoporation: efficient gene transduction technique for chick embryos. *Genesis* 37(2):91–101.

Opas M, Davies JR, Zhou Y, Dziak E (2001). Formation of retinal pigment epithelium in vitro by transdifferentiation of neural retina cells. *Int J Dev Biol* 45(4):633–642.

Opas M, Dziak E (1994). bFGF-induced transdifferentiation of RPE to neuronal progenitors is regulated by the mechanical properties of the substratum. *Devel Biol* 161(2):440–454.

Ooto S, Haruta M, Honda Y, Kawasaki H, Sasai Y, Takahashi M (2003). Induction of the differentiation of lentoids from primate embryonic stem cells. *Invest Ophthalmol Vis Sci* 44(6): 2689–2693.

Park CM, Hollenberg MJ (1989). Basic fibroblast growth factor induces retinal regeneration in vivo. *Dev Biol* 134(1):201–205.

Park CM, Hollenberg MJ (1991). Induction of retinal regeneration in vivo by growth factors. *Dev Biol* 1 48(1):322–333.

Pekarik V, Bourikas D, Miglino N, Joset P, Preiswerk S, Stoeckli ET (2003). Screening for gene function in chicken embryo using RNAi and electroporation. *Nat Biotechnol.* 21(1):93–96. Erratum in: Nat Biotechnol. 2003 Feb; 21:199.

Perron M, Boy S, Amato MA, Viczian A, Koebernick K, Pieler T, Harris WA (2003). A novel function for Hedgehog signalling in retinal pigment epithelium differentiation. *Development* 130(8):1565–1577.

Pimentel B, Sanz C, Varela-Nieto I, Rapp UR, De Pablo F, de La Rosa EJ (2000). c-Raf regulates cell survival and retinal ganglion cell morphogenesis during neurogenesis. *J Neurosci* 20(9):3254–3262.

Pittack C, Jones M, Reh TA (1991). Basic fibroblast growth factor induces retinal pigment epithelium to generate neural retina *in vitro*. *Development (Cambridge, England)* 113(2):577–588.

Planque N, Leconte L, Coquelle FM, Martin P, Saule S (2001). Specific Pax-6/microphthalmia transcription factor interactions involve their DNA-binding domains and inhibit transcriptional properties of both proteins. *J Biol Chem* 3: 276(31):29330–29337.

Planque N, Raposo G, Leconte L, Anezo O, Martin P, Saule S (2004). Microphthalmia transcription factor induces both retinal pigmented epithelium and neural crest melanocytes from neuroretina cells. *J Biol Chem* 279(40):41911–41917.

Planque N, Turque N, Opdecamp K, Bailly M, Martin P, Saule S (1999). Expression of the microphthalmia-associated basic helix-loop-helix leucine zipper transcription factor Mi in avian neuroretina cells induces a pigmented phenotype. *Cell Growth Differ* 10(7):525–536.

Prada C, Puga J, Pérez-Méndez L, López R, Ramírez G (1991). Spatial and Temporal Patterns of Neurogenesis in the Chick Retina. *Eur J Neurosci* 3(6):559–569.

Rao M, Baraban JH, Rajaii F, Sockanathan S (2004). In vivo comparative study of RNAi methodologies by in ovo electroporation in the chick embryo. *Dev Dyn* 231(3):592–600.

Repka A, Adler R (1992a). Differentiation of retinal precursor cells born in vitro. *Dev Biol* 153(2):242–249.

Repka AM, Adler R (1992b). Accurate determination of the time of cell birth using a sequential labeling technique with [3H]-thymidine and bromodeoxyuridine ("window labeling"). *J Histochem Cytochem* 40(7):947–953.

Riddle RD, Johnson RL, Laufer E, Tabin C (1993). Sonic hedgehog mediates the polarizing activity of the ZPA. *Cell* 75(7):1401–1416.

Rosenblum CI, Chen HY (1995). In ovo transfection of chicken embryos using cationic liposomes. *Transgenic Res* 4(3):192–198.

Rothermel A, Volpert K, Burghardt M, Lantzsch C, Robitzki AA, Layer PG (2006). Knock-down of GFRalpha4 expression by RNA interference affects the development of retinal cell types in three-dimensional histiotypic retinal spheres. *Invest Ophthalmol Vis Sci* 47(6):2716–2725.

Rothermel A, Willbold E, Degrip WJ, Layer PG (1997). Pigmented epithelium induces complete retinal reconstitution from dispersed embryonic chick retinae in reaggregation culture. *Proc Biol Sci* 264(1386):1293–1302.

Rowan S, Chen CM, Young TL, Fisher DE, Cepko CL (2004). Transdifferentiation of the retina into pigmented cells in ocular retardation mice defines a new function of the homeodomain gene Chx10. *Development* 131(20):5139–5152.

Saito M, Mazda O, Takahashi KA, Arai Y, Kishida T, Shin-Ya M, Inoue A, Tonomura H, Sakao K, Morihara T, Imanishi J, Kawata M, Kubo T (2007). Sonoporation mediated transduction of pDNA/siRNA into joint synovium in vivo. *J Orthop Res* 25(10):1308–1316.

Sakami S, Etter P, Reh TA (2008). Activin signaling limits the competence for retinal regeneration from the pigmented epithelium. *Mech Dev* 125(1-2):106–116.

Sakami S, Hisatomi O, Sakakibara S, Liu J, Reh TA, Tokunaga F (2005). Downregulation of Otx2 in the dedifferentiated RPE cells of regenerating newt retina. *Brain Res Dev Brain Res.* 155(1):49–59.

Sakuta H, Suzuki R, Takahashi H, Kato A, Shintani T, Iemura S, Yamamoto TS, Ueno N, Noda M (2001). Ventroptin: a BMP-4 antagonist expressed in a double-gradient pattern in the retina. *Science* 293(5527):111–115.

Sato Y, Kasai T, Nakagawa S, Tanabe K, Watanabe T, Kawakami K, Takahashi Y (2007). Stable integration and conditional expression of electroporated transgenes in chicken embryos. *Dev Biol* 305(2):616–624.

Sato N, Matsuda K, Sakuma C, Foster DN, Oppenheim RW, Yaginuma H (2002). Regulated gene expression in the chicken embryo by using replication-competent retroviral vectors. *J Virol* 76(4):1980–1985.

Schulte D, Furukawa T, Peters MA, Kozak CA, Cepko CL (1999). Misexpression of the Emx-related homeobox genes cVax and mVax2 ventralizes the retina and perturbs the retinotectal map. *Neuron* 24(3):541–553.

Sehgal R, Andres DJ, Adler R, Belecky-Adams TL (2006). Bone morphogenetic protein 7 increases chick photoreceptor outer segment initiation. *Invest Ophthalmol Vis Sci* 47(8):3625–3634.

Spence JR, Aycinena JC, Del Rio-Tsonis K (2007). Fibroblast growth factor-hedgehog interdependence during retina regeneration. *Dev Dyn* 236(5):1161–1174.

Spence JR, Madhavan M, Aycinena JC, Del Rio-Tsonis K (2007b). Retina regeneration in the chick embryo is not induced by spontaneous Mitf downregulation but requires FGF/FGFR/MEK/Erk dependent upregulation of Pax6. *Mol Vis* 13:57–65.

Spence JR, Madhavan M, Ewing JD, Jones DK, Lehman BM, Del Rio-Tsonis K (2004). The hedgehog pathway is a modulator of retina regeneration. *Development* 131(18):4607–4621.

Stern CD (2005). The chick; a great model system becomes even greater. *Dev cell* 8(1):9–17.

Swartz M, Eberhart J, Mastick GS, Krull CE (2001). Sparking new frontiers: using in vivo electroporation for genetic manipulations. *Dev Biol* 233(1):13–21.

Takahashi M, Haruta M (2006). Derivation and characterization of lentoid bodies and retinal pigment epithelial cells from monkey embryonic stem cells in vitro. *Methods Mol Biol* 330:417–429.

Thanos S, Mey J, Dutting D, Hummler E (1996). Positional determination of the naso-temporal retinal axis coincides with asymmetric expression of proteins along the anterior-posterior axis of the eye primordium. *Exp Eye Res* 63(5):479–492.

Timmer J, Johnson J, Niswander L (2001). The use of in ovo electroporation for the rapid analysis of neural-specific murine enhancers. *Genesis* 29(3):123–132.

Trousse F, Esteve P, Bovolenta P (2001). Bmp4 mediates apoptotic cell death in the developing chick eye. *J Neurosci* 21(4):1292–1301.

Toy J, Yang JM, Leppert GS, Sundin OH (1998). The optx2 homeobox gene is expressed in early precursors of the eye and activates retina-specific genes. *Proc Natl Acad Sci USA* 95(18): 10643–10648.

Tsonis PA, Del Rio-Tsonis K (2004). Lens and retina regeneration: transdifferentiation, stem cells and clinical applications. *Exp Eye Res* 78(2):161–172.

Uchikawa M, Takemoto T, Kamachi Y, Kondoh H (2004). Efficient identification of regulatory sequences in the chicken genome by a powerful combination of embryo electroporation and genome comparison. *Mech Dev* 121(9):1145–1158.

Uemonsa T, Sakagami K, Yasuda K, Araki M (2002). Development of dorsal-ventral polarity in the optic vesicle and its presumptive role in eye morphogenesis as shown by embryonic transplantation and *in ovo* explant culturing. *Dev Biol* 248(2):319–330.

Vogel-Höpker A, Momose T, Rohrer H, Yasuda K, Ishihara L, Rapaport DH (2000 Jun). Multiple functions of fibroblast growth factor-8 in chick eye development. *Mech Dev* 94(1–2):25–36.

Vugler A, Lawrence J, Walsh J, Carr A, Gias C, Semo M, Ahmado A, da Cruz L, Andrews P, Coffey P (2007). Embryonic stem cells and retinal repair. *Mech Dev* 124(11-12):807–829.

Waid DK, McLoon SC (1998). Ganglion cells influence the fate of dividing retinal cells in culture. *Development (Cambridge, England)* 125(6):1059–1066.

Wallis JW et al. (2004). A physical map of the chicken genome. *Nature* 432(7018):761–764.

Watanabe T, Saito D, Tanabe K, Suetsugu R, Nakaya Y, Nakagawa S, Takahashi Y (2007). Tet-on inducible system combined with in ovo electroporation dissects multiple roles of genes in somitogenesis of chicken embryos. *Dev Biol* 305(2):625–636.

Weng J, Belecky-Adams T, Adler R, Travis GH (1998). Identification of two rds/peripherin homologs in the chick retina. *Invest Ophthalmol Vis Sci* 39(2):440–443.

Wilson JM, Sato K, Chernoff EA, Belecky-Adams TL (2007). Expression patterns of chick Musashi-1 in the developing nervous system. *Gene Expr Patterns* 7(7):817–825.

Xie W, Yan RT, Ma W, Wang SZ (2004). Enhanced retinal ganglion cell differentiation by ath5 and NSCL1 coexpression. *Invest Ophthalmol Vis Sci.* 45(9):2922–2928.

Yamagata M, Sanes JR (1995). Target-independent diversification and target-specific projection of chemically defined retinal ganglion cell subsets. *Development* 121(11):3763–3776.

Yamaguchi S, Katagiri S, Hirose N, Fujimoto Y, Mori M, Fujii-Taira I, Takano T, Matsushima T, Homma KJ (2007). In-vivo gene transfer into newly hatched chick brain by electroporation. *Neuroreport* 18(8):735–739.

Yan RT, Ma W, Liang L, Wang SZ (2005). bHLH genes and retinal cell fate specification. *Mol Neurobiol* 32(2):157–171.

Yan RT, Ma WX, Wang SZ (2001). neurogenin2 elicits the genesis of retinal neurons from cultures of nonneural cells. *Proc Natl Acad Sci USA* 98(26):15014–15019.

Yan RT, Wang SZ (1998). NeuroD induces photoreceptor cell overproduction in vivo and de novo generation *in vitro*. *J Neurobiol* 36(4):485–496.

Yan RT, Wang SZ (2000a). Differential induction of gene expression by basic fibroblast growth factor and neuroD in cultured retinal pigment epithelial cells. *Vis Neurosci.* 17(2):157–164.

Yan RT, Wang SZ (2000b). Expression of an array of photoreceptor genes in chick embryonic retinal pigment epithelium cell cultures under the induction of neuroD. *Neurosci Lett* 280(2):83–86.

Yang XJ (2004). Roles of cell-extrinsic growth factors in vertebrate eye pattern formation and retinogenesis. *Semin Cell Dev Biol* 15(1):91–103.

Yasuda K, Momose T, Takahashi Y (2000). Applications of microelectroporation for studies of chick embryogenesis. *Dev Growth Differ* 42(3):203–206.

Yasugi S, Nakamura H (2000). Gene transfer into chicken embryos as an effective system of analysis in developmental biology. *Dev Growth Differ* 42(3):195–197.

Yeh JR, Crews CM (2003). Chemical genetics: adding to the developmental biology toolbox. *Dev Cell* 5(1):11–19.

Zhang XM, Yang XJ (2001 May 15). Temporal and spatial effects of Sonic hedgehog signaling in chick eye morphogenesis. *Dev Biol* 233(2):271–290.

Zhao X, Liu J, Ahmad I (2002). Differentiation of embryonic stem cells into retinal neurons. *Biochem Biophys Res Commun* 297(2):177–184.

Zhao X, Liu J, Ahmad I (2006). Differentiation of embryonic stem cells to retinal cells *in vitro*. *Methods Mol Biol, Clifton, NJ* 330:401–416.

Zhou Y, Opas M (1994). Cell shape, intracellular pH, and fibroblast growth factor responsiveness during transdifferentiation of retinal pigment epithelium into neuroepithelium in vitro. *Biochem Cell Biol* 72(7–8):257–265.

Zinn KM (1970). Changes in corneal ultrastructure resulting from early lens removal in the developing chick embryo. *Invest Ophthalmol* 9(3):165–182.

CHAPTER

9

Eye Development Using Mouse Genetics

Ni Song[1,2], *Richard A. Lang*[1,2]
[1]Divisions of Pediatric Ophthalmology and Developmental Biology, Children's Hospital Research Foundation, Cincinnati, OH 45229, USA
[2]Department of Ophthalmology, and Graduate Program of Molecular and Developmental Biology, College of Medicine, University of Cincinnati, Cincinnati, OH 45229, USA

OUTLINE

INTRODUCTION

There is a relationship between the age of a research field and the sophistication of the models used. Eye development is no exception. In the beginning when Spemann and Mencl were investigating lens induction (Spemann, 1901; Mencl, 1903), the model was an amphibian capable of external development and the experimental tools were the hot needles and cutting tools that allowed physical manipulation of the embryo. Now, the state-of-the-art amphibian experiment can involve fate-mapping (Moody, 1987; Huang and Moody, 1993), Morpholino® oligonucleotide-mediated gene suppression (Heasman, 2002), *in situ* hybridization detection of mRNAs (Hirsch and Harris, 1997), antibody detection of gene products (Hemmati-Brivanlou *et al.*, 1992), and the use of overexpressed (Chow *et al.*, 1999) and inducible (Kolm and Sive, 1995) protein activities.

At the time of Spemann and Mencl, the mouse was not an especially useful experimental system for studying development. Only recently, as we have understood how to establish genetically modified lines, has the mouse become a powerful experimental model useful for all kinds of analysis. We are strongly influenced to use the mouse by the need for a good model for the human. Being a mammal, the mouse fulfills this role. The purpose of this chapter is to summarize the genetic methods that have made the mouse a powerful tool in developmental analysis. We illustrate the unique features of the mouse using examples of analysis that have advanced our understanding of eye development.

NATURALLY OCCURRING MUTANTS

The mouse began its career as an eye development research subject with the identification of naturally occurring mutants with eye defects. These typically emerged from a mouse colony as a mouse with an eyeless or aphakic (lensless) phenotype after the astute investigator realized their value and propagated a line. Early, the type of analysis performed was usually simple histology at different developmental stages. This led to an understanding of the developmental basis of the defect and often an idea of which cells carried the primary defect. Typical examples of this type of mouse include *Dysgenetic lens* (Sanyal and Hawkins, 1979) and *Aphakia* (Varnum and Stevens, 1968) both of which have defects in the early stages of lens development. In more recent years, the *Dysgenetic lens* mutation has been assigned to the *FoxE3* gene (Blixt *et al.*, 2000; Brownell *et al.*, 2000), a transcription factor of the forkhead class. Similarly *Aphakia* encodes *PitX3*, a homeodomain transcription factor (Semina *et al.*, 2000; Rieger *et al.*, 2001).

For the eye development field, arguably the most important naturally occurring mouse mutants identified are the *Small eye* series of alleles. These were identified as semi-dominant mutations that gave small eyes in the heterozygous state. There are four different alleles of *Small eye* isolated from different laboratories at different times (Hill *et al.*, 1991; Ton *et al.*, 1992). Examination of embryos at early stages of development revealed that although the early steps in forming an optic vesicle occurred in *Small eye* mice, the final outcome was a complete absence of eyes (Fig. 9.1) (Hogan *et al.*, 1986b). We now know that *Small eye* encodes a paired and homeodomain transcription factor Pax6 that has a critical role at multiple stages of eye development in both humans and mice (Hill *et al.*, 1991; Glaser *et al.*, 1992; Jordan *et al.*, 1992). Furthermore, the discovery that the *Drosophila eyeless* gene was a Pax6 orthologue (Quiring *et al.*, 1994) was exciting because it indicated that the same gene was

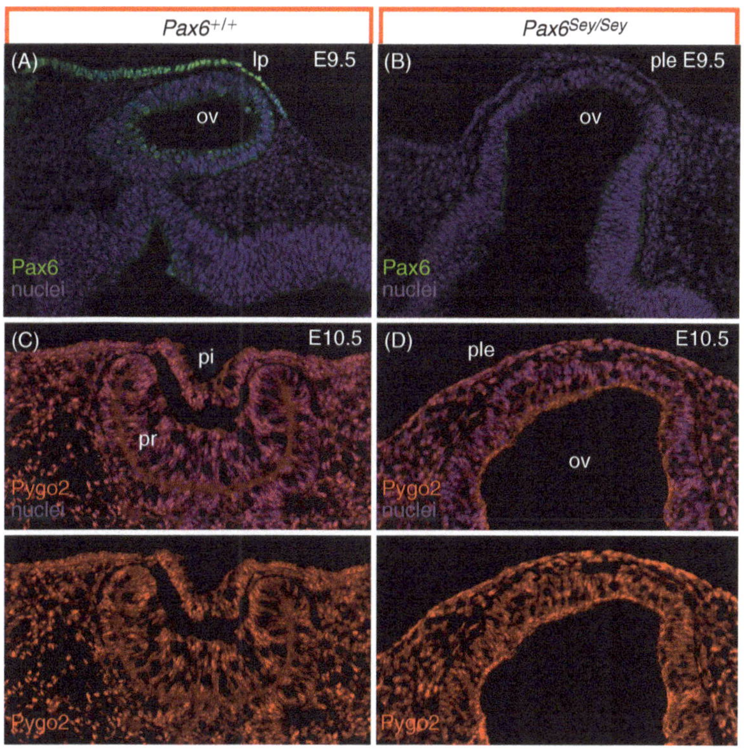

FIGURE 9.1 Failure of lens induction in $Pax6^{sey/sey}$ embryos. (A–D) Eye region cryosections labeled for nuclei with Hoechst 33258 (A–D, blue), Pax6 (A, B, green), Pygo2 (C, D, red). Note: absence of lens pit formation and abnormal opening of proximal optic vesicle in the *Small eye* mutants. Ov, optic vesicle; lp, lens placode, ple, presumptive lens ectoderm; pi, lens pit; pr, presumptive retina.

required for eye development in both insects and flies. This realization has been used to argue that evolution of different complex eye types occurred from a simple, primordial form and was monophyletic (Gehring, 2002), but see (Salvini-Plawen and Mayr, 1977). The observation that in flies and frogs, misexpression of *eyeless/Pax6* could induce the formation of ectopic eyes also reinforced the idea that Pax6 lies at the apex of a genetic hierarchy regulating eye development (Halder *et al.*, 1995; Chow *et al.*, 1999). The mouse $Pax6^{Small\ eye}$ ($Pax6^{Sey}$) alleles continue to be an important resource in the analysis of mammalian eye development, Fig. 9.2.

TRANSGENIC MOUSE LINES

The method to introduce DNA into the zygote of developing mice – the generation of transgenic mice – was established by the early 1980s. The first paper (Gordon *et al.*, 1980) described the insertion of a pBR332-derived plasmid that carried SV40 and Herpes *thymidine kinase* (*TK*) sequences. Subsequently, a group of papers described transgenic lines in which incorporated recombinant genes were expressed. In one case, the *TK* gene was expressed from the *metallothionein-I* promoter (Brinster *et al.*, 1981) in another, the β-*globin* gene was introduced (Costantini and Lacy, 1981). Subsequently, it became clear that introduction of foreign genes into the mouse germ line could be a routine procedure and that this method could be used to study gene function and regulation (Hogan *et al.*, 1986a).

A key finding in the early days of transgenic mouse generation was that the regulatory elements of the introduced gene could to a large degree control the tissue type expression pattern. For example, the *metallothionein-I* promoter gave gene expression in the liver and kidney (Brinster *et al.*, 1981) while that for *αA-crystallin* gave expression in the lens (Overbeek *et al.*, 1985). This feature of transgenic mouse constructs has been thoroughly exploited for many different purposes. The high expression levels and tissue specificity of the crystallin promoters have helped to make the lens a favorite tissue in which to address different biological questions (Lok *et al.*, 1985; Breitman *et al.*, 1987; Chow *et al.*, 1995; Reneker and Overbeek, 1996; Lovicu *et al.*, 1999; Zhao and Overbeek, 2001; Faber *et al.*, 2002). As other expression specificities have been discovered, the variety of experimental possibilities has expanded. For example, we now have the ability to target gene expression to either the presumptive lens ectoderm (Williams *et al.*, 1998; Xu *et al.*, 1999; Ashery-Padan *et al.*, 2000) or the optic vesicle (Swindell *et al.*, 2006) the two main interacting tissues from which the eye is constructed. These expression patterns allow the mechanisms of presumptive retina–presumptive lens interactions to be examined.

A major use of transgenic mouse technology in the eye development field has been to over-express a protein with

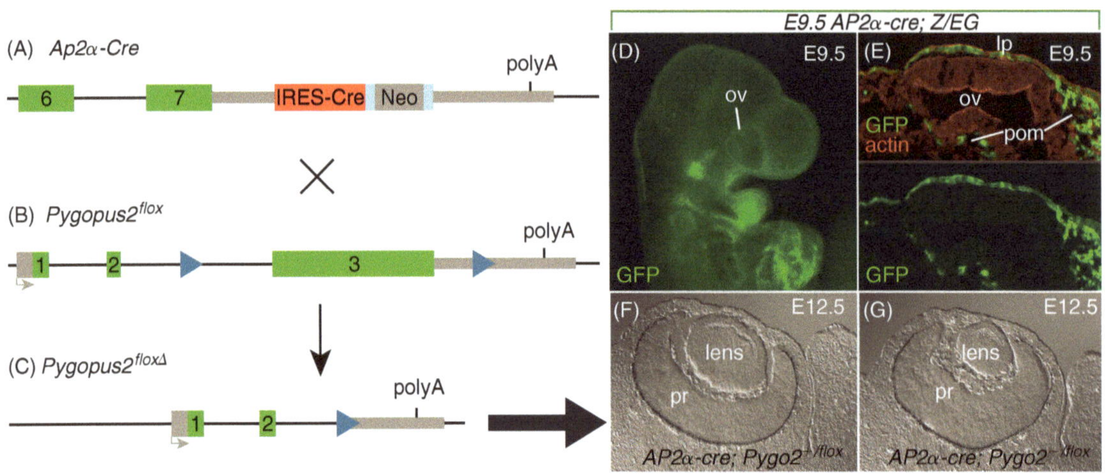

FIGURE 9.2 A small lens defect in *AP2α-cre; Pygo2*$^{flox/-}$ embryos. (A–C) Schematic of somatic mutation of *Pygo2* conditional allele. By combining *AP2α-cre* (A) and *Pygo2*^{flox} (B) alleles, Pygo2 function is eliminated in cells of interest. Green and gray boxes represent, respectively, coding and non-coding exons. Red box indicates *Cre* cDNA linked with internal ribosome entry sequence (IRES). The box marked Neo represents the positive selectable gene *Neomycin*. Light blue bars show *frt* sites used for deletion of *Neo* in the final allele. Blue arrowheads denote *LoxP* sites. (D–G) Genotypes as labeled. (D) Whole mount embryos visualized for GFP. (E–G) Cryosections labeled for nuclei with GFP (E, green) and F-actin (E, red) or unlabeled and DIC illuminated (F, G). Ple, presumptive lens ectoderm; ov, optic vesicle; pom, periocular mesenchyme; lp, lens placode; pr, presumptive retina.

a particular biological function as a means of determining, whether a particular signaling pathway or process is involved. A good example is the fibroblast growth factor (FGF) signaling pathway, as it has an important role in development of both lens and retina. In one set of experiments, the Overbeek laboratory misexpressed various forms of FGF ligands using the *αA-crystallin* promoter that gives high expression in the developing lens fiber cells (Robinson *et al.*, 1995b; Robinson *et al.*, 1998; Lovicu *et al.*, 1999). In some cases, this stimulated the differentiation of fiber cells from the overlying lens epithelium (Robinson *et al.*, 1995b). In the case of FGF7 overexpression, ectopic periocular glands were stimulated to develop (Lovicu *et al.*, 1999). By making *αA-crystallin* promoter transgenic mice with many FGF ligands, the Overbeek laboratory was able to define which ligands could stimulate fiber cell differentiation.

During the execution of this work, a number of laboratories took an alternative approach for investigating the role of the FGF pathway in lens development and expressed a truncated, dominant negative form of FGFR1 in the lens, again using the *αA-crystallin* promoter. This resulted in the diminished elongation and differentiation of lens fiber cells (Chow *et al.*, 1995; Robinson *et al.*, 1995a). Combined, these data argued that the FGF pathway was an important signaling event in fiber cell differentiation. A similar experimental strategy was used in combination with the *Pax6* ectoderm enhancer (*EE*) to demonstrate that the FGF pathway also had a role during the inductive phases of lens development (Faber *et al.*, 2001).

One of the more recent applications of transgenic mice has been the generation of mouse lines in which cre recombinase is expressed in different tissue types. Cre recombinase is a site-specific recombinase derived from phage P1 that normally regulates lysogeny of the phage genome through DNA recombination reactions at *loxP* sites (Shaikh and Sadowski, 1997). It has been successfully adapted for mouse genetic analysis and is often used to perform somatic mutation of so called conditional alleles in which a critical region of a gene is flanked by loxP sites (a "floxed" allele) through gene targeting (Lewandoski, 2001; Yu and Bradley, 2001). This has been particularly important to allow the study of genes that give a lethal phenotype when deleted in the germ line or where cell–cell interactions are the object of study. Table 9.1 lists a series of cre recombinase expressing mouse lines that have proven useful in the study of eye development. With all these lines, there is a fairly dense coverage of the different tissue types and developmental stages for eye development. The somatic mutation options will no doubt become greater as more cre recombinase expressing mouse lines are derived.

GENE TARGETING

Gene targeting is arguably the single most important technological advance in the history of the mouse as an experimental organism (Joyner, 1995). The ability to modify a mouse gene in a way that can be precisely defined to the base pair has had broad-reaching consequences. For all kinds of biologists, gene targeting in the mouse has become an essential tool. For developmental biologists, gene targeting is critical in performing the analysis that allows genetic and biochemical pathways to be defined. Human geneticists can generate models of human diseases by mimicking naturally occurring human mutational events using this method. At last count about 4000 genes had been targeted in mice and another 7000 were mutated by gene-trapping techniques in ES cells (Collins *et al.*, 2007b). Eliminating overlap, this means that about 9000 unique mouse genes have been targeted (Collins *et al.*, 2007b). This is a sizable portion of the approximately 22,000 genes in the mouse genome and a wonderful resource for the experimentalist. To build on this, the International Mouse Knockout Consortium plans to target, either with germ line null, conditional null or gene-trap alleles, of all the genes in the mouse genome (Collins *et al.*, 2007a).

The Germ Line Null Allele

When gene targeting was first being established, targeting schemes were simple (Joyner, 1993). This often meant that a straightforward loss-of-function allele of the gene was generated by deleting a region that was critical. In many cases, the *Neo* gene expressed from the *PGK* promoter was used as a positive selectable marker and after homologous recombination, remained in place. Subsequently it was shown that *PGK-Neo* could modify expression of genes in the region, sometimes giving a false genotype–phenotype link. To overcome this, gene targeting now generally uses a *PGK-Neo* expression unit that is flanked by *frt* sites and can therefore be deleted with the site-specific recombinase flippase (Joyner, 1995). The *frt-PGK-Neo-frt* cassette can either be deleted in ES cells by transfecting them with a flippase expressing plasmid or by crossing gene targeted mice with those that express flippase in the germ line (Joyner, 1995). The generation of germ line null alleles has resulted in many wonderful advances in vision research. A few examples will suffice.

The bone morphogenetic proteins (BMPs) are a family of signaling molecules that function through serine-threonine receptor kinases to activate Smad family

TABLE 9.1 Cre recombinase mouse lines useful for studies of eye development

Line (type)	Activity	Origin	Usage
Le-cre(transgene)	Lens placode, lens vesicle, periocular ectoderm and glands, lens epithelium, lens fiber cells, corneal epithelium	(Ashery-Padan *et al.*, 2000)	(Lakso *et al.*, 1992; Davis-Silberman *et al.*, 2005; Dwivedi *et al.*, 2005; Garcia *et al.*, 2005; Smith *et al.*, 2005; Yoshimoto *et al.*, 2005; Liu *et al.*, 2006; Song *et al.*, 2007; Swamynathan *et al.*, 2007)
MLR10 (transgene)	Lens vesicle, lens fiber cells	(Zhao *et al.*, 2004)	(Ebong *et al.*, 2004; Zhao *et al.*, 2006)
AP2-cre (knock-in)	Dorsal neural tube, neural crest including periocular mesenchyme, head surface ectoderm including presumptive lens, all lens cells, corneal mesenchyme	(Macatee *et al.*, 2003)	(Macatee *et al.*, 2003; Song *et al.*, 2007)
Wnt1-cre (transgene)	Dorsal neural tube, neural crest including periocular mesenchyme, corneal mesenchyme	(Danielian *et al.*, 1998a,b)	(Brault *et al.*, 2001; Jaskoll *et al.*, 2002; Smith *et al.*, 2005; Yoshida *et al.*, 2006; Song *et al.*, 2007)
Six3-cre (transgene)	Retina from ~E9.5, ventral forebrain	(Furuta *et al.*, 2000)	(Murali *et al.*, 2005; Fu *et al.*, 2006)
TRP1-cre	Retinal pigment epithelium from E10.5	(Mori *et al.*, 2002)	(Marneros *et al.*, 2005)
L7/pcp-2:cre	Retinal bipolar neurons and Purkinje cells	(Barski *et al.*, 2000)	(Barski *et al.*, 2003)
Nestin-cre	Retinal neurons; distal retina, ciliary epithelium, iris and lens in adults	(Haigh *et al.*, 2003)	(Calera *et al.*, 2006)
M-opsin-cre	M-cone photoreceptors	(Akimoto *et al.*, 2004)	
S-opsin-cre	S-cone photoreceptors	(Akimoto *et al.*, 2004)	
Chx10-cre	Retinal progenitors, Muller glia subset	(Rowan and Cepko, 2004)	(Jadhav *et al.*, 2006)
SMOPC1-cre (transgene)	Rod photoreceptor cells	(Le *et al.*, 2006)	(Jimeno *et al.*, 2006; Zheng *et al.*, 2006)
RHO-Cre-8 (transgene)	Rod photoreceptor cells	(Jimeno *et al.*, 2006)	(Jimeno *et al.*, 2006)
α-cre (transgene)	Anterior and peripheral retina	(Marquardt *et al.*, 2001)	(Fu *et al.*, 2006)
Keratin12-cre (Knock-in)	Cornea epithelium	(Chikama *et al.*, 2005)	
Keratocan-cre (transgene)	Stromal keratocytes in adult and neural crest cells in embryos	(Kao and Liu, 2003)	

proteins and in turn, regulate gene transcription. The germ line null allele of *Bmp7* was generated in the mouse and found to have a variable defect in eye development (Luo *et al.*, 1995; Jena *et al.*, 1997). Further study showed that when $Bmp7^{-/-}$ embryos showed anophthalmia, the likely primary defect was a failure of lens placode formation (Wawersik *et al.*, 1999). A failure of lens formation was consistent with the absence of *Pax6* expression in the presumptive lens. In this way Bmp7 was established as the first lens induction signaling molecule. This observation was the beginning of a whole series of molecular genetic investigations of lens induction that have led to complex models for regulation of the process (Chow and Lang, 2001; Lang, 2004).

Math5 encodes a basic helix-loop-helix transcription factor that is expressed in the very early stages of retinal neurogenesis. Generation of a germ line null allele in the Glaser laboratory resulted in mice without an optic nerve because the ganglion cells that produce ganglion cell axon are mostly missing. This observation has been one of many in which mutant mice have been used to understand mechanisms of retinal neurogenesis (Brown *et al.*, 2001).

The Wnt pathway repressor Dkk2 is expressed in the cornea. Apparently, mice with a germ line disruption of *Dkk2* developed skin instead of cornea (Mukhopadhyay *et al.*, 2006). This suggested that suppression of the canonical Wnt pathway was involved in the development of non-keratinizing stratified epithelium characteristic of the cornea. Other investigations of the Wnt pathway function in eye development have also suggested that it must be suppressed if eye development is to proceed (Smith *et al.*, 2005; Cho and Cepko, 2006).

The three examples given above demonstrate the value of germ line gene disruption. However, one of the disadvantages of a germ line null allele is that it is only possible to study its role in development up to the point where it has a critical function. For example, germ line deletion of β-*catenin*, encoding a factor important in Wnt pathway signaling and cadherin-mediated cell–cell adhesion, results in lethality at the stage of gastrulation (Haegel *et al.*, 1995). Germ line deletion of *Sox2*, encoding an HMG-box transcription factor, similarly results in lethality (Avilion *et al.*, 2003). To overcome this limitation, it is possible to design genetic analyses so that only some cells in the mouse become mutant. This somatic mutation method employs the so-called conditional allele described below.

The Conditional Allele

The most popular method for producing a conditional allele is to place *LoxP* sites on either side of a critical gene region and then delete this region using cre recombinase in a desired tissue type or stage of development. If deletion, rather than inversion, is to be the outcome of cre recombinase activity in such an allele, the *LoxP* sites must be in direct repeat orientation. To generate this type of allele, the *LoxP* sites are placed in pre-planned locations during a normal gene-targeting procedure. Since the *frt-PGK-Neo-frt* expression unit is usually used for positive selection, these types of alleles actually employ two different types of site-specific recombination systems – flp/frt and cre/loxP-borrowed from other systems. Sometimes the presence of *LoxP* sites can interfere with normal gene expression prior to recombination, but usually, placing them in non-conserved regions of the gene avoids this problem. Well-designed LoxP conditional alleles (often referred to as "floxed" alleles) offer great versatility for analysis.

One option after the generation of a floxed allele is to cross the mouse line with one of a few lines that express cre recombinase in the germ line and therefore generate a germ line null. The *Sox2-cre* (Hayashi *et al.*, 2002) or *CMV-cre* (Schwenk *et al.*, 1995) lines can be used for this purpose. A germ line null allele is often used in combination with a floxed allele when performing somatic deletions as this means that cre recombinase needs only to recombine a single allele to generate the null. This can lead to phenotypes that appear earlier in development and show less variability.

The most common means of recombining a floxed allele is to use it in combination with a cre expressing transgene or "knock-in" line so that null cells are restricted to the tissue of interest. In the eye development field this strategy has been used very effectively. For example, a cell-autonomous function for *Pax6* in lens development was demonstrated by deleting a conditional allele, *Pax6*^{flox}, using a transgene, *Le-cre*, driven by the *Pax6* ectoderm enhancer (EE) that has activity in the presumptive lens ectoderm (Ashery-Padan *et al.*, 2000). This was genetic confirmation of the conclusions of earlier studies that used tissue recombination (Fujiwara *et al.*, 1994) and chimeric mouse (Collinson *et al.*, 2000) techniques.

Investigations of the role of periocular mesenchyme in eye development have become possible in the mouse because the *Wnt1-cre* transgene (Brault *et al.*, 2001) is expressed in the dorsal neural tube that is the origin of neural crest. This can be demonstrated when *Wnt1-cre* is combined with the Nagy *Z/EG* reporter (that is converted from *Lacz* to *GFP* expression upon cre recombinase action (Novak *et al.*, 2000)); GFP expressing cells are found surrounding the optic vesicle and at embryonic day (E) 8.5, between optic vesicle and presumptive lens. Use of *Wnt1-cre* to conditionally delete the *Pygopus2* gene shows that Pygopus2 activity in the periocular mesenchyme is required for development of a full-size lens (Song *et al.*, 2007). Cre recombinase expressing transgenes can also be used in combination. A mouse embryo with the genotype *Le-cre; Wnt1-cre; Pygo2*$^{-/flox}$ has a smaller lens that either *Le-cre; Pygo2*$^{-/flox}$, or *Wnt1-cre; Pygo2*$^{-/flox}$ demonstrating that Pygopus2 has a lens development role in both the lens placode (*Le-cre*) and periocular mesenchyme (*Wnt1-cre*) (Song *et al.*, 2007).

Heterologous Gene Expression via "Knock-in"

The generation of transgenic mice (using pronuclear injection of an expression unit) carries some risk that the transgene will insert in a genomic location that will not allow expression or will modify the desired tissue-specific expression pattern. Clearly, if trying to generate cre recombinase mouse lines for use in somatic mutation, this could be a problem. One way around this is to perform gene targeting of a cre recombinase expression unit into an existing gene. This has the advantage that since the larger genomic context is unchanged; one can generally expect to get the tissue-specific expression pattern of the target gene.

A good example of this type of gene targeted line is *AP2*α*-cre* (Macatee *et al.*, 2003). In this case, cre recombinase was targeted to the 3′ untranslated region of *AP2α* and placed behind an internal ribosome entry sequence to allow efficient translation. *AP2α* is normally expressed in the embryonic head surface ectoderm (including presumptive lens) from E8.0 and in the periocular mesenchyme (Macatee *et al.*, 2003). The expression of the *Z/EG* reporter in these tissues reflects the faithfully reproduced expression pattern of the *AP2α-cre* locus. When *AP2α-cre* is used to conditionally delete *Pygo2*^{flox}, the result is a severe defect in lens formation (Song *et al.*, 2007). This indicates that ectodermal and mesenchymal (neural crest) Pygopus2 cooperate in lens formation (Song *et al.*, 2007).

TEMPORAL CONTROL

Over the years in which transgenic mice have been a useful tool for developmental analysis, there have been many attempts to identify a system for temporal regulation of gene expression. Only recently, however, have these systems reached the point where they are routinely useful. Several of the more popular temporal regulation systems are described below and are illustrated schematically in Fig. 9.3.

Hormone-regulated Protein Activity

Fusion of many different proteins to the hormone binding domain (HBD) of the estrogen receptor (ER) or progesterone receptor (PR) renders them hormone inducible (Fig. 9.3(A)). This type of fusion protein was first used in oncogene studies (Eilers *et al.*, 1991; Jackson *et al.*, 1993). Why the HBD of hormone receptors should inactivate protein function is not understood mechanistically but may be related to the binding of heat shock proteins (HSPs) 70 and 90; this could sterically inhibit protein activity or possibly maintain partial unfolding. It has also been observed that ER fusion proteins increase their half-life in the presence of estradiol and this too may explain fusion protein inducibility.

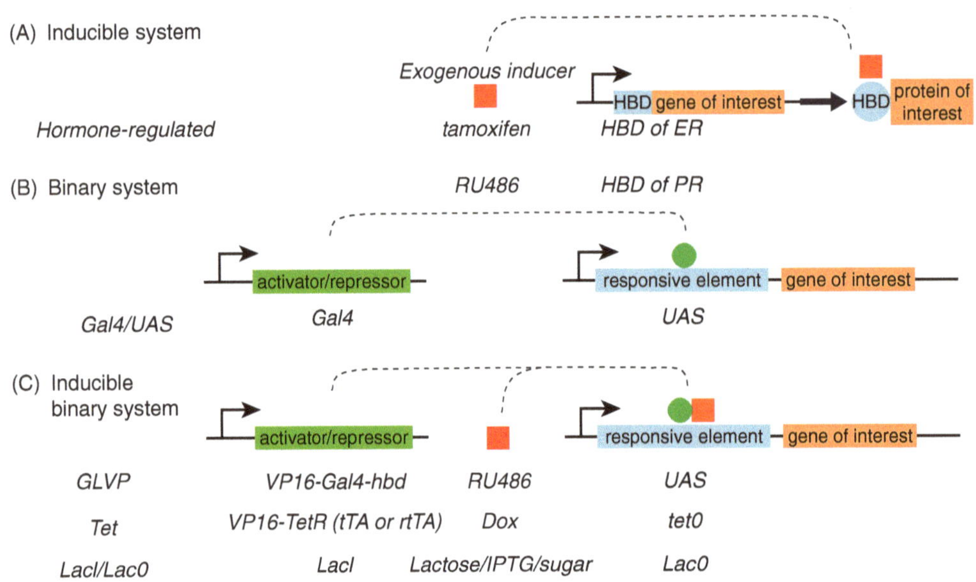

FIGURE 9.3 Schematic of an inducible and/or binary system for reversible, temporal and quantitative control of gene expression. (A) An inducible system such as a hormone-regulated system contains an exogenous inducer hormone (tamoxifen or RU486) and a fusion transgene containing the HBD of ER or PR and gene of interest. The protein is activated upon application of the hormone. (B) A binary system such as the Gal4/UAS is composed of a transgene encoding an activator (or repressor) protein driven by a promoter and another transgene encoding a protein of interest driven by a responsive element for the activator (*UAS*) and a minimal promoter. Gene expression is activated or repressed when a cell contains both transgenes. (C) An inducible binary system is composed of three elements: a transgene encoding an activator (or repressor) protein driven by a promoter, an exogenous inducer and another transgene encoding a protein of interest driven by a responsive element and a minimal promoter. Gene expression is activated by the responsive element only when both the activator protein and exogenous inducer are present. Examples shown include the GLVP, Tet and LacI/LacO systems. HBD, hormone binding domain; ER, estrogen receptor; PR, progesterone receptor; Dox, doxycycline; *tetO*, tetracycline operator sequences; VP16, *herpes simplex virus* VP16 activation domain; tTA, tetracycline controlled transactivator; rtTA, the reverse tTA, *lacO*, the *lac* operon of *E. coli*, lacR, the inhibitor protein lacR which can bind to lacO.

The HBDs used for hormone-regulated fusion proteins are variant forms that are selectively activated by the synthetic antiprogestin RU486 (Kellendonk *et al.*, 1996) or by tamoxifen (Feil *et al.*, 1996) in preference to endogenous progesterone and estrogen, respectively. This minimizes unwanted effects of the natural hormones. Building on this information, the Chambon laboratory generated an ER fusion form of cre recombinase and showed that its activity could be hormone regulated (Feil *et al.*, 1996). Although it is not in widespread use, the cre-ERT2 fusion protein has been used successfully in conditional deletion of floxed alleles (Kimmel *et al.*, 2000; Monvoisin *et al.*, 2006). So far, there are few examples in which hormone inducible cre recombinase has been used to analyze eye development. In one of these, the MerCreMer fusion protein that has both N- and C-terminal ER HBD fusions was used to examine the mechanism of choroideremia pathogenesis (Tolmachova *et al.*, 2006).

The GAL4/UAS System for Transcriptional Control

The GAL4/UAS system popular in studies of *Drosophila* has also been proved to be useful in mice. The key components of the system (Fig. 9.3(B)) include a transcriptional activator Gal4 from *Saccharomyces cerevisiae* (yeast) and a Gal4-responsive element – upstream activator sequences (*UASs*) (Lewandoski, 2001).

The earliest application of this system is made in studying the role of Shh in the dorsal-ventral patterning of the mouse central nervous system. The use of *Wnt1-Gal4* and *UAS-shh* bitransgenic system wins a ~24 h delay of expression and thus bypasses the embryonic lethality compared to *Wnt1-shh* transgene (Ornitz *et al.*, 1991).

A hybrid system combining the Gal4/UAS and hormone-regulated system has been generated, known as the inducible Gal4/UAS system (GLVP system; Fig. 9.3(C)). This system employs a chimeric protein (GLVP) consisting of the herpes simplex virus VP16 activation domain, the Gal4 DNA-binding domain and the ligand-binding domain of the progesterone receptor (LBD 42). GLVP is believed to be sequestered in the cytoplasm by binding to heat-shock proteins (HSPs) 70 and 90. In the presence of the synthetic steroids RU486 or ZK98.734, GLVP is released and translocated to the nucleus and transactvates the *UAS*-bearing target genes. This system has been used in live studies (Pierson *et al.*, 2000; Chaisson *et al.*, 2002).

Tetracycline-regulated Transcriptional Control

The tetracycline regulation system is becoming increasingly popular. In this inducible binary system (Fig. 9.3(C)), one mouse strain contains and fusion of the herpes simplex virus VP16 transactivation domain and the *Escherichia coli* tetracycline repressor (TetR) designated tTA. A second strain carries a gene of interest under the control of the 19-bp operator sequences (*tetO*) of the tet operon, which is activated upon the presence of both the transactivation protein and tetracycline (Gossen and Bujard, 1992; Kistner *et al.*, 1996). Two versions of this system have been developed to either activate or suppress target gene expression conditionally. The tet-on system contains the tetracycline controlled transactivator (tTA) which cannot bind DNA when the inducer is present. The tet-off system contains the reverse tTA (rtTA) which binds DNA when the inducer is present (Kistner *et al.*, 1996).

One application of the tet system is to inducibly express a transgene. This bitransgenic system contains a mouse strain carrying a tissue specific promoter driven rtTA with a second strain carrying the *tet-O* driven gene of interest. For example, the *Keratocan-rtTA/tet-O-FGF7* bitransgenic mice are used to direct inducible expression of Fgf7 in the corneal stroma. This results in enhanced cell proliferation, but otherwise fails to cause pathology in corneal epithelium (Hayashi *et al.*, 2005). A *tyrosinase-rtTA/tet-O-tyrosinase* bitransgenic system is used to study the role of tyrosinase in regulation of abnormal chiasmatic projections found in albinism (Gimenez *et al.*, 2004). A tet-off system containing *VE-cadherin-rtTA* and *tet-myrAkt* transgenes is applied to study the role of endothelial Akt in vascular lesion formation. The data from this study suggest that enhancing endothelial Akt activity alone could have therapeutic benefits after injury (Mukai *et al.*, 2006). Combining the tet system with the cre/flox system, one can inducibly and tissue-specifically delete a gene of interest. For example in eye development, the *Keratocan*- or Keratin12-rtTA (knock-in)/*tet-O-cre* lines are used to study corneal development and wound healing (Kao, 2006).

The LacO/LacIR System for Transcriptional Control

The LacO/LacIR system utilizes the *lac* operon of *E. coli* and functions in an analogous way to the tet system described above (Cronin *et al.*, 2001). In this system (Fig. 9.3(C)), the inhibitor protein LacIR can bind to *lacO* regulatory sites in the promoter and turn off transcription. By adding lactose, IPTG or sugar, the LacIR protein dissociates to relieve repression.

For example, this system has been used to identify the developmental window in which tyrosinase activity is critical for ganglion axon cell pathfinding, an important issue in oculocutaneous albinism (Cronin *et al.*, 2003). A transgene ubiquitously expressing lacI was combined with one composed of the *lacO* sequences as well as the *Tyrosinase* minimal promoter and open reading frame. In combination, these were used to direct inducible tyrosinase expression during discrete periods of visual system development in an albino background. This showed that there was a period of neuroblast cell division in which tyrosinase activity was critical for ganglion cell axonal pathfinding (Cronin *et al.*, 2001, 2003).

FORWARD GENETICS IN THE MOUSE

In a distinct experimental approach from most of what is described above, some investigative groups, often consortia, have chosen to use the mouse for forward genetics. Two popular approaches to using the mouse for forward genetics include gene-trapping methods and chemical mutagenesis screens.

Gene Trapping

Gene trapping is an efficient system to introduce enhancer-, gene- or promoter-trap vectors to the mouse germ line. This is typically done by electroporating suitable constructs into ES cells, generating chimeric mice and screening those mice for construct expression and phenotypes that are the result of insertional mutagenesis (Stanford *et al.*, 2001). There are three distinct trapping vector designs. The enhancer-trap vector contains a minimal promoter and a reporter gene open reading frame (Fig. 9.4(A)). It needs to be inserted near to a *cis*-acting enhancer element to produce expression of the reporter gene. Typically, enhancer trap constructs produce loss-of-function mutations at low efficiency presumably because there is no requirement for insertion into the transcription unit. The gene-trap vector (Fig. 9.4(B)) has the distinct design feature of a splice acceptor site immediately upstream of a promoterless reporter. This arrangement means that to be expressed, it needs to be inserted into an intron. This type of vector design often results in the generation of hypomorphic alleles as the splicing pattern is disrupted but not eliminated. Finally, promoter-trap vectors (Fig. 9.4(C)) require a promoterless reporter gene and a selectable marker. In some promoter trap vectors, these two requirements are combined in a single open reading frame. To be expressed, this type of vector needs to be inserted into an exon and as a consequence usually results in the generation of null alleles.

A hypomorphic mutant allele of *Crim1* (Cysteine-rich, motor neurons (Kolle *et al.*, 2000)) designated *KST264* has been generated from a gene-trap screen designed to identify novel proteins containing signal sequences (Leighton *et al.*, 2001). Analysis of the $Crim1^{KST264/KST264}$ mice reveals that *Crim1* might be involved in multiple organogenesis including the eye (Pennisi *et al.*, 2007). Recently, the lens intrinsic membrane protein-2 (Lim-2) deficient mice have been derived from the Omnibank library of gene-trap embryonic stem (ES) cells (Zambrowicz *et al.*, 1998; Shiels *et al.*, 2007). The refractive defects and cataracts detected in these mice provide direct evidence of the crucial role of Lim-2 in establishing the correct internal refractive properties of the crystalline lens (Shiels *et al.*, 2007).

Chemical Mutagenesis Screens

Chemical mutagenesis screens in the mouse are phenotype-driven. This approach has the appeal that the screens should be largely unbiased and with current technology the prospects of identifying mutated genes within a reasonable length of time are improved (Nolan, 2000). The chemical mutagens used primarily generate point mutations, and occasionally very small deletions (20–50 bps).

Recently, *N*-ethyl-*N*-nitrosourea (ENU) mutagenesis was employed to identify mutations causing neural tube closure defects (Kasarskis *et al.*, 1998; Zohn *et al.*, 2005). Two mutants obtained from this screen harbor eye defects. One is the *droopy eye* (*drey*) mutant with a hypomorphic mutation in a p38-interacting protein (P38IP), which has an RPE abnormality in addition to the neural tube closure defects. It was shown that P38IP down-regulates E-cadherin protein expression downstream from NCK-interacting kinase (NIK) during gastrulation (Zohn *et al.*, 2005). In addition, the *humpty dumpty* (*Humdy*) mouse mutant, carries a null mutation in Phactr4, an uncharacterized protein phosphatase 1 (PP1) and actin regulator family member. *Humdy* mutants fail to close the optic fissure and the neural tube perhaps due to the requirement of Phactr4 in cell-cycle progression (Kim *et al.*, 2007). Recently, a genome-wide screen using ENU has been performed,

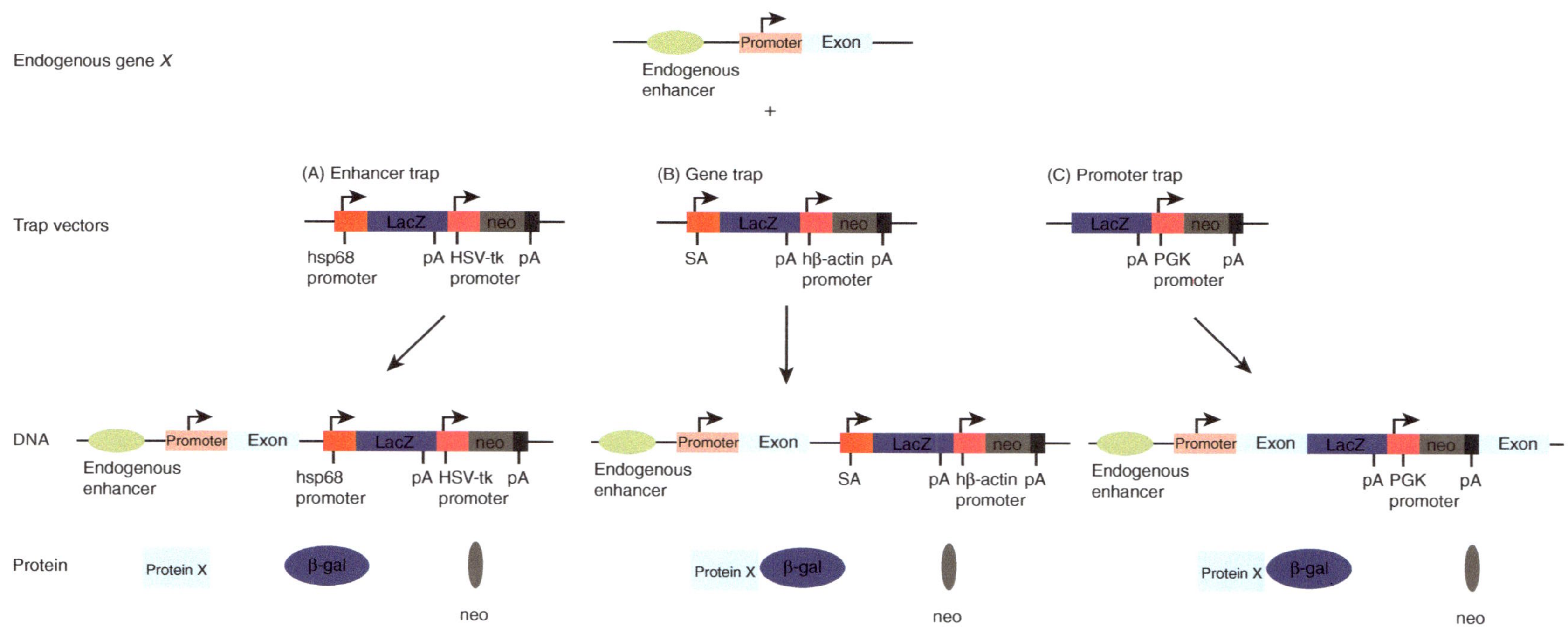

FIGURE 9.4 The three basic gene-trapping methods. An endogenous gene *X* is shown to be trapped by Enhancer- (A), gene- (B) and promoter- (C) trap vectors. The modified endogenous locus after vector integration (grey arrows) and resulting translated protein are shown. (A) An enhancer-trap vector is shown containing two transcription units. The first is a *lacZ* reporter cDNA with a polyadenylation signal driven by a heat-shock inducible minimal (*hsp68*) promoter. The second transcription unit is driven by the HSV-tk minimal promoter and encodes the *Neomycin* resistance gene (*Neo*). Both transcription units terminate with a polyadenylation signal (pA). Integration of the enhancer-trap vector anywhere within the range of an endogenous enhancer activity will lead to the transcription and translation of the *lacZ* reporter. A low frequency of loss-of-function mutations is produced with this method. (B) A gene-trap vector is shown containing two transcription unit. The first is a promoterless *lacZ* gene immediately downstream of a splice acceptor (SA). The second encodes *Neo* and is driven by *hβ-actin* promoter. A fusion protein of protein X and β-gal is generated only if the vector is inserted into an intron. (C) A promoter-trap vector is shown containing a promoterless *lacZ* gene combined with a *PGK* promoter-*neo* transcription unit. A fusion protein of protein X and β-gal is generated only if the vector is inserted into the coding sequence of gene *X*. *β-gal*, β-galactosidase; *HSV-tk*, herpes simplex virus thymidine kinase, *hβ-actin*, human β-actin; pA, polyadenylation, *PGK*, phosphoglycerate kinase 1. (*Source*: Stanford *et al.*, 2001 with permission.)

aiming at identifying novel mutations that give rise to eye and vision abnormalities in the mouse (Thaung *et al.*, 2002). They have identified new loci which are required for formation of a normal visual system and valuable for future study.

CONCLUDING COMMENTS

From characterization of the *Small eye* mouse (Hogan *et al.*, 1986b), to identification of the *Pax6* gene (Hill *et al.*, 1991; Ton *et al.*, 1992), the use of gene targeting, conditional genetics (Joyner, 1995), trapping techniques (Stanford *et al.*, 2001) and chemical mutagenesis (Zohn *et al.*, 2005), the various tools of the mouse geneticist's trade, are proving powerful for the analysis of developmental events including those that build the visual system. While there are alternative experimental strategies available in some systems, it seems likely that the various forms of mouse genetic analysis will form the backbone of visual system developmental analysis for some years. In particular, as the questions we ask become more closely aimed at answering mechanistic questions, our ability to generate subtle mutations with gene targeting techniques will become all-the-more important.

REFERENCES

Akimoto M, Filippova E, Gage PJ, Zhu X, Craft CM, Swaroop A (2004). Transgenic mice expressing Cre-recombinase specifically in M- or S-cone photoreceptors. *Invest Ophthalmol Vis Sci* 45:42–47.

Ashery-Padan R, Marquardt T, Zhou X, Gruss P (2000). Pax6 activity in the lens primordium is required for lens formation and for correct placement of a single retina in the eye. *Genes Dev.* 14:2701–2711.

Avilion AA, Nicolis SK, Pevny LH, Perez L, Vivian N, Lovell-Badge R (2003). Multipotent cell lineages in early mouse development depend on SOX2 function. *Genes Dev.* 17:126–140.

Barski JJ, Dethleffsen K, Meyer M (2000). Cre recombinase expression in cerebellar Purkinje cells. *Genesis* 28:93–98.

Barski JJ, Hartmann J, Rose CR, Hoebeek F, Morl K, Noll-Hussong M, De Zeeuw CI, Konnerth A, Meyer M (2003). Calbindin in cerebellar Purkinje cells is a critical determinant of the precision of motor coordination. *J Neurosci* 23:3469–3477.

Blixt A, Mahlapuu M, Aitola M, Pelto-Huikko M, Enerback S, Carlsson P (2000). A forkhead gene, FoxE3, is essential for lens epithelial proliferation and closure of the lens vesicle. *Genes Dev* 14:245–254.

Brault V, Moore R, Kutsch S, Ishibashi M, Rowitch DH, McMahon AP, Sommer L, Boussadia O, Kemler R (2001). Inactivation of the beta-catenin gene by Wnt1-Cre-mediated deletion results in dramatic brain malformation and failure of craniofacial development. *Development* 128:1253–1264.

Breitman ML, Clapoff S, Rossant J, Tsui LC, Glode LM, Maxwell IH, Bernstein A (1987). Genetic ablation: targeted expression of a toxin gene causes microphthalmia in transgenic mice. *Science* 238:1563–1566.

Brinster RL, Chen HY, Trumbauer M, Senear AW, Warren R, Palmiter RD (1981). Somatic expression of herpes thymidine kinase in mice following injection of a fusion gene into eggs. *Cell* 27:223–231.

Brown NL, Patel S, Brzezinski J, Glaser T (2001). Math5 is required for retinal ganglion cell and optic nerve formation. *Development* 128:2497–2508.

Brownell I, Dirksen M, Jamrich M (2000). Forkhead Foxe3 maps to the dysgenetic lens locus and is critical in lens development and differentiation. *Genesis* 27:81–93.

Calera MR, Topley HL, Liao Y, Duling BR, Paul DL, Goodenough DA (2006). Connexin43 is required for production of the aqueous humor in the murine eye. *J Cell Sci* 119:4510–4519.

Chaisson ML, Brooling JT, Ladiges W, Tsai S, Fausto N (2002). Hepatocyte-specific inhibition of NF-κB leads to apoptosis after TNF treatment, but not after partial hepatectomy. *J Clin Invest* 110:193–202.

Chikama T, Hayashi Y, Liu CY, Terai N, Terai K, Kao CW, Wang L, Hayashi M, Nishida T, Sanford P, Doestchman T, Kao WW (2005). Characterization of tetracycline-inducible bitransgenic Krt12rtTA/+/tet-O-LacZ mice. *Invest Ophthalmol Vis Sci* 46:1966–1972.

Cho SH, Cepko CL (2006). Wnt2b/beta-catenin-mediated canonical Wnt signaling determines the peripheral fates of the chick eye. *Development* 133:3167–3177.

Chow RL, Altmann CR, Lang RA, Hemmati-Brivanlou A (1999). Pax6 induces ectopic eyes in a vertebrate. *Development* 126:4213–4222.

Chow RL, Lang RA (2001). Early eye development in vertebrates. *Annu Rev Cell Dev Biol* 17:255–296.

Chow RL, Roux GD, Roghani M, Palmer MA, Rifkin DB, Moscatelli DA, Lang RA (1995). FGF suppresses apoptosis and induces differentiation of fibre cells in the mouse lens. *Development* 121:4383–4393.

Collins FS, Finnell RH, Rossant J, Wurst W (2007a). A new partner for the international knockout mouse consortium. *Cell* 129:235.

Collins FS, Rossant J, Wurst W (2007b). A mouse for all reasons. *Cell* 128:9–13.

Collinson JM, Hill RE, West JD (2000). Different roles for Pax6 in the optic vesicle and facial epithelium mediate early morphogenesis of the murine eye. *Development* 127:945–956.

Costantini F, Lacy E (1981). Introduction of a rabbit beta-globin gene into the mouse germ line. *Nature* 294:92–94.

Cronin CA, Gluba W, Scrable H (2001). The lac operator-repressor system is functional in the mouse. *Genes Dev* 15:1506–1517.

Cronin CA, Ryan AB, Talley EM, Scrable H (2003). Tyrosinase expression during neuroblast divisions affects later pathfinding by retinal ganglion cells. *J Neurosci* 23:11692–11697.

Danielian PS, Muccino D, Rowitch DH, Michael SK, McMahon AP (1998a). Modification of gene activity in mouse embryos in utero by a tamoxifen- inducible form of Cre recombinase. *Curr Biol* 8:1323–1326.

Danielian PS, Muccino D, Rowitch DH, Michael SK, McMahon AP (1998b). Modification of gene activity in mouse embryos in utero by a tamoxifen-inducible form of Cre recombinase. *Curr Biol* 8:1323–1326.

Davis-Silberman N, Kalich T, Oron-Karni V, Marquardt T, Kroeber M, Tamm ER, Ashery-Padan R (2005). Genetic dissection of Pax6 dosage requirements in the developing mouse eye. *Hum Mol Genet* 14:2265–2276.

Dwivedi DJ, Pontoriero GF, Ashery-Padan R, Sullivan S, Williams T, West-Mays JA (2005). Targeted deletion of AP-2alpha leads to disruption in corneal epithelial cell integrity and defects in the corneal stroma. *Invest Ophthalmol Vis Sci* 46:3623–3630.

Ebong S, Chepelinsky AB, Robinson ML, Zhao H, Yu CR, Egwuagu CE (2004). Characterization of the roles of STAT1 and STAT3 signal transduction pathways in mammalian lens development. *Mol Vis* 10:122–131.

Eilers M, Schirm S, Bishop JM (1991). The MYC protein activates transcription of the alpha-prothymosin gene. *Embo J* 10: 133–141.

Faber SC, Dimanlig P, Makarenkova HP, Shirke S, Ko K, Lang RA (2001). Fgf receptor signaling plays a role in lens induction. *Development* 128:4425–4438.

Faber SC, Robinson ML, Makarenkova HP, Lang RA (2002). Bmp signaling is required for development of primary lens fiber cells. *Development* 129:3727–3737.

Feil R, Brocard J, Mascrez B, LeMeur M, Metzger D, Chambon P (1996). Ligand-activated site-specific recombination in mice. *P Natl Acad Sci USA* 93:10887–10890.

Fu X, Sun H, Klein WH, Mu X (2006). Beta-catenin is essential for lamination but not neurogenesis in mouse retinal development. *Dev Biol* 299:424–437.

Fujiwara M, Uchida T, Osumi-Yamashita N, Eto K (1994). Uchida rat (rSey): a new mutant rat with craniofacial abnormalities resembling those of the mouse Sey mutant. *Differentiation* 57:31–38.

Furuta Y, Lagutin O, Hogan BL, Oliver GC (2000). Retina- and ventral forebrain-specific Cre recombinase activity in transgenic mice. *Genesis* 26:130–132.

Garcia CM, Yu K, Zhao H, Ashery-Padan R, Ornitz DM, Robinson ML, Beebe DC (2005). Signaling through FGF receptor-2 is required for lens cell survival and for withdrawal from the cell cycle during lens fiber cell differentiation. *Dev Dynam* 233:516–527.

Gehring WJ (2002). The genetic control of eye development and its implications for the evolution of the various eye-types. *Int J Dev Biol* 46:65–73.

Gimenez E, Lavado A, Giraldo P, Cozar P, Jeffery G, Montoliu L (2004). A transgenic mouse model with inducible Tyrosinase gene expression using the tetracycline (Tet-on) system allows regulated rescue of abnormal chiasmatic projections found in albinism. *Pigm Cell Res* 17:363–370.

Glaser T, Walton DS, Maas RL (1992). Genomic structure, evolutionary conservation and aniridia mutations in the human PAX6 gene. *Nat Genet* 2:232–239.

Gordon JW, Scangos GA, Plotkin DJ, Barbosa JA, Ruddle FH (1980). Genetic transformation of mouse embryos by microinjection of purified DNA. *P Natl Acad Sci USA* 77:7380–7384.

Gossen M, Bujard H (1992). Tight control of gene expression in mammalian cells by tetracycline-responsive promoters. *P Natl Acad Sci USA* 89:5547–5551.

Haegel H, Larue L, Ohsugi M, Fedorov L, Herrenknecht K, Kemler R (1995). Lack of beta-catenin affects mouse development at gastrulation. *Development* 121:3529–3537.

Haigh JJ, Morelli PI, Gerhardt H, Haigh K, Tsien J, Damert A, Miquerol L, Muhlner U, Klein R, Ferrara N, Wagner EF, Betsholtz C, Nagy A (2003). Cortical and retinal defects caused by dosage-dependent reductions in VEGF-A paracrine signaling. *Dev Biol* 262:225–241.

Halder G, Callaerts P, Gehring WJ (1995). Induction of ectopic eyes by targeted expression of the eyeless gene in *Drosophila*. *Science* 267:1788–1792.

Hayashi M, Hayashi Y, Liu CY, Tichelaar JW, Kao WW (2005). Over expression of FGF7 enhances cell proliferation but fails to cause pathology in corneal epithelium of Kerapr-rtTA/FGF7 bitransgenic mice. *Mol Vis* 11:201–207.

Hayashi S, Lewis P, Pevny L, McMahon AP (2002). Efficient gene modulation in mouse epiblast using a Sox2Cre transgenic mouse strain. *Gene Expr Patterns* 2:93–97.

Heasman J (2002). Morpholino oligos: making sense of antisense? *Dev Biol* 243:209–214.

Hemmati-Brivanlou A, Mann RW, Harland RM (1992). A protein expressed in the growth cones of embryonic vertebrate neurons defines a new class of intermediate filament protein. *Neuron* 9:417–428.

Hill RE, Favor J, Hogan BL, Ton CC, Saunders GF, Hanson IM, Prosser J, Jordan T, Hastie ND, van Heyningen V (1991). Mouse small eye results from mutations in a paired-like homeobox-containing gene. *Nature* 354:522–525.

Hirsch N, Harris WA (1997). *Xenopus* Pax-6 and retinal development. *J Neurobiol* 32:45–61.

Hogan B, Costantini F, Lacy E (1986a). Manipulating the mouse embryo: a laboratory manual.

Hogan BL, Horsburgh G, Cohen J, Hetherington CM, Fisher G, Lyon MF (1968b). Small eyes (Sey): a homozygous lethal mutation on chromosome 2 which affects the differentiation of both lens and nasal placodes in the mouse. *J Embryol Exp Morphol* 97:95–110.

Huang S, Moody SA (1993). The retinal fate of *Xenopus* cleavage stage progenitors is dependent upon blastomere position and competence: studies of normal and regulated clones. *J Neurosci* 13:3193–3210.

Jackson P, Baltimore D, Picard D (1993). Hormone-conditional transformation by fusion proteins of c-Abl and its transforming variants. *Embo J* 12:2809–2819.

Jadhav AP, Mason HA, Cepko CL (2006). Notch 1 inhibits photoreceptor production in the developing mammalian retina. *Development* 133:913–923.

Jaskoll T, Zhou YM, Chai Y, Makarenkova HP, Collinson JM, West JD, Hajihosseini MK, Lee J, Melnick M (2002). Embryonic submandibular gland morphogenesis: stage-specific protein localization of FGFs, BMPs, Pax6 and Pax9 in normal mice and abnormal SMG phenotypes in FgfR2-IIIc(+/Δ), BMP7(−/−) and Pax6(−/−) mice. *Cells Tissues Organs* 170:83–98.

Jena N, Martin-Seisdedos C, McCue P, Croce CM (1997). BMP7 null mutation in mice: developmental defects in skeleton, kidney, and eye. *Exp Cell Res* 230:28–37.

Jimeno D, Feiner L, Lillo C, Teofilo K, Goldstein LS, Pierce EA, Williams DS (2006). Analysis of kinesin-2 function in photoreceptor cells using synchronous *Cre*-loxP knockout of *Kif3a* with *RHO-Cre*. *Invest Ophthalmol Vis Sci* 47:5039–5046.

Jordan T, Hanson I, Zaletayev D, Hodgson S, Prosser J, Seawright A, Hastie N, van Heyningen V (1992). The human PAX6 gene is mutated in two patients with aniridia. *Nat Genet* 1:328–332.

Joyner A (1993). *Gene Targeting A Practical Approach*. Oxford University Press, New York.

Joyner AL (1995) Gene targeting: a practical approach. In: Joyner AL (ed.) *The Practical Approach Series*. IRL Press at Oxford University Press, New York, pp. 234.

Kao WW (2006). Ocular surface tissue morphogenesis in normal and disease states revealed by genetically modified mice. *Cornea* 25:S7–S19.

Kao WW, Liu CY (2003). The use of transgenic and knock-out mice in the investigation of ocular surface cell biology. *Ocul Surf* 1:5–19.

Kasarskis A, Manova K, Anderson KV (1998). A phenotype-based screen for embryonic lethal mutations in the mouse. *P Natl Acad Sci USA* 95:7485–7490.

Kellendonk C, Tronche F, Monaghan AP, Angrand PO, Stewart F, Schutz G (1996). Regulation of Cre recombinase activity by the synthetic steroid RU 486. *Nucleic Acids Res* 24:1404–1411.

Kim TH, Goodman J, Anderson KV, Niswander L (2007). Phactr4 regulates neural tube and optic fissure closure by controlling PP1-, Rb-, and E2F1-regulated cell-cycle progression. *Dev Cell* 13:87–102.

Kimmel RA, Turnbull DH, Blanquet V, Wurst W, Loomis CA, Joyner AL (2000). Two lineage boundaries coordinate vertebrate apical ectodermal ridge formation. *Genes Dev* 14:1377–1389.

Kistner A, Gossen M, Zimmermann F, Jerecic J, Ullmer C, Lubbert H, Bujard H (1996). Doxycycline-mediated quantitative and tissue-specific control of gene expression in transgenic mice. *P Natl Acad Sci USA* 93:10933–10938.

Kolle G, Georgas K, Holmes GP, Little MH, Yamada T (2000). CRIM1, a novel gene encoding a cysteine-rich repeat protein, is developmentally regulated and implicated in vertebrate CNS development and organogenesis. *Mech Dev* 90:181–193.

Kolm PJ, Sive HL (1995). Efficient hormone-inducible protein function in *Xenopus laevis*. *Dev Biol* 171:267–272.

Lakso M, Sauer B, Mosinger B, Jr, Lee EJ, Manning RW, Yu SH, Mulder KL, Westphal H (1992). Targeted oncogene activation by site-specific recombination in transgenic mice. *P Natl Acad Sci USA* 89:6232–6236.

Lang RA (2004). Pathways regulating lens induction in the mouse. *Int J Dev Biol* 48:783–791.

Le YZ, Zheng L, Zheng W, Ash JD, Agbaga MP, Zhu M, Anderson RE (2006). Mouse opsin promoter-directed Cre recombinase expression in transgenic mice. *Mol Vis* 12:389–398.

Leighton PA, Mitchell KJ, Goodrich LV, Lu X, Pinson K, Scherz P, Skarnes WC, Tessier-Lavigne M (2001). Defining brain wiring patterns and mechanisms through gene trapping in mice. *Nature* 410:174–179.

Lewandoski M (2001). Conditional control of gene expression in the mouse. *Nat Rev Genet* 2:743–755.

Liu W, Lagutin OV, Mende M, Streit A, Oliver G (2006). Six3 activation of Pax6 expression is essential for mammalian lens induction and specification. *Embo J* 25:5383–5395.

Lok S, Breitman ML, Chepelinsky AB, Piatigorsky J, Gold RJM, Tsui LC (1985). Lens-specific promoter activity of a mouse γ-crystallin gene. *Mol Cell Biol* 5:2221–2230.

Lovicu FJ, Kao WW, Overbeek PA (1999). Ectopic gland induction by lens-specific expression of keratinocyte growth factor (FGF-7) in transgenic mice. *Mech Dev* 88:43–53.

Luo G, Hofmann C, Bronckers AL, Sohocki M, Bradley A, Karsenty G (1995). BMP-7 is an inducer of nephrogenesis, and is also required for eye development and skeletal patterning. *Genes Dev* 9:2808–2820.

Macatee TL, Hammond BP, Arenkiel BR, Francis L, Frank DU, Moon AM (2003). Ablation of specific expression domains reveals discrete functions of ectoderm- and endoderm-derived FGF8 during cardiovascular and pharyngeal development. *Development* 130:6361–6374.

Marneros AG, Fan J, Yokoyama Y, Gerber HP, Ferrara N, Crouch RK, Olsen BR (2005). Vascular endothelial growth factor expression in the retinal pigment epithelium is essential for choriocapillaris development and visual function. *Am J Pathol* 167:1451–1459.

Marquardt T, Ashery-Padan R, Andrejewski N, Scardigli R, Guillemot F, Gruss P (2001). Pax6 is required for the multipotent state of retinal progenitor cells. *Cell* 105:43–55.

Mencl E (1903). Ein fall von beiderseitiger augenlinsenausbildung wahrend der abwesenheit von augenblasen. *Arch EntwMech Org* 16:328–339.

Monvoisin A, Alva JA, Hofmann JJ, Zovein AC, Lane TF, Iruela-Arispe ML (2006). VE-cadherin-CreER^{T2} transgenic mouse: a model for inducible recombination in the endothelium. *Dev Dynam* 235:3413–3422.

Moody SA (1987). Fates of the blastomeres of the 32-cell-stage *Xenopus* embryo. *Dev Biol* 122:300–319.

Mori M, Metzger D, Garnier JM, Chambon P, Mark M (2002). Site-specific somatic mutagenesis in the retinal pigment epithelium. *Invest Ophthalmol Vis Sci* 43:1384–1388.

Mukai Y, Rikitake Y, Shiojima I, Wolfrum S, Satoh M, Takeshita K, Hiroi Y, Salomone S, Kim HH, Benjamin LE, Walsh K, Liao JK (2006). Decreased vascular lesion formation in mice with inducible endothelial-specific expression of protein kinase Akt. *J Clin Invest* 116:334–343.

Mukhopadhyay M, Gorivodsky M, Shtrom S, Grinberg A, Niehrs C, Morasso MI, Westphal H (2006). Dkk2 plays an essential role in the corneal fate of the ocular surface epithelium. *Development* 133:2149–2154.

Murali D, Yoshikawa S, Corrigan RR, Plas DJ, Crair MC, Oliver G, Lyons KM, Mishina Y, Furuta Y (2005). Distinct developmental programs require different levels of Bmp signaling during mouse retinal development. *Development* 132:913–923.

Nolan PM (2000). Generation of mouse mutants as a tool for functional genomics. *Pharmacogenomics* 1:243–255.

Novak A, Guo C, Yang W, Nagy A, Lobe CG (2000). Z/EG, a double reporter mouse line that expresses enhanced green fluorescent protein upon Cre-mediated excision. *Genesis* 28:147–155.

Ornitz DM, Moreadith RW, Leder P (1991). Binary system for regulating transgene expression in mice: targeting int-2 gene expression with yeast GAL4/UAS control elements. *P Natl Acad Sci USA* 88:698–702.

Overbeek PA, Chepelinsky AB, Khillan JS, Piatigorsky J, Westphal H (1985). Lens-specific expression and developmental regulation of the bacterial chloramphenicol acetyltransferase gene driven by the murine alpha A-crystallin promoter in transgenic mice. *P Nal Acad Sci USA* 82:7815–7819.

Pennisi DJ, Wilkinson L, Kolle G, Sohaskey ML, Gillinder K, Piper MJ, McAvoy JW, Lovicu FJ, Little MH (2007). Crim1KST264/KST264 mice display a disruption of the Crim1 gene resulting in perinatal lethality with defects in multiple organ systems. *Dev Dynam* 236:502–511.

Pierson TM, Wang Y, DeMayo FJ, Matzuk MM, Tsai SY, Omalley BW (2000). Regulable expression of inhibin A in wild-type and inhibin null mice. *Mol Endocrinol* 14:1075–1085.

Quiring R, Walldorf U, Kloter U, Gehring WJ (1994). Homology of the eyeless gene of *Drosophila* to the Small eye gene in mice and Aniridia in humans. *Science* 265:785–789.

Reneker LW, Overbeek PA (1996). Lens-specific expression of PDGF-A alters lens growth and development. *Dev Biol* 180:554–565.

Rieger DK, Reichenberger E, McLean W, Sidow A, Olsen BR (2001). A double-deletion mutation in the Pitx3 gene causes arrested lens development in aphakia mice. *Genomics* 72:61–72.

Robinson ML, MacMillan-Crow LA, Thompson JA, Overbeek PA (1995a). Expression of a truncated FGF receptor results in defective lens development in transgenic mice. *Development* 121:3959–3967.

Robinson ML, Overbeek PA, Verran DJ, Grizzle WE, Stockard CR, Friesel R, Maciag T, Thompson JA (1995b). Extracellular FGF-1 acts as a lens differentiation factor in transgenic mice. *Development* 121:505–514.

Robinson ML, Ohtaka-Maruyama C, Chan CC, Jamieson S, Dickson C, Overbeek PA, Chepelinsky AB (1998). Disregulation of ocular

morphogenesis by lens-specific expression of FGF-3/int-2 in transgenic mice. *Dev Biol* 198:13–31.
Rowan S, Cepko CL (2004). Genetic analysis of the homeodomain transcription factor Chx10 in the retina using a novel multifunctional BAC transgenic mouse reporter. *Dev Biol.* 271:388–402.
Salvini-Plawen L, Mayr E (1977). On the evolution of photoreceptors and eyes. *Evol Biol* 10:207–263.
Sanyal S, Hawkins RK (1979). Dysgenetic lens (dyl) – a new gene in the mouse. *Invest Ophthalmol Vis Sci* 18:642–645.
Schwenk F, Baron U, Rajewsky K (1995). A cre-transgenic mouse strain for the ubiquitous deletion of loxP-flanked gene segments including deletion in germ cells. *Nucleic Acids Res* 23:5080–5081.
Semina EV, Murray JC, Reiter R, Hrstka RF, Graw J (2000). Deletion in the promoter region and altered expression of Pitx3 homeobox gene in aphakia mice. *Hum Mol Genet* 9:1575–1585.
Shaikh AC, Sadowski PD (1997). The Cre recombinase cleaves the lox site in trans. *J Biol Chem* 272:5695–5702.
Shiels A, King JM, Mackay DS, Bassnett S (2007). Refractive defects and cataracts in mice lacking lens intrinsic membrane protein-2. *Invest Ophthalmol Vis Sci* 48:500–508.
Smith AN, Miller LA, Song N, Taketo MM, Lang RA (2005). The duality of beta-catenin function: a requirement in lens morphogenesis and signaling suppression of lens fate in periocular ectoderm. *Dev Biol* 285:477–489.
Song N, Schwab KR, Patterson LT, Yamaguchi T, Lin X, Potter SS, Lang RA (2007). pygopus 2 has a crucial, Wnt pathway-independent function in lens induction. *Development* 134:1873–1885.
Spemann H (1901). Uber Correlationen in der Entwickelung des Auges. *Verh Anat Ges* 15:61–79.
Stanford WL, Cohn JB, Cordes SP (2001). Gene-trap mutagenesis: past, present and beyond. *Nat Rev Genet* 2:756–768.
Swamynathan SK, Katz JP, Kaestner KH, Ashery-Padan R, Crawford MA, Piatigorsky J (2007). Conditional deletion of the mouse Klf4 gene results in corneal epithelial fragility, stromal edema, and loss of conjunctival goblet cells. *Mol Cell Biol* 27: 182–194.
Swindell EC, Bailey TJ, Loosli F, Liu C, Amaya-Manzanares F, Mahon KA, Wittbrodt J, Jamrich M (2006). Rx-Cre, a tool for inactivation of gene expression in the developing retina. *Genesis* 44:361–363.
Thaung C, Arnold K, Jackson IJ, Coffey PJ (2002). Presence of visual head tracking differentiates normal sighted from retinal degenerate mice. *Neurosci Lett* 325:21–24.
Tolmachova T, Anders R, Abrink M, Bugeon L, Dallman MJ, Futter CE, Ramalho JS, Tonagel F, Tanimoto N, Seeliger MW, Huxley C, Seabra MC (2006). Independent degeneration of photoreceptors and retinal pigment epithelium in conditional knockout mouse models of choroideremia. *J Clin Invest* 116:386–394.
Ton CC, Miwa H, Saunders GF (1992). Small eye (Sey): cloning and characterization of the murine homolog of the human aniridia gene. *Genomics* 13:251–256.
Varnum DS, Stevens LC (1968). Aphakia, a new mutation in the mouse. *J Hered* 59:147–150.
Wawersik S, Purcell P, Rauchman M, Dudley AT, Robertson EJ, Maas R (1999). BMP7 acts in murine lens placode development. *Dev Biol* 207:176–188.
Williams SC, Altmann CR, Chow RL, Hemmati-Brivanlou A, Lang RA (1998). A highly conserved lens transcriptional control element from the Pax-6 gene. *Mech Dev* 73:225–229.
Xu PX, Zhang X, Heaney S, Yoon A, Michelson AM, Maas RL (1999). Regulation of Pax6 expression is conserved between mice and flies. *Development* 126:383–395.
Yoshida S, Shimmura S, Nagoshi N, Fukuda K, Matsuzaki Y, Okano H, Tsubota K (2006). Isolation of multipotent neural crest-derived stem cells from the adult mouse cornea. *Stem Cells* 24:2714–2722.
Yoshimoto A, Saigou Y, Higashi Y, Kondoh H (2005). Regulation of ocular lens development by Smad-interacting protein 1 involving Foxe3 activation. *Development* 132:4437–4448.
Yu Y, Bradley A (2001). Engineering chromosomal rearrangements in mice. *Nat Rev Genet* 2:780–790.
Zambrowicz BP, Friedrich GA, Buxton EC, Lilleberg SL, Person C, Sands AT (1998). Disruption and sequence identification of 2,000 genes in mouse embryonic stem cells. *Nature* 392:608–611.
Zhao H, Yang Y, Partanen J, Ciruna BG, Rossant J, Robinson ML (2006). Fibroblast growth factor receptor 1 (Fgfr1) is not essential for lens fiber differentiation in mice. *Mol Vis* 12:15–25.
Zhao H, Yang Y, Rizo CM, Overbeek PA, Robinson ML (2004). Insertion of a Pax6 consensus binding site into the alphaA-crystallin promoter acts as a lens epithelial cell enhancer in transgenic mice. *Invest Ophthalmol Vis Sci* 45:1930–1939.
Zhao S, Overbeek PA (2001). Elevated TGFbeta signaling inhibits ocular vascular development. *Dev Biol* 237:45–53.
Zheng L, Anderson RE, Agbaga MP, Rucker EB, III, Le YZ (2006). Loss of BCL-XL in rod photoreceptors: increased susceptibility to bright light stress. *Invest Ophthalmol Vis Sci* 47:5583–5589.
Zohn IE, Anderson KV, Niswander L (2005). Using genomewide mutagenesis screens to identify the genes required for neural tube closure in the mouse. *Birth Defects Res A* 73:583–590.

CHAPTER

10

Epithelial Explants and Their Application to Study Developmental Processes in the Lens

F.J. Lovicu, J.W. McAvoy

Department of Anatomy and Histology, Save Sight Institute and Discipline of Ophthalmology, The University of Sydney, and The Vision Cooperative Research Centre, Sydney, NSW, Australia

OUTLINE

INTRODUCTION

Explant cultures have long been used in developmental biology. Experimental embryologists frequently used explant or transplant assays to ask fundamental questions about when cells became committed to a certain lineage and what cell interactions were involved in establishing commitment. In these studies, pieces of embryos were explanted into culture dishes (in the early days these were cultured "in glass" hence the term "in vitro"). Explantation was also used to see how embryonic cells responded to disruptions and perturbations in their environment. These studies led to recognition of the critical role played by tissue interactions in mediating developmental processes. The logical progression was then to identify the bioactive molecules involved in these inductive interactions. Lens developmental biology has followed a similar pattern from the early days in the late 19th and early 20th century when experimentalists such as Spemann and Mangold initiated studies with amphibians into the role of inductive interactions in ocular development (see Spemann, 1901). Since then, much work has gone into identifying the mechanisms of lens induction. Various explantation experiments by Jacobson (1966) were important for identifying stages and the possible roles of different tissue interactions in lens induction and morphogenesis. Following on from this concept, the series of explantation and transplantation studies by Henry and Grainger (1990) identified key early inductive tissue interactions and led to the guiding principle that presumptive lens ectodermal cells go through several stages, including competence and bias, before induction culminates in lens morphogenesis and differentiation (Fisher and Grainger, 2004). As with other developmental systems, emphasis is now placed on identifying the molecular basis of these inductive interactions.

Whilst mouse mutant and transgenic models that are described in other chapters of this volume have been very important in identifying transcription and growth factors involved in regulating some of the earlier lens inductive processes, tissue explants have been fundamental to the identification of factors that determine and modulate the differentiated state of lens cells once they have formed. This has been the focus of our research and the remainder of this chapter will describe the mammalian lens epithelial explant system and how it has provided a unique and important model for identifying key factors that determine the fate and behavior of lens cells.

LENS MORPHOGENESIS, DIFFERENTIATION AND GROWTH

The lens develops from head ectoderm that is associated with the optic vesicle (Figure 10.1(A)). Growth and thickening of presumptive lens ectoderm gives rise to the lens placode (Figure 10.1(B)). Further placode growth and invagination results in the lens pit (Figure 10.1(C)) which subsequently closes to form the lens vesicle (Figure 10.1(D)). Cells in the posterior hemisphere of the vesicle elongate to form the primary lens fibers (Figure 10.1(E)) whereas cells in the anterior hemisphere of the vesicle differentiate into lens epithelial cells (Figure 10.1(F)). These divergent fates of embryonic lens cells give the lens its distinctive polarity (McAvoy, 1981). From this stage onwards the lens grows by continued proliferation of epithelial cells and differentiation of fiber cells. Proliferation is restricted to the lens epithelium and progeny of divisions elongate in the transitional zone at the lens equator to give rise to secondary fibers (McAvoy, 1978a,b). These growth patterns ensure that lens polarity is maintained as new fibers continue to differentiate throughout life. This is crucial for the maintenance of the ordered lens cellular architecture that contributes to its transparency and optical properties.

To elucidate how lens polarity and growth patterns are generated, the Coulombre's carried out their classical lens inversion experiment (Coulombre and Coulombre, 1963). They turned the lens of the chicken eye through 180° so that the lens epithelium faced the retina. In this environment, the epithelial cells elongated and formed a new fiber mass. This experiment showed that the optic cup environment facilitates fiber differentiation and as a result research began to focus on the identification of the factor(s) involved.

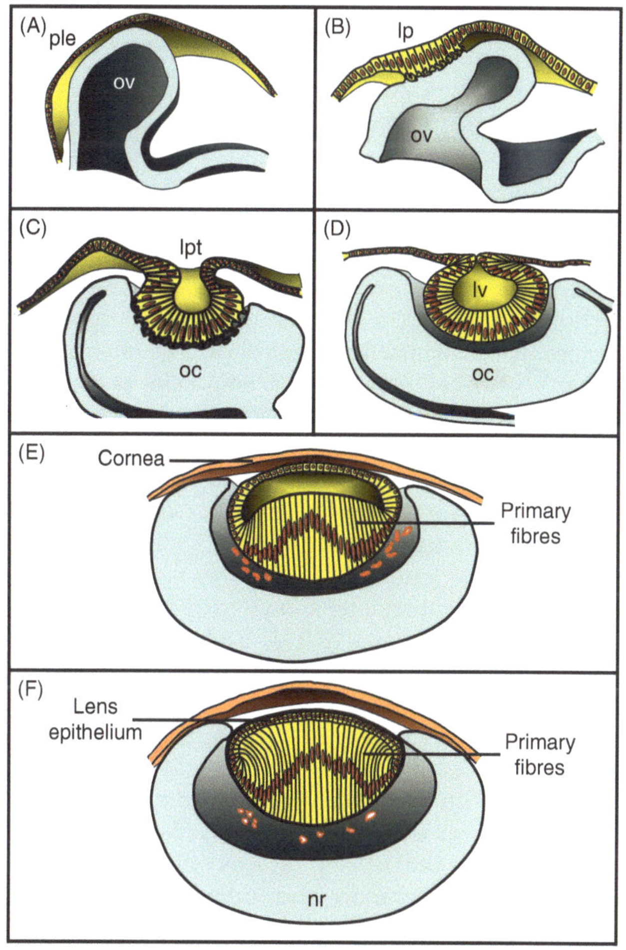

FIGURE 10.1 Schematic diagram representing sections of the developing embryonic rodent eye, from 8.5 to 13.5 d.p.c. At 8.5 d.p.c. (A), the optic vesicle (ov, blue) evaginates from the developing forebrain approaching the region of presumptive lens ectoderm (ple, yellow). By 9.5 d.p.c. (B), the optic vesicle associates with head ectoderm making direct contact via basal cellular extensions (see McAvoy, 1981). The ectoderm thickens to form the lens placode (lp). Coordinated invagination of the placode and optic vesicle occurs at 10.5 d.p.c. (C), leading to the formation of the lens pit (lpt) and optic cup (oc), respectively. By 11.5 d.p.c. (D), the lens pit deepens to form the lens vesicle (lv). At 12.5 d.p.c. (E), the lens vesicle has completely closed and detached from the optic cup. The posterior lens vesicle cells elongate to form the primary lens fiber cells. By 13.5 d.p.c (F), the lumen of the lens vesicle is lost as the primary lens fibers make contact with the anterior overlying lens vesicle cells that differentiate to form the lens epithelium. The vitreous humor and hyaloid vasculature (orange, E, F) develop between the lens and neural retina (nr, which arises from the optic cup). The ectoderm that forms over the developing lens differentiates to give rise to the cornea (pink).

DEVELOPMENT OF EXPLANT MODELS

It was in the 1960s that the first major approaches were made toward developing a lens epithelial explant system to study lens fiber differentiation. Philpott and Coulombre (1965) developed an *in vitro* system whereby cells of the embryonic chick lens epithelium (still attached to their lens capsule) could be isolated from the fiber cells and induced by serum to elongate in tissue culture. The early studies using this chick system were primarily focused on the mechanism of early fiber cell elongation (Piatigorsky *et al.*, 1970; Piatigorsky and Rothschild, 1971; Piatigorsky *et al.*, 1972a; Piatigorsky and Rothschild, 1972; Piatigorsky *et al.*, 1972b). Insulin was soon after shown to substitute for serum in inducing lens fiber cell elongation in these explants (Piatigorsky, 1973). Further studies identified "lentropin" from the vitreous (Beebe *et al.*, 1980), a protein later shown to be related to insulin-like growth factor-1 (IGF-1; Beebe *et al.*, 1987), as a promoter of epithelial cell elongation in chick lens explants.

A mammalian lens epithelial explant system was also introduced around this time (McAvoy, 1980). Consistent with earlier studies identifying neural retina as a key regulator of lens growth, co-culture experiments showed that cells in rat lens epithelial explants underwent proliferation and differentiation in the presence of neural retina (McAvoy, 1980; McAvoy and Fernon, 1984). The observation that retina-conditioned media induced these same effects in rat lens explants (Campbell and McAvoy, 1984; Walton and McAvoy, 1984) led to fractionation studies that identified a number of proteins running between 23 and 27 kD that were classed as the lens "fiber differentiation factor" (Campbell and McAvoy, 1986). At around that time, several laboratories had identified the retina as a source of eye-derived growth factors (EDGF) I and II (Courty *et al.*, 1985), and β- and α-retina-derived growth factors (D'Amore and Klagsbrun, 1984; Baird *et al.*, 1985). These turned out to be basic fibroblast growth factor (FGF) and acidic FGF, respectively (now known as FGF2 and FGF1, respectively). Whilst the Courtois group had shown that their EDGF I and II were mitogenic for lens cells, there was no evidence that these factors promoted fiber differentiation. However, at this stage it was not known, even given the appropriate stimulus, if dissociated and cultured mammalian lens cells could undergo a fiber differentiation response. In contrast, the rat lens epithelial explant system had already been shown to be capable of a fiber differentiation response that faithfully reiterated many of the morphologic and molecular processes that occur *in vivo*. When EDGFs and the FGFs, purified from both retina and brain, were tested on explants, they both induced a strong fiber differentiation response (Chamberlain and McAvoy, 1987, 1989). Since these early days, numerous *in vitro* and *in vivo* studies from a range of laboratories have provided compelling support for the conclusion that members of the FGF growth factor family play a key role in inducing lens fiber differentiation (Lovicu and McAvoy, 2005; Robinson, 2006).

Explant studies have also shown that although other growth factors such as IGF (Klok *et al.*, 1998) and PDGF (Kok *et al.*, 2002) are capable of potentiating the FGF-induced response, FGF is the only factor with the ability to induce mammalian lens epithelial cells to undergo many of the morphologic (see Figure 10.2; Lovicu and McAvoy, 1989, 1992) and molecular (Chamberlain and McAvoy, 1989; McAvoy and Chamberlain, 1989; Lovicu *et al.*, 2004) changes characteristic of fiber differentiation *in situ*. Consistent with a role for FGF in the eye, both prototype FGFs (FGF1 and FGF2; de Iongh and McAvoy, 1992, 1993; Schulz *et al.*, 1993; Lovicu and McAvoy, 1993; Lovicu *et al.*, 1997) and their high affinity receptors (de Iongh *et al.*, 1996, 1997) are expressed throughout the eye, and in particular, in the lens. Other FGFs (altogether there are 22 family members) are now known to be present in the eye and some of these have been shown to be capable of inducing fiber differentiation (Lovicu and Overbeek, 1998; Robinson, 2006). The presence of several FGFs with fiber-differentiating capability has left it open as to whether the effects of one FGF predominates, or whether fiber differentiation is the result of the effects of several members of the FGF family that are bioavailable to lens cells. As many factors in the lens cell environment such as heparan sulfate proteoglycans (Schulz *et al.*, 1997) influence FGF bioavailability, potency and stability, this is a very challenging question to resolve.

Therefore, the first insights into the role of FGF in the induction of fiber differentiation hinged on the development of the explant method for culturing mammalian lens epithelial cells during the 1980s. With the exception of the chick studies, up to this time the standard method for culturing lens cells, as with other cell types, was to dissociate them and culture them in medium containing fetal calf serum. This was clearly an inappropriate system to identify factors that control fiber differentiation as serum is essentially a broth containing many growth factors. Explanting lens epithelial cells but leaving them on their natural substratum, the lens capsule, is critical because the

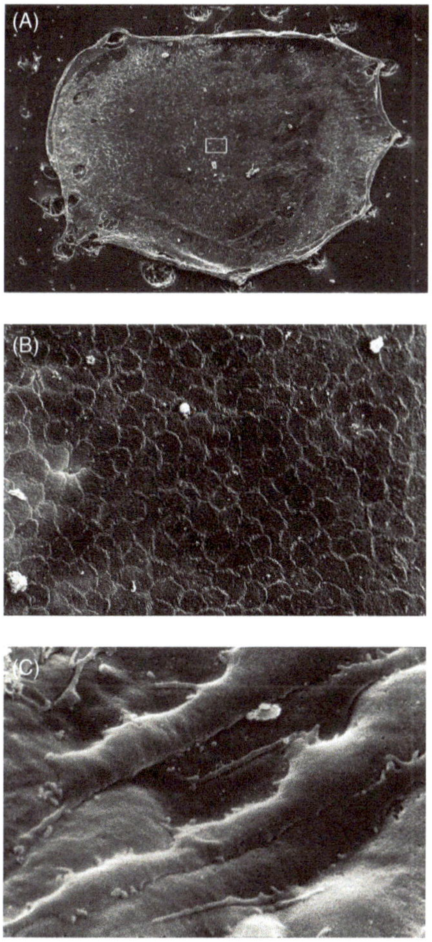

FIGURE 10.2 Scanning electron micrographs demonstrating a low power view (A) of a rat lens epithelial explant on the base of a tissue culture dish. (B) Higher magnification of the explant (represented by box in A) shows a tightly packed sheet of "cobble-stone"-shaped cells. These cells undergo a dramatic change in morphology, elongating and differentiating into lens fibre cells (C), when cultured in the presence of FGF (FGF2) for up to 7 days.

cells remain viable (without the need for serum) and maintain many normal phenotypic features. The fact that these cells remain associated with the lens capsule also provides an ideal system to examine the role of this basement membrane, as well as better assess how molecules that normally influence lens cells *in situ*, that need to traverse the lens capsule, are presented to the cells. The inability of dissociated cultured lens cells in other *in vitro* systems to undergo fiber differentiation is consistent with recent studies showing that they exhibit marked changes in gene expression patterns, compared with freshly explanted lens epithelial cells; that is, cells that are not subjected to culture and transformation (see for example, Wang-Su *et al.*, 2003). Finally, the other critical factor that has contributed to the usefulness of mammalian lens explants was that β- and γ-crystallins, which are strongly expressed in lens fiber cells but undetected in epithelial cells (McAvoy, 1978a,b), have provided key fiber differentiation markers for many studies.

PREPARATION OF LENS EPITHELIAL EXPLANTS

The preparation of mammalian lens epithelial explants was originally described in brief by McAvoy and Fernon in 1984. Since this time, this model has been used extensively and little has changed in the way the explants are prepared. This section of this chapter provides for the first time a detailed account of the dissection process, covering many of the finer details never before published, as well as a description of some of the different applications and types of explants that can be prepared.

Choice of Animals

As alluded to earlier, depending on the nature of the experimentation, lens epithelial explants can be readily prepared from different vertebrate species, with the only limitation being the size of the lens. For example, it is very challenging to prepare explants from lenses collected from early stage murine embryos. Having said this, with the appropriate equipment, fine surgical instruments, and a steady hand, it is possible. When selecting an animal model one must also take into consideration what the resultant tissue will ultimately be used for. Larger lenses used for explants will ultimately yield a lot more material for analysis (e.g., one rat lens epithelial explant will provide significantly more protein than a lens explant from the equivalent age-matched mouse).

In light of this, rats have proven to be an ideal animal model for use in preparation of lens epithelial explants. Although different strains of rats can be used, owing largely to their bigger size, albino Wistar rats (*Rattus norvegicus*) have been routinely adopted as a source of mammalian lens tissue for explant preparation. Postnatal ages are primarily used in experiments due to the ease of their collection and preventing the need to sacrifice healthy reproductive females (if embryonic tissue is required). Female Wistar rats also deliver relatively large litters (up to 20 pups in some cases) on an approximate 4 week cycle. If electing to use newborn rat pups, they are relatively large (as is their lens) and very easy to handle. An added advantage of using rodents for lens explants is that they have a large lens relative to the overall size of the eye.

Selection of the age of donor animals will depend on the nature of the experiments but there are several issues that need to be considered. First and foremost, in the case of rats at least, lenses from younger postnatal animals are usually a lot more amenable to dissection and recover and fare better in tissue culture. For Wistar rats, this may be due to the fact that the lens epithelial cells from younger donors are a lot more responsive to stimulatory factors than lens epithelial cells from more mature donors (Lovicu and McAvoy, 1992). With regards to amenability of dissection, as the explant procedure is largely dependent on pressing the lens capsule at select points into the base of the tissue culture dish to immobilize it (see later), the considerably thicker anterior capsule of lenses from mature animals does not assist this process. To overcome this problem, some lens epithelial explants from mature animals have been required to be physically pinned to the base of the culture dish using shortened, sterile, entomological pins. Although lens tissue from younger donors is better suited to the explanting procedure, again depending on the nature of the experiments, in the case of rodents it may be best to consider using young postnatal animals just prior to eye opening (approximately 14 days). The reason for this being that up until this stage, the lens is surrounded by the closely adherent fine capillary mesh of the tunica vasculosa lentis. Unlike humans, who lose this capillary network before birth, in rodents this vascular net covers much of the lens capsule at birth but by postnatal days 10–12 has mostly regressed. If the presence of such vessels has the potential to impact on the type of experimentation planned, it may be best to use slightly older animals. In most cases; however, unlike the lens epithelial cells, the fine capillaries are not maintained in prolonged tissue culture and are readily lost. In most instances, their presence has little effect on lens cells in culture and with appropriate controls in place, the use of younger animals as donors should be considered for preparation of lens epithelial explants.

Setting Up for the Explant Procedure

Preparation of lens epithelial explants requires very little in the way of equipment. To maintain aseptic technique, a laminar flow cabinet is ideal but not essential. If this is not readily available, providing that the appropriate anti-fungal reagents and antibiotics are included in the tissue culture media, explants can successfully be prepared without a laminar flow cabinet, in a room with minimal traffic, as well as providing that extra care is taken to avoid contaminating the media or tissue. Given the small size of the lens, all procedures are routinely carried out using a dissecting microscope with the ability to magnify at least 6–12 times. The light source can be supplied by external fiber-optics or a light source from the base of the microscope. If the dissecting microscope is fitted with dark field illumination, this can assist with the explant procedure, although it is not essential providing you have at least a dark (ideally black) background to work against. The transparency of the lens tissue does not readily lend itself to working on a white or light colored background.

All tissue dissection is carried out using a pair of stainless steel, fine watchmakers forceps (at least grade number 5). Although new forceps have a very fine point and are very sharp, this is not essential for the dissection procedure. In fact, slightly dull forceps are more suitable for the dissection process and lend themselves favorably to the "pinning down" of the explants to the base of the tissue culture dish (see later). It is important that forceps be kept sterile when in contact with the tissue or culture media and that any adherent ocular or lens tissue be routinely removed from the tips of the forceps. This is especially important if you are flaming your forceps to maintain sterility. The alcohol wash involved in this process will dehydrate any adherent soft tissue and the subsequent flaming will only further cake this tissue onto the forceps. This will compromise the functionality of the tips of the forceps, making them less effective in their use and will make the cleaning of the forceps a lot more difficult. On the subject of forceps, it is important that the tips meet to allow better handling of tissues. Given the fine nature of the tips of the forceps, they are easily distorted if not handled properly (simply hitting any hard surface will readily bend the tips). If this is the case, before proceeding with any dissection, the tips of the forceps will need to be straightened and sharpened, for example, with an emery stone, to ensure the tips once again meet and are level.

Other than that highlighted above, there is no more specialist equipment required to prepare lens epithelial explants. Lens explants are prepared in dedicated tissue culture dishes. In most cases, providing you can easily manipulate the tissue within the dish, any culture dish will suffice. We routinely use 35 mm × 10 mm tissue culture dishes. If this is not suitable, one option may be to first prepare explants on sterile plastic coverslips, which can then be transferred to a specific plate or dish of choice. The portability of a coverslip may also be advantageous with the subsequent processing of the tissue. Some general things to keep in mind are that the explant, once prepared in a dish, will more than likely need to be processed within the same dish for subsequent analysis. Secondly, each explant will take at least up to 2 min to prepare, so direct explanting into multi-well plates may not be ideal if you are setting up multiple explants in the one plate. The length of time fresh tissue can be left out in one sitting will depend on how quickly the tissue culture media cools down and how readily gas exchange takes place (subsequently altering the pH of the media). These changes usually can take effect within 5 min, before the tissues are required to be returned to the incubator for the media to be equilibrated and warmed up. Because of this, it is beneficial to include an inert pH indicator in the tissue culture media, such as phenol red. This will allow you to instantly see changes in pH and act accordingly.

All explants are prepared in pre-incubated (37°C in 5% CO_2/air) minimal essential media. As mentioned earlier, a major advantage of this system is that the explants are cultured without the need for serum. We routinely use Medium 199 with Earle's salts containing phenol red, and this is supplemented with 0.1% bovine serum albumin (BSA), 50 IU/ml penicillin, 50 μg/ml streptomycin, 2.5 μg/ml Fungizone and 0.1 μg/ml L-glutamine. Note that L-glutamine is not very stable so needs to be added fresh, just before the media is ready to be used.

Collection of Lens Tissue

Once you have dispensed the media into the tissue culture dishes and these have equilibrated (37°C in 5% CO_2/air) in the incubator, it is time to collect the fresh tissue to commence the preparation of the lens epithelial explants. Once the donor animal has been humanely sacrificed, the eyeballs are collected. If using postnatal rodents prior to day 14, the overlying eyelids will first need to be removed to expose the eyes. This is readily done by pinching the eyelid with a pair of dull forceps (this will automatically result in the eyeball sinking into the eye socket) and with sharp scissors, cut underneath the point of the pinch and remove the overlying skin. This cut will readily expose the eyeball. With a pair of small curved scissors, place these over the eyeball in an open position and press on the exposed region. This will force the eyeball to surface above the level of the scissor blades. At this time the scissors are gradually closed but not to the extent of severing the optic nerve or cutting into the eyeball. Using the scissors like a claw hammer, the eyeball is easily removed from the eye socket as the optic nerve detaches with gentle pressure. To begin with, collect two eyeballs at any one time. As you become more proficient with the explanting procedure, you can collect anywhere from 6 to 10 eyeballs at the one time. These eyeballs are placed in pre-equilibrated tissue culture media and immediately dissected under the microscope to isolate the lenses.

Care needs to be taken when isolating the lens so as to avoid prematurely rupturing the lens capsule. The most effective way to cleanly remove the lens from the eyeball is to use fine forceps to pinch a point of the outer sclera where it meets the cornea. Care must be taken not to puncture the lens by stabbing the eyeball at this time (the lens sits relatively close to this point of contact). Once you have a good grasp of the eyeball, using both forceps, it is torn open. The immediate release of intraocular pressure is usually sufficient to cleanly expel and separate the lens from all of the surrounding ocular tissues. If this is not the case, it will most likely come out of the eye with a skirt of ciliary body attached around the lens equator. This ciliary body is readily removed by grasping it with the forceps and teasing it off, usually as one string. If puncturing the eyeball at this sensitive spot is too problematic, the best approach is to puncture it through the vitreal chamber tearing through the back of the eye, through the retina. The advantage of this is that you reduce the likelihood of damaging the lens, which sits quite anteriorly. The disadvantage is that in all instances the lens will be isolated with a lot of adherent surrounding ocular tissues. These are readily removed as described above with forceps, but will simply add more time to the isolation process and may increase the risk of damaging the lens through over handling. Once the lenses have been cleanly isolated and all the extra-lenticular tissues removed, the lenses can be transferred to tissue culture dishes containing fresh media. These are the dishes that will be used to isolate the lens epithelia and prepare the lens explant. It is at this time that you need to decide how many explants you would like to have in any given dish. We routinely set up two explants per 35 mm culture dish, hence would add two lenses to the dish. This number allows both lenses to be set up in the one sitting. Any more lenses would increase the time that the tissue is left out of the incubator. If more than two lenses are required per dish, this is easily overcome if subsequent explants are set up after re-equilibrating the media, once the first set of explants are prepared. We would not recommend more than four explants per 35 mm dish. When setting up the explants in a dish, also keep in mind what the tissues will subsequently be used for. If they are to be immunolabeled, it will be beneficial to prepare the explants close to each other so that when applying expensive reagents, small volumes will readily cover both explants at any one time. If they are to be immunolabeled differently, it may be best to leave some space between the explants to prevent any cross contamination of reagents during this process. While you are preparing one set of explants in one dish, the other lenses are kept in the incubator (at 37°C in 5% CO_2/air) for approximately 5–10 min. This preculture assists with the ease of separation of the lens epithelium from the fiber cell mass during the dissection process.

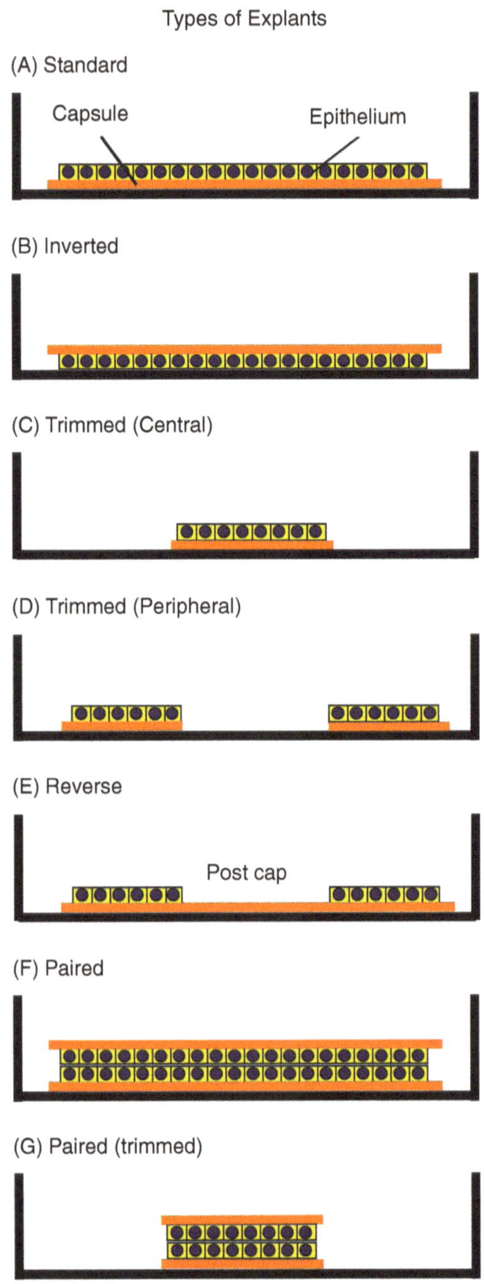

FIGURE 10.3 Schematic diagram demonstrating the different types of lens epithelial explants that can be prepared, all supported on the base of a tissue culture dish. Explants are shown in section and their size relative to the dish are not to scale. Abbreviation: Post cap; posterior capsule.

Orientating the lens

One of the first important tasks to undertake before dissecting the lens is to determine its orientation. Given the distinct biconvex shape of the lens (see Figure 10.4(C)), once placed in the tissue culture dish, the lens will either sit on its anterior pole (epithelial side) or its posterior pole (fiber side). Depending on what type of explant (see Figure 10.3) you need to prepare, you will need to know which pole of the lens is facing the base of the dish. When preparing "standard" lens epithelial explants (Figure 10.3(A)), defined as explants primarily made up of anterior lens capsule completely covered by the lens epithelium, you will need the anterior pole to be facing the base of the culture dish. That is, the posterior pole is facing up and this is the pole that you will be looking at through the dissecting microscope. One of the more difficult tasks in explanting is to determine the polarity of the lens and this tends to become even more difficult as the age of donor increases. When dealing with lenses from younger donors, there are three distinct features of the lens that will assist with its orientation, including (i) the presence of a distinct suture plane; (ii) predominance of capillaries on the posterior pole of the lens; and (iii) subtle differences in convexity of the anterior and posterior poles:

1. *Presence of sutures*: In a young postnatal lens, the maturing secondary fibers will meet at distinct suture planes along the midline of the lens. As the lens grows with age, these suture planes become a lot more elaborate as they begin to extensively branch (see Kuszak and Costello, 2004). In the young rodent lens, the suture planes appear as a distinct Y. Both anterior and posterior poles will display such a Y-suture but given the fact that the anterior Y-suture is obscured by the lens epithelium and its thicker anterior lens capsule, it is not as obvious as the Y-suture of the posterior pole (Figure 10.4(A)). Hence, this is the first and most obvious marker observed when comparing the poles of the lens. One potential complication that will obscure the ability to readily see the Y-suture is the presence of a "cold" cataract. If the tissue culture media is allowed to cool prior to the dissection process, the intact lens

will also cool and the first signs of this is reflected by protein changes in the lens nuclear fibers. The core or nucleus of the lens, made up of the primary fiber cells, will readily become opaque. As the lens nucleus is in direct alignment with the Y-sutures, they will no longer be visible. If this occurs, the process is easily reversible by placing the intact lens back in the incubator to warm, or simply transferring it into a fresh tissue culture dish containing warm media. Within seconds, the transparency of the lens is restored and the orientation and dissection procedure can be resumed.

2. *Presence of capillaries*: As mentioned earlier, the young postnatal lens is surrounded by a rich network of fine capillaries, collectively known as the tunica vasculosa lentis. These vessels are distinctly arranged, with the main hyaloid artery branching into the vasa hyaloidea propria which anastomoses over the entire posterior pole of the lens to form the posterior vascular capsule (Figure 10.4(B), arrows). These blood vessels extend over the lens equator into the lateral part of the vascular capsule as they straighten into the capsulo-pupillary vessels. These straight lateral vessels at the lens equator again anastomose with loops of the anterior vascular capsule which are not as densely placed as those on the posterior lens pole, and also do not necessarily extend to cover the immediate center of the anterior pole. Based on this distinction, examination of the capillary network surrounding the lens will provide a clear idea of the polarity of the lens. This marker can only be used in younger rodents, because as mentioned earlier, as the donor rodent ages (closer to postnatal day 14 with eye opening), these vessels have mostly regressed and are no longer apparent, hence polarity will mostly be based on the shape of the lens.
3. *Lens shape*: In the younger donors, the shape of the lens is the confirmatory marker determining lens polarity. As mentioned, the lens is biconvex; however, the degree of this convexity allows you to distinguish the anterior from the posterior pole. The anterior pole of a younger rodent lens appears less convex than the posterior pole (Figure 10.4(C)). The anterior pole hence may seem a little flatter than the posterior pole. As the donor age increases, the rodent lens tends to lose this characteristic and appears uniformly spherical.

One additional indicator of the polarity of the lens, irrespective of age, is the thickness of the lens capsule. The anterior lens capsule is significantly thicker than the posterior capsule but unfortunately the best way to readily determine this is invasive; either by puncturing or making a tear in the capsule. Tearing the thicker anterior capsule is most apparent as the fold of the tear tends to sit idle. If a similar tear were to be made in the thinner posterior capsule, this would readily roll up tight as if it were recoiling. This distinction is always apparent once you have commenced the dissection process. If the tear is small enough and you find that you have started at the opposite pole required, it may still be possible to invert the lens and re-commence the explanting procedure.

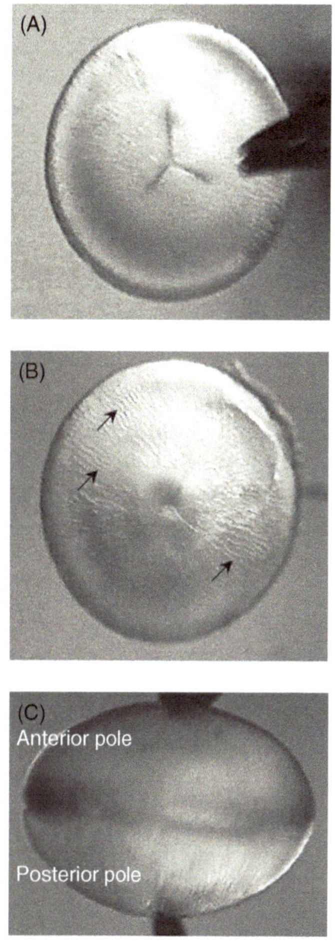

FIGURE 10.4 Characteristic features of the young postnatal rat lens that assist with determining orientation. (A) Distinct Y-suture on posterior pole. (B) Numerous fine capillaries on posterior pole of lens. (C) Subtle differences in convexity between anterior and posterior poles of the lens. In (A) and (C), tips of forceps used to handle lens are evident.

Isolating the Lens Epithelium

Once the orientation of the lens is established, with the anterior pole facing the base of the dish (for a "standard" explant), you are ready to commence isolating the lens epithelium. It is important to note that you are in fact manipulating and attempting to isolate the lens capsule that will contain the adherent monolayer of lens epithelial cells. The beauty of this system is that the fiber cell mass will readily dissociate, not only from the overlying epithelial cells but also from the posterior lens capsule. To commence the procedure, the forceps are used to gently pinch the posterior lens capsule, close to the center of the lens. It is important not to stab the lens too deeply as the disruption to the lens fiber mass will compromise this early step. Once you have a firm hold of the lens, the forceps are gently separated in opposite directions to place a substantial tear in the posterior capsule, remembering to not extend this tear beyond the lens equator. This single tear will readily expose the underlying fiber mass but will still not permit you to separate it from the epithelium. It should be kept in mind that the tearing process is to facilitate opening the posterior of the lens so that the anterior portion can eventually be pinned out flat on the base of the culture dish. If insufficient tears are made, it will be similar to flattening a capsular bag which is not ideal for the purpose of preparing an epithelial explant. To continue with the procedure, after the first tear is made, use the forceps to tear one half of the separated posterior capsule. Again, ensure this tear does not extend beyond the lens equator. Repeat this step for the other half of the posterior lens capsule. If the tears were carried out as described, you should now have the posterior capsule opened up as quadrants, with each quadrant recoiling (see above) over itself toward the lens equator, exposing the complete posterior surface of the fiber cell mass. If this is not the case, make the necessary additional tears until you have reached this point.

The next stage of the process requires the separation of the fiber cell mass from the capsule (hence the lens epithelium). To do this, with one pair of forceps at the lens equator, hold the lens capsule firmly at one point onto the base of the culture dish (Figure 10.5(A)). This will tilt the lens to an angle. A thing to note here is to carry out this step in the position on the dish where you expect to have the final explant. This is because all the subsequent steps will be carried out in one line of action. Once the lens is secured by the forceps to the base of the dish, using your other forceps take a hold of the fiber cell mass in a position close to where you are holding the lens capsule with the other forceps. Once this is secured, gently roll off the fiber cell mass (Figure 10.5(B)), away from the point where you are holding down the lens capsule. It is important to roll off the fiber mass in one motion, if possible, and

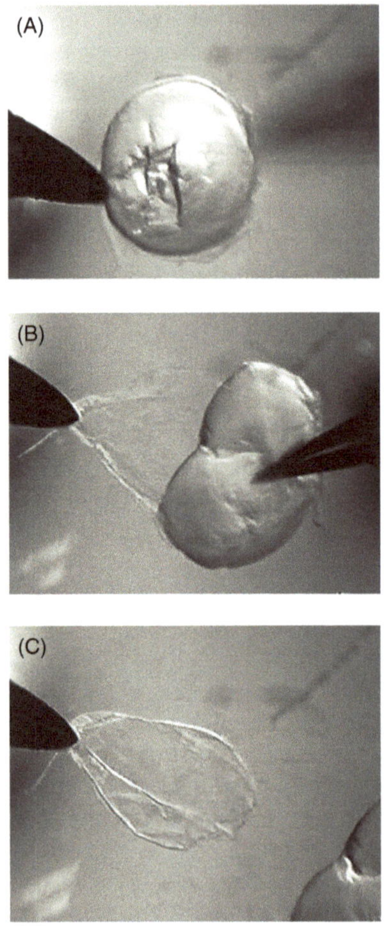

FIGURE 10.5 Early stages in the preparation of rat lens epithelial explants demonstrating the separation of lens epithelial sheet from the fibre cell mass. Using fine forceps to hold down one spot of the peeled back posterior capsule (A), the second pair of forceps is used to grab the fibre cell mass, and in one motion roll it off the underlying lens epithelial sheet (B), until the transparent capsule supporting the lens epithelial cells is free (C).

not pull at it which will only encourage the teasing off and separation of individual groups of lens fibers. Separating the fibers as one mass makes the procedure a lot cleaner and will prevent any fiber cell contamination of your epithelial preparation. With the successful separation of the fiber cell mass, the epithelial cells should be facing uppermost with the capsule side of the explant facing the base of the dish. At this point you are now securing the lens capsule at one point with a pair of forceps (Figure 10.5(C)). As the focal plane has changed, from the surface of the posterior pole of the intact lens to a thin monolayer of cells, it is best to adjust the focus of the microscope to better visualize the lens capsule, which will now be attached to the base of the culture dish. If possible, do not lose the position of the transparent lens capsule by releasing it.

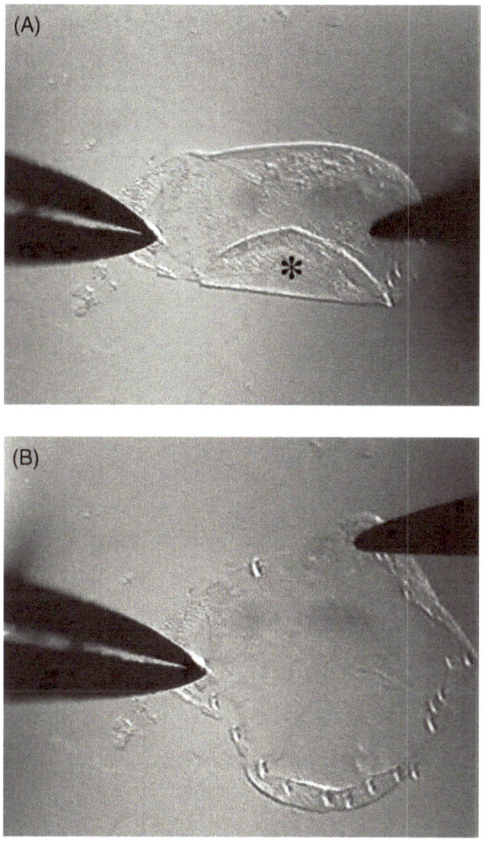

FIGURE 10.6 Later stages in the preparation of rat lens epithelial explants demonstrating the "pinning down" or securing of the explant to the base of the tissue culture dish. (A) The isolated lens epithelial explant is secured to the base of the culture with a pair of forceps. Until it is "pinned" to the dish, it will have a tendency to roll under itself (asterisk denotes fold). This is readily rectified by applying gentle pressure around the circumference of the explant with fine forceps, flattening and pinning as you proceed.

If this occurs, you can still determine which side of the capsule the epithelial cells are facing (see below).

Securing the Lens Explant

With one edge of the lens capsule held in place on the base of the dish with forceps, use the other pair of forceps to simultaneously flatten the lens capsule and secure it to the base of the dish, as you pin it along its circumference. Note that the lens capsule will have a tendency to roll up under itself, like a scroll (see Figure 10.6(A)). This is beneficial to some degree as it will confirm the orientation of the lens epithelial cells on the capsule. The epithelial cells are always on the uppermost face of the capsule, opposite to the side that it is scrolling under. At this point you will need to decide whether you would like the cells to be in direct contact with the base of the culture dish (capsule facing uppermost; inverted explant, see Figure 10.3(B)) or for the cells to be facing uppermost (with the capsule in direct contact with the base of the culture dish; standard explant, see Figure 10.3(A)). Again, the positioning of the explant will depend on how you plan to utilize the tissue at the completion of the culture period. In most cases, the explants are prepared so that the epithelial cells face up. The reverse of this, having the epithelial cells sheltered by the overlying lens capsule does not effect their ability to be cultured, nor their ability to respond to exogenously applied factors. Most of these, such as growth factors, will readily traverse the lens capsule, as they do *in situ*, to influence the lens epithelial cells. The major disadvantage of positioning of the cells face down is evident during subsequent processing of the tissue for analysis. If the explants are "inverted" so that the capsule faces uppermost, it is difficult to immunolabel the cells as the antibodies and some reagents do not effectively penetrate the lens capsule.

To adhere the explant to the base of the tissue culture dish, as mentioned above, forceps are used to apply gentle pressure around the explant periphery. With the explant in position after removal of the fiber cell mass, gently stretch to flatten the explant and place gentle pressure with the forceps on the opposite side to the first holding point on the explant (Figure 10.6(A)). For this, pressure has to be firm enough to physically allow the capsule to lightly embed itself into the base of the tissue culture dish, like a "press button." Once the explant is secured on two opposing sides, the same "pinning" process is applied around the remainder of the explant periphery, unscrolling any parts of the capsule that may have rolled underneath, ensuring a flat epithelial preparation (Figure 10.6(B)). In most cases, up to 10 points of adherence are sufficient; however, if you foresee that the analysis of the tissue will undergo some rigorous processing, a few more points of adherence may be required. Once you have set up an explant, another explant can be set up in close proximity to it, completing the explanting procedure for the one dish.

Variations on a Theme

Reverse explants. Earlier we discussed the principle of "inverted" explants, with the epithelial cells facing the base of the culture dish, as opposed to "standard" explants where the lens epithelial cells face uppermost. Another variation on this theme are "reverse" explants which are similar to the standard explants with the exception that they are prepared to better visualize the most peripheral lens epithelial cells that are in closer association with the posterior capsule (see Figure 10.3(E)). The center of these explants (made up of posterior capsule) is therefore primarily devoid of cells, with epithelial cells only found around the periphery of the explant on anterior capsule. These "reverse" explants allow visualization of the leading edge of the peripheral epithelial cells as they migrate and/or differentiate from the anterior capsule, over the posterior lens capsule (see Chen *et al.*, 2006). Preparation of these is very similar to setting up "standard" explants as described in detail above, only now instead of tearing the capsule through the posterior pole of the lens, it is torn through the anterior pole. These explants are hence "heavier" around the edges owing to the thicker anterior capsule containing the epithelial cells. Unlike "standard" explants that have the most posterior (closest to the lens equator) epithelial cells situated around the explant periphery with the most anterior epithelial cells throughout the center of the explant, in "reverse" explants the anterior epithelial cells are on the periphery of the explant with the posterior cells innermost, all surrounding a bare posterior lens capsule that makes up the explant center.

Trimmed explants. It is well recognized that the lens epithelium is comprised of a relatively heterogenous population of cells. And this is no more evident than when comparing the anterior central lens epithelial cells with the more posterior peripheral epithelial cells. As these two regions of the epithelium have been shown to differentially respond to external stimuli (see e.g., Lovicu and McAvoy, 1992), to ensure that the responsiveness of the different regions of the explant are totally independent of each other, both central and peripheral preparations of the explant can be prepared by trimming and removing a specific region of the explant. These "trimmed" explants are relatively easy to prepare once you have first set up the explant. Using a small scalpel blade, a central square with sides approximately equivalent in length to a third of the original explant diameter, is made in the center of the explant. For central trimmed explants, the peripheral cells outside of this square are removed (see Figure 10.7 and 10.3(C)). For a peripheral trimmed explant, the central cells within the central square are removed (see Figure 10.3(D)). These procedures result in isolating either the central or peripheral lens epithelial cells.

Paired explants. More recent studies have made attempts to reconstitute an intact lens *in vitro* by using lens epithelial explants. Earlier experiments by Coulombre and Coulombre (1971) surgically removed the lens from an embryonic chick eye and replaced it

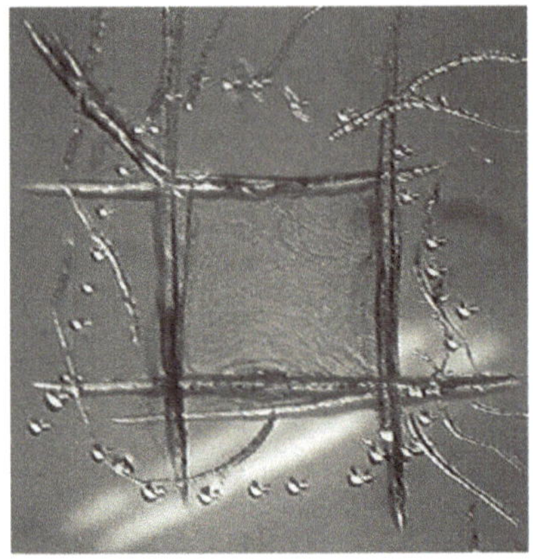

FIGURE 10.7 A central trimmed lens epithelial explant, produced by scoring out a square with sides approximately equivalent in length to a third of the original explant diameter, using a scalpel. In this instance, the peripheral cells outside of this square are removed (see also Figure 10.3(C)).

with an isolated lens epithelium (attached to its lens capsule). This small lens epithelial explant, with an exposed epithelium was shown to form a lens vesicle *in situ* and subsequently form an intact lens with normal polarity; the cells of the lens vesicle facing the cornea formed the lens epithelium while those facing the vitreous elongated and differentiated into lens fiber cells. *In vitro*, by sandwiching two lens epithelial explants together, so that the epithelial cells are innermost and the lens capsules are outermost, under the right stimulation (in this case, in response to vitreous humor), long-term tissue culture of this "paired" explant (see Figure 10.3(F), (G)) results in the formation of a refractive transparent lens with a distinct polarity (O'Connor and McAvoy, 2007). Epithelial cells are confined to one side of this newly formed lens and overlie a mass of regularly elongated fiber cells that express fiber-specific cell markers.

PROCESSING EXPLANTS FOR ANALYSIS

As mentioned earlier, depending on what the resultant lens material is required for, much of the processing of explants can be carried out in the original tissue culture dish. If the resultant explants are required for determining protein or gene expression levels, the explants can simply be gently lifted off the base off the culture dish with fine forceps, and transferred to the appropriate reagents for processing. On the other hand, if you would like to stain or immunolabel the lens cells as a wholemount, this is all carried out in the dish. First and foremost, the explants will need to be fixed. Given that the explants start as a monolayer of cells that may multilayer (dependent on the treatment of the explants) over the culture period, these tissues are relatively thin, so will not require a lengthy fixation period. Twenty minutes usually suffices for aldehyde fixatives and methanol fixation usually takes only a few minutes. Depending on the proteins to be identified, the type of fixative usually needs to be experimented with, as it may influence the labeling process, as for any other cell type. Following rinses to wash out the fixative, tissue can be partially dehydrated and stored in 70% ethanol at 4°C until ready for use.

Light Microscopy Applications

If explants are required to be sectioned for staining or labeling purposes at the light microscope level, as for all small and delicate tissues being prepared for paraffin wax embedding, it is recommended that they are first pre-embedded in agar. Following fixation in the dish, explants are rinsed and left in phosphate buffered saline (PBS), which we use to dissolve our agar. Once in PBS, excess PBS is drained and the explants are gently unpinned from the base of the culture dish using fine forceps. Successful unpinning of the explant can be tested by gently moving the fixed explant over the residual PBS lining the base of the culture dish. Approximately, 1.5ml of molten 2.5% Noble agar is added to the dish around the explants. By gently moving the explant at this time, you will ensure that the agar lifts the explants off the base of the dish as it begins to set. Once the agar is set, the embedded explants can be cut out, as cubes, using a scalpel blade. This agar cube protects the explant which is still visibly sandwiched through the midline of the cube. The agar cube containing the embedded explant is then transferred to a graded series of ethanol to begin its dehydration, and processed accordingly for wax embedding.

Electron Microscopy Applications

For preparation of explants for transmission electron microscopy (TEM), the explants are fixed, post-fixed and dehydrated in their dishes. Although the tissues will tolerate acetone, the dishes will not, so the explants must be transferred for this step, or alternatively directly transferred to resin for embedding. Although the explants are thin and brittle at this stage, with care they are easily manipulated with forceps and readily transferred. For scanning electron microscopy (SEM), the explants will remain attached to the base of the tissue culture dish throughout the whole processing stage, through to visualization under the electron microscope. If you are preparing explants for SEM analysis, be sure to pin the tissue down in multiple places around its periphery, as the dehydration and critical point drying of the explants will encourage shrinkage and their possible dislodgement from the supporting dish. It is important that the tissue does not dislodge as the orientation of the explant (important for SEM that cells are facing up) is very difficult to determine once the tissue is at this stage of processing. Once tissues reach 70% ethanol as part of their dehydration step for SEM analysis, using a hot scalpel blade, cut out the small portion of the tissue culture dish containing the explant. This is then transferred to 100% ethanol to continue the dehydration process. If different treated explants are being examined, these can be color coded by using permanent markers to label the underside of each of the dish pieces. These permanent markers are indelible and tolerate the final stages of tissue processing for SEM.

FUTURE PERSPECTIVES

Clearly, studies with epithelial explants have contributed much to the identification and understanding of factors that regulate the differentiated state and behavior of lens cells. For example, the fiber-differentiating influence of members of the FGF growth factor family was first illustrated in rat lens epithelial explants and this provided the impetus for further investigations into the function of FGF using other more complex experimental models, notably *in vivo* transgenic mice. Lens explants have also been useful for investigating signaling pathways, activated by the ocular media and different growth factors, as they are amenable to the use of pharmacological inhibitors that are specific for these pathways. Undoubtedly, this *in vitro* system will yield further insights into FGF signaling, as well as the role of other growth factor-mediated pathways, in modulating the behavior of lens cells. The use of viral vectors and gene delivery techniques, to differentially express genes of interest into lens cells, is also likely to be an important and sought-after application of future research.

Besides contributing information to understanding normal developmental processes, explants have also led to the identification of factors that promote aberrant growth and differentiation of lens cells, such as occurs in subcapsular cataracts. For example, it was studies with rat lens explants that first identified the TGFβ family as key inducers of epithelial mesenchymal transition (EMT) and fibrosis in lens cells (Liu *et al.*, 1994; Hales *et al.*, 1994). An important innovation from these cataract studies with explants was the development of the human lens capsular bag model in the late George Duncan's laboratory (Wormstone, 2002). Explanting the capsular bag, after removal of the fibers, provided an *in vitro* model that closely mimicked the situation *in vivo* after cataract surgery where the lens epithelial cells had the exposed posterior capsule to migrate along. Studies with this model confirmed that TGFβ promoted a similar EMT/fibrosis in human lens cells to that described in rat explants. In addition, this model mimicked many of the processes seen in development of posterior subcapsular cataract (PCO) that is a common complication of modern cataract surgery. This system has been, and will continue to be, useful in identifying features of aberrant behavior of human lens cells, as well as identifying useful inhibitors and strategies for prevention of PCO (see e.g., Duncan *et al.*, 2007). Recently, chick capsular bags have also been used in this context (Walker *et al.*, 2007). Clearly these capsular bags/explants will be a quick way of identifying useful molecules for PCO prevention.

Finally, recent studies have shown that if given appropriate conditions, explants can be used, not only to study cellular responses, but also to understand how the three-dimensional lens cell architecture develops. These studies with "paired" explants have shown that lentoids can develop, which have both epithelial and fiber cell populations in relatively normal arrangements (O'Connor and McAvoy, 2007). These lentoids are also transparent and capable of focusing light. Achieving the "holy grail" of successful regeneration of lens structure and function after cataract surgery will depend on our understanding not only of the factors that regulate the behavior of lens cells but also how this relates to the generation of their appropriately ordered arrangements. These questions provide enormous challenges for the future but we believe that lens epithelial explants will play an important part in providing the answers, as they have over the last two decades.

REFERENCES

Baird A, Esch F, Gospodarowicz D, Guillemin R (1985). Retina- and eye-derived endothelial cell growth factors: partial molecular characterization and identity with acidic and basic fibroblast growth factors. *Biochemistry* 24:7855–7860.

Beebe DC, Feagans DE, Jebens HA (1980). Lentropin: a factor in vitreous humor which promotes lens fiber cell differentiation. *Proc Natl Acad Sci USA* 77:490–493.

Beebe DC, Silver MH, Belcher KS, Van Wyk JJ, Svoboda ME, Zelenka PS (1987). Lentropin, a protein that controls lens fiber formation, is related functionally and immunologically to the insulin-like growth factors. *Proc Natl Acad Sci USA* 84:2327–2330.

Campbell MT, McAvoy JW (1984). Onset of fibre differentiation in cultured rat lens epithelium under the influence of neural retina-conditioned medium. *Exp Eye Res* 39:83–94.

Campbell MT, McAvoy JW (1986). A lens fibre differentiation factor from calf neural retina. *Exp Cell Res* 163:453–466.

Chamberlain CG, McAvoy JW (1987). Evidence that fibroblast growth factor promotes lens fibre differentiation. *Curr Eye Res* 6:1165–1169.

Chamberlain CG, McAvoy JW (1989). Induction of lens fibre differentiation by acidic and basic fibroblast growth factor (FGF). *Growth Factors* 1:125–134.

Chen Y, Stump RJ, Lovicu FJ, McAvoy JW (2006). A role for Wnt/planar cell polarity signaling during lens fiber cell differentiation? *Semin Cell Dev Biol* 17:712–725.

Coulombre JL, Coulombre AJ (1963). Lens development: fiber elongation and lens orientation. *Science* 142:1489–1490.

Coulombre JL, Coulombre AJ (1971). Lens development. V. Histological analysis of the mechanism of lens reconstitution from implants of lens epithelium. *J Exp Zool* 176:15–24.

Courty J, Loret C, Moenner M, Chevallier B, Lagente O, Courtois Y, Barritault D (1985). Bovine retina contains three growth factor activities with different affinity to heparin: eye derived growth factor I, II, III. *Biochimie* 67:265–269.

D'Amore PA, Klagsbrun M (1984). Endothelial cell mitogens derived from retina and hypothalamus: biochemical and biological similarities. *J Cell Biol* 99:1545–1549.

de Iongh R, McAvoy JW (1992). Distribution of acidic and basic fibroblast growth factors (FGF) in the foetal rat eye: implications for lens development. *Growth Factors* 6:159–177.

de Iongh R, McAvoy JW (1993). Spatio-temporal distribution of acidic and basic FGF indicates a role for FGF in rat lens morphogenesis. *Dev Dyn* 198:190–202.

de Iongh RU, Lovicu FJ, Hanneken A, Baird A, McAvoy JW (1996). FGF receptor-1 (*flg*) expression is correlated with fibre differentiation during rat lens morphogenesis and growth. *Dev Dyn.* 206:412–426.

de Iongh RU, Lovicu FJ, Chamberlain CG, McAvoy JW (1997). Differential expression of fibroblast growth factor receptors during rat lens morphogenesis and growth. *Invest Ophthalmol Vis Sci* 38:1688–1699.

Duncan G, Wang L, Neilson GJ, Wormstone IM (2007). Lens cell survival after exposure to stress in the closed capsular bag. *Invest Ophthalmol Vis Sci* 48:2701–2707.

Fisher M, Grainger RM (2004). Lens induction and determination. In: Lovicu FJ, Robinson ML (eds), *Development of the Ocular Lens.* Cambridge University Press, New York, pp. 27–47.

Hales AM, Schulz MW, Chamberlain CG, McAvoy JW (1994). TGF-beta 1 induces lens cells to accumulate alpha-smooth muscle actin, a marker for subcapsular cataracts. *Curr Eye Res* 13:885–890.

Henry JJ, Grainger RM (1990). Early tissue interactions leading to embryonic lens formation in Xenopus laevis. *Dev Biol* 141:149–163.

Jacobson AG (1966). Inductive processes in embryonic development. *Science* 152:25–34.

Klok E, Lubsen NH, Chamberlain CG, McAvoy JW (1998). Induction and maintenance of differentiation of rat lens epithelium by FGF-2, insulin and IGF-1. *Exp Eye Res* 67:425–431.

Kok A, Lovicu FJ, Chamberlain CG, McAvoy JW (2002). Influence of platelet-derived growth factor on lens epithelial cell proliferation and differentiation. *Growth Factors* 20:27–34.

Kuszak JR, Costello MJ (2004). The structure of the vertebrate lens. In: Lovicu FJ, Robinson ML (eds), *Development of the Ocular Lens.* Cambridge University Press, New York. 2004:71–118.

Liu J, Hales AM, Chamberlain CG, McAvoy JW (1994). Induction of cataract-like changes in rat lens epithelial explants by transforming growth factor beta. *Invest Ophthalmol Vis Sci* 35:388–401.

Lovicu FJ, McAvoy JW (1989). Structural analysis of lens epithelial explants induced to differentiate into fibres by fibroblast growth factor (FGF). *Exp Eye Res* 49:479–494.

Lovicu FJ, McAvoy JW (1992). The age of rats affects the response of lens epithelial explants to fibroblast growth factor. An ultrastructural analysis. *Invest Ophthalmol Vis Sci* 33:2269–2278.

Lovicu FJ, McAvoy JW (1993). Localization of acidic fibroblast growth factor, basic fibroblast growth factor, and heparan sulphate proteoglycan in rat lens: implications for lens polarity and growth patterns. *Invest Ophthalmol Vis Sci* 34:3355–3365.

Lovicu FJ, Overbeek PA (1998). Overlapping effects of different members of the FGF family on lens fiber differentiation in transgenic mice. *Development* 125:3365–3377.

Lovicu FJ, McAvoy JW (2005). Growth factor regulation of lens development. *Dev Biol* 280:1–14.

Lovicu FJ, de Iongh RU, McAvoy JW (1997). Expression of FGF-1 and FGF-2 mRNA during lens morphogenesis, differentiation and growth. *Curr Eye Res* 16:222–230.

Lovicu FJ, Ang S, Chorazyczewska M, McAvoy JW (2004). Deregulation of lens epithelial cell proliferation and differentiation during the development of TGFbeta-induced anterior subcapsular cataract. *Dev Neurosci* 26:446–455.

McAvoy JW (1978a). Cell division, cell elongation and distribution of alpha-, beta- and gamma-crystallins in the rat lens. *J Embryol Exp Morphol* 44:149–165.

McAvoy JW (1978b). Cell division, cell elongation and the co-ordination of crystallin gene expression during lens morphogenesis in the rat. *J Embryol Exp Morphol* 45:271–281.

McAvoy JW (1980). Beta- and gamma-crystallin synthesis in rat lens epithelium explanted with neural retinal. *Differentiation* 17:85–91.

McAvoy JW (1981). The spatial relationship between presumptive lens and optic vesicle/cup during early eye morphogenesis in the rat. *Exp Eye Res* 33:447–458.

McAvoy JW, Fernon VT (1984). Neural retinas promote cell division and fibre differentiation in lens epithelial explants. *Curr Eye Res* 3:827–834.

McAvoy JW, Chamberlain CG (1989). Fibroblast growth factor (FGF) induces different responses in lens epithelial cells depending on its concentration. *Development* 107:221–228.

O'Connor MD, McAvoy JW (2007). In vitro generation of functional lens-like structures with relevance to age-related nuclear cataract. *Invest Ophthalmol Vis Sci* 48:1245–1252.

Philpott GW, Coulombre AJ (1965). Lens development II. The differentiation of embryonic chick lens epithelial cells in vitro and in vivo. *Exp Cell Res* 38:635–644.

Piatigorsky J, Rothschild SS (1971). Effect of serum on the synthesis of RNA and of DNA in the cultured lens epithelium of the chick embryo: initiation of lens fiber formation in vitro. *Biochim Biophys Acta* 238:86–98.

Piatigorsky J, Rothschild SS (1972). Loss during development of the ability of chick embryonic lens cells to elongate in culture: inverse relationship between cell division and elongation. *Dev Biol* 28:382–389.

Piatigorsky J, Webster H deF, Craig SP (1970). Ultrastructural and biochemical aspects of lens fiber formation in vivo and in vitro. *J Cell Biol* 47:158a.

Piatigorsky J, Webster H deF, Craig SP (1972a). Protein synthesis and ultrastructure during the formation of embryonic chick lens fibers in vivo and in vitro. *Dev Biol* 27:176–189.

Piatigorsky J, Webster H deF, Wollberg M (1972c). Cell elongation in the cultured embryonic chick lens epithelium with and without protein synthesis. Involvement of microtubules. *J Cell Biol* 55:82–92.

Piatigorsky J (1973). Insulin initiation of lens fiber differentiation in culture: elongation of embryonic lens epithelial cells. *Dev Biol* 30:214–216.

Robinson ML (2006). An essential role for FGF receptor signaling in lens development. *Semin Cell Dev Biol* 17:726–740.

Schulz MW, Chamberlain CG, de Iongh RU, McAvoy JW (1993). Acidic and basic FGF in ocular media and lens: implications for lens polarity and growth patterns. *Development* 118:117–126.

Schulz MW, Chamberlain CG, McAvoy JW (1997). Binding of FGF-1 and FGF-2 to heparan sulphate proteoglycans of the mammalian lens capsule. *Growth Factors* 14:1–13.

Spemann H (1901). Uber Korrelationen in der Entwicklung des Auges. *Verh Anat Ges* 15(Vers. Bonn.):61–79.

Walker JL, Wolff IM, Zhang L, Menko AS (2007). Activation of SRC kinases signals induction of posterior capsule opacification. *Invest Ophthalmol Vis Sci* 48:2214–2223.

Walton J, McAvoy J (1984). Sequential structural response of lens epithelium to retina-conditioned medium. *Exp Eye Res* 39:217–229.

Wang-Su ST, McCormack AL, Yang S, Hosler MR, Mixon A, Riviere MA, Wilmarth PA, Andley UP, Garland D, Li H, David LL, Wagner BJ (2003). Proteome analysis of lens epithelia, fibers, and the HLE B-3 cell line. *Invest Ophthalmol Vis Sci* 44:4829–4836.

Wormstone IM (2002). Posterior capsule opacification: a cell biological perspective. *Exp Eye Res* 74:337–347.

CHAPTER

11

Mouse Models of the Cornea and Lens: Understanding Ocular Disease

Satori A. Marchitti[1], *J. Bronwyn Bateman*[2], *J. Mark Petrash*[3], *Vasilis Vasiliou*[1]

[1]Molecular Toxicology and Environmental Health Sciences Program, Department of Pharmaceutical Sciences, University of Colorado Health Sciences Center at Denver and Aurora, CO 80262, USA

[2]Ophthalmology and Pediatrics, Rocky Mountain Lions Eye Institute, The Children's Hospital, University of Colorado Health Sciences Center at Denver and Aurora, CO 80262, USA

[3]Department of Ophthalmology and Visual Science, Washington University School of Medicine, St. Louis, MO 63110, USA

OUTLINE

INTRODUCTION

The cornea is an avascular, transparent tissue located at the anterior most surface of the eye that covers the iris, pupil, and anterior chamber. It is a highly specialized structure, providing greater than 60% of the optical power of the eye to refract and focus incident light on the retina (Land and Fernald, 1992). In addition, the cornea serves as a protective physical and biochemical barrier against environmental insults. The human cornea has five distinct layers. The most anterior, the corneal epithelium, is a thin multi-cellular layer of fast-growing and easily regenerated epithelial cells. Tight junctions between neighboring cells act as a physical barrier against noxious environmental agents. Irregularity or edema of the epithelial cell layer can disrupt the smoothness of the air-tear film interface, the most significant factor in the total refractive power of the eye. The next layer, Bowman's layer, is composed of irregularly arranged collagen fibers that protect the underlying corneal stroma, which is a thick transparent middle layer consisting of regularly arranged type I collagen fibers along with sparsely populated karatocytes. Descemet's membrane is a thin acellular layer that serves as a modified basement membrane of the corneal endothelium. The most posterior layer of the cornea is the corneal endothelium, a simple squamous monolayer of mitochondria-rich cells responsible for regulating fluid and solute transport between the aqueous and corneal stromal compartments. In humans, the corneal endothelium is essentially a non-divided monolayer; however, it helps preserve corneal optical properties by eliciting a net fluid transport outward from the stroma into the anterior chamber, thus preventing stromal swelling and preserving corneal clarity. Corneal transparency and optical refraction are dependent both on the continuous renewal of the anterior epithelial layer and on the fluid transport of the endothelial layer. The lens, located behind the cornea, is a transparent, avascular, flexible, biconvex structure that refracts light onto the retina while also changing the refractive index to allow the focusing on objects at various distances. Unlike the multiple layers of the cornea, the lens consists entirely of an encapsulated tissue comprising an anterior layer of epithelial cells and a posterior array of elongated fiber cells. While the curvature of the cornea is fixed, that of the lens is flexible and can be adjusted in a process called accommodation to fine-tune the focus depending upon an object's distance. The lens central fiber cells lack organelles, including their cell nuclei, and are instead primarily composed of a high concentration (>90%) of transparent water-soluble proteins, which have a key role in maintaining the refractive properties of the lens while at the same time allowing most light to pass through (Andley, 2007). These proteins, called crystallins, are believed to be evolutionarily related to stress proteins and are also found in the cornea where they can make up approximately 40–50% of total water-soluble protein (Piatigorsky, 2000). Studies of the various crystallins, long believed to be merely structural proteins, have revealed a variety of diverse functions including roles as molecular chaperones (Horwitz, 1992), as cellular defense systems against oxidative stress (Lassen *et al.*, 2007), and as direct absorbers of UV light (Estey *et al.*, 2007a). Many of these previously unknown functions have been demonstrated using gene-altered mice (Lassen *et al.*, 2007). Indeed, mouse models of human ocular abnormalities and pathologies, including numerous transgenic and knockout mutant mouse lines, have been instrumental in advancing many areas of eye research and elucidating the consequences of altered gene function in regards to ocular physiology and disease. In this regard, normal vision depends on the ability of the cornea and lens to maintain transparency, appropriate curvature and protective defense systems, thus, genetic or environmental factors that affect these properties can lead to impaired visual acuity and disease states. The purpose of this chapter is to summarize important mouse models of the cornea and lens. First, an overview of several corneal mouse models of various ocular disease states and developmental processes is given, followed by a focused discussion on the role of corneal crystallins in ocular disease as evidenced by relevant mouse models. Second, mouse models of the lens will be reviewed including a summary of several transgene mouse models, followed by a detailed discussion of single gene mouse models of lens cataract formation.

MOUSE MODELS OF THE CORNEA

Mouse models investigating critical functions of the cornea have led to an increased understanding of mechanisms involving the maintenance of the cornea and its role in protecting the underlying structures of the eye. The following sections will, first, highlight several corneal disease states and physiological processes in which mouse models have proven instrumental followed by, second, a focused discussion of crystallin proteins of the cornea including significant mouse models that have contributed to our understanding of eye physiology and disease.

Mouse Models of Corneal Development and Disease

Many transgenic and knockout mice exhibit phenotypes resembling human ocular disease. As such, mouse models have proven instrumental in the investigation of human ocular physiological processes and the development of disease states. Table 11.1 summarizes several important mouse models of the cornea and includes relevant phenotypes and investigational diseases or processes.

Cornea Plana

Cornea plana (CNA2), a human disease often associated with glaucoma in which the forward convex curvature of the cornea is flattened leading to a decrease in light refraction, has been directly linked to mutations in the human keratocan gene (*KERA*) (Pellegata *et al.*, 2000). *Kera* knockout mice (*Kera*$^{-/-}$), generated

TABLE 11.1 Mouse models of the cornea

Model	Ocular phenotype	Relevant disease/process	Reference
Kera$^{-/-}$	Thin corneal stroma, narrow cornea-iris angles, disorganized corneal collagen fibers	Cornea plana	Liu *et al.* (2003)
C57BL/6	Resistant to HSV-1 corneal infection	HSV-1 corneal infection	Norose *et al.* (2002)
DBA/2	Susceptible to HSV-1 corneal infection	HSV-1 corneal infection	Metcalf and Michaelis (1984)
Tnf$^{-/-}$	Susceptible to HSV-1 corneal infection	HSV-1 corneal infection	Minagawa *et al.* (2004)
Il4$^{-/-}$	Reduced severity and inflammatory response of onchocercal keratitis	Onchocercal keratitis	Pearlman *et al.* (1995)
BALB/c	CD4$^+$ Th1 responder, resistant to corneal bacterial infection	Microbial keratitis	Hazlett *et al.* (2000)
C57BL/6	CD4$^+$ Th2 responder, susceptible to corneal bacterial infection	Microbial keratitis	Hazlett *et al.* (2000)
Pax6$^{+/-}$	Small eyes, aniridia, atrophic corneal epithelium, goblet cells in cornea	Aniridia-related keratopathy	Ramaesh *et al.* (2005)
Plg$^{-/-}$	Increased fibrin deposition, opacity, inflammation, and neovascularization of injured corneas	Corneal wound healing	Kao *et al.* (1998)
Krt12$^{-/-}$	Thin, fragile corneal epithelial layers	Corneal wound healing	Kao *et al.* (1996)
Aqp1$^{-/-}$	Delayed recovery of corneal transparency and thickness after edema	Corneal wound healing	Thiagarajah and Verkman (2002)
Klf4$^{-/-}$	Corneal epithelial fragility, stromal edema	Corneal morphogenesis	Swamynathan *et al.* (2007)
Tgfβ2$^{-/-}$	Thin corneal stroma, decreased ECM, absence of corneal endothelium, fusion of cornea to lens	Corneal morphogenesis	Saika *et al.* (2001)
KGF over-expression	Hyperproliferation of embryonic corneal epithelial cells with subsequent differentiation into lacrimal gland-like tissues	Corneal morphogenesis	Lovicu *et al.* (1999)
SWR/J	Increased UV-induced corneal opacification	Corneal opacification	Downes *et al.* (1994)
Aldh1a1$^{-/-}$	Increased lens opacification, sensitivity to UV light, corneal edema	Cataract formation, ocular oxidative stress	Lassen *et al.* (2007)
Aldh3a1$^{-/-}$	Increased lens opacification, sensitivity to UV light, corneal edema	Cataract formation, ocular oxidative stress	Lassen *et al.* (2007)
Aldh1a1$^{-/-}$/ *Aldh3a1*$^{-/-}$	Increased lens opacification, sensitivity to UV light, corneal edema	Cataract formation, ocular oxidative stress	Lassen *et al.* (2007)

by gene targeting, exhibit thinner corneal stroma, narrower cornea-iris angles, and have large diameters of disorganized corneal collagen fibers in comparison to wild-type littermates, demonstrating that keratocan plays a key role in maintaining the structural integrity of collagenous matrix and appropriate corneal shape (Liu *et al.*, 2003).

Herpes Simplex Virus

Herpes simplex virus type 1 (HSV-1) infection of the eye is one of the world's leading infectious causes of blindness (Liesegang, 2001). Utilizing animal models of HSV-1 corneal infection, it has been determined that the genetic background of the host, the host adaptive immune system response, and the strain of virus all interact to determine disease severity (Brandt, 2005). Indeed, inbred mouse strains show different susceptibility and innate immune response to HSV-1 infection with C57BL/6 mice being the most resistant and DBA/2 mice being the most susceptible (Metcalf and Michaelis, 1984). Studies of C57BL/6 mice have revealed the *igh* locus on chromosome 12 to be a factor inferring innate resistance to HSV-1 (Norose *et al.*, 2002). Tumor necrosis factor (TNF) is an important cytokine in the innate immune response against various infections and is frequently a target of anti-cytokine treatment in certain inflammatory diseases. However, a *Tnf* null mouse model (*Tnf*$^{-/-}$) has shown that the absence of TNF results in increased susceptibility to acute corneal HSV-1 infection, as compared to wild-type mice (Minagawa *et al.*, 2004). These results indicate that treatment with TNF antagonists may facilitate the exacerbation and propagation of infectious HSV-1. Studies to determine a role for TNF in HSV-1 infection utilizing TNF receptor null mice, have revealed a genetic locus closely linked to the p55 TNF receptor gene termed *Hrl* (herpes resistance locus) on mouse chromosome 6 that determines resistance or susceptibility to HSV-1 depending on whether the allele derives from the resistant C57BL/6 or susceptible 129 strain background (Lundberg *et al.*, 2003).

Onchocercal Keratitis

Onchocercal keratitis, or river blindness as it is commonly called, is the world's second leading infectious cause of blindness, affecting approximately 18 million persons worldwide (World Health Organization, 1995). Mouse models that partially reproduce the clinical features of the human disease have led to a greater understanding of its immunopathogenesis. These data have indicated that disease severity is regulated in part by the degree of inflammatory response and inflammatory cell recruitment to the cornea by the cytokine IL4 (Hall and Pearlman, 1999). Indeed, *Il4* knockout mice ($Il4^{-/-}$) mice do not develop severe onchocercal keratitis and fewer inflammatory cells, including eosinophils, are observed in the corneas of these animals, as compared to wild-type animals (Pearlman *et al.*, 1995).

Microbial Keratitis

Recent advances have also been made in the field of microbial keratitis, specifically bacterial infections of the cornea due to *Pseudomonas aeruginosa*, a common Gram-negative pathogen (Hazlett, 2007). BALB/c mice are naturally resistant (cornea heals) to *P. aeruginosa* while C57BL/6 mice are susceptible (cornea perforates) to infection (Hazlett *et al.*, 2000). Susceptibility to *P. aeruginosa* in different strains of mice has been shown to involve $CD4^+$ T cells and their respective cytokine response. In this regard, it has been shown that strains of mice that favor a Th1 response (C57BL/6) are more susceptible to *P. aeruginosa* infection of the cornea while those strains that favor a Th2 responsiveness (BALB/c) are more resistant (Hazlett *et al.*, 2000, p. 805). Based on these models, it has been proposed that resistance in Th2 responder mice to *P. aeruginosa* may arise from their ability to downregulate the inflammatory response. In support of this, resistant BALB/c mouse corneas have lower levels of the chemokine macrophage inflammatory protein-2 (Mip-2) and the cytokine Il1, as compared to susceptible C57BL/6 mice, which directly correlate with decreased persistence of polymorphonuclear neutrophil (PMN) recruitment and ocular pathology including stromal scarring and corneal perforation (Steuhl *et al.*, 1987; Rudner *et al.*, 2000).

Aniridia-related Keratopathy

Aniridia-related keratopathy (ARK) is a bilateral panocular condition that affects the cornea and can lead to such corneal changes as vascular pannus formation, conjunctival invasion and epithelial erosion ultimately resulting in corneal opacification and loss of vision (Ramaesh *et al.*, 2005). Heterozygous mutation of human *PAX6*, normally widely expressed in the developing eye where it regulates cell proliferation, differentiation and apoptosis, has been found to result in ARK (Jordan *et al.*, 1992). Heterozygous $Pax6^{+/-}$ (small-eye) mice have similar corneal abnormalities as human aniridia, including atrophic corneal epithelium, infiltration of goblet cells, corneal vascular pannus formation and central corneal nebulae, and have been widely accepted as an animal model of ARK (Ramaesh *et al.*, 2003). Studies of $Pax6^{+/-}$ mice have suggested that the mechanism of corneal pathology in ARK may be related to an abnormality within the limbal stem cell niche, as opposed to an intrinsic deficiency of limbal stem cells, as was previously thought (Ramaesh *et al.*, 2005).

Corneal Wound Healing

Corneal wound healing is an important function of the cornea following injury. Corneal epithelium must be rapidly resurfaced to avoid microbial infection and further damage to the underlying stroma. Corneal epithelial renewal is essential to this process and further facilitates the maintenance of the smooth optical surface of the cornea. *In vivo* wound healing models, consisting of making a defined central epithelial wound and characterizing the healing response, have provided insight into how corneal renewal and repair occurs. Indeed, transgenic and knockout mouse models have identified many key factors in the corneal response to injury (Lu *et al.*, 2001). Utilizing *plasminogen-* (*Plg*) and *fibrinogen-* (*Fib*) null mice, it has been revealed that cross-linking of the expanding fibrin network is essential for corneal wound healing and that the central role of plasmin in corneal wound healing is fibrinolysis (Kao *et al.*, 1998). In this regard, $Plg^{-/-}$ mice have significant fibrin deposition in injured corneas, which are further characterized by prolonged inflammation, cloudiness, and neovascularization;

however, apparently normal corneal wound repair occurs in double knockout mice deficient for both *Plg* and *Fib* ($Plg^{-/-}/Fib^{-/-}$). Cytokeratin 12 (KRT12), an intermediate filament component, has been found to play an important role in the maintenance of corneal epithelial integrity during wound healing and *Krt12*-deficient mice display thin and fragile corneal epithelial layers in comparison to wild-type animals (Kao *et al.*, 1996). Edema of the cornea often occurs as a result of injury. Aquaporin proteins, a family of homologous water channels expressed in the cornea, appear to be important in the restoration of corneal water content and transparency following edema (Verkman, 2002, p. 617). Indeed, transgenic mice lacking *aquaporin 1* (*Aqp1*), normally expressed in the corneal endothelium, have remarkably delayed recovery of corneal transparency and thickness after hypotonic swelling, providing evidence for the involvement of AQP1 in the active extrusion of fluid from the corneal stroma across the corneal endothelium (Thiagarajah and Verkman, 2002).

Development and Maintenance of the Cornea

Studies utilizing mouse models of transcription and growth factors have also greatly contributed to our understanding of the development, maintenance, and physiology of the cornea. The Krüppel-like transcription factor (Klf4) is one of the most highly expressed transcription factors in the mouse cornea (Norman *et al.*, 2004). While conditional (using a Cre-lox approach) corneal and lens ablation of mouse *Klf4* results in no visible phenotype at postnatal day 1, by 8 weeks corneal epithelial fragility, stromal edema and defective lens are apparent indicating that KLF4 has a critical role in postnatal ocular surface maturation and maintenance (Swamynathan *et al.*, 2007). Indeed, the structural integrity of the corneal epithelium, maintenance of stromal hydration levels and the development of conjunctival goblet cells are all affected in the absence of *Klf4*. In this regard, the *Klf4* conditional null mouse may be a useful model for investigating ocular surface pathologies such as dry eye, Meesmann's dystrophy, and Stevens-Johnson syndrome. Transforming growth factor (TGF)-β2 has also been shown to be essential for corneal and lens morphogenesis using animal models (Kao, 2006). $Tgfb2^{-/-}$ mice have abnormal ocular morphogenesis characterized by thin corneal stroma resulting from decreased extracellular matrix synthesis, absence of corneal endothelium, fusion of cornea to lens and accumulation of hyaline cells in vitreous (Saika *et al.*, 2001). Keratinocyte growth factor (KGF) is produced in stromal cells and is thought to play a role in mediating epithelial cell behavior. Transgenic mice over-expressing human KGF in the eye have proven to be a valuable tool in studying mechanisms of cell fate decisions during ocular morphogenesis (Lovicu *et al.*, 1999). These mice exhibit hyperproliferation of embryonic corneal epithelial cells with subsequent differentiation into lacrimal gland-like tissues, indicating that early stimulation of the KGF receptor is critical in altering cellular developmental fate.

Corneal Crystallins and Relevant Mouse Models

Genetically modified mice have greatly increased our understanding of the mechanisms driving corneal development and disease, and these models will continue to advance eye research. An extremely important class of structural proteins in the eye is the eye crystallins located in both the cornea and lens. Mouse models of corneal crystallins have revealed many surprising and previously unknown cellular functions of these proteins including their role as molecular chaperones and in the maintenance and defense of eye structures.

While it has been well established that the major water-soluble proteins in the lens, termed crystallins, are responsible for the optical properties of the lens; the notion that corneal cells may also express similar enzymes for maintaining the transparency and optical power of the cornea was proposed only relatively recently (Cooper *et al.*, 1991). Prior to this it was commonly accepted that corneal transparency was due to the precise structure of the tightly packed, orthogonally arranged collagen fibers of the extracellular stroma (Maurice, 1957). However, utilizing non-invasive optical imaging techniques and confocal microscopy, which allows the identification of light-reflecting and scattering structures within living cornea (Li *et al.*, 1997), it was revealed that swollen, fixed, and wounded corneas have decreased corneal transparency that correlates with a marked increase in light scattering from corneal stromal cells (keratocytes) (Jester *et al.*, 1992). This cell-related loss of corneal transparency suggested a unique and previously unknown mechanism of maintaining corneal clarity, similar to that for the transparent lens of the eye (Jester *et al.*, 1999). In the lens, transparency and refractive properties are attributed to the high accumulation of a select number of soluble crystalline proteins and the subsequent minimizing of refractive index fluctuations by short range interactions within the cellular cytoplasm (Benedek, 1983). The first evidence of the existence of corneal crystallins came from early studies of the bovine cornea in which unusually high-cytosolic expression of aldehyde dehydrogenase 3A1 (ALDH3A1) was found (Holt and Kinoshita, 1973). Indeed, ALDH3A1 comprises 20–40% of the water-soluble protein in the bovine cornea (Alexander *et al.*, 1981) and approximately 50% of the major fraction of water-soluble protein in normal mouse cornea (Nees *et al.*, 2002), leading investigators to speculate that ALDH3A1 may contribute to corneal transparency (Silverman *et al.*, 1981) and indicating that the cornea, like the lens, may contain crystallins. Later studies revealed that both ALDH3A1 and aldehyde dehydrogenase 1A1 (ALDH1A1), identified as a major crystallin in the lens (Wistow and Kim, 1991), have high levels of cytosolic expression in the human cornea, although ALDH1A1 is expressed at lower concentrations than ALDH3A1 (King and Holmes, 1998). ALDH proteins comprise a superfamily of $NAD(P)^+$-dependent enzymes that are primarily known for their cellular detoxification role in the oxidative conversion of highly reactive aldehyde species to their corresponding carboxylic acids (Vasiliou *et al.*, 2004). ALDH3A1 and ALDH1A1 are the predominant ALDH isozymes localized to mouse cornea, specifically in corneal epithelial cells and keratocytes but not endothelial cells (Pappa *et al.*, 2003). Expression of ALDHs in ocular structures is not surprising given the presence of aldehydes generated by UV light-induced oxidative stress and lipid peroxidation (Feeney and Berman, 1976). Along with ALDH1A1, transketolase (TKT), a ubiquitous enzyme comprising up to 10% of the soluble protein in mouse cornea, also has unexpectedly high expression in keratocytes of the rabbit cornea and levels of these proteins were found to be decreased in highly reflective keratocytes after injury, suggesting ALDH1A1 and TKT contributed to a previously unrecognized cellular component of corneal transparency (Jester *et al.*, 1999). TKT is an enzyme in the non-oxidative branch of the pentose phosphate pathway, which produces pentoses and NADP (Kochetov, 1982). Similar to ALDH3A1 and ALDH1A1, TKT mRNA, protein and enzyme activity were found to be unusually high in cornea compared to other tissues, indicating they may have additional functions aside from their enzymatic roles (Sax *et al.*, 1996). Indeed, high-corneal concentrations of these enzymes are now believed to be required for a structural role in corneal transparency and refraction, as is the case with lens crystallins (Sax *et al.*, 2000). Further evidence that ALDH3A1, ALDH1A1, and TKT are corneal crystallins came from the fact that their expression was determined to be taxon-specific, similar to crystallins present in the lens (Wistow and Piatigorsky, 1988). Indeed, while ALDH3A1 is abundantly expressed in the corneas of numerous mammals, it is not present in all species including the rabbit (Jester *et al.*, 1999). Accordingly, in contrast to the human cornea where ALDH3A1 is predominant, ALDH1A1 appears to be the major corneal crystalline present in rabbit cornea where it constitutes approximately 30% of the water-soluble protein (Jester *et al.*, 1999). Nonetheless, ALDH3A1, ALDH1A1, and TKT have all been found in higher levels in human and mouse cornea than would be expected for them serving strictly enzymatic roles, thus supporting their putative role as corneal crystallins in mammals (Piatigorsky, 2000). It has now been shown that the corneal epithelial cells of vertebrates, including humans, accumulate different enzymes in a taxon-specific manner at concentrations similar to those for crystallins in the lens (Cuthbertson *et al.*, 1992). These water-soluble proteins of the cornea are enzymes and, like lens crystallins, are believed to have stress-protective functions and structural as well as enzymatic roles (Piatigorsky, 1998). In addition to ALDH3A1, ALDH1A1, and TKT, to date, other putative mammalian corneal crystallins that have been identified include isocitrate dehydrogenase (Sun *et al.*, 1999), serum albumin (Nees *et al.*, 2003), α-enolase (Jester *et al.*, 2005), glutathione-s-transferase (Cuthbertson *et al.*, 1992), lactate and glyceraldehyde-3-phosphate dehydrogenases (Jester *et al.*, 2005), and actin (Swamynathan *et al.*, 2003). Mouse models of several corneal crystallins have been developed, however, to date, few gene ablation models of these proteins are viable and display corneal phenotypes. Indeed, *Tkt*-null embryos created by gene targeting are not viable and, while $Tkt^{+/-}$ mice do survive and display growth retardation, reduced adipose tissue levels and female fertility as compared to wild-type animals, no corneal phenotype has been observed (Xu *et al.*, 2002). Similarly, early recessive embryonic lethality occurs in mice in which the α-enolase gene has been disrupted by retroviral gene trapping and, like $Tkt^{+/-}$ mice, mice heterozygous for the α-enolase mutation have no obvious phenotype (Couldrey *et al.*, 1998). In contrast, mouse models of ALDH3A1 and ALDH1A1 deficiency are viable and display ocular phenotypes, thus, these models, including a double knockout $Aldh1a1^{-/-}/Aldh3a1^{-/-}$ mouse line, will be the focus of the remaining discussion of corneal crystallins.

One of the earliest mouse models of ALDH3A1 deficiency came about serendipitously from a study of inbred albino mouse strains subjected to ultraviolet radiation (UVR), and examined for ALDH activity and soluble protein content (Downes *et al.*, 1994). In this study, the SWR/J strain of mice exhibited more extensive corneal clouding and opacification after UV exposure than did other mouse strains. Untreated SWR/J mice exhibited a "low-activity" variant of ALDH3A1 as well as overall decreased levels of soluble corneal protein. It was discovered that most inbred mouse strains code for a "high-activity" ALDH3A1 through the Aldh3a1a allele, while the SWR/J strain of mice have "low-activity" ALDH3A1 encoded by a variant allele termed Aldh3a1c. Accordingly, SWR/J mice have only trace levels of ALDH3A1 activity in various tissues compared to all other inbred strains examined and this "low-activity" variant is associated with extensive corneal clouding and opacification after a single exposure to UV light (Downes *et al.*, 1994). From these data, it was hypothesized that ALDH3A1 has a major role in the cornea as a UVR-defense system. Later studies from our laboratory revealed 13 nucleotide changes in the Aldh3a1c allele as compared to the Aldh3a1a allele (Shiao *et al.*, 1999). Four of these changes resulted in amino acid substitutions (G88R, I154N, H305R, and I352V) while nine changes were determined to be silent. The I154N and H305R substitutions were specifically proposed to affect critical regions involved in co-factor binding and catalysis. These mutation-mediated structural changes are believed to be the molecular basis for the loss of ALDH3A1 enzymatic activity in SWR/J mice and may also be responsible for the corneal sensitivity to UV light that is demonstrated in these mice (Shiao *et al.*, 1999).

An Aldh3a1-deficient (*Aldh3a1*$^{-/-}$) mouse model has been generated by utilizing *Aldh3a1* gene targeting (Nees *et al.*, 2002). These mice were found to be viable and fertile, displaying a corneal epithelium water-soluble protein content approximately half that of wild-type mice. However, despite the significant loss of Aldh3a1 protein from the cornea, surprisingly, *Aldh3a1*$^{-/-}$ mouse corneas appeared structurally normal, transparent, and were indistinguishable from wild-type corneas upon histological analysis, electron microscopy, and light and slit lamp microscopy. No compensatory increase in the amounts of water-soluble corneal proteins was found. However, no other ocular structures of the eye including the lens of these mice appear to have been evaluated for effects relating to ALDH3A1 deficiency. Thus, these data led investigators to conclude that ALDH3A1 is unnecessary for corneal transparency and maintenance (Nees *et al.*, 2002). Nevertheless, despite no obvious difference in corneal clarity or structure in these *Aldh3a1*$^{-/-}$ mice, corneal functions of ALDH3A1 including directly absorbing UV light, enzymatic detoxification of UV-derived lipid peroxidation products (Abedinia *et al.*, 1990) and functions analogous to those of lens crystallins such as thiol regulation (Uma *et al.*, 1996a) and molecular chaperone activity (Uma *et al.*, 1996b) were not ruled out. Similar to *Aldh3a1*$^{-/-}$ mice, *Aldh1a1*$^{-/-}$ mice, created by gene targeting to determine the role of Aldh1a1 in the retina where it is highly expressed, are viable but show no obvious phenotype in the intended target tissue, however, again, other ocular tissues such as the cornea or lens were not evaluated (Fan *et al.*, 2003).

Recent data from our laboratory indicate that, while ALDH3A1 is undetectable in the mouse lens, ALDH3A1 present in the cornea may actually serve a critical role in the protection of the lens against environmentally induced oxidative damage (Lassen *et al.*, 2007). Indeed, the use of *Aldh3a1*$^{-/-}$ and *Aldh1a1*$^{-/-}$ single and double knockout mouse models have recently revealed that both ALDH3A1 and ALDH1A1 are instrumental and have additive functions in protecting the lens against cataract formation via multiple mechanisms (Lassen *et al.*, 2007). In this study, *Aldh3a1*$^{-/-}$ and *Aldh1a1*$^{-/-}$ single knockout (Nees *et al.*, 2002; Fan *et al.*, 2003) F3 homozygous hybrids were bred to generate an *Aldh1a1*$^{-/-}$/*Aldh3a1*$^{-/-}$ double knockout mouse model (Lassen *et al.*, 2007). *Aldh3a1*$^{-/-}$ and *Aldh1a1*$^{-/-}$ single knockout and *Aldh1a1*$^{-/-}$/*Aldh3a1*$^{-/-}$ double knockout mouse lines were evaluated for biochemical changes and premature lens cataract formation (lens opacification) by *in vivo* slit lamp biomicroscopy. By 1 month of age, *Aldh3a1*$^{-/-}$ single and *Aldh1a1*$^{-/-}$/*Aldh3a1*$^{-/-}$ double knockout mice develop lens cataracts in the anterior and posterior subcapsular regions while *Aldh1a1*$^{-/-}$ single knockout mice also develop lens cataracts, albeit later in life (6–9 months of age), as compared to wild-type mice. The percentage of mice that develop cataracts is significantly higher in the double knockout mouse model and this trend increases with age, as compared to age-matched wild-type mice. Additionally, mice (1–3 months of age) exposed to UVB exhibit accelerated anterior lens opacification, with this effect being more prominent in *Aldh3a1*$^{-/-}$ single and *Aldh1a1*$^{-/-}$/*Aldh3a1*$^{-/-}$ double knockout mice, as compared to *Aldh1a1*$^{-/-}$ single knockout and wild-type animals. One mechanism by which both ALDH3A1 and ALDH1A1 protect the corneal epithelium from UV-induced oxidative damage is the detoxification of lipid peroxidation-derived aldehydes such as 4-hydroxy-2-nonenal (4-HNE) and malondialdehyde (MDA) (Manzer *et al.*, 2003; Pappa *et al.*, 2003). Accordingly, cataract formation in *Aldh1a1*$^{-/-}$/*Aldh3a1*$^{-/-}$ double knockout mice is associated with increased corneal and lens levels of 4-HNE- and MDA-protein adducts and increased protein carbonyl content, all measures of protein oxidation, indicating the importance of both ALDH3A1 and ALDH1A1 as cellular defense mechanisms against ocular toxicity. Aldehydes are generally long-lived compounds that can diffuse to sites some distance from their origin (Esterbauer *et al.*, 1991), thus, aldehydes accumulating in the cornea due to ALDH3A1 deficiency may diffuse into the lens leading to reduced lens clarity. Lens opacification is known to result from aggregation, cross-linking, and denaturation of proteins. This can occur as aldehyde-mediated adduction of proteins, thus, aldehyde detoxification by ALDH3A1 and ALDH1A1 is believed to be an important cellular defense against lens opacification (Lassen *et al.*, 2007). *Aldh1a1*$^{-/-}$/*Aldh3a1*$^{-/-}$ double knockout mice also display decreased proteosome function, as compared to wild-type mice. Proper proteosome function is critical for the removal of damaged proteins and the prevention of protein aggregation. Indeed, impaired proteosome function is associated with cataract formation (Zetterberg *et al.*, 2003). Interestingly, *Aldh3a1*$^{-/-}$ and *Aldh1a1*$^{-/-}$ single and *Aldh1a1*$^{-/-}$/*Aldh3a1*$^{-/-}$ double knockout mice all show increased levels of reduced GSH that correlate with increased γ-glutamylcysteine synthase (GCS) content in the lens and cornea, as compared to wild-type animals, with the most significant increase occurring in *Aldh1a1*$^{-/-}$/*Aldh3a1*$^{-/-}$ double knockout mice (Lassen *et al.*, 2007). These results are not surprising given that cellular adaptation to oxidative stress often involves the upregulation of GSH synthesis. Indeed, GSH is a major regulator of the redox environment in both cornea and lens (Ganea and Harding, 2006), and its synthesis is known to be inducible by electrophiles such as 4-HNE (Iles and Liu, 2005). ALDH3A1 appears to have additional functions aside from aldehyde oxidation including the direct absorption of UV radiation, chaperone-like activity, and scavenging of UV-generated reactive oxygen species (ROS) via –SH groups of cysteine and methionine residues (Estey *et al.*, 2007a). Considering that ALDH3A1 is not present in the lens, the primary role of ALDH3A1 in protecting the lens is most likely due to the ability of ALDH3A1 to act as a UV-filter in the cornea by absorbing light, thus minimizing the amount of UV radiation that reaches the lens (Estey *et al.*, 2007b). In support of this hypothesis, ALDH3A1 has an enrichment of UV-absorbing tryptophan residues (Mitchell and Cenedella, 1995) and *in vitro* data has shown that ALDH3A1 has a structural role in the cornea by protecting proteins from UV-mediated inactivation (Estey *et al.*, 2007a). On the other hand, the primary role of ALDH1A1 in protecting the lens against cataract formation is most likely the detoxification of reactive aldehydes in both the cornea and lens (Lassen *et al.*, 2007). In support of this hypothesis, the drug chloroquine, known to bind and inhibit ALDH1A1 (Graves *et al.*, 2002), causes distinct anterior cataracts in rats (Drenckhahn and Lullmann-Rauch, 1977). Taken together, mouse models of the corneal crystallins, including SWR/J mice and *Aldh3a1*$^{-/-}$ and *Aldh1a1*$^{-/-}$ single and *Aldh1a1*$^{-/-}$/*Aldh3a1*$^{-/-}$ double knockout mice have proven instrumental in elucidating the role of ALDH3A1 and ALDH1A1 in safeguarding the optical properties of the cornea and lens through both enzymatic and nonenzymatic mechanisms.

MOUSE MODELS OF THE LENS

Animal models created using transgene or targeted gene knock out strategies have been invaluable tools to probe mechanisms leading to lens abnormalities. Several features make the lens an ideal target tissue for studies of disease mechanisms. The lens is readily accessible to light and therefore to imaging instruments, so it is possible to obtain detailed information about lens structure and transparency in the living animal without use of deep anesthesia. This makes it possible to conduct time course studies of changes in parameters such as the onset and progression of lens opacities using slit lamp ophthalmoscopy (Seeberger *et al.*, 2004), alterations in lens proteins using dynamic light scattering (Simpanya *et al.*, 2005), and refractive changes (Bantseev *et al.*, 2004). Still more detailed measurements can be made on freshly dissected lenses to probe for changes in optical quality that result from targeted mutations introduced via a transgene or gene ablation approach (Shiels *et al.*, 2007). Much is known about lens development and the patterns of change that occur when lens epithelial cells develop into elongated fiber cells. Consequently, alterations in lens differentiation can be readily appreciated using simple histochemical procedures (Bassnett, 2002). The following sections will, first, highlight several recent studies that have utilized transgenic mouse models of the lens followed by, second, a detailed discussion of single gene mouse models of cataract formation.

Transgenic Lens Models

For studies based on a transgene strategy, one of the most important elements is the choice of promoter to control tissue and developmental specificity for expression of the desired protein. Not surprisingly, promoter sequences derived from the α*A-crystallin* gene have been shown to drive expression of many different transgenes almost exclusively in cells of the lens outer cortex (Zhao *et al.*, 2004). Although the core α promoter shows strong preference for supporting transgene expression exclusively in lens, little or no expression is typically observed in the epithelial layer which plays a critical role in development (Hsu *et al.*, 2006). By combining enhancer sequences from *Pax6* or *delta-crystallin* with the core α*A-crystallin* promoter, it has been possible to create chimeric promoters that support transgene expression not only in lens cortex but also in lens epithelium (Reneker *et al.*, 2004; Zhao *et al.*, 2004). Hybrid promoters of this type have been used for transgenic expression of signaling pathway components such as growth factors (Xie *et al.*, 2007) and offer promise as drivers for cell type-specific expression of Cre recombinase for conditional deletion of targeted genes. Due to a wealth of tools available to direct expression of transgenes to targeted tissues, and clinically validated imaging devices to monitor their biological effects in the eye, the lens and cornea appear to be very attractive platforms for the use of gene manipulation as an experimental approach to mechanistic studies of disease. The following section will highlight several recent studies that have utilized transgenic mouse models of the lens to investigate various eye processes and disease states.

Transgenic Mouse Models of Ion Homeostasis Defects

Transgenic as well as gene-targeted mice have been extremely useful experimental tools for studies of gap junction proteins in lens development and ion homeostasis. Gap junction coupling is thought to play an essential role in keeping intracellular calcium at low (nM) concentrations (Gao *et al.*, 2004). Movement of ions and small molecules between coupled cells occurs through channels formed by gap junction proteins called connexins. *Connexin* gene mutations, which have been linked to cataracts, most likely result in disruption of ion homeostasis. A number of laboratories have used a targeted gene knock out approach to show that Cx46, which is responsible for gap junction coupling in inner cortical fiber cells, most likely plays a major role as a conduit for movement of calcium from inner to outer cortical fiber cells (Gong *et al.*, 1998). Loss of Cx46-mediated gap junctions results in accumulation of abnormal calcium levels in the deep cortical fiber cells and activation of Lp82, a calcium-dependent protease (Baruch *et al.*, 2001). Protease activation then leads to degradation of lens crystallins and opacification of the lens nucleus. In contrast, targeted deletion of Cx50, which is expressed in cells of the outer lens cortex, results in impaired lens growth and development and zonular cataracts (White *et al.*, 1998). Clever use of mouse genetics and replacement of *Cx50* sequences by *Cx46* using a knock in approach has led to a new appreciation of the functional diversity of connexins and their roles in ionic and biochemical coupling across lens cells (White *et al.*, 2001; White, 2002; Xia *et al.*, 2006a).

Transgenic Mouse Models of Autosomal Dominant Cataract

Familial cataracts in humans have recently been associated with mutations associated with the three major crystallin families: the α-, β-, and γ-crystallins (Shiels and Hejtmancik, 2007). Of these, the best studied are mutations in the α*A-crystallin* gene that give rise to autosomal dominant cataracts (Litt *et al.*, 1998; Mackay *et al.*, 2003). Given the autosomal dominant pattern of the phenotype in affected families, it was suggested that the cataract mechanism involves a deleterious gain of function in the mutant protein (Cobb and Petrash, 2000, 2002). Biochemical studies of mutant crystallins provided clues for possible mechanisms to explain why individuals with the corresponding mutations develop cataract at an early age (Cobb and Petrash, 2000, 2002; Koteiche and Mchaourab, 2006). While *in vitro* studies at the biochemical level can lead to hypotheses regarding disease mechanisms, these must be validated in animal models. The transgenic mouse lens appears to be ideally suited as an experimental system for mechanistic studies of autosomal dominant cataract. Using well characterized α*-crystallin* promoters, it is possible to direct biosynthesis of wild-type (as control) or mutant-human α*-crystallins* to lens fiber cells which will also contain their normal complement of endogenous crystallins. Using this transgene approach, Hsu and coworkers observed cataract phenotypes in a transgenic mouse model of human autosomal dominant cataract caused by a R116C missense mutation in the α*A-crystallin* gene (Hsu *et al.*, 2006). This approach appeared to be quite robust, as lens abnormalities were observable among different founder lines characterized by large differences in transgene expression levels. Furthermore, no abnormalities were observed in transgenic mice engineered to express wild-type αA-crystallin. Because relevant phenotypes were observed in transgenic lines that express relatively low levels of transgene product, it will be possible to obtain lens tissue with obvious cataract-related changes but without extensive protein aggregation and tissue disorganization that could preclude mechanistic studies at the molecular level. Thus, the transgene approach often carries the opportunity to select founder lines with distinct tissue abnormalities that are suitable for follow up study at the biochemical level.

Transgenic Mouse Models of Cataracts Linked to Metabolism and Oxidative Stress

Chronic oxidative stress associated with metabolic imbalances is thought to be a risk factor for human cataract formation. Because long duration of exposure to cataract-inducing conditions (e.g. diabetes mellitus, UV irradiation) is typically required before appearance of lens opacities in humans, it has been difficult to study the relevant mechanisms using traditional laboratory models. Now it appears that availability of transgenic animals may substantially lower the barrier to studies of this kind. Using transgenic mice engineered for lens specific expression of aldose reductase, Chung and coworkers demonstrated the importance of oxidative stress associated with accelerated polyol pathway metabolism in a humanized mouse model of diabetic cataract (Lee and Chung, 1999). Similarly, Monnier and coworkers have used a transgenic mouse model to accelerate the formation and accumulation of protein modifications associated with senile cataract formation (Fan *et al.*, 2006). By creating transgenic mice designed for over-expression of the vitamin C co-transporter, these investigators were able to compress the time course required for accumulation of changes to lens proteins from decades (in humans) to months (in the mouse model). Cataracts produced in this mouse model will provide the lens tissue necessary to conduct mechanistic studies of senile cataract formation.

Aside from transgene approaches, genetic ablation can be used to probe the functional roles of targeted genes and their cognate metabolic pathways. In this way, as mentioned above, Vasiliou and coworkers showed that animals that are null for activity for two different aldehyde dehydrogenases, ALDH1A1 and ALDH3A1, are at significantly higher risk for cataract development (Lassen *et al.*, 2007). These results are consistent with roles for the ALDH family as both catalysts responsible for detoxification of deleterious oxidants as well as structural proteins in lens and cornea with potential roles as UV filters (Lassen *et al.*, 2007). Similarly, Reddy and coworkers demonstrated the importance of glutathione peroxidase as a defense enzyme against oxidative damage in the lens. GPX-1-deficient mice were found to develop significantly higher amounts of lens opacities and markers of oxidative stress as compared to age-matched controls (Reddy *et al.*, 2001).

Single Gene Mouse Models of Cataract Formation

Mouse models provide important insight into the mechanisms of cataract formation; such studies are relevant to human cataract formation, because ocular gene sequences are similar between the two species. Multiple models of cataracts in mice have been developed and studied with single gene forms having been available for the longest period of time. Although some single gene models have occurred spontaneously, most have been induced by irradiation or chemical mutagenesis. Several valuable reviews have been published (Graw *et al.*, 1984; Graw, 1999a,b, 2004; Graw and Loster, 2003) and the website Mouse Genome Informatics (www.informatics.jax.org) is a useful resource. The following section will provide a detailed discussion of single gene mouse models of cataract formation, including human corollaries, and the role of mouse models in elucidating mechanisms of cataract.

There are 20 genes that produce cataracts in humans as a result of mutations of these genes expressed in the lens. Many genetic loci have been identified by linkage and mutational analyses for human hereditary cataracts; over 30 independent chromosomal regions for human ADC have been mapped. Mutations in genes encoding many protein groups expressed in the lens have been reported to cause autosomal dominant and recessive patterns of inheritance. Mutations in the human genes encoding connexins (gap junction α8 and gap junction α3), lens membrane junction proteins (major intrinsic protein of lens fiber), crystallins (γC, γD, γS, βA1, βB2, βB1, αB, and αA), structural proteins (beaded filament structural protein 2), intracellular storage proteins (ferritin light chain), transcription factors (paired-like homeodomain transcription factor 3 and paired box gene 6) and heat shock proteins (heat shock transcription factor 4) have been reported. Crystallins are the major protein group in the lens and are responsible for transparency. The α crystallins belong to the category of heat shock proteins and are molecular chaperones. The β and γ crystallins are related and belong to a superfamily. Although the range of genes and mutations has not reached the diversity found in retinal degenerations, the numbers have increased exponentially over the past 6–8 years. The specific genes and mutations are summarized by Shiels and Hejtmancik (Shiels and Hejtmancik, 2007). Many single genes have been identified as causative of cataracts in mice, and most of these genes have multiple mutations (alleles); Table 11.2 summarizes the cataract mouse models, spontaneously occurring or induced by irradiation or chemicals.

TABLE 11.2 Single gene mouse models of lens cataract

Mouse gene/ chromosome[a]		Cataract phenotype[b]	Mode[c]	Inheritance[d]	Mouse Strain[e]	Reference
Acc	U	Anterior capsular cataract	I	S-D		Kratochvilova (1981)
act	U	Adult cataract	S	R		Trigg (1972)
Apo	U	Anterior polar opacity	C	D		Kratochvilova (1978)
Asc1	U	Anterior suture cataract 1	I	S-D		Kratochvilova and Favor (1988); Graw *et al.* (1990b)
Asc2	U	Anterior suture cataract 2	I	S-D		Kratochvilova and Favor (1988)
bs	2	Unknown	S	R		Varnum (1983); Li *et al.* (2002)
cac	U	Cataract	S	R		Konyukhov and Wachtel (1963)
Cad	4	Congenital cataract	S	D		Tissot and Cohen (1972)
Casr	16	Central embryonal flecks, more in homozygote	C	S-D	$Casr^{Nuf}$	Favor and Neuhauser-Klaus (2000); Hough *et al.* (2004)
Cat3	10	Anterior; osmotic	I	D	$Cat3^{vao}$	Kratochvilova and Favor (1988); Graw *et al.* (1990a); Loster *et al.* (1997)
		Vacuolated lens	I	S-D	$Cat3^{vl}$	Kratochvilova and Favor (1988); Loster *et al.* (1997)
Cat4	8	Anterior polar	I/S	D	$Cat4^{Apcat1-1}$	Favor *et al.* (1997); Grimes *et al.* (1998); Wakefield *et al.* (2007)
			I/S	S-D	$Cat4^{Apcat1-2}$	Favor *et al.* (1997); Grimes *et al.* (1998); Wakefield *et al.* (2007)
			I/S	S-D	$Cat4^{Apcat1-3}$	Favor *et al.* (1997); Grimes *et al.* (1998); Wakefield *et al.* (2007)
			I/S	S-D	$Cat4^{Apcat1-4}$	Favor *et al.* (1997); Grimes *et al.* (1998); Wakefield *et al.* (2007)
Cat5	10	Total cataract	C	D		Everett *et al.* (1994)
Ccw	4	Progressive cataract, curly whiskers		D		Kerscher *et al.* (1996); Lyon *et al.* (2000)
Coc	16	Nuclear fleck ("coralliform") opacities	I	D		Kratochvilova and Favor (1988); Sidjanin *et al.* (1997)
cgct	4		S	D		Bennett and Charlton (1992); Argeson *et al.* (1994)
Col4a1	8	Vacuolar	C	S-D	$Col4a1^{Bru}$	Lyon *et al.* (1984); Van *et al.* (2005)
		Vacuolar	C	D	$Col4a1^{Svc}$	Thaung *et al.* (2002); Van *et al.* (2005)
Cryaa	17	Nuclear, subcortical zonular; denser with microspherophakia in heterozygote	C	R	$Cryaa^{Aey7}$	Graw *et al.* (2001a)
		White nuclear, mild cortical	S	R	$Cryaa^{lop18}$	Chang *et al.* (1996); Chang *et al.* (1999)
		Nuclear, denser in homozygote	C	S-D	$Cryaa^{Lin}$	Xia *et al.* (2006b)
		Nearly total, microspherophakia, microphthalmia	S	R	$Cryaa^{2J}$	Xia *et al.* (2006b)

TABLE 11.2 Continued

Mouse gene/ chromosome[a]		Cataract phenotype[b]	Mode[c]	Inheritance[d]	Mouse Strain[e]	Reference
Cryba1	11	Minimal nuclear, zonular heterozygote; total, microphthalmia in homozygote	C	S-D	$Cryba1^{Po1}$	Graw *et al.* (1999)
Crybb2	5	Faint anterior cataract to dense nuclear and mild anterior and posterior opacity	S	R	$Crybb2^{Phil}$	Kador *et al.* (1980); Chambers and Russell (1991)
Crybb2	5	Progressive cortical and anterior suture opacity; same in heterozygote/homozygote	C	D	$Crybb2^{Aey2}$	Graw *et al.* (2001d)
Cryga	1	Nuclear	C	D	$Cryga^{1Neu}$	Favor (1983); Favor (1984); Klopp *et al.* (1998)
		Total, vacuoles	C	S-D	$Cryga^{tol}$	Ehling *et al.* (1982); Graw *et al.* (2004)
Crygb	1	Nuclear	S	S-D	$Crygb^{nop}$	Ehling *et al.* (1982); Graw *et al.* (1984); Klopp *et al.* (1998); Graw *et al.* (2004)
		Nuclear; denser in homozygote	C	S-D	$Crygb^{Clapper}$	Liu *et al.* (2005)
Crygc	1	Central with radial spokes into cortex; denser with mild microphthalmia in homozygote	C	S-D	$Crygc^{Chl3}$	Graw *et al.* (2002b)
		Dense nuclear, subcortical with vacuoles; denser in homozygote	C	S-D	$Cryc^{MNU8}$	Graw (2004)
Crygd	1	Central fetal dense opacity; progresses to total cataract; cataract similar in heterozygotes and homozygotes; microphthalmia evident	S	S-D	$Crygd^{Lop12}$	Smith *et al.* (2000)
		Nuclear with mild cortical opacities; denser with microspherophakia in homozygote	C	D	$Crygd^{Aey4}$	Graw *et al.* (2002a)
		Dense nuclear with less dense cortical opacities; denser in homozygote	C	S-D	$Crygd^{ENU4011}$	Graw *et al.* (2004)
		Mild, diffuse opacity in homozygote and heterozygote	C	S-D	$Crygd^{ENU910}$	Graw *et al.* (2004)
		Dense nuclear with less dense cortical opacities; denser in homozygote	S	S-D	$Cryd^{K10}$	Graw *et al.* (2004)
Cryge	1	Total	I	D	$Cryge^{t}$	Klopp *et al.* (1998); Kratochvilova and Favor (1988)
		Defective lens development; microphthalmia	S	D	$Cryge^{Elo}$	Oda *et al.* (1980); Cartier *et al.* (1992)
		Sutural	I	S-D	$Cryge^{Ns}$	Graw (1999b)
		Nuclear, zonular	I	S-D	$Cryge^{Nz}$	Kratochvilova (1981); Klopp *et al.* (1998)
		Nuclear, lamellar	C	S-D	$Cryge^{ENU418}$	Graw *et al.* (2002a)
		Diffuse	C	S-D	$Cryge^{ENU449}$	Graw *et al.* (2004)

(Continued)

TABLE 11.2 Continued

Mouse gene/ chromosome[a]		Cataract phenotype[b]	Mode[c]	Inheritance[d]	Mouse Strain[e]	Reference
		Nuclear, zonular	C	S-D	$Cryge^{Aey1}$	Hrabe de Angelis *et al.* (2000); Graw *et al.* (2001b)
		Total, lamellar	S	S-D	$Cryge^{Z2}$	Graw *et al.* 2004
		Zonular	C	S-D	$Cryge^{ADD15306}$	Graw *et al.* (2004)
		Nuclear	S/RV	S-D	$Cryge^{No3}$	Nag *et al.* (2007)
Crygf	1	Central fetal, lamellar cortex; denser in homozygote	C	S-D	$Crygf^{Rop}$	Graw *et al.* (2002a)
Cts	7	Cataract, small eye; more severe in homozygote	S	S-D		Ohotori *et al.* (1968); Kimura *et al.* (1998)
dcm	U	Progressive mild opacity with vacuoles to complete cataract; iris abnormalities/microphthalmia	S	D	dcm	Kohale *et al.* (2004)
dwg	10	Unknown	S	R	dwg	Harris and Davisson (1990)
dwg	10	Unknown	S	R	dwg^{Bayer}	Davis (2005)
Em	U	Progressive anterior cortical to total opacity; early and late onset strains	S	D		Kuck *et al.* (1981); Kuck (1990)
Enc	U	Embryonic nucleus cataract	I	D		Kratochvilova (1983); Kratochvilova and Favor (1988)
Foxe3	4	Corneal opacity with progressive cataract	S	R	$Foxe3^{dyl}$	Sanyal and Hawkins (1979)
GalKO	U	White pinhead opacity progressing to total cataract	U	D		Eyssens (1999)
Gja8	3	Nuclear; more severe in homozygote	C	S-D	$Gja8^{No2}$	Favor (1983); Steele *et al.* (1998)
		Cortical opacity with clear central fetal lens, snowflake central opacity, dense central opacity depending upon cross breed; homozygote dense cataract with microphthalmia	S	S-D	$Gja8^{Lop10}$	Runge *et al.* (1992); Chang *et al.* (2002)
		Progressive nuclear and zonular; same in hetero- and homozygotes	C	S-D	$Gja8^{Ae5}$	Graw *et al.* (2001a)
		Complete cataract with microphakia and microphthalmia in hetero- and homozygotes	C	S-D	$Gja8^{S50P}$	Xia *et al.* (2006a)
Iac	U	Iris anomaly with cortical or total cataract; homozygote not viable	I	D		Kratochvilova (1981); Kratochvilova and Favor (1988)
Idc	U	Iris anomaly with cortical or total cataract with cloudy cornea and microphthalmia; homozygote not viable	I	D		Kratochvilova (1981); Kratochvilova and Favor (1988)
Icl	U	Unknown	C	U		Thaung *et al.* (2002)
jrc	7	Progressive to total	S	R		Cargill *et al.* (2001)
Lim2	7	Complete cataract; with microphthalmia in homozygote	C	D	$Lim2^{To3}$	Ehling *et al.* (1982); Favor (1983); Favor (1984); Kerscher *et al.* (1996); Steele *et al.* (1997)
lop2	U	Unknown	C	R		West *et al.* (1985)

TABLE 11.2 Continued

Mouse gene/ chromosome[a]		Cataract phenotype[b]	Mode[c]	Inheritance[d]	Mouse Strain[e]	Reference
Maf	8	Central embryonal pulverulent	C	S-D	Maf^{ENU424}	Copeland *et al.* (1993); Perveen *et al.* (2007)
		Pulverulent cataract entire lens; microphthalmia	S	D	Maf^{Ofl}	Lyon *et al.* (2003)
Mip	10	Unknown	S	S-D	Mip^{CAT}	Paget (1953)
		Progressive central opacity; more severe in homozygote	S	S-D	Mip$^{CAT\text{-}FR}$	Fraser and Schabtach (1962); Koniukhov and Kolesova (1976); Verrusio and Fraser (1966); Zwaan and Williams (1968); Zwaan and Williams (1969); Shiels and Bassnett (1996)
		Progressive central opacity; microphthalmia; more severe in homozygote	S	S-D	Mip$^{CAT\text{-}LOP}$	Lyon *et al.* (1981); Shiels and Bassnett (1996)
		Progressive cataract; microphthalmia	S	S-D	Mip$^{Cat\text{-}Tohm}$	Okamura *et al.* (2003)
		Cortical cataract in heterozygote; total cataract in homozygote	I	S-D	Mip^{Hfi}	Kratochvilova and Favor (1988); Kratochvilova and Favor (1992); Sidjanin *et al.* (2001)
			S	R	Mip$^{Cat\text{-}AA}$	Magon and Erickson (1983)
Nhs	X	Progressive cataract	I	S-D	Nhs^{Xcat}	Favor *et al.* (1987); Grimes *et al.* (1993); Huang *et al.* (2006); [Huang *et al.* (2007)
nct	16	Nakano cataract; dense central cataract and progressive diffuse cataract, depending upon strain; osmotic cataract; microspherophakia	S	R		Brown *et al.* (1970); Hamai *et al.* (1974); Piatigorsky *et al.* (1978); Hara *et al.* (1999); Iida *et al.* (1997); Takehana (1990); Narita *et al.* (2002)
Nuca	U	Dominant nuclear cataract	I	S-D		Kratochvilova (1981)
Pax6	2	Anterior cataract	C	R	Pax6^{Leca4}	Thaung *et al.* (2002)
		Anterior cataract	C	R	Pax6 ^{Leca2}	Thaung *et al.* (2002)
		Central; microphthalmia	C	D	Pax6^{Gsfaey11}	Graw *et al.* (2005)
		Central; microphthalmia	C	D	Pax6^{Aey18}	Haubst *et al.* (2004); Graw *et al.* (2005)
		Central; microphthalmia	C	D	Pax6^{ADD4802}	Graw *et al.* (2005)
		Anterior polar; corneal opacities; microphthalmia	C	S-D	Pax6^{Neu2}	Favor and Neuhauser-Klaus (2000)
		Anterior polar; corneal opacities; microphthalmia	C	S-D	Pax6^{Neu4}	Favor and Neuhauser-Klaus (2000)
		Anterior polar; corneal opacities; microphthalmia	C	S-D	Pax6^{Neu5}	Favor (1986)
		Anterior polar; corneal opacities; microphthalmia	C	S-D	Pax6^{Neu6}	Favor and Neuhauser-Klaus (2000)
		Anterior polar; corneal opacities; microphthalmia	C	S-D	Pax6^{Neu7}	Favor (1986)

(Continued)

TABLE 11.2 Continued

Mouse gene/ chromosome[a]		Cataract phenotype[b]	Mode[c]	Inheritance[d]	Mouse Strain[e]	Reference
		Anterior polar; corneal opacities; microphthalmia	C	S-D	$Pax6^{Neu8}$	Favor and Neuhauser-Klaus (2000)
		Anterior polar; corneal opacities; microphthalmia	C	S-D	$Pax6^{Neu10}$	Favor *et al.* (2001)
		Unknown	C	R	$Pax6^{1Jrt}$	Rossant (2003)
		Complete cataract with Coloboma/microspherophakia/ aphakia/microphthalmia	S	S-D	$Pax6^{Sey\text{-}Dey}$	Varnum and Stevens (1974); Theiler *et al.* (1978); Theiler *et al.* (1980); Theiler and Varnum (1981); Hogan (1987)
		Progressive cataract with vacuolization	S	S-D	$Pax6^{Sey}$	Roberts (1967); Hogan (1987); Hill *et al.* (1991); Hill *et al.* (1992);
Pcs1	U	Unknown	R	U		Kratochvilova (1983); Kratochvilova and Favor (1988)
rct/mrct	4,5	No cataract, or early or late onset based on a two allele system; cataract with vacuoles; microphthalmia	S	R	rct/mrct	Maeda *et al.* (2001)
rlc	14	Progressive to complete opacity	S	R		Matsushima *et al.* (1996); Iida *et al.* (1997); Song *et al.* (1997); Kim *et al.* (2007)
Tcm	4	Total cataract; coloboma; microphthalmia	R	S-D		Kratochvilova and Favor (1988); Adler (1990); Graw *et al.* (1990b); Zhou *et al.* (1997)
vl	1	Vacuolated lens	S	R		Dickie (1967)

[a] Mouse gene names are abbreviated using standard genomic abbreviations. Chromosomal location is represented by chromosome number or unknown (U).

[b] Cataract phenotypes are listed as described in the reference or as evaluated by Dr. Bronwyn Bateman from published photographs.

[c] The mode is abbreviated as follows: irradiation-induced (I), chemically induced (C), spontaneous (S), radiation-induced/spontaneous (I/S), or spontaneous/retrovirus insertion (S/RV).

[d] Inheritance is abbreviated as follows: dominant (D), semi-dominant (S-D), recessive (R), or unknown (U).

[e] When applicable, the mouse strain is given, otherwise it is left blank.

Biochemical studies and crystallography are useful methods for identifying the mechanisms of human cataract formation. However, methods that require fresh tissue including histology, immunohistochemistry, expression studies such as Western blotting, and *in vitro* transfection necessitate the use of animal models. For example, a range of mechanisms for human cataract formation as a result of *CRYD* mutations have been established for some mutations using these techniques (Kmoch *et al.*, 2000; Pande *et al.*, 2000, 2001, 2005; Basak *et al.*, 2003; Wang *et al.*, 2007). The molecular differences among the reported mutations in the spontaneously occurring and induced mouse models for cataracts provide opportunities to study mechanisms *in vivo* and *in vitro*, with fresh tissue techniques.

Spontaneous mouse mutations and those caused by irradiation or chemical induction have been studied for decades. Documentation of cataracts in mouse models have been reported in structural proteins including crystallins, cytoskeletal and membrane proteins, and DNA-binding proteins including transcription factors (Table 11.2). Effective chemical methods for induction of single gene mutations include ethylnitrosourea (ENU), methylnitrosourea (MNU), iso-propylmethanesufonate (iPMS), 3-aminobezamide (3-AB) and chlorambucil; irradiation is usually Gy (γ irradiation). The molecular differences among the reported mutations in the spontaneously occurring and induced mouse models permit study of phenotypic expression and the mechanisms of cataract formation. Although most cataracts are caused by protein effects in the cytosol or alteration of membrane proteins, some are caused by interruption of denucleation (Wang *et al.*, 2007). And, different mutations of the same gene can cause cataracts by different mechanisms.

Some genes that cause cataracts in humans are not represented in mouse models; examples include *GCNT2*, *CRYAB*, *CRYBB1*, *CRYBB3*, *HSF4*, and *FTL*. Similarly, there are genes in spontaneously occurring and induced mouse models that are not have been reported in human disease including *Cryge*, *Cryga*, *Crygb*, *Crygf*, and *Six5*. For reasons that are, yet, unclear, some mouse genes are more prone to mutations. For example, mutations in the *Cryg* genes are not equally distributed and there are numerous polymorphisms, particularly for the *Crygd* and *Cryge* genes (Graw et al., 2004); *Cryg* has more mutations than *Cryb* (Graw *et al.*, 2004). Polymorphisms of the *Cryg* group are not strain-specific. Curiously, ENU has induced the identical mutation in *Crygd* and *Cryge* with different phenotypes (Graw *et al.*, 2001a, 2004). The identical mutation has been reported in the *Crygd* (Smith *et al.*, 2000) gene of the mouse and the *CRYGD* (Santhiya *et al.*, 2002) gene in the human; the mouse cataract (based on published photograph) and the human (based on description) were similar, a central (embryonal) opacity. Such observations of mutation frequency are not reliable in human populations because of scientific ascertainment biases and cultural differences in reproductive patterns.

Phenotypes: Human Corollaries

The terminology used to describe mouse cataract phenotypes is similar to the human. Historically, human cataract phenotypes have been described in an inconsistent and, frequently, incomplete fashion. Phenotypic classifications have been based on morphology, size, color, and location in the lens using slit lamp biomicroscopy, and/or the name of the author describing the cataract or affected family. Wide variations of cataract phenotypes among human families have been described. In early publications, cataracts were documented using drawings, and some intrafamilial variability was evident (Nettleship and Ogilvie, 1906; Lutman and Neel, 1945). Most recent published reports of hereditary cataracts document the phenotype in one or two phakic (unoperated eye) individuals, usually because most affected members of the family have had surgery, the family is relatively small, and/or an ophthalmologist has not participated in the study. Phenotypes among human families have been described as nuclear, anterior polar (Jaafar and Robb, 1984), posterior polar (Tulloh, 1955), coralliform (Gunn, 1895; Nettleship and Ogilvie, 1906), blue dot cerulean (Kivlin *et al.*, 1985), pulverulent (Mackay *et al.*, 1997), cortical (Berry *et al.*, 1999), zonular (Basti *et al.*, 1996), aculeiform (Heon *et al.*, 1998), pouch-like (Vanita *et al.*, 2001), and sutural cataracts (Vanita *et al.*, 2001). Progression is frequently not documented and asymmetry has been documented rarely (Scott *et al.*, 1994; Shafie *et al.*, 2006). In publications of specific gene defects, descriptions and photos are limited in most and, in those with detail, there is consistency within the family with respect to density and morphology of the cataract (Ferrini *et al.*, 2004; Addison *et al.*, 2005). There has been recent recognition of variability within a family (Shafie *et al.*, 2006). As in humans, most reports of mouse models report a single phenotype. Phenotype can be documented in mice photographically using various techniques (Brady *et al.*, 1997; Hsu *et al.*, 2006; Xia *et al.*, 2006b; Lassen *et al.*, 2007; Shiels and Hejtmancik, 2007). Anatomically, the lens of the mouse is more spherical and relatively larger in comparison to the size of the eye than in humans; therefore, the vitreous space is relatively smaller in the mouse eye than in the human.

Mutations involving *Cryaa* are illustrative of phenotypic variability among a range of mutations of a single gene. Autosomal dominant ($Cryaa^{V124E}$, $Cryaa^{aey7}$, and $Cryaa^{Y118D}$) (Graw *et al.*, 2001c; Xia *et al.*, 2006b) and autosomal recessive ($Cryaa^{R54H}$, $Cryaa^{lop18}$, $Cryaa^{V124E}$, $Cryaa^{aey7}$, $Cryaa^{R54C}$, $Cryaa^{Y118D}$, and $Cryaa^{-/-}$) (Brady *et al.*, 1997; Chang *et al.*, 1999; Graw *et al.*, 2001c; Xia *et al.*, 2006b) models have been studied. Of note, both the heterozygous and homozygous $Cryaa^{V124E}$ mice were described as having nuclear and posterior sutural opacities at postnatal day 12 that progressed and stabilized by 2 months of age to a nuclear cataract and a "zonular" opacity in the "subcortical" region (Graw *et al.*, 2001c). Based on the published postmortem photographs of the 2-month adult, the embryonal nuclear opacities consisted of white, relatively round opacities of various sizes that formed an irregular fetal opacity in the heterozygous mouse; a denser, white embryonal cataract with a "zonular" (lamellar or ring) opacity in the cortical region of a small lens was evident in the homozygous model. The eyes were described as small in both the heterozygous and the homozygous mice with the homozygous being more microphthalmic (Graw *et al.*, 2001c). In contrast, the cataract in the heterozygote mouse with the $Cryaa^{Y118D}$ mutation is located in the embryonal nucleus with sharp margins in a circular shape and irregular density within; in the homozygous state, the lens is of normal size with a dense, central, white circular opacity in the embryonal nucleus with a less dense, diffuse opacity in the juvenile lens as well as a less dense annular (zonular) ring in the subcapsular region (Xia *et al.*, 2006b). Progression was not described. In the two mouse models with mutations in codon 54 (*Cryaa*^{R54H} and *Cryaa*^{R54C}) the inheritance is autosomal recessive (Chang *et al.*, 1999; Xia *et al.*, 2006b). The homozygous *Cryaa*^{R54H} mutation results in progressive white opacification of the central (embryonal) region of the lenses with mild cortical changes in a lens of normal size (Chang *et al.*, 1996). In the homozygous *Cryaa*^{R54C} mouse, significant microspherophakia, microphthalmia and dense, central white fetal and cortical opacities with irregular margins and less dense haze in the remaining peripheral cortex was evident (Xia *et al.*, 2006b). Other models with mutations of the *Cryaa* gene have demonstrated a variety of cataracts with microspherophakia in some. Thus, the clinical similarities and differences between the human and mouse provide further evidence of the usefulness of mouse models.

Phenotypic observations of mouse models and comparisons with human disease are instructive. Some mouse models, such as those with mutations of the *Mip* gene, have a consistent pattern of a more severe phenotype in the homozygous mouse, including the Mip^{Hfi} (Sidjanin *et al.*, 2001), Mip^{Cat-Fr} (Shiels and Bassnett, 1996) and Mip^{Lop} models (Lyon *et al.*, 1981). Conversely, the heterozygote and the homozygote mouse with the *Crybb2*^{Aey2} mutation have a similar phenotype (Graw *et al.* 2001d). Different PITX3 mutations cause a relatively consistent posterior polar cataract in humans (Berry *et al.*, 2004; Finzi *et al.*, 2005; Bidinost *et al.*, 2006; Burdon *et al.*, 2006) but have not caused cataracts in mice, to date; severe microphthalmia with aphakia occurs in the mouse models (Varnum and Stevens, 1968; Varnum and Stevens, 1975; Semina *et al.*, 2000; Rieger *et al.*, 2001). Similarly, the *Pax6* mouse phenotype is remarkably similar amongst many different mutations (Favor *et al.*, 2001); the human disease is variable within families and among different mutations. Although progressive cataracts are common (Table 11.1), some, such as the heterozygote *Gja8*^{Lop10}, are stable (Runge *et al.*, 1992); most human cataracts are progressive.

Although there are exceptions, the phenotypes of most mouse models are not similar to their human counterpart. One exception is the "knockout" mouse model in which the cataract formation is similar to humans; the *Cryaa*$^{-/-}$(Brady *et al.*, 1997) mouse has no expression of the gene and cataracts develop soon after birth, similar to the autosomal recessive human cataract, described as developing in the first few weeks of life (Pras *et al.*, 2000). Although most publications do not comment on laterality in mouse models, asymmetry between eyes has been reported rarely (Favor *et al.*, 2001). Gender can result in phenotypic differences in the late-onset cataract of the Emory (Shang *et al.*, 2002) and Nakano models (Matsuzawa and Wada, 1988).

Many reports of mouse models include a single photograph of an affected lens from a heterozygote and a homozygote, and descriptions of phenotypic consistency within and among litters are uncommon. As an example, in the *Gja8*^{Lop10} mouse, cataracts in the homozygote are consistent. In the heterozygote *Gja8*^{Lop10} mouse, the phenotype varies from no cataract, to only cortical opacities, nuclear snow-flake opacities, or a small central (fetal) opacity to a dense central (fetal) nuclear opacity, depending on the cross strain; this heterozygote mouse is consistent within the cross strain (Runge *et al.*, 1992). In other mutations of the *Gja8* gene, there is no difference between the heterozygotes and the homozygote (Graw *et al.*, 2001b; Xia *et al.*, 2006c).

The phenotypic variability of cataracts among mice from a single litter has not been studied. But, there is evidence in mouse models to support the concept that modifier genes influence the cataract phenotype (see below). Mouse models offer valuable opportunities to correlate the cataract phenotype with the genotype.

Molecular Bases: Mechanistic Understanding

The inheritance patterns of mutations of specific genes can differ by mutation and are probably related to the mechanism of cataract formation. Different mutations of some genes can cause either dominant or recessive forms of cataract. For example, mutations of the *Crybb2* cause either autosomal recessive cataract in the *Crybb2*^{Phil} (Kador *et al.*, 1980; Chambers and Russell, 1991) form or autosomal dominant disease in the *Crybb2*^{Aey} mouse (Graw *et al.*, 2001d). *Pax6* mutations can cause a cataract with an autosomal dominant, semidominant, or recessive pattern of inheritance; mutations of *Mip* and *Cryaa* can be inherited in autosomal semidominant and recessive patterns. Based on a late truncation of the protein, a *Crygs* mutation is an autosomal recessive cataract (Bu *et al.*, 2002); in humans, the cataracts caused by mutations of this gene are inherited in an autosomal dominant pattern (Sun *et al.*, 2005).

Mouse models have been useful in studying the mechanisms of cataract formation. Protein–protein interactions are the bases of cataract formation in some forms, such as *Gja8*^{Lop10}, in which the semidominant inheritance has been shown to be based on a loss of function as well as a dominant negative effect. The mutated Gja8 protein (Gja8^{Lop10}) interferes with normal function of the Gja3 protein (Chang *et al.*, 2002); in the absence of the Gja3 protein product, the cataract phenotype of the mouse is altered and the secondary fibers (cortex) are clear (Chang *et al.*, 2002). In this model, the cataract is formed by loss of function of the Gja8 protein and a negative dominant caused by reduced phosphorylated Gja3 protein (Chang *et al.*, 2002), based on both the histology and immunohistochemistry of fresh tissue. A similar model was studied with a different *Gja3* mutation; both the heterozygote and homozygote exhibit microspherophakia and complete cataract with the *Gja8*^{S50P} mutation. The heterozygote mice with this mutation demonstrate inhibition of elongation of the primary lens cells (Xia *et al.*, 2006c); in the homozygote mouse, the secondary fibers do not elongate (Xia *et al.*, 2006c). The experiment was repeated with a knockout *Gja3* mouse and the affected mice developed a cataract in the central (embryonal) region (primary lens fibers) of the lens with a clear cortex (secondary lens fibers). Thus, Gja8^{S50P} protein interacted with the wild-type Gja8 protein to inhibit the primary fibers and with the wild-type Gja3 protein to inhibit the secondary fibers. Some cataracts are osmotic including the Asc-1 (Graw *et al.*, 1990b), Cat-3vao (Graw *et al.*, 1990b), nct (Nakano) (Takehana, 1990), and Tcm mouse models (Graw *et al.*, 1990b). A gain of function mutation is the basis for the cataract in the Mip^{Tohn} mouse; the protein localization shifts from the plasma membrane of the fiber cells to the intracellular and perinuclear spaces (Okamura *et al.*, 2003). A dominant negative effect is caused by a mutation of the *Cryge* start codon (*Cryge*^{Aey1}) creating a novel protein that is the basis of the cataract. The autosomal recessive *Crygs* mutation results in absence of the protein due to premature truncation (Bu *et al.*, 2002).

Mouse models of the same mutation but in different genetic backgrounds are probably the bases for much of the reported phenotypic variability. For example, the phenotype in the homozygous (Graw *et al.*, 2002a) model of a *Crygd* mutation is similar to the phenotype in a heterozygous (Wang *et al.*, 2007) model with a different strain background. Background strains influence the phenotype of mutations in connexin46 (*Gja3*) (Runge *et al.*, 1992; Gong *et al.*, 1997; Chang *et al.*, 2002) and connexin50 (*Gja8*) mouse models (Gerido *et al.*, 2003). Gong *et al.* (1999) found that the variability of the cataract phenotype of *Gja3* (*Cx46*) null mice was γ crystallin solubility. Background strains also influence the severity of the cataract phenotype in knockout mice for *Gja8* (*Cx50*; connexin50) but does not influence the microphthalmia phenotype (Gerido *et al.*, 2003). The Nakano mouse is an example with different phenotypes in mice with different genetic backgrounds (Lipman *et al.*, 1981; Narita *et al.*, 2002).

Modifier genes may be the bases for the phenotypic variability among different background strains. Although not identified as such, genes controlling connexin expression and/or genetic connexin polymorphisms may act as modifiers as *Crygd* mutant proteins alter connexin levels (Wang *et al.*, 2007). For the mrct cataract, the causative gene locus is on chromosome 4 and an unknown modifier locus is on chromosome 5 (Maeda *et al.*, 2001). For the Nakano mouse, an autosomal recessive cataract caused by a gene on chromosome 16, linkage analysis has demonstrated two phenotypic subtypes with modifier genes on chromosomes 3 and 10. However, for some forms such as the jrc, the cataract is identical in different background strains (Cargill *et al.*, 2001).

CONCLUDING REMARKS

Diseases affecting the cornea are a major cause of blindness worldwide, second only to cataract in overall importance. Indeed, cataract of the lens causes upwards of 50% of blindness and affects nearly 20 million people worldwide. In this regard, mouse models of the cornea and lens have proven invaluable in the investigation of both human ocular physiology and disease mechanisms. With the advent of new mouse transgene and gene ablation knockout strategies and technologies, these models will continue to contribute to the future of eye research.

ACKNOWLEDGMENTS

Grant support: EY11490 (V.V.); EY05856 (J.M.P.); EY02687 (Core Grant for Vision Research to Washington University). Satori A. Marchitti was supported by NIH/NIAAA Pre-doctral Fellowship AA016875.

REFERENCES

Abedinia M, Pain T, Algar EM, Holmes RS (1990). Bovine corneal aldehyde dehydrogenase: The major soluble corneal protein with a possible dual protective role for the eye. *Exp Eye Res* 51:419–426.

Addison PK, Berry V, Ionides AC, Francis PJ, Bhattacharya SS, Moore AT (2005). Posterior polar cataract is the predominant consequence of a recurrent mutation in the PITX3 gene. *Br J Ophthalmol* 89:138–141.

Adler ID (1990). New stocks of mice carrying reciprocal translocations. *Mouse Genome* 86:251–253.

Alexander RJ, Silverman B, Henley WL (1981). Isolation and characterization of BCP 54, the major soluble protein of bovine cornea. *Exp Eye Res* 32:205–216.

Andley UP (2007). Crystallins in the eye: Function and pathology. *Prog Retin Eye Res* 26:78–98.

Argeson AC, Buchberg AM, Siracusa LD (1994). Molecular genetic mapping of D6Kcc1, a novel DNA marker on mouse chromosome 6. *Mouse Genome* 94:488–490.

Bantseev V, Oriowo OM, Giblin FJ, Leverenz VR, Trevithick JR, Sivak JG (2004). Effect of hyperbaric oxygen on guinea pig lens optical quality and on the refractive state of the eye. *Exp Eye Res* 78:925–931.

Baruch A, Greenbaum D, Levy ET, Nielsen PA, Gilula NB, Kumar NM, Bogyo M (2001). Defining a link between gap junction communication, proteolysis, and cataract formation. *J Biol Chem* 276:28999–29006.

Basak A, Bateman O, Slingsby C, Pande A, Asherie N, Ogun O, Benedek GB, Pande J (2003). High-resolution X-ray crystal structures of human gammaD crystallin (1.25 A) and the R58H mutant (1.15 A) associated with aculeiform cataract. *J Mol Biol* 328:1137–1147.

Bassnett S (2002). Lens organelle degradation. *Exp Eye Res* 74:1–6.

Basti S, Hejtmancik JF, Padma T, Ayyagari R, Kaiser-Kupfer MI, Murty JS, Rao GN (1996). Autosomal dominant zonular cataract with sutural opacities in a four-generation family. *Am J Ophthalmol* 121:162–168.

Benedek G (1983). Why the eye lens is transparent. *Nature* 302:383–384.

Bennett MK, Charlton HM (1992). A PCR based assay for the murine testicular feminisation mutation (Tfm). *Mouse Genome* 90:216–217.

Berry V, Mackay D, Khaliq S, Francis PJ, Hameed A, Anwar K, Mehdi SQ, Newbold RJ, Ionides A, Shiels A *et al.* (1999). Connexin 50 mutation in a family with congenital "zonular nuclear" pulverulent cataract of Pakistani origin. *Hum Genet* 105:168–170.

Berry V, Yang Z, Addison PK, Francis PJ, Ionides A, Karan G, Jiang L, Lin W, Hu J, Yang R et al. (2004). Recurrent 17 bp duplication in PITX3 is primarily associated with posterior polar cataract (CPP4). *J Med Genet* 41:e109.

Bidinost C, Matsumoto M, Chung D, Salem N, Zhang K, Stockton DW, Khoury A, Megarbane A, Bejjani BA, Traboulsi EI (2006). Heterozygous and homozygous mutations in PITX3 in a large Lebanese family with posterior polar cataracts and neurodevelopmental abnormalities. *Invest Ophthalmol Vis Sci* 47:1274–1280.

Brady JP, Garland D, Duglas-Tabor Y, Robison WG, Jr, Groome A, Wawrousek EF (1997). Targeted disruption of the mouse alpha A-crystallin gene induces cataract and cytoplasmic inclusion bodies containing the small heat shock protein alpha B-crystallin. *Proc Natl Acad Sci USA* 94:884–889.

Brandt CR (2005). The role of viral and host genes in corneal infection with herpes simplex virus type 1. *Exp Eye Res* 80:607–621.

Brown ER, Nakano T, Vankin GL (1970). Early devlopment of an inherited cataract in mice. *Jikken Dobutsu* 19:95–99.

Bu L, Yan S, Jin M, Jin Y, Yu C, Xiao S, Xie Q, Hu L, Xie Y, Solitang Y *et al.* (2002). The gamma S-crystallin gene is mutated in autosomal recessive cataract in mouse. *Genomics* 80:38–44.

Burdon KP, McKay JD, Wirth MG, Russell-Eggit IM, Bhatti S, Ruddle JB, Dimasi D, Mackey DA, Craig JE (2006). The PITX3 gene in posterior polar congenital cataract in Australia. *Mol Vis* 12:367–371.

Cargill EJ, Happold TR, Bertani G, Rocha J, Lou MF, Pomp D, Nielsen MK (2001). Localization of a recessive juvenile cataract mutation to proximal chromosome 7 in mice. *Hum Hered* 52:77–82.

Cartier M, Breitman ML, Tsui LC (1992). A frameshift mutation in the gamma E-crystallin gene of the Elo mouse. *Nat Genet* 2:42–45.

Chambers C, Russell P (1991). Deletion mutation in an eye lens beta-crystallin. An animal model for inherited cataracts. *J Biol Chem* 266:6742–6746.

Chang B, Hawes NL, Roderick TH, Smith RS, Heckenlively JR, Horwitz J, Davisson MT (1999). Identification of a missense mutation in the alphaA-crystallin gene of the lop18 mouse. *Mol Vis* 5:21.

Chang B, Hawes NL, Smith RS, Heckenlively JR, Davisson MT, Roderick TH (1996). Chromosomal localization of a new mouse lens opacity gene (lop18). *Genomics* 36:171–173.

Chang B, Wang X, Hawes NL, Ojakian R, Davisson MT, Lo WK, Gong X (2002). A Gja8 (Cx50) point mutation causes an alteration of alpha 3 connexin (Cx46) in semi-dominant cataracts of Lop10 mice. *Hum Mol Genet* 11:507–513.

Cobb BA, Petrash JM (2000). Structural and functional changes in the alpha A-crystallin R116C mutant in hereditary cataracts. *Biochemistry* 39:15791–15798.

Cobb BA, Petrash JM (2002). alpha-Crystallin chaperone-like activity and membrane binding in age-related cataracts. *Biochemistry* 41:483–490.

Cooper DL, Baptist EW, Enghild JJ, Isola NR, Klintworth GK (1991). Bovine corneal protein 54K (BCP54) is a homologue of the tumor-associated (class 3) rat aldehyde dehydrogenase (RATALD). *Gene* 98:201–207.

Copeland NG, Jenkins NA, Gilbert DJ, Eppig JT, Maltais LJ, Miller JC, Dietrich WF, Weaver A, Lincoln SE, Steen RG (1993). A genetic linkage map of the mouse: Current applications and future prospects. *Science* 262:57–66.

Couldrey C, Carlton MB, Ferrier J, Colledge WH, Evans MJ (1998). Disruption of murine alpha-enolase by a retroviral gene trap results in early embryonic lethality. *Dev Dyn* 212:284–292.

Cuthbertson RA, Tomarev SI, Piatigorsky J (1992). Taxon-specific recruitment of enzymes as major soluble proteins in the corneal epithelium of three mammals, chicken, and squid. *Proc Natl Acad Sci USA* 89:4004–4008.

Davis AC (2005). Dwarf Grey Bayer, A Remutation of the dwg Locus. MGI direct data submission.

Dickie MM (1967). Vacuolated lens. *Mouse News Lett*:39–40.

Downes JE, Swann PG, Holmes RS (1994). Differential corneal sensitivity to ultraviolet light among inbred strains of mice. Correlation of ultraviolet B sensitivity with aldehyde dehydrogenase deficiency. *Cornea* 13:67–72.

Drenckhahn D, Lullmann-Rauch R (1977). Lens opacities associated with lipidosis-like ultrastructural alterations in rats treated with chloroquine, chlorphentermine, or iprindole. *Exp Eye Res* 24:621–632.

Ehling UH, Favor J, Kratochvilova J, Neuhauser-Klaus A (1982). Dominant cataract mutations and specific-locus mutations in mice induced by radiation or ethylnitrosourea. *Mutat Res* 92:181–192.

Esterbauer H, Schaur RJ, Zollner H (1991). Chemistry and biochemistry of 4-hydroxynonenal, malonaldehyde and related aldehydes. *Free Radic Biol Med* 11:81–128.

Estey T, Cantore M, Weston PA, Carpenter JF, Petrash JM, Vasiliou V (2007a). Mechanisms involved in the protection of UV-induced protein inactivation by the corneal crystallin ALDH3A1. *J Biol Chem* 282:4382–4392.

Estey T, Piatigorsky J, Lassen N, Vasiliou V (2007b). ALDH3A1: A corneal crystallin with diverse functions. *Exp Eye Res* 84:3–12.

Everett CA, Glenister PH, Taylor DM, Lyon MF, Kratochvilova-Loester J, Favor J (1994). Mapping of six dominant cataract genes in the mouse. *Genomics* 20:429–434.

Eyssens D (1999). Cataract development in the GalKO mouse. *Clin Exp Optom* 82:119–123.

Fan X, Molotkov A, Manabe S, Donmoyer CM, Deltour L, Foglio MH, Cuenca AE, Blaner WS, Lipton SA, Duester G (2003). Targeted disruption of Aldh1a1 (Raldh1) provides evidence for a complex mechanism of retinoic acid synthesis in the developing retina. *Mol Cell Biol* 23:4637–4648.

Fan X, Reneker LW, Obrenovich ME, Strauch C, Cheng R, Jarvis SM, Ortwerth BJ, Monnier VM (2006). Vitamin C mediates chemical aging of lens crystallins by the Maillard reaction in a humanized mouse model. *Proc Natl Acad Sci USA* 103:16912–16917.

Favor J (1983). A comparison of the dominant cataract and recessive specific-locus mutation rates induced by treatment of male mice with ethylnitrosourea. *Mutat Res* 110:367–382.

Favor J (1984). Characterization of dominant cataract mutations in mice: Penetrance, fertility and homozygous viability of mutations recovered after 250 mg/kg ethylnitrosourea paternal treatment. *Genet Res* 44:183–197.

Favor J (1986). The frequency of dominant cataract and recessive specific-locus mutations in mice derived from 80 or 160 mg ethylnitrosourea per kg body weight treated spermatogonia. *Mutat Res* 162:69–80.

Favor J, Neuhauser-Klaus A (2000). Saturation mutagenesis for dominant eye morphological defects in the mouse Mus musculus. *Mamm Genome* 11:520–525.

Favor J, Neuhauser-Klaus A, Ehling UH (1987). Radiation-induced forward and reverse specific locus mutations and dominant cataract mutations in treated strain BALB/c and DBA/2 male mice. *Mutat Res* 177:161–169.

Favor J, Grimes P, Neuhauser-Klaus A, Pretsch W, Stambolian D (1997). The mouse Cat4 locus maps to chromosome 8 and mutants express lens-corneal adhesion. *Mamm Genome* 8:403–406.

Favor J, Peters H, Hermann T, Schmahl W, Chatterjee B, Neuhauser-Klaus A, Sandulache R (2001). Molecular characterization of Pax6(2Neu) through Pax6(10Neu): An extension of the Pax6 allelic series and the identification of two possible hypomorph alleles in the mouse Mus musculus. *Genetics* 159:1689–1700.

Feeney L, Berman ER (1976). Oxygen toxicity: Membrane damage by free radicals. *Invest Ophthalmol* 15:789–792.

Ferrini W, Schorderet DF, Othenin-Girard P, Uffer S, Heon E, Munier FL (2004). CRYBA3/A1 gene mutation associated with suture-sparing autosomal dominant congenital nuclear cataract: A novel phenotype. *Invest Ophthalmol Vis Sci* 45:1436–1441.

Finzi S, Li Y, Mitchell TN, Farr A, Maumenee IH, Sallum JM, Sundin O (2005). Posterior polar cataract: Genetic analysis of a large family. *Ophthalmic Genet* 26:125–130.

Fraser FC, Schabtach G (1962). "Shrivelled": A hereditray degeneration of the lens in the house mouse. *Genet Res* 3:383–387.

Ganea E, Harding JJ (2006). Glutathione-related enzymes and the eye. *Curr Eye Res* 31:1–11.

Gao J, Sun X, Martinez-Wittinghan FJ, Gong X, White TW, Mathias RT (2004). Connections between connexins, calcium, and cataracts in the lens. *J Gen Physiol* 124:289–300.

Gerido DA, Sellitto C, Li L, White TW (2003). Genetic background influences cataractogenesis, but not lens growth deficiency, in Cx50-knockout mice. *Invest Ophthalmol Vis Sci* 44:2669–2674.

Gong X, Li E, Klier G, Huang Q, Wu Y, Lei H, Kumar NM, Horwitz J, Gilula NB (1997). Disruption of alpha3 connexin gene leads to proteolysis and cataractogenesis in mice. *Cell* 91:833–843.

Gong X, Agopian K, Kumar NM, Gilula NB (1999). Genetic factors influence cataract formation in alpha3 connexin knockout mice. *Dev Genet* 24:27–32.

Gong X, Baldo GJ, Kumar NM, Gilula NB, Mathias RT (1998). Gap junctional coupling in lenses lacking alpha3 connexin. *Proc Natl Acad Sci USA* 95:15303–15308.

Graves PR, Kwiek JJ, Fadden P, Ray R, Hardeman K, Coley AM, Foley M, Haystead TA (2002). Discovery of novel targets of quinoline drugs in the human purine binding proteome. *Mol Pharmacol* 62:1364–1372.

Graw J (1999a). Cataract mutations and lens development. *Prog Retin Eye Res* 18:235–267.

Graw J (1999b). Mouse models of congenital cataract. *Eye* 13(Pt 3b):438–444.

Graw J (2004). Congenital hereditary cataracts. *Int J Dev Biol* 48:1031–1044.

Graw J, Loster J (2003). Developmental genetics in ophthalmology. *Ophthalmic Genet* 24:1–33.

Graw J, Kratochvilova J, Summer KH (1984). Genetical and biochemical studies of a dominant cataract mutant in mice. *Exp Eye Res* 39:37–45.

Graw J, Bors W, Gopinath PM, Merkle S, Michel C, Reitmeir P, Schaffer E, Summer KH, Wulff A (1990a). Characterization of Cat-2t, a radiation-induced dominant cataract mutation in mice. *Invest Ophthalmol Vis Sci* 31:1353–1361.

Graw J, Reitmeir P, Wulff A (1990b). Osmotic state of lenses in three dominant murine cataract mutants. *Graefes Arch Clin Exp Ophthalmol* 228:252–254.

Graw J, Jung M, Loster J, Klopp N, Soewarto D, Fella C, Fuchs H, Reis A, Wolf E, Balling R *et al.* (1999). Mutation in the betaA3/A1-crystallin encoding gene Cryba1 causes a dominant cataract in the mouse. *Genomics* 62:67–73.

Graw J, Klopp N, Loster J, Soewarto D, Fuchs H, Becker-Follmann J, Reis A, Wolf E, Balling R, Habre de AM (2001a). Ethylnitrosourea-induced mutation in mice leads to the expression of a novel protein in the eye and to dominant cataracts. *Genetics* 157:1313–1320.

Graw J, Loster J, Soewarto D, Fuchs H, Meyer B, Reis A, Wolf E, Balling R, Hrabe de AM (2001b). Characterization of a mutation in the lens-specific MP70 encoding gene of the mouse leading to a dominant cataract. *Exp Eye Res* 73:867–876.

Graw J, Loster J, Soewarto D, Fuchs H, Meyer B, Reis A, Wolf E, Balling R, Hrabe de AM (2001c). Characterization of a new, dominant V124E mutation in the mouse alphaA-crystallin-encoding gene. *Invest Ophthalmol Vis Sci* 42:2909–2915.

Graw J, Loster J, Soewarto D, Fuchs H, Reis A, Wolf E, Balling R, Hrabe de AM (2001d). Aey2, a new mutation in the betaB2-crystallin-encoding gene of the mouse. *Invest Ophthalmol Vis Sci* 42:1574–1580.

Graw J, Loster J, Soewarto D, Fuchs H, Reis A, Wolf E, Balling R, Hrabe de AM (2002a). V76D mutation in a conserved gD-crystallin region leads to dominant cataracts in mice. *Mamm Genome* 13:452–455.

Graw J, Neuhauser-Klaus A, Loster J, Favor J (2002b). A 6-bp deletion in the Crygc gene leading to a nuclear and radial cataract in the mouse. *Invest Ophthalmol Vis Sci* 43:236–240.

Graw J, Loster J, Puk O, Munster D, Haubst N, Soewarto D, Fuchs H, Meyer B, Nurnberg P, Pretsch W et al. (2005). Three novel Pax6 alleles in the mouse leading to the same small-eye phenotype caused by different consequences at target promoters. *Invest Ophthalmol Vis Sci* 46:4671–4683.

Graw J, Neuhauser-Klaus A, Klopp N, Selby PB, Loster J, Favor J (2004). Genetic and allelic heterogeneity of Cryg mutations in eight distinct forms of dominant cataract in the mouse. *Invest Ophthalmol Vis Sci* 45:1202–1213.

Grimes PA, Favor J, Koeberlein B, Silvers WK, Fitzgerald PG, Stambolian D (1993). Lens development in a dominant X-linked congenital cataract of the mouse. *Exp Eye Res* 57:587–594.

Grimes PA, Koeberlein B, Favor J, Neuhauser-Klaus A, Stambolian D (1998). Abnormal eye development associated with Cat4a, a dominant mouse cataract mutation on chromosome 8. *Invest Ophthalmol Vis Sci* 39:1863–1869.

Gunn R (1895). Peculiar coralliform cataract with crystals in the lens. *Trans Ophthalmol Soc UK*:19.

Hall LR, Pearlman E (1999). Pathogenesis of onchocercal keratitis (River blindness). *Clin Microbiol Rev* 12:445–453.

Hamai Y, Fukui HN, Kuwabara T (1974). Morphology of hereditary mouse cataract. *Exp Eye Res* 18:537–546.

Hara A, Matsumoto M, Uga S (1999). Morphological study on cataractogenesis of the Nakano mouse lens. *Graefes Arch Clin Exp Ophthalmol* 237:249–255.

Harris B, Davisson MT (1990). Dwarf grey (dwg). *Mouse Genome* 86:238.

Haubst N, Berger J, Radjendirane V, Graw J, Favor J, Saunders GF, Stoykova A, Gotz M (2004). Molecular dissection of Pax6 function: The specific roles of the paired domain and homeodomain in brain development. *Development* 131: 6131–6140.

Hazlett LD (2007). Bacterial infections of the cornea (*Pseudomonas aeruginosa*). *Chem Immunol Allergy* 92:185–194.

Hazlett LD, McClellan S, Kwon B, Barrett R (2000). Increased severity of *Pseudomonas aeruginosa* corneal infection in strains of mice designated as Th1 versus Th2 responsive. *Invest Ophthalmol Vis Sci* 41:805–810.

Heon E, Liu S, Billingsley G *et al.* (1998). Gene localization for aculeiform cataract, on chromosome 2q33-35. *Am J Hum Genet* 921-926. [letter] [published erratum appears in *Am J Hum Genet* 1999 Jan;64(1):334].

Hill RE, Favor J, Hogan BL, Ton CC, Saunders GF, Hanson IM, Prosser J, Jordan T, Hastie ND, van Heyningen V (1991). Mouse small eye results from mutations in a paired-like homeobox-containing gene. *Nature* 354:522–525.

Hill RE, Favor J, Hogan BL, Ton CC, Saunders GF, Hanson IM, Prosser J, Jordan T, Hastie ND, van Heyningen V (1992). Mouse small eye results from mutations in a paired-like homeobox-containing gene. *Nature* 355:750.

Hogan BL (1987). Allelism of small eyes (Sey) with Dickie's small eye (Dey) on chromosome 2. *Mouse News Lett* 77:135–138.

Holt WS, Kinoshita JH (1973). The soluble proteins of the bovine cornea. *Invest Ophthalmol* 12:114–126.

Horwitz J (1992). Alpha-crystallin can function as a molecular chaperone. *Proc Natl Acad Sci USA* 89:10449–10453.

Hough TA, Bogani D, Cheeseman MT, Favor J, Nesbit MA, Thakker RV, Lyon MF (2004). Activating calcium-sensing receptor mutation in the mouse is associated with cataracts and ectopic calcification. *Proc Natl Acad Sci USA* 101:13566–13571.

Hrabe de Angelis MH, Flaswinkel H, Fuchs H, Rathkolb B, Soewarto D, Marschall S, Heffner S, Pargent W, Wuensch K, Jung M et al. (2000). Genome-wide, large-scale production of mutant mice by ENU mutagenesis. *Nat Genet* 25:444–447.

Hsu CD, Kymes S, Petrash JM (2006). A transgenic mouse model for human autosomal dominant cataract. *Invest Ophthalmol Vis Sci* 47:2036–2044.

Huang KM, Wu J, Duncan MK, Moy C, Dutra A, Favor J, Da T, Stambolian D (2006). Xcat, a novel mouse model for Nance-Horan syndrome inhibits expression of the cytoplasmic-targeted Nhs1 isoform. *Hum Mol Genet* 15:319–327.

Huang KM, Wu J, Brooks SP, Hardcastle AJ, Lewis RA, Stambolian D (2007). Identification of three novel NHS mutations in families with Nance-Horan syndrome. *Mol Vis* 13:470–474.

Iida F, Matsushima Y, Hiai H, Uga S, Honda Y (1997). Rupture of lens cataract: A novel hereditary recessive cataract model in the mouse. *Exp Eye Res* 64:107–113.

Iles KE, Liu RM (2005). Mechanisms of glutamate cysteine ligase (GCL) induction by 4-hydroxynonenal. *Free Radic Biol Med* 38:547–556.

Jaafar MS, Robb RM (1984). Congenital anterior polar cataract: A review of 63 cases. *Ophthalmology* 91:249–254.

Jester JV, Budge A, Fisher S, Huang J (2005). Corneal keratocytes: Phenotypic and species differences in abundant protein expression and *in vitro* light-scattering. *Invest Ophthalmol Vis Sci* 46:2369–2378.

Jester JV, Moller-Pedersen T, Huang J, Sax CM, Kays WT, Cavangh HD, Petroll WM, Piatigorsky J (1999). The cellular basis of corneal transparency: Evidence for "corneal crystallins". *J Cell Sci* 112(Pt 5):613–622.

Jester JV, Petroll WM, Garana RM, Lemp MA, Cavanagh HD (1992). Comparison of *in vivo* and ex vivo cellular structure in rabbit eyes detected by tandem scanning microscopy. *J Microsc* 165:169–181.

Jordan T, Hanson I, Zaletayev D, Hodgson S, Prosser J, Seawright A, Hastie N, van Heyningen V (1992). The human PAX6 gene is mutated in two patients with aniridia. *Nat Genet* 1:328–332.

Kador PF, Fukui HN, Fukushi S, Jernigan HM, Jr, Kinoshita JH (1980). Philly mouse: A new model of hereditary cataract. *Exp Eye Res* 30:59–68.

Kao WW (2006). Ocular surface tissue morphogenesis in normal and disease states revealed by genetically modified mice. *Cornea* 25:S7–S19.

Kao WW, Liu CY, Converse RL, Shiraishi A, Kao CW, Ishizaki M, Doetschman T, Duffy J (1996). Keratin 12-deficient mice have fragile corneal epithelia. *Invest Ophthalmol Vis Sci* 37: 2572–2584.

Kao WW, Kao CW, Kaufman AH, Kombrinck KW, Converse RL, Good WV, Bugge TH, Degen JL (1998). Healing of corneal epithelial defects in plasminogen- and fibrinogen-deficient mice. *Invest Ophthalmol Vis Sci* 39:502–508.

Kerscher S, Glenister PH, Favor J, Lyon MF (1996). Two new cataract loci, Ccw and To3, and further mapping of the Npp and Opj cataracts in the mouse. *Genomics* 36:17–21.

Kim E, Rhee SD, Bae Y, Okumoto M, Yoon JB, Song CW, Yoon SK (2007). Genomic organization of the region spanning D14Mit262 and D14Mit86 on mouse chromosome 14 and exclusion of Adam28 and Adamdec1 as the cataract-causing gene, lr2. *Cytogenet Genome Res* 116:12–17.

Kimura S, Hirasawa T, Nakamura K, Yagi H, Matsumoto M, Makino S, Harada M (1998). Genetic control of peripheral T-cell deficiency in the cataract Shionogi (CTS) mouse linked to chromosome 7. *Immunogenetics* 47:278–280.

King G, Holmes R (1998). Human ocular aldehyde dehydrogenase isozymes: Distribution and properties as major soluble proteins in cornea and lens. *J Exp Zool* 282:12–17.

Kivlin JD, Lovrien E, George C *et al.* (1985). E. Linkage between cerulean cataract and PGP. *Cytogenet Cell Genet*:669.

Klopp N, Favor J, Loster J, Lutz RB, Neuhauser-Klaus A, Prescott A, Pretsch W, Quinlan RA, Sandilands A, Vrensen GF *et al.* (1998). Three murine cataract mutants (Cat2) are defective in different gamma-crystallin genes. *Genomics* 52:152–158.

Kmoch S, Brynda J, Asfaw B, Bezouska K, Novak P, Rezacova P, Ondrova L, Filipec M, Sedlacek J, Elleder M (2000). Link between a novel human gammaD-crystallin allele and a unique cataract phenotype explained by protein crystallography. *Hum Mol Genet* 9:1779–1786.

Kochetov GA (1982). Transketolase from yeast, rat liver, and pig liver. *Methods Enzymol* 90(Pt E):209–223.

Kohale K, Ingle A, Kelkar A, Parab P (2004). Dense cataract and microphthalmia – new spontaneous mutation in BALB/c mice. *Comp Med* 54:275–279.

Koniukhov BV, Kolesova NA (1976). Ultrastructural analysis of the effects of dominant cataract-Fr gene in mouse embryos. *Ontogenez* 7:271–277.

Konyukhov BV, Wachtel AW (1963). Electrophoretic studies of proteins in normal lenses and cataracts of inbred and mutant mice. *Exp Eye Res* 45:325–330.

Koteiche HA, Mchaourab HS (2006). Mechanism of a hereditary cataract phenotype. Mutations in alphaA-crystallin activate substrate binding. *J Biol Chem* 281:14273–14279.

Kratochvilova J (1978). Evaluation of pre-implantation loss in dominant-lethal assay in the mouse. *Mutat Res* 54:47–54.

Kratochvilova J (1981). Dominant cataract mutations detected in offspring of gamma-irradiated male mice. *J Hered* 72:302–307.

Kratochvilova J (1983). New cataract mutants. *Mouse News Lett* 68:86.

Kratochvilova J, Favor J (1988). Phenotypic characterization and genetic analysis of twenty dominant cataract mutations detected in offspring of irradiated male mice. *Genet Res* 52:125–134.

Kratochvilova J, Favor J (1992). Allelism tests of 15 dominant cataract mutations in mice. *Genet Res* 59:199–203.

Kuck JF (1990). Late onset hereditary cataract of the emory mouse. A model for human senile cataract. *Exp Eye Res* 50:659–664.

Kuck JF, Kuwabara T, Kuck KD (1981). The Emory mouse cataract: An animal model for human senile cataract. *Curr Eye Res* 1:643–649.

Land MF, Fernald RD (1992). The evolution of eyes. *Annu Rev Neurosci* 15:1–29.

Lassen N, Bateman JB, Estey T, Kuszak JR, Nees DW, Piatigorsky J, Duester G, Day BJ, Huang J, Hines LM *et al.* (2007). Multiple and additive functions of ALDH3A1 and ALDH1A1: Cataract phenotype and ocular oxidative damage in Aldh3a1(−/−)/Aldh1a1(−/−) KNOCK-OUT MICE. *J Biol Chem* 282: 25668–25676.

Lee AY, Chung SS (1999). Contributions of polyol pathway to oxidative stress in diabetic cataract. *FASEB J* 13:23–30.

Li B, Dai Q, Li L, Nair M, Mackay DR, Dai X (2002). Ovol2, a mammalian homolog of Drosophila ovo: Gene structure, chromosomal mapping, and aberrant expression in blind-sterile mice. *Genomics* 80:319–325.

Li HF, Petroll WM, Moller-Pedersen T, Maurer JK, Cavanagh HD, Jester JV (1997). Epithelial and corneal thickness measurements by *in vivo* confocal microscopy through focusing (CMTF). *Curr Eye Res* 16:214–221.

Liesegang TJ (2001). Herpes simplex virus epidemiology and ocular importance. *Cornea* 20:1–13.

Lipman RD, Muggleton-Harris AL, Aroian MA (1981). Phenotypic variation of cataractogenesis in the Nakano mouse. *Exp Eye Res* 32:255–256.

Litt M, Kramer P, LaMorticella DM, Murphey W, Lovrien EW, Weleber RG (1998). Autosomal dominant congenital cataract associated with a missense mutation in the human alpha crystallin gene CRYAA. *Hum Mol Genet* 7:471–474.

Liu CY, Birk DE, Hassell JR, Kane B, Kao WW (2003). Keratocan-deficient mice display alterations in corneal structure. *J Biol Chem* 278:21672–21677.

Liu H, Du X, Wang M, Huang Q, Ding L, McDonald HW, Yates JR, III, Beutler B, Horwitz J, Gong X (2005). Crystallin {gamma}B-I4F mutant protein binds to {alpha}-crystallin and affects lens transparency. *J Biol Chem* 280:25071–25078.

Loster J, Immervoll T, Schmitt-John T, Graw J (1997). Cat3vl and Cat3vao cataract mutations on mouse chromosome 10: Phenotypic characterization, linkage studies and analysis of candidate genes. *Mol Gen Genet* 257:97–102.

Lovicu FJ, Kao WW, Overbeek PA (1999). Ectopic gland induction by lens-specific expression of keratinocyte growth factor (FGF-7) in transgenic mice. *Mech Dev* 88:43–53.

Lu L, Reinach PS, Kao WW (2001). Corneal epithelial wound healing. *Exp Biol Med (Maywood)* 226:653–664.

Lundberg P, Welander P, Openshaw H, Nalbandian C, Edwards C, Moldawer L, Cantin E (2003). A locus on mouse chromosome 6 that determines resistance to herpes simplex virus also influences reactivation, while an unlinked locus augments resistance of female mice. *J Virol* 77:11661–11673.

Lutman FC, Neel JV (1945). Inherited cataract in the B geneology. Occurrence of diverse types of cataract in the descendants of one person. *Arch Ophthalmol*:341–357.

Lyon MF, Jarvis SE, Sayers I, Holmes RS (1981). Lens opacity: A new gene for congenital cataract on chromosome 10 of the mouse. *Genet Res* 38:337–341.

Lyon MF, Glenister PH, West JD (1984). Buised (Bru). *Mouse News Lett* 71:26.

Lyon MF, Bogani D, Boyd Y, Guillot P, Favor J (2000). Further genetic analysis of two autosomal dominant mouse eye defects, Ccw and Pax6(coop). *Mol Vis* 6:199–203.

Lyon MF, Jamieson RV, Perveen R, Glenister PH, Griffiths R, Boyd Y, Glimcher LH, Favor J, Munier FL, Black GC (2003). A dominant mutation within the DNA-binding domain of the bZIP transcription factor Maf causes murine cataract and results in selective alteration in D NA binding. *Hum Mol Genet* 12:585–594.

Mackay D, Ionides A, Berry V, Moore A, Bhattacharya S, Shiels A (1997). A new locus for dominant "zonular pulverulent" cataract, on chromosome 13. *Am J Hum Genet* 60:1474–1478.

Mackay DS, Andley UP, Shiels A (2003). Cell death triggered by a novel mutation in the alphaA-crystallin gene underlies autosomal dominant cataract linked to chromosome 21q. *Eur J Hum Genet* 11:784–793.

Maeda YY, Funata N, Takahama S, Sugata Y, Yonekawa H (2001). Two interactive genes responsible for a new inherited cataract (RCT) in the mouse. *Mamm Genome* 12:278–283.

Magon A, Erickson R (1983). Autosomal recessive mutation for cataracts and male sterility. *Mouse News Lett* 68:78.

Manzer R, Qamar L, Estey T, Pappa A, Petersen DR, Vasiliou V (2003). Molecular cloning and baculovirus expression of the rabbit corneal aldehyde dehydrogenase (ALDH1A1) cDNA. *DNA Cell Biol* 22:329–338.

Matsushima Y, Kamoto T, Iida F, Abujiang P, Honda Y, Hiai H (1996). Mapping of rupture of lens cataract (rlc) on mouse chromosome 14. *Genomics* 36:553–554.

Matsuzawa A, Wada E (1988). Retarded and distinct progress of lens opacification in congenic hereditary cataract mice, Balb/c-nct/nct. *Exp Eye Res* 47:705–711.

Maurice DM (1957). The structure and transparency of the cornea. *J Physiol* 136:263–286.

Metcalf JF, Michaelis BA (1984). Herpetic keratitis in inbred mice. *Invest Ophthalmol Vis Sci* 25:1222–1225.

Minagawa H, Hashimoto K, Yanagi Y (2004). Absence of tumour necrosis factor facilitates primary and recurrent herpes simplex virus-1 infections. *J Gen Virol* 85:343–347.

Mitchell J, Cenedella RJ (1995). Quantitation of ultraviolet light-absorbing fractions of the cornea. *Cornea* 14:266–272.

Nag N, Peterson K, Wyatt K, Hess S, Ray S, Favor J, Bogani D, Lyon M, Wistow G (2007). Endogenous retroviral insertion in Cryge in the mouse No3 cataract mutant. *Genomics* 89:512–520.

Narita M, Wang Y, Kita A, Omi N, Yamada Y, Hiai H (2002). Genetic analysis of Nakano Cataract and its modifier genes in mice. *Exp Eye Res* 75:745–751.

Nees DW, Wawrousek EF, Robison WG, Jr, Piatigorsky J (2002). Structurally normal corneas in aldehyde dehydrogenase 3a1-deficient mice. *Mol Cell Biol* 22:849–855.

Nees DW, Fariss RN, Piatigorsky J (2003). Serum albumin in mammalian cornea: Implications for clinical application. *Invest Ophthalmol Vis Sci* 44:3339–3345.

Nettleship E, Ogilvie FM (1906). A peculiar form of hereditary cataract. *Trans Ophthalmol Soc UK*:191–207.

Norman B, Davis J, Piatigorsky J (2004). Postnatal gene expression in the normal mouse cornea by SAGE. *Invest Ophthalmol Vis Sci* 45:429–440.

Norose K, Yano A, Zhang XM, Blankenhorn E, Heber-Katz E (2002). Mapping of genes involved in murine herpes simplex virus keratitis: Identification of genes and their modifiers. *J Virol* 76:3502–3510.

Oda SI, Watanabe T, Kondo K (1980). A new mutation, eye lens obsolescence, Elo, on chromosome 1 in the mouse. *Jpn J Genet* 55:71–75.

Ohotori H, Yoshida T, Inuta T (1968). Small eye and cataract, a new dominant mutation in the mouse. *Jikken Dobutsu* 17:91–96.

Okamura T, Miyoshi I, Takahashi K, Mototani Y, Ishigaki S, Kon Y, Kasai N (2003). Bilateral congenital cataracts result from a gain-of-function mutation in the gene for aquaporin-0 in mice. *Genomics* 81:361–368.

Paget OE (1953). Hereditary subcapsular cataract; a new dominant allelomorph in house mouse. *Z Indukt Abstamm Vererbungsl* 85:238–244.

Pande A, Pande J, Asherie N, Lomakin A, Ogun O, King JA, Lubsen NH, Walton D, Benedek GB (2000). Molecular basis of a progressive juvenile-onset hereditary cataract. *Proc Natl Acad Sci USA* 97:1993–1998.

Pande A, Pande J, Asherie N, Lomakin A, Ogun O, King J, Benedek GB (2001). Crystal cataracts: Human genetic cataract caused by protein crystallization. *Proc Natl Acad Sci USA* 98:6116–6120.

Pande A, Annunziata O, Asherie N, Ogun O, Benedek GB, Pande J (2005). Decrease in protein solubility and cataract formation caused by the Pro23 to Thr mutation in human gamma D-crystallin. *Biochemistry* 44:2491–2500.

Pappa A, Estey T, Manzer R, Brown D, Vasiliou V (2003). Human aldehyde dehydrogenase 3A1 (ALDH3A1): Biochemical characterization and immunohistochemical localization in the cornea. *Biochem J* 376:615–623.

Pearlman E, Lass JH, Bardenstein DS, Kopf M, Hazlett FE, Jr, Diaconu E, Kazura JW (1995). Interleukin 4 and T helper type 2 cells are required for development of experimental onchocercal keratitis (river blindness). *J Exp Med* 182:931–940.

Pellegata NS, Deguez-Lucena JL, Joensuu T, Lau S, Montgomery KT, Krahe R, Kivela T, Kucherlapati R, Forsius H, de la Chapelle A (2000). Mutations in KERA, encoding keratocan, cause cornea plana. *Nat Genet* 25:91–95.

Perveen R, Favor J, Jamieson RV, Ray DW, Black GC (2007). A heterozygous c-Maf transactivation domain mutation causes congenital cataract and enhances target gene activation. *Hum Mol Genet* 16:1030–1038.

Piatigorsky J (1998). Multifunctional lens crystallins and corneal enzymes. More than meets the eye. *Ann N Y Acad Sci* 842:7–15.

Piatigorsky J (2000). Review: A case for corneal crystallins. *J Ocul Pharmacol Ther* 16:173–180.

Piatigorsky J, Fukui HN, Kinoshita JH (1978). Differential metabolism and leakage of protein in an inherited cataract and a normal lens cultured with ouabain. *Nature* 274:558–562.

Pras E, Frydman M, Levy-Nissenbaum E, Bakhan T, Raz J, Assia EI, Goldman B, Pras E (2000). A nonsense mutation (W9X) in CRYAA causes autosomal recessive cataract in an inbred Jewish Persian family. *Invest Ophthalmol Vis Sci* 41:3511–3515.

Ramaesh K, Ramaesh T, Dutton GN, Dhillon B (2005). Evolving concepts on the pathogenic mechanisms of aniridia related keratopathy. *Int J Biochem Cell Biol* 37:547–557.

Ramaesh T, Collinson JM, Ramaesh K, Kaufman MH, West JD, Dhillon B (2003). Corneal abnormalities in Pax6+/− small eye mice mimic human aniridia-related keratopathy. *Invest Ophthalmol Vis Sci* 44:1871–1878.

Reddy VN, Giblin FJ, Lin LR, Dang L, Unakar NJ, Musch DC, Boyle DL, Takemoto LJ, Ho YS, Knoernschild T *et al.* (2001). Glutathione peroxidase-1 deficiency leads to increased nuclear light scattering, membrane damage, and cataract formation in gene-knockout mice. *Invest Ophthalmol Vis Sci* 42:3247–3255.

Reneker LW, Chen Q, Bloch A, Xie L, Schuster G, Overbeek PA (2004). Chick delta1-crystallin enhancer influences mouse alphaA-crystallin promoter activity in transgenic mice. *Invest Ophthalmol Vis Sci* 45:4083–4090.

Rieger DK, Reichenberger E, McLean W, Sidow A, Olsen BR (2001). A double-deletion mutation in the Pitx3 gene causes arrested lens development in aphakia mice. *Genomics* 72:61–72.

Roberts RC (1967). Small-eyes, a new dominant mutant in the mouse. *Genet Res* 9:121–122.

Rossant J (2003). A new allele at the Pax6 locus from the Center of Modeling Human Disease. MGI direct data submission.

Rudner XL, Kernacki KA, Barrett RP, Hazlett LD (2000). Prolonged elevation of IL-1 in Pseudomonas aeruginosa ocular infection regulates macrophage-inflammatory protein-2 production, polymorphonuclear neutrophil persistence, and corneal perforation. *J Immunol* 164:6576–6582.

Runge PE, Hawes NL, Heckenlively JR, Langley SH, Roderick TH (1992). Autosomal dominant mouse cataract (Lop-10). Consistent differences of expression in heterozygotes. *Invest Ophthalmol Vis Sci* 33:3202–3208.

Saika S, Saika S, Liu CY, Azhar M, Sanford LP, Doetschman T, Gendron RL, Kao CW, Kao WW (2001). TGFbeta2 in corneal morphogenesis during mouse embryonic development. *Dev Biol* 240:419–432.

Santhiya ST, Shyam MM, Rawlley D, Vijayalakshmi P, Namperumalsamy P, Gopinath PM, Loster J, Graw J (2002). Novel mutations in the gamma-crystallin genes cause autosomal dominant congenital cataracts. *J Med Genet* 39:352–358.

Sanyal S, Hawkins RK (1979). Dysgenetic lens (dyl) – a new gene in the mouse. *Invest Ophthalmol Vis Sci* 18:642–645.

Sax CM, Salamon C, Kays WT, Guo J, Yu FX, Cuthbertson RA, Piatigorsky J (1996). Transketolase is a major protein in the mouse cornea. *J Biol Chem* 271:33568–33574.

Sax CM, Kays WT, Salamon C, Chervenak MM, Xu YS, Piatigorsky J (2000). Transketolase gene expression in the cornea is influenced by environmental factors and developmentally controlled events. *Cornea* 19:833–841.

Scott MH, Hejtmancik JF, Wozencraft LA, Reuter LM, Parks MM, Kaiser-Kupfer MI (1994). Autosomal dominant congenital cataract. Interocular phenotypic variability. *Ophthalmology* 101:866–871.

Seeberger TM, Matsumoto Y, Alizadeh A, Fitzgerald PG, Clark JI (2004). Digital image capture and quantification of subtle lens opacities in rodents. *J Biomed Opt* 9:116–120.

Semina EV, Murray JC, Reiter R, Hrstka RF, Graw J (2000). Deletion in the promoter region and altered expression of Pitx3 homeobox gene in aphakia mice. *Hum Mol Genet* 9:1575–1585.

Shafie SM, Barria von-Bischhoffshausen FR, Bateman JB (2006). Autosomal dominant cataract: Intrafamilial phenotypic variability, interocular asymmetry, and variable progression in four Chilean families. *Am J Ophthalmol* 141:750–752.

Shang F, Nowell T, Jr, Gong X, Smith DE, Dallal GE, Khu P, Taylor A (2002). Sex-linked differences in cataract progression in Emory mice. *Exp Eye Res* 75:109–111.

Shiao T, Tran P, Siegel D, Lee J, Vasiliou V (1999). Four amino acid changes are associated with the Aldh3a1 locus polymorphism in mice which may be responsible for corneal sensitivity to ultraviolet light. *Pharmacogenetics* 9:145–153.

Shiels A, Bassnett S (1996). Mutations in the founder of the MIP gene family underlie cataract development in the mouse. *Nat Genet* 12:212–215.

Shiels A, Hejtmancik JF (2007). Genetic origins of cataract. *Arch Ophthalmol* 125:165–173.

Shiels A, King JM, Mackay DS, Bassnett S (2007). Refractive defects and cataracts in mice lacking lens intrinsic membrane protein-2. *Invest Ophthalmol Vis Sci* 48:500–508.

Sidjanin DJ, Grimes PA, Pretsch W, Neuhauser-Klaus A, Favor J, Stambolian DE (1997). Mapping of the autosomal dominant cataract mutation (Coc) on mouse chromosome 16. *Invest Ophthalmol Vis Sci* 38:2502–2507.

Sidjanin DJ, Parker-Wilson DM, Neuhauser-Klaus A, Pretsch W, Favor J, Deen PM, Ohtaka-Maruyama C, Lu Y, Bragin A, Skach WR et al. (2001). A 76-bp deletion in the Mip gene causes autosomal dominant cataract in Hfi mice. *Genomics* 74:313–319.

Silverman B, Alexander RJ, Henley WL (1981). Tissue and species specificity of BCP 54, the major soluble protein of bovine cornea. *Exp Eye Res* 33:19–29.

Simpanya MF, Ansari RR, Suh KI, Leverenz VR, Giblin FJ (2005). Aggregation of lens crystallins in an *in vivo* hyperbaric oxygen guinea pig model of nuclear cataract: Dynamic light-scattering and HPLC analysis. *Invest Ophthalmol Vis Sci* 46:4641–4651.

Smith RS, Hawes NL, Chang B, Roderick TH, Akeson EC, Heckenlively JR, Gong X, Wang X, Davisson MT (2000). Lop12, a mutation in mouse Crygd causing lens opacity similar to human Coppock cataract. *Genomics* 63:314–320.

Song CW, Okumoto M, Mori N, Yamate J, Sakuma S, Kim JS, Han SS, Hilgers J, Esaki K (1997). A new hereditary cataract mouse with lens rupture. *Lab Anim* 31:248–253.

Steele EC, Jr, Kerscher S, Lyon MF, Glenister PH, Favor J, Wang J, Church RL (1997). Identification of a mutation in the MP19 gene, Lim2, in the cataractous mouse mutant To3. *Mol Vis* 3:5.

Steele EC, Jr, Lyon MF, Favor J, Guillot PV, Boyd Y, Church RL (1998). A mutation in the connexin 50 (Cx50) gene is a candidate for the No2 mouse cataract. *Curr Eye Res* 17:883–889.

Steuhl KP, Doring G, Henni A, Thiel HJ, Botzenhart K (1987). Relevance of host-derived and bacterial factors in *Pseudomonas aeruginosa* corneal infections. *Invest Ophthalmol Vis Sci* 28:1559–1568.

Sun H, Ma Z, Li Y, Liu B, Li Z, Ding X, Gao Y, Ma W, Tang X, Li X *et al.* (2005). Gamma-S crystallin gene (CRYGS) mutation causes dominant progressive cortical cataract in humans. *J Med Genet* 42:706–710.

Sun L, Sun TT, Lavker RM (1999). Identification of a cytosolic NADP+-dependent isocitrate dehydrogenase that is preferentially expressed in bovine corneal epithelium. A corneal epithelial crystallin. *J Biol Chem* 274:17334–17341.

Swamynathan SK, Crawford MA, Robison WG, Jr, Kanungo J, Piatigorsky J (2003). Adaptive differences in the structure and macromolecular compositions of the air and water corneas of the "four-eyed" fish (Anableps anableps). *FASEB J* 17: 1996–2005.

Swamynathan SK, Katz JP, Kaestner KH, Ashery-Padan R, Crawford MA, Piatigorsky J (2007). Conditional deletion of the mouse Klf4 gene results in corneal epithelial fragility, stromal edema, and loss of conjunctival goblet cells. *Mol Cell Biol* 27:182–194.

Takehana M (1990). Hereditary cataract of the Nakano mouse. *Exp Eye Res* 50:671–676.

Thaung C, West K, Clark BJ et al. (2002). Novel ENU-induced eye mutations in the mouse: Models for human eye disease. *Hum Mol Genet* 11:755–767.

Theiler K, Varnum DS, Stevens LC (1978). Development of Dickie's small eye, a mutation in the house mouse. *Anat Embryol (Berl)* 155:81–86.

Theiler K, Varnum DS, Stevens LC (1980). Development of Dickie's small eye: An early lethal mutation in the house mouse. *Anat Embryol (Berl)* 161:115–120.

Theiler K, Varnum DS (1981). Development of coloboma (Cm/+), a mutation with anterior lens adhesion. *Anat Embryol (Berl)* 162:121–126.

Thiagarajah JR, Verkman AS (2002). Aquaporin deletion in mice reduces corneal water permeability and delays restoration of transparency after swelling. *J Biol Chem* 277:19139–19144.

Tissot RG, Cohen C (1972). A new congenital cataract in the mouse. *J Hered* 63:197–201.

Trigg M (1972). New Mutations-provisional gene symbol act (adult cataract) and provisional gene symbol fp (flopsy). *Mouse News Lett* 47:19.

Tulloh CG (1955). Hereditary posterior polar cataract with report of a pedigree. *Br J Ophthalmol* 39:374–379.

Uma L, Hariharan J, Sharma Y, Balasubramanian D (1996a). Corneal aldehyde dehydrogenase displays antioxidant properties. *Exp Eye Res* 63:117–120.

Uma L, Hariharan J, Sharma Y, Balasubramanian D (1996b). Effect of UVB radiation on corneal aldehyde dehydrogenase. *Curr Eye Res* 15:685–690.

Van AT, Schlotzer-Schrehardt U, McKie L, Brownstein DG, Lee AW, Cross SH, Sado Y, Mullins JJ, Poschl E, Jackson IJ (2005). Dominant mutations of Col4a1 result in basement membrane defects which lead to anterior segment dysgenesis and glomerulopathy. *Hum Mol Genet* 14:3161–3168.

Vanita A, Singh JR, Sarhadi VK, Singh D, Reis A, Rueschendorf F, Becker-Follmann J, Jung M, Sperling K (2001). A novel form of "central pouchlike" cataract, with sutural opacities, maps to chromosome 15q21-22. *Am J Hum Genet* 68: 509–514.

Varnum DS (1983). Blind-sterile: A new mutation on chromosome 2 of the house mouse. *J Hered* 74:206–207.

Varnum DS, Stevens LC (1968). Aphakia, a new mutation in the mouse. *J Hered* 59:147–150.

Varnum DS, Stevens LC (1974). Rachiterata: A new skeletal mutation on chromosome 2 of the mouse. *J Hered* 65:91–93.

Varnum DS, Stevens LC (1975). Aphakia linked with brachymorphic. *Mouse News Lett*:35.

Vasiliou V, Pappa A, Estey T (2004). Role of human aldehyde dehydrogenases in endobiotic and xenobiotic metabolism. *Drug Metab Rev* 36:279–299.

Verkman AS (2002). Aquaporin water channels and endothelial cell function. *J Anat* 200:617–627.

Verrusio AC, Fraser FC (1966). Identity of mutant genes "Shrivelled" and cataracta congenita subcapsularis in the mouse. *Genet Res* 8:377–378.

Wakefield L, Long H, Lack N, Sim E (2007). Ocular defects associated with a null mutation in the mouse arylamine N-acetyltransferase 2 gene. *Mamm Genome* 18:270–276.

Wang K, Cheng C, Li L, Liu H, Huang Q, Xia CH, Yao K, Sun P, Horwitz J, Gong X (2007). GammaD-crystallin associated protein aggregation and lens fiber cell denucleation. *Invest Ophthalmol Vis Sci* 48:3719–3728.

West JD, Fisher G, Loutit JF, Marshall MJ, Nisbet NW, Perry VH (1985). A new allele of microphthalmia induced in the mouse: Microphthalmia – defective iris (midi). *Genet Res* 46:309–324.

White TW (2002). Unique and redundant connexin contributions to lens development. *Science* 295:319–320.

White TW, Goodenough DA, Paul DL (1998). Targeted ablation of connexin50 in mice results in microphthalmia and zonular pulverulent cataracts. *J Cell Biol* 143:815–825.

White TW, Sellitto C, Paul DL, Goodenough DA (2001). Prenatal lens development in connexin43 and connexin50 double knockout mice. *Invest Ophthalmol Vis Sci* 42:2916–2923.

Wistow G, Kim H (1991). Lens protein expression in mammals: Taxon-specificity and the recruitment of crystallins. *J Mol Evol* 32:262–269.

Wistow GJ, Piatigorsky J (1988). Lens crystallins: The evolution and expression of proteins for a highly specialized tissue. *Annu Rev Biochem* 57:479–504.

World Health Organization (1995). Onchocerciasis and its control. *World Health Organ Tech Rep Ser* 852:1–103.

Xia CH, Cheung D, DeRosa AM, Chang B, Lo WK, White TW, Gong X (2006a). Knock-in of alpha3 connexin prevents severe cataracts caused by an alpha8 point mutation. *J Cell Sci* 119:2138–2144.

Xia CH, Liu H, Chang B, Cheng C, Cheung D, Wang M, Huang Q, Horwitz J, Gong X (2006b). Arginine 54 and Tyrosine 118 residues of {alpha}A-crystallin are crucial for lens formation and transparency. *Invest Ophthalmol Vis Sci* 47:3004–3010.

Xia CH, Liu H, Cheung D, Cheng C, Wang E, Du X, Beutler B, Lo WK, Gong X (2006c). Diverse gap junctions modulate distinct mechanisms for fiber cell formation during lens development and cataractogenesis. *Development* 133:2033–2040.

Xie L, Chen H, Overbeek PA, Reneker LW (2007). Elevated insulin signaling disrupts the growth and differentiation pattern of the mouse lens. *Mol Vis* 13:397–407.

Xu ZP, Wawrousek EF, Piatigorsky J (2002). Transketolase haploinsufficiency reduces adipose tissue and female fertility in mice. *Mol Cell Biol* 22:6142–6147.

Zetterberg M, Petersen A, Sjostrand J, Karlsson J (2003). Proteasome activity in human lens nuclei and correlation with age, gender and severity of cataract. *Curr Eye Res* 27:45–53.

Zhao H, Yang Y, Rizo CM, Overbeek PA, Robinson ML (2004). Insertion of a Pax6 consensus binding site into the alphaA-crystallin promoter acts as a lens epithelial cell enhancer in transgenic mice. *Invest Ophthalmol Vis Sci* 45:1930–1939.

Zhou E, Grimes P, Favor J, Koeberlein B, Pretsch W, Neuhauser-Klaus A, Sidjanin D, Stambolian D (1997). Genetic mapping of a mouse ocular malformation locus, Tcm, to chromosome 4. *Mamm Genome* 8:178–181.

Zwaan J, Williams RM (1968). Morphogenesis of the eye lens in a mouse strain with hereditary cataracts. *J Exp Zool* 169:407–421.

Zwaan J, Williams RM (1969). Cataracts and abnormal proliferation of the lens epithelium in mice carrying the CatFr gene. *Exp Eye Res* 8:161–167.

CHAPTER

12

Deciphering Irradiance Detection in the Mammalian Retina

Robert J. Lucas[1], *Daniela Vallone*[2], *Nicholas S. Foulkes*[2]

[1]Faculty of Life Sciences, University of Manchester, Manchester M13 9PT, UK

[2]Institute of Toxicology and Genetics, Forschungszentrum Karlsruhe, Hermann-von-Helmholtz Platz 1, Eggenstein-Leopoldshafen 76344, Germany

OUTLINE

INTRODUCTION

Instead of focusing on a single animal model and its contribution to eye research, the theme of this chapter is a complementary collection of animal models. Based on 150 years of research, until relatively recently almost everyone agreed that the eye's light detecting function relied exclusively upon photoreceptive rods and cones in the outer layers of the retina. These provide a photoreceptive surface while the inner layers of retinal cells contribute to the first stages of signal processing. However, as the result of some very active research during the past decade, it has become clear that there is an additional class of photoreceptive cells located in the inner retina (intrinsically photosensitive retinal ganglion cells or ipRGCs) that rely on a newly discovered photoreceptive molecule (melanopsin). ipRGCs measure irradiance (the gross amount of light in the environment) and contribute to the well known ability of mammals to respond to light exposure by resetting their circadian clock, down-regulating the production of the hormone melatonin, regulating pupil size and controlling aspects of their behavior.

Apart from the obvious biological interest, the study of irradiance detection is an excellent illustration of how a set of animal models can enable the cellular and molecular basis of eye function to be deciphered. More specifically, mutant mouse models have allowed researchers to genetically eliminate subsets of retinal cells or disrupt key elements of phototransduction pathways and thus probe the retina for the origin of irradiance detection. Furthermore, comparative studies with other non-mammalian vertebrates as well as invertebrates have proven particularly valuable.

In order to explain the logic behind the various studies, it is important to briefly introduce the fields that initially lead to the interest in irradiance detection.

Circadian Clock Entrainment

The key adaptation enabling animals to anticipate day–night changes in the environment is the circadian clock. This endogenous timing mechanism is able to generate rhythms independently of the environmental day–night cycle, and in turn regulates most behavioral and physiological processes (Pittendrigh, 1993). The molecular components of the core clock mechanism are expressed in nearly all cell types and so most tissues possess their own endogenous clocks. However, in vertebrates, one clock located in the hypothalamic suprachiasmatic nucleus (SCN) appears to play a privileged role in coordinating the phase of the other multiple tissue clocks or peripheral clocks (Schibler and Sassone-Corsi, 2002).

Left to their own devices, circadian clocks have periods slightly shorter or longer than 24h (hence the term circa-diem; approximately 1 day) and, as a result, drift out of synchrony with local time (free run). To stop this happening they rely upon daily resetting by signals that are diagnostic of the time of day (Roenneberg *et al.*, 2003). These environmental signals are termed as zeitgebers (time-givers) and are relayed to the clock mechanism via the so-called clock "input" pathways (Menaker *et al.*, 1978). The most important zeitgeber is light. When light falls at a time when the clock is expecting darkness (subjective night), it causes an adjustment in the phase of the clock (a phase shift). The nature of the phase shift depends on whether the light is experienced in the early part of the subjective night, in which case it induces a delay, or the late part of the subjective night when it produces an advance. In this way, the shift drives the clock toward a phase when it could expect to experience light (subjective day). The amplitude of the phase shift is determined by the intensity, and duration of any light exposure, as well as its spectral composition (Roenneberg and Foster, 1997; Roenneberg *et al.*, 2003). In mammals, enucleation experiments have demonstrated that the circadian clock is totally dependent on the eyes for photoentrainment (Foster, 1998). Complete loss of the eye structures results in a free-running clock.

The Pupillary Light Reflex (PLR)

The association between pupil size and environmental light intensity is a mechanism for regulating the amount of light reaching the retina (Kardon, 1995; Lucas *et al.*, 2001). In mammals, it is achieved by a well-known reflex pathway originating in a retinal projection to the olivary pretectal nucleus (OPN), in the pretectal region of the midbrain. The OPN innervates the Edinger-Westphal nucleus whose axons run along both the left and right oculomotor nerves. The oculomotor nerve axons subsequently form synapses with ciliary ganglion neurons whose axons innervate the constrictor muscle of the iris. This reflex is commonly used as a diagnostic tool in medicine, providing a rapid way to gauge the function of the brain stem.

Masking

In rodents such as mice, the period of highest locomotor activity coincides with the night. The circadian clock plays a key role in timing this nocturnal activity period. However, in addition, exposure to light has a direct inhibitory effect on locomotion. This acute behavioral response to light has been termed as "Masking" and serves to complement the clock function to ensure that locomotor activity is restricted to the night (Aschoff, 1960; Mrosovsky, 1999).

Melatonin Suppression

One of the key endocrine output pathways of the circadian clock is the nocturnal production of the hormone melatonin (Arendt, 1995; Korf *et al.*, 1998). Circulating melatonin is synthesized in the pineal gland via a multistep enzymatic pathway. Norepinephrine release from sympathetic nerve fibers in the pineal gland and subsequent activation of the cyclic adenosine monophosphate (cAMP) signaling pathway via the β-adrenergic receptor represent the key signals driving melatonin synthesis. cAMP in turn upregulates protein kinase A activity that induces arylalkylamine-*N*-acetyl transferase (AA-NAT) activity acting at the transcriptional or posttranslational levels depending on the species. AA-NAT catalyzes the penultimate and rate-limiting step in melatonin synthesis and serves as a central control point. The sympathetic innervation of the pineal is controlled indirectly by the SCN and so represents the link between the clock and melatonin levels. In addition to this clock regulation, exposure to light during the night period leads to a characteristic rapid shut down of melatonin synthesis (Klein and Weller, 1972).

In mammals, although pinealocytes share some similarities with photoreceptor cells, the adult pineal gland is not able to respond directly to light. Instead the retina appears crucial for light to block the pineal gland synthesis of melatonin (Klein and Weller, 1972; Lucas and Foster, 1999). Photic information reaches the pineal via a multisynaptic pathway that originates in the retina and passes through the SCN region of the hypothalamus (Arendt, 1995). Many of the more commonly used inbred laboratory strains of mice are genetically deficient in melatonin synthesis (Roseboom *et al.*, 1998). In these cases, changes in expression of AA-NAT have been used as an assay of photic regulation of the pineal.

IRRADIANCE DETECTION IN OTHER VERTEBRATES

Early attempts to trace the origins of these irradiance-dependent responses confirmed that, in mammals, the integrity of the eye is crucial (Foster, 1998). It was therefore assumed that rod and/or cones acted as photoreceptors for these pathways. However, critical evidence that it did not have to be this way came from non-mammalian vertebrates, which have several extraretinal photoreceptive structures. In fish, amphibia, birds and reptiles, the pineal (and, where present, parapineal complex and parietal eye) and also cells lining the third ventricle of the brain represent additional sites of photoreception (Fig. 12.1) (Menaker *et al.*, 1997; Shand and Foster, 1999; Bertolucci and Foa, 2004). Many ablation experiments have confirmed that

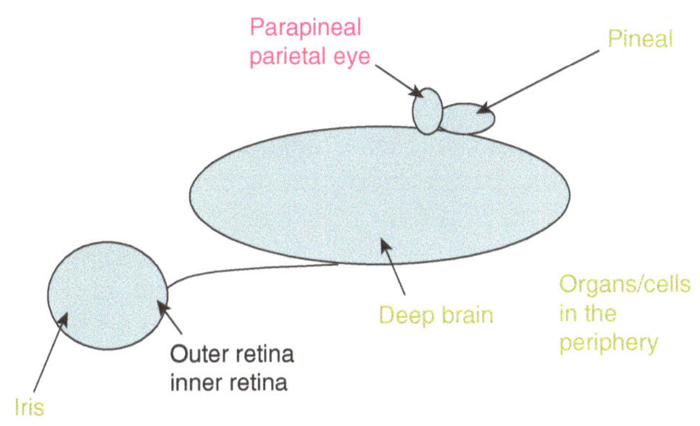

FIGURE 12.1 A schematic view of the vertebrate eye and brain showing the location of photoreceptors responsible for irradiance detection. In mammals these are restricted to the inner and outer retina, while in non-mammalian vertebrates photoreception is also found in structures that are not photosensitive (green), or indeed even absent (purple), in mammals.

these extraretinal photoreceptive sites in non-mammalian vertebrates do not contribute to image forming photoreception – but are instead involved in irradiance detection. The widespread presence of dedicated irradiance detection photoreceptors in non-mammalian vertebrates implies a distinct evolutionary advantage in separating this function from image forming photoreception. In which case, are mammals not subject to this selection pressure, or could they perhaps also have separate irradiance detectors?

RODLESS CONELESS MICE

Among the first direct approaches to determining the necessity of rods and/or cones for mammalian irradiance detection were studies of retinally degenerate mouse models. Several strains were used in these initial experiments:

1. Mice carrying the *rd* mutation. This affects the *Pde6b* gene that encoding the β-subunit of the rod-specific cGMP phosphodiesterase (Bowes *et al.*, 1990; Pittler and Baehr, 1991). The *rd* mutation consists of a murine viral insertion and a second nonsense mutation in exon 7 of *Pde6b*. In homozygous mutants, there is a constitutively high level of cGMP in the rod photoreceptors and so these cells are unable to respond to stimulation by light. This leads to an attrition of the rods which subsequently also affects the cones (Carter-Dawson *et al.*, 1978). By the time the mice reach 85 to 90-days-old, the cell bodies of the rods are completely absent and the number of cones is reduced by around 50%.
2. The *rds* mutation is an insertion in the peripherin gene that encodes a key structural component of photoreceptor cells' outer segments (Travis *et al.*, 1989, 1991; Connell *et al.*, 1991). In homozygous *rds* mutant mice, outer segments of photoreceptors are never formed and by 90-days-old the total number of cells in the outer retinal layer is reduced by around 50% (Sanyal *et al.*, 1980).
3. The transgenic line *rdta* expresses an attenuated form of the diptheria toxin A chain-encoding gene under the control of the rhodopsin promoter (McCall *et al.*, 1996). From the time that rhodopsin starts to be expressed during postnatal development (P5), there is a progressive cell loss from the outer nuclear layer of the retina. By P17, rods are completely absent, with the remaining cell bodies corresponding to cones.
4. The *cl* transgenic line, like the *rdta* transgenic expresses the attenuated diptheria toxin A-chain

but here it is expressed under the control of the human red cone opsin promoter (Wang *et al.*, 1992; Soucy *et al.*, 1998). This results in a substantial loss of green cones in the mouse retina and, although a significant proportion of UV cones do survive, it is possible to exclude their contribution to evoked-responses by using long wavelength stimuli (Lucas *et al.*, 1999).

rd/rd, *rds/rds*, *rdta* and *cl* mice all showed circadian clock entrainment and unimpaired sensitivity for circadian phase-shifting responses (Foster *et al.*, 1991; Argamaso *et al.*, 1995; Freedman *et al.*, 1999; Lupi *et al.*, 1999). Similarly, when studied in the C3H/He genetic background that is capable of melatonin production, *rd/rd*, *rds/rds* and *cl* mice all showed photic suppression of pineal melatonin (Lucas and Foster, 1999; Lucas *et al.*, 1999). This body of data suggested that irradiance detection survives substantial loss of rod or cone photoreception. However, the implications of this work were ambiguous. Could the small numbers of surviving rods or cones in these degeneration mutants still be sufficient to fulfill irradiance detection tasks, or was there a non-rod non-cone photoreceptor responsible for driving these responses? The obvious solution to answering this question was to generate rodless plus coneless double mutants. To this end, the *cl* transgene was introduced into the *rd/rd* mouse in the C3H/He genetic background to generate *rd/rd cl* mice that completely lack photoreceptive cells of the outer nuclear layer (Lucas *et al.*, 1999). At the same time, double transgenic *rdta/cl* mice were also generated to achieve a similar retinal phenotype (Freedman *et al.*, 1999). Both rodless + coneless genotypes retained circadian entrainment and *rd/rd cl* animals also exhibited normal melatonin suppression (Freedman *et al.*, 1999; Lucas *et al.*, 1999). *rd/rd cl* animals have since been shown to also retain masking and a pupil light reflex (Lucas *et al.*, 2001; Mrosovsky *et al.*, 2001).

The photosensitivity of rodless + coneless mice implied the presence of non-rod, non-cone photoreceptors, but perhaps the first positive evidence for their existence came from the action spectra defined for the PLR (pupillary light reflex) in the *rd/rd cl* animals, and phase-shifting in some *rd/rd* animals (Yoshimura and Ebihara, 1996; Lucas *et al.*, 2001). These revealed that pupil constriction and photoentrainment in these mice originates with light absorption by a single opsin/vitamin A based photopigment whose spectral sensitivity profile (peak sensitivity around 479nm) is distinct from those of the known murine rod and cone opsins (Lucas *et al.*, 2001).

ipRGCs

If rods and cones are not required for irradiance detection, then which retinal cells perform this function? The first answer to this question came from retrograde labeling experiments in rats. The aim of these experiments was to inject a fluorescent tracer into the SCN and then exploit the subsequent retrograde transport of tracer to the retina to study the specific population of retinal ganglion cells responsible for photoentrainment of the clock. Amazingly, a series of single cell recordings revealed light-induced depolarization of these labeled cells that was retained even when they were pharmacologically or physically isolated from the rest of the retina (Berson *et al.*, 2002). It was these experiments that first revealed the presence of a small number of directly photosensitive ganglion cells and led to the term ipRGCs being used to describe them. Since then, others have adopted similar approaches in rats and primates, and also identified ipRGCs on the basis of calcium imaging/multielectrode plate recordings in *rd/rd cl*, *rd/rd* and neonatal mouse retinas (Sekaran *et al.*, 2003, 2005; Tu *et al.*, 2005); of autofluorescence in the primate retina (Dacey *et al.*, 2005); and using viral tract tracing (Viney *et al.*, 2007). From these experiments a view of the physiological features of ipRGCs is emerging. Their spectral sensitivity matches that originally described for the *rd/rd cl* pupil light response, they require relatively high irradiances to be activated and are slow to react to changes in illumination. On the other hand, they respond continuously to extended stimuli without obvious adaptation.

XENOPUS MELANOPHORES: THE DISCOVERY OF MELANOPSIN

How do ipRGCs attain their photosensitivity? The expression of rod and cone opsins was not detected in these cells. Furthermore, the action spectrum for the PLR in *rd/rd cl* mutant mice suggested the involvement of an opsin/vitamin A-based photopigment separate from those in mouse rods and cones (Lucas *et al.*, 2001). Therefore the search was on for additional opsins. In several non-mammalian vertebrates, non-rod or -cone opsins have shown to be expressed in extraretinal photoreceptor sites (Foster and Hankins, 2002). Indeed the first clue as to what the ipRGC opsin might be, came from a study of *Xenopus laevis* dermal melanophores by Provencio *et al.* (1998). Even when these dermal pigment cells are cultured *in vitro*, direct exposure to light causes melanosomes to migrate to the cell periphery. The clear prediction was that these cells must express their own photopigment leading Provencio *et al.* to screen a melanophore cDNA library for opsin-like cDNAs. The result of this work was the cloning of melanopsin, an opsin that actually shared slightly more homology with invertebrate opsins than vertebrate rod or cone opsins (Provencio *et al.*, 1998). The authors discovered that melanopsin was also expressed in regions of the brain predicted to contain deep brain photoreceptors, as well as in the iris which is directly photosensitive in *Xenopus* and most interestingly, in cells within the retina (Provencio *et al.*, 1998). Melanopsin positive cells were localized in the outermost lamina of the inner nuclear layer, the site of horizontal cells. Subsequently, melanopsin homologs were cloned in other mammals and were found to be expressed in a subset of cells within the ganglion and amacrine cell layers of the primate and murine retinas (Provencio *et al.*, 2000). Importantly, it was not expressed in the rod and cone photoreceptors. This lead to speculation that melanopsin might function as the photopigment of ipRGCs.

The next step was to test whether there was any link between melanopsin expression and the ipRGC cells. By retrograde labeling experiments, it was shown that the majority of ipRGCs that project to the SCN (predicted to be around 1–2% of the RGC population) are also melanopsin positive and that a majority of melanopsin positive RGCs also project to the SCN (Gooley *et al.*, 2001; Hannibal *et al.*, 2002). More detailed studies showed that melanopsin immunoreactivity was present in cell bodies, dendrites and axons, predominantly at the cell surface (Hattar *et al.*, 2002). These melanopsin-expressing cells had properties that lend them very well to the task of detecting light intensity rather than image formation. For a start they have very long melanopsin-immunoreactive dendrites that form a photoreceptive web through the retina. Thus single cells can integrate light signals over a relatively large area of the retina when compared with rods and cones that respond to light restricted to relatively small surface areas.

In order to gain more precise information on the axonal projections of these cells, a mouse model was generated where a hybrid *tau-lacZ* gene was inserted into the melanopsin locus (Hattar *et al.*, 2002). This gene encodes the bacterial β-galactosidase enzyme fused to the signal sequence of the actin-associated protein, tau. β-galactosidase activity is thereby transported down the axons of melanopsin expressing cells toward the presynaptic terminals and can be visualized by *in situ* X-gal labeling (Fig. 12.2) (Mombaerts *et al.*, 1996). Mice heterozygous for this targeted construct revealed that axons from melanopsin positive RGCs project not only to the SCN, but also to the intergeniculate leaflet (IGL), the olivary pretectal nucleus (OPN) and to a much lesser extent the ventral lateral geniculate. Neurons in the IGL and OPN had been shown previously to encode ambient lighting levels. Furthermore, the IGL, like the SCN have been implicated in photoentrainment of the circadian clock while the OPN is a key regulatory site of the pupillary response (Hattar *et al.*, 2002). Together, these results pointed strongly to melanopsin positive ipRGCs being the cell type responsible for irradiance detection.

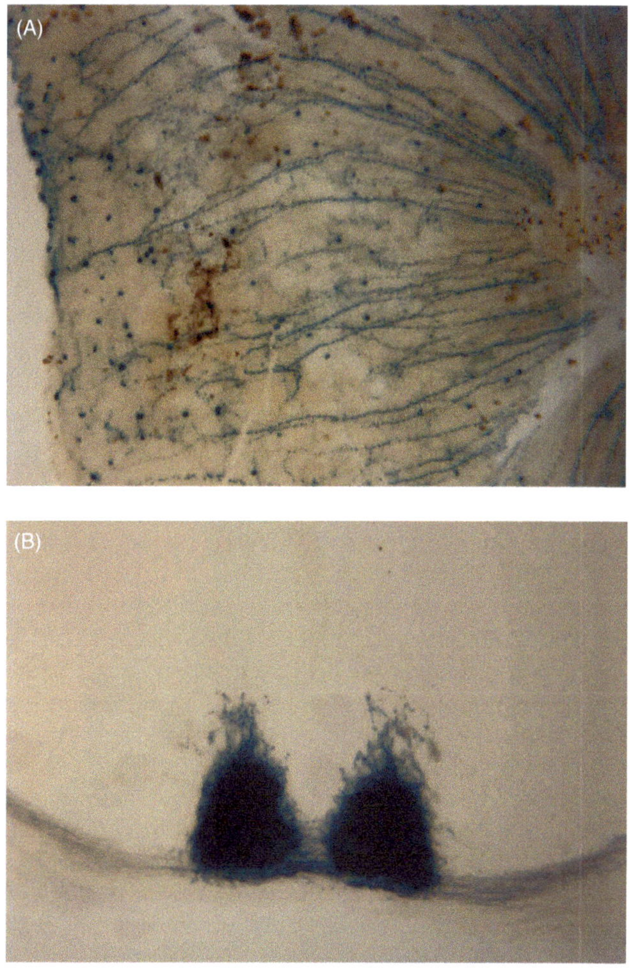

FIGURE 12.2 ipRGCs in the mouse retina (A), and their axonal projections to the suprachiasmatic nuclei (B) are revealed by β-galactosidase staining in the *Opn4 tau-lacZ* knockin mouse. Blue cell bodies and in some cases their associated dendritic arborization can be identified sparsely distributed across the *en face* retina. Blue stained axons converge on the optic nerve head. Most of these blue fibers terminate in the bilateral suprachiasmatic nuclei shown here (B) in a coronal section of the mouse brain at the level of the optic chiasm. Courtesy of Emma Tarttelin.

MELANOPSIN-KNOCKOUT MICE

The important direct test of the contribution of melanopsin to irradiance detection came with the generation of mice models carrying targeted disruption of the melanopsin locus (*Opn4*). Three different lines were generated by (i) the insertion of a neomycin resistance cassette into exon 1 (Panda *et al.*, 2002); (ii) the insertion of an *IRES-lacZ-Neo* casstette into exon 3 (Ruby *et al.*, 2002) and (iii) the insertion of a *tau-lacZ* expression cassette (previously mentioned above) (Hattar *et al.*, 2002). In all cases, the viability and development of homozygous mutants was normal. Furthermore, the *tau-lacZ*-knockin mouse revealed that the morphology, number and projections of ipRGCs are not altered following melanopsin loss. However electrophysiological recordings showed that ipRGCs were no longer intrinsically light responsive in knockout mice confirming that melanopsin constitutes an essential component of their photoreceptive machinery (Lucas *et al.*, 2003).

However, it came as a big surprise that all these knockouts retained irradiance-dependent responses. Thus they entrained their locomotor activity rhythms to LD cycles, showed masking behavior (Panda *et al.*, 2002), exhibited light induced *c-fos* expression in the SCN and phase-shifts of the circadian clock (Ruby *et al.*, 2002) and retained a pupil light reflex (Lucas *et al.*, 2003). Several of these responses did show impairments compared with wild types, but none was sufficiently dramatic to render them non-functional.

ROLE OF RODS AND CONES IN IRRADIANCE DETECTION

As melanopsin loss abolishes ipRGC photosensitivity, the light responses of melanopsin-knockout mice must originate with some other photoreceptors. Could these be the rods and/or cones, or was there yet another new photoreceptor waiting to be discovered? In order to address this question, transgenic mice lacking rods, cones and melanopsin were generated by two groups. In the simpler strategy, Panda *et al.* (2002) bred the *rd* mutation into their melanopsin-knockout mice. Although some cones would be expected to survive in these *Opn4*$^{-/-}$; *rd/rd* mice the researchers hoped that their residual activity would be insufficient to drive irradiance responses. Indeed, these animals lacked circadian photoentrainment, pupillary light responses, masking and light induced reduction of AA-NAT transcript levels (Panda *et al.*, 2003). In parallel experiments, Hattar *et al.* (2003) generated a triple-knockout mouse in which the melanopsin-null mutant was combined with knockouts of the rod transducin α-subunit (*Gnat1*) and the cone cyclic GMP gated channel A-subunit 3 (*Cnga3*) genes which encode critical elements of the rod and cone phototransduction cascades (Hattar *et al.*, 2003). These *Opn4*$^{-/-}$; *Gnat1*$^{-/-}$; *Cnga3*$^{-/-}$ mice lack phototransduction but their rods, cones and ipRGCs remain physically intact. In these triple mutant mice, circadian clock photic entrainment, the PLR and the masking response to light were all absent.

Thus, by studying increasingly sophisticated retinally degenerate and knockout mice, it became clear that irradiance detection can survive loss of rods, cones or melanopsin, but not all three. This sets the challenge to determine the exact significance of each photoreceptor class for irradiance detection and how their signals are integrated in efferent pathways. Addressing the former issue is a substantial undertaking and will probably require a whole new array of transgenic models. Thus far, work on the PLR of *rd/rd cl* and *Opn4*$^{-/-}$ mice suggests that there is a division of labor between the three photoreceptor classes based upon their unique sensory capabilities, with rather little overlap in their contributions to encoding irradiance (Lucas *et al.*, 2003). With regards to the mechanism of signal integration, there is anatomical and physiological evidence for convergence of rod, cone and ipRGC output pathways (Sollars *et al.*, 2003; Dacey *et al.*, 2005). It has been shown that stimulation of short-wavelength cones seems to attenuate the response of ipRGCs while rods and medium and long-wavelength cones provide excitatory input for ipRGCs. Furthermore, in the absence of melanopsin, visual photoreceptors can clearly contribute to irradiance detection tasks (Hattar *et al.*, 2003; Panda *et al.*, 2003).

IS MELANOPSIN A PHOTOPIGMENT?

One of the clear consequences of the melanopsin-knockout is the loss of photosensitivity in the ipRGCs. What is the basis for this phenotype? Could it be that melanopsin itself functions as a photopigment or alternatively, does it perform a supporting function for an as yet unknown photoreceptive opsin (Bellingham and Foster, 2002; Bellingham *et al.*, 2002). To address this issue and to explore how melanopsin might function, several studies reported the heterologous expression of melanopsin in non-photoreceptive cells: in microinjected *Xenopus laevis* oocytes, as well as the HEK293 (Fig. 12.3) and Neuro2a cell lines (Melyan

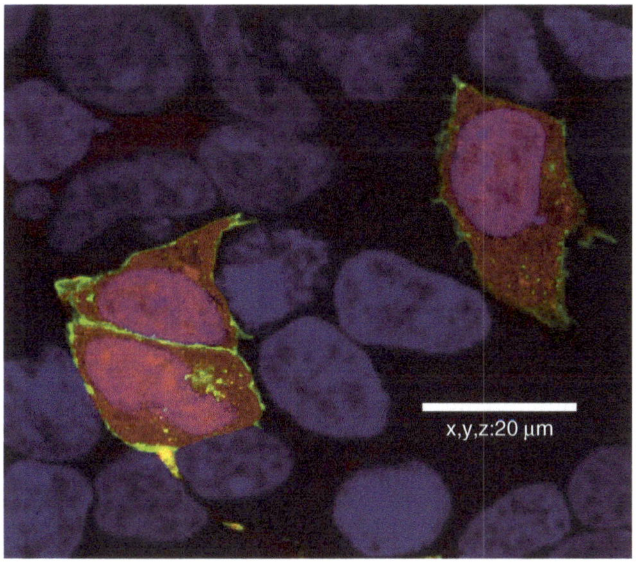

FIGURE 12.3 Confocal image of HEK293 cells (nuclei stained with DAPI), transiently transfected with a bicistronic vector driving expression of a human melanopsin, eGFP fusion construct and a cytoplasmic RFP. As expected for an opsin protein, melanopsin is located primarily in the plasma membrane. Courtesy of Jim Bellingham

et al., 2005; Panda *et al.*, 2005; Qiu *et al.*, 2005). In all cases, melanopsin expression rendered these cells photoresponsive, with light triggering membrane depolarization and increases in intracellular levels of calcium, confirming that melanopsin is capable of acting as a photopigment. Opsins are G-protein-coupled receptors, and the researchers were able to show that in all the three cell types, melanopsin activation was based upon G-protein signaling cascades. Two of the studies were also able to confirm that melanopsin's activity was reliant on the presence of retinaldehyde (Melyan *et al.*, 2005; Panda *et al.*, 2005). This is entirely expected, as all known opsin photopigments use 11-*cis*-retinaldehyde or closely related isomers as a chromophore. However, it dose pose a particular problem for melanopsin. In opsins, the first step in photoreception is the photoisomerization of *cis*-retinaldehyde isoforms to all-*trans*-retinaldehyde, meaning that they require a reliable source of *cis*-retinaldehyde to retain photosensitivity. Rod and cone opsins release their bleached all-*trans*-retinaldehyde chromophore, which is subsequently converted back to *cis*-retinaldehyde enzymatically via the so-called "visual cycle" located primarily in the retinal pigment epithelium (RPE) (Fig. 12.4). Melanopsin's location in the inner retina makes it ill placed to take advantage of this source of chromophore. Interestingly, invertebrate opsins use an alternative strategy for obtaining *cis*-retinaldehyde

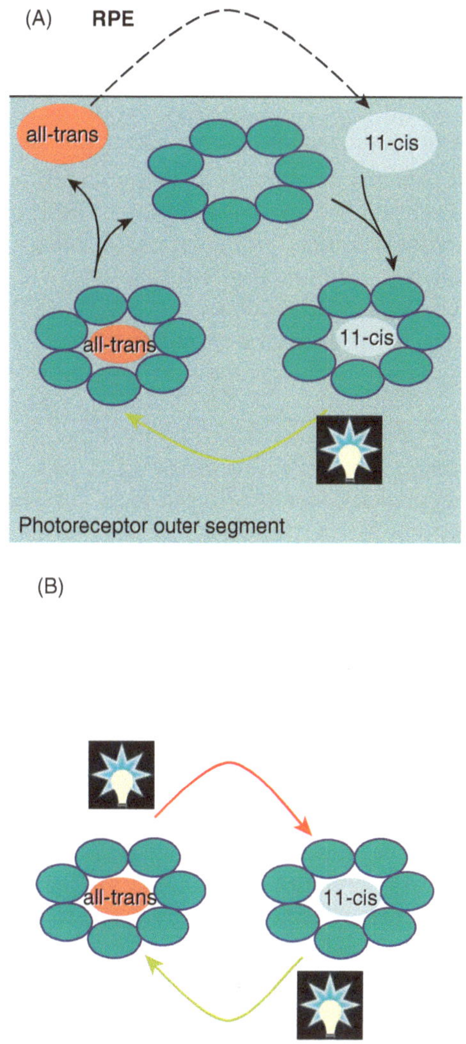

FIGURE 12.4 Mechanisms of chromophore regeneration for vertebrate rod opsin (A) and invertebrate opsin (B) In both cases the heptahelical opsin protein binds a *cis*-retinal isoform which, upon absorption of light, is isomerized to all-*trans*. The vertebrate photopigment then releases its chromophore and is insensitive to light until a new molecule of 11-*cis* retinal is presented. The conversion of all-*trans* to 11-*cis*-retinal occurs in the neighboring retinal pigment epithelium (RPE). By contrast, invertebrate opsins (B) form a stable association with all-*trans* retinal, allowing a second photon to regenerate the *cis*-isoform. There is commonly, but not always, a difference between the spectral sensitivity of the pigment when binding *cis*- or *trans*-chromophores.

based upon their ability to form a stable association with both *cis*- and *trans*-isoforms. This bistability enables them to hold on to bleached chromophore which may then be re-isomerized to a *cis*-isoform by absorption of a second photon (Fig. 12.4). Several observations suggest that melanopsin might also be bistable and consequently have an intrinsic bleach recovery mechanism. One important finding from the cell culture studies was that melanopsin could drive light responses if *cis*-retinaldehyde in the culture media was replaced by all-*trans*-retinaldehyde (Melyan *et al.*, 2005; Panda *et al.*, 2005). Furthermore, pre-exposure of cells expressing human melanopsin treated with all-*trans*-retinaldehyde to longer wavelengths of light (520nm) subsequently enhanced the response of cells to a 420-nm wavelength of light suggesting a long-wavelength bleach recovery event (Melyan *et al.*, 2005). More direct evidence in favor of such a mechanism was published for a melanopsin-like protein in the protochordate *Amphioxus* (Koyanagi *et al.*, 2005). *Amphioxus* melanopsin *in vitro* does not bleach to light but appears to form photointerconvertable stable states.

Further support for the hypothesis that melanopsin possesses its own photoisomerase activity comes from the use of mouse models, where the "visual cycle" in the RPE of the retina has been genetically inactivated. Specifically, in the work of Doyle *et al.* (2006), Tu *et al.* (2006) and Fu *et al.* (2005) irradiance detection was tested in two mutant mice strains lacking Rpe65 or lecithin-retinal acyl transferase (Lrat), critical components of the visual cycle pathway. The *Rpe65* gene encodes the retinoid isomerohydrolase and null mutants loose all cone as well as almost completely rod function. Lecithin-retinol acyl transferase (Lrat) acylates all-*trans*-retinaldehyde with a fatty acid ester tail and thereby traps it within the RPE for subsequent enzymatic processing steps. In *Lrat* mouse mutants, chromophore regeneration is also severely impaired with a consequent block of rod and cone phototransduction activity. Strangely however, photic entrainment of the clock, the pupillary light reflex and ipRGC activity are also strongly impaired (Fu *et al.*, 2005; Doyle *et al.*, 2006; Tu *et al.*, 2006). Thus for example, in both mutants the PLR is around 1000-fold less sensitive than in wild type controls. Normal photosensitivity can be rescued in the case of the *Rpe65* mutant by the administration of exogenous 9-*cis*-retinal (Fu *et al.*, 2005). Superficially, these results would tend to argue that melanopsin does indeed depend on the RPE and the visual cycle to regenerate its chromophore. However, this reduced sensitivity of irradiance detection seems to be more likely a secondary consequence of visual cycle inactivation. For a start, treatment of the wild-type retina with all-*trans*-retinylamine which acutely blocks the visual cycle, does not affect ipRGC activity (Tu *et al.*, 2006) suggesting that the ipRGCs do not rely on the RPE visual cycle for chromophore regeneration. In addition, when the visual cycle mutants are crossed with the *rd* mutant and *rdta* transgenic lines of mice to generate *Rpe65*$^{-/-}$; *rdta* and *Lrat*$^{-/-}$; *rd/rd* mutants, there is paradoxically an increase in the photosensitivity for the irradiance detection responses (Doyle *et al.*, 2006; Tu *et al.*, 2006). Thus it would seem that the presence of an intact outer retina renders the ipRGC cells sensitive to lesions in the visual cycle. Various explanations for this have been proposed to explain such an interaction within the retina (Lucas, 2006). One possibility is that the rods and cones make high demands for active chromophore and so compete with the ipRGCs for this limiting resource. Alternatively, the inactive outer retina may inhibit ipRGC function or signaling. In support of this idea is growing evidence for interactions between the ipRGCs and the rod and cone photoreceptors. However, ipRGC activity is normal when measured in the context of the *Cnga3*$^{-/-}$ *Gnat1*$^{-/-}$ double mutant mice, and addition of the *Rpe65* mutation to this genetic background still interferes with ipRGC function (Fu *et al.*, 2005). An alternative possibility is that the *Rpe65* or *Lrat* mutations actually impair the development and/or the viability of the ipRGCs. Supporting this possibility, it has been shown that the number of ipRGCs is actually reduced in *Rpe65*$^{-/-}$ mice (Doyle *et al.*, 2006). Furthermore the expression of melanopsin is altered in the surviving cells, with less being detected in the outer–inner plexiform layer (Doyle *et al.*, 2006).

ipRGCS, MELANOPSIN AND EARLY DEVELOPMENT

One of the more intriguing discoveries relates to the development of the melanopsin expressing ipRGCs. These cells seem to gain their photoreceptive function well before the other photoreceptors in the retina. In mice, at birth, the RGCs are not synaptically connected with the rods and cones. Subsequently vertical synaptic connections are established via bipolar cells in time for the rods and cones to start their phototransduction function around P12. The situation for the ipRGCs seems to be very different. Melanopsin is already expressed halfway through gestation (embryonic day 18) in contrast to the much later appearance of mRNA for UV cone opsin (P1), rod opsin (P5) and green cone opsin (P7) (Tarttelin *et al.*, 2003). Melanopsin expression is first detected in the inner neuroblast layer and then migrates outwards as its levels of expression increase to coincide with the RGC layer at birth (Fahrenkrug *et al.*, 2004; Hannibal and Fahrenkrug, 2004). Already at birth, the melanopsin expressing cells show light induced *c-fos* expression – as is also the case in the SCN, a sign that the mechanism for photic entrainment of the clock is established (Hannibal and Fahrenkrug, 2004). *In vivo* calcium imaging has allowed light induced ipRGC activity to be visualized directly at birth (Sekaran *et al.*, 2005). What is fascinating is that at P0, around 14% of RGCs are light responsive, at P4–P5 this is reduced to 5.4% and then in the adult the proportion of ipRGCs is only 2.7%. Thus there is a 70% reduction in the number of ipRGCs that occurs between P4 and P14 (Sekaran *et al.*, 2005). What can we conclude from these striking observations? It would seem that an independent irradiance detection system is required well before the eye is able to form and process images, possibly to ensure independence of the offspring from their mother. It has even been speculated that at birth, the ipRGC network may actually also constitute a very basic image forming system (Hattar *et al.*, 2002; Dacey *et al.*, 2005). Indeed, some of these cells do project to visual centers in the brain. Alternatively, they may assist the maturation of rods and cones and subsequently contribute to conveying global changes in images. The apparent overproduction of ipRGCs could help to consolidate development of their retinal projections. Indeed it has been shown that from late embryogenesis until P15/P21 in the mouse, immature RGCs spontaneously fire in waves across the developing retina (Sernagor *et al.*, 2001). This is thought to contribute to the so-called Hebbian strengthening of connections.

EVOLUTIONARY PERSPECTIVES AND CONCLUDING REMARKS

The discovery of ipRGCs and their melanopsin photopigment reinforces the conclusion drawn from the multiple extraretinal photoreceptors of non-mammalian vertebrates that there is a selective advantage to maintaining dedicated irradiance detection photoreceptors. This presumably reflects the particular sensory abilities required to accurately measure environmental irradiance, which are likely quite different from those required for classical image forming vision. However, it also begs the question of why mammals lack extraretinal photoreceptors. In fact, early mammalian evolution seems to have been characterized by a general reduction in photoreceptive machinery, perhaps as a result of a "nocturnal bottleneck" (Menaker *et al.*, 1997). In this scenario, the early ancestors of mammals lived underground during the day to avoid predation and came out to feed only under the cover of darkness. Under such light-limiting circumstances, it is assumed that the benefits of complexity in both irradiance and image forming photoreception would have been reduced. Whatever the origin of the reduction in photoreceptive capacity, it seems that even melanopsin itself did not escape this evolutionary event, as it is now clear that the counterpart of the original *Xenopus* melanopsin gene has in fact been lost from the mammalian genome. The extant mammalian melanopsin is actually an orthologe of a second, more recently discovered *Xenopus* gene that is also present across non-mammalian genomes (Bellingham *et al.*, 2006). We do not yet know what advantage non-mammals gain from having two separate melanopsin genes or, perhaps more importantly, how having only one limits our own sensory capabilities.

In addition to the challenge of understanding how irradiance detection evolved in vertebrates, many basic questions still remain concerning the molecular biology of melanopsin as well as how ipRGCs interact with rod and cone outputs. Clearly the various mouse models that have been described in this chapter represent key tools that should help us to gain new insight into these important issues.

ACKNOWLEDGMENTS

We would like to acknowledge the support from the BBSRC (RJL), Wellcome Trust (RJL) and the Forschungszentrum, Karlsruhe (NSF and DV).

REFERENCES

Arendt J (1995). *Melatonin and the Mammalian Pineal Gland*. Chapman and Hall, Cambridge.

Argamaso SM, Froehlich AC, McCall MA *et al.* (1995). Photopigments and circadian systems of vertebrates. *Biophys Chem* 56(1–2):3–11.

Aschoff J (1960). Exogenous and endogenous components in circadian rhythms. *Cold Spring Harb Sym* 25:11–28.

Bellingham J, Chaurasia SS, Melyan Z *et al.* (2006). Evolution of melanopsin photoreceptors: discovery and characterization of a new melanopsin in nonmammalian vertebrates. *PLoS Biol* 4(8):e254.

Bellingham J, Foster RG (2002). Opsins and mammalian photoentrainment. *Cell Tissue Res* 309(1):57–71.

Bellingham J, Whitmore D, Philp AR *et al.* (2002). Zebrafish melanopsin: isolation, tissue localisation and phylogenetic position. *Brain Res Mol Brain Res* 107(2):128–136.

Berson DM, Dunn FA, Takao M (2002). Phototransduction by retinal ganglion cells that set the circadian clock. *Science* 295(5557):1070–1073.

Bertolucci C, Foa A (2004). Extraocular photoreception and circadian entrainment in nonmammalian vertebrates. *Chronobiol Int* 21(4–5):501–519.

Bowes C, Li T, Danciger M *et al.* (1990). Retinal degeneration in the *rd* mouse is caused by a defect in the β subunit of rod cGMP-phosphodiesterase. *Nature* 347(6294):677–680.

Carter-Dawson LD, LaVail MM, Sidman RL (1978). Differential effect of the *rd* mutation on rods and cones in the mouse retina. *Invest Ophthalmol Vis Sci* 17(6):489–498.

Connell G, Bascom R, Molday L *et al.* (1991). Photoreceptor peripherin is the normal product of the gene responsible for retinal degeneration in the rds mouse. *P Natl Acad Sci USA* 88(3):723–726.

Dacey DM, Liao HW, Peterson BB *et al.* (2005). Melanopsin-expressing ganglion cells in primate retina signal colour and irradiance and project to the LGN. *Nature* 433(7027):749–754.

Doyle SE, Castrucci AM, McCall M *et al.* (2006). Nonvisual light responses in the Rpe65 knockout mouse: rod loss restores sensitivity to the melanopsin system. *P Natl Acad Sci USA* 103(27):10432–10437.

Fahrenkrug J, Nielsen HS, Hannibal J (2004). Expression of melanopsin during development of the rat retina. *Neuroreport* 15(5):781–784.

Foster RG (1998). Shedding light on the biological clock. *Neuron* 20(5):829–832.

Foster RG, Hankins MW (2002). Non-rod, non-cone photoreception in the vertebrates. *Prog Retin Eye Res* 21(6):507–527.

Foster RG, Provencio I, Hudson D *et al.* (1991). Circadian photoreception in the retinally degenerate mouse (rd/rd). *J Comp Physiol A* 169(1):39–50.

Freedman MS, Lucas RJ, Soni B *et al.* (1999). Regulation of mammalian circadian behavior by non-rod, non-cone, ocular photoreceptors. *Science* 284(5413):502–504.

Fu Y, Zhong H, Wang MH *et al.* (2005). Intrinsically photosensitive retinal ganglion cells detect light with a vitamin A-based photopigment, melanopsin. *P Natl Acad Sci USA* 102(29):10339–10344.

Gooley JJ, Lu J, Chou TC *et al.* (2001). Melanopsin in cells of origin of the retinohypothalamic tract. *Nat Neurosci* 4(12):1165.

Hannibal J, Fahrenkrug J (2004). Melanopsin containing retinal ganglion cells are light responsive from birth. *Neuroreport* 15(15):2317–2320.

Hannibal J, Hindersson P, Knudsen SM *et al.* (2002). The photopigment melanopsin is exclusively present in pituitary adenylate cyclase-activating polypeptide-containing retinal ganglion cells of the retinohypothalamic tract. *J Neurosci* 22(1):RC191.

Hattar S, Liao HW, Takao M *et al.* (2002). Melanopsin-containing retinal ganglion cells: architecture, projections, and intrinsic photosensitivity. *Science* 295(5557):1065–1070.

Hattar S, Lucas RJ, Mrosovsky N *et al.* (2003). Melanopsin and rod-cone photoreceptive systems account for all major accessory visual functions in mice. *Nature* 424(6944):76–81.

Kardon R (1995). Pupillary light reflex. *Curr Opin Ophthalmol* 6(6):20–26.

Klein DC, Weller JL (1972). Rapid light-induced decrease in pineal serotonin *N*-acetyltransferase activity. *Science* 177(48):532–533.

Korf HW, Schomerus C, Stehle JH (1998). The pineal organ, its hormone melatonin, and the photoneuroendocrine system. *Adv Anat Embryol Cell Biol* 146:1–100.

Koyanagi M, Kubokawa K, Tsukamoto H *et al.* (2005). Cephalochordate melanopsin: evolutionary linkage between invertebrate visual cells and vertebrate photosensitive retinal ganglion cells. *Curr Biol* 15(11):1065–1069.

Lucas RJ (2006). Chromophore regeneration: melanopsin does its own thing. *P Natl Acad Sci USA* 103(27):10153–10154.

Lucas RJ, Douglas RH, Foster RG (2001). Characterization of an ocular photopigment capable of driving pupillary constriction in mice. *Nat Neurosci* 4(6):621–626.

Lucas RJ, Foster RG (1999). Neither functional rod photoreceptors nor rod or cone outer segments are required for the photic inhibition of pineal melatonin. *Endocrinology* 140(4):1520–1524.

Lucas RJ, Freedman MS, Munoz M *et al.* (1999b). Regulation of the mammalian pineal by non-rod, non-cone, ocular photoreceptors. *Science* 284(5413):505–507.

Lucas RJ, Hattar S, Takao M *et al.* (2003). Diminished pupillary light reflex at high irradiances in melanopsin-knockout mice. *Science* 299(5604):245–247.

Lupi D, Cooper HM, Froehlich A *et al.* (1999). Transgenic ablation of rod photoreceptors alters the circadian phenotype of mice. *Neuroscience* 89(2):363–374.

McCall MA, Gregg RG, Merriman K *et al.* (1996). Morphological and physiological consequences of the selective elimination of rod photoreceptors in transgenic mice. *Exp Eye Res* 63(1):35–50.

Melyan Z, Tarttelin EE, Bellingham J *et al.* (2005). Addition of human melanopsin renders mammalian cells photoresponsive. *Nature* 433(7027):741–745.

Menaker M, Moreira LF, Tosini G (1997). Evolution of circadian organization in vertebrates. *Braz J Med Biol Res* 30(3):305–313.

Menaker M, Takahashi JS, Eskin A (1978). The physiology of circadian pacemakers. *Annu Rev Physiol* 40:501–526.

Mombaerts P, Wang F, Dulac C *et al.* (1996). Visualizing an olfactory sensory map. *Cell* 87(4):675–686.

Mrosovsky N (1999). Masking: history, definitions, and measurement. *Chronobiol Int* 16(4):415–429.

Mrosovsky N, Lucas RJ, Foster RG (2001). Persistence of masking responses to light in mice lacking rods and cones. *J Biol Rhythm* 16(6):585–588.

Panda S, Nayak SK, Campo B *et al.* (2005). Illumination of the melanopsin signaling pathway. *Science* 307(5709):600–604.

Panda S, Provencio I, Tu DC *et al.* (2003). Melanopsin is required for non-image-forming photic responses in blind mice. *Science* 301(5632):525–527.

Panda S, Sato TK, Castrucci AM *et al.* (2002). Melanopsin (Opn4) requirement for normal light-induced circadian phase shifting. *Science* 298(5601):2213–2216.

Pittendrigh CS (1993). Temporal organization: reflections of a Darwinian clock-watcher. *Annu Rev Physiol* 55:16–54.

Pittler SJ, Baehr W (1991). Identification of a nonsense mutation in the rod photoreceptor cGMP phosphodiesterase beta-subunit gene of the rd mouse. *P Natl Acad Sci USA* 88(19):8322–8326.

Provencio I, Jiang G, De Grip WJ *et al.* (1998). Melanopsin: an opsin in melanophores, brain, and eye. *P Natl Acad Sci USA* 95(1):340–345.

Provencio I, Rodriguez IR, Jiang G *et al.* (2000). A novel human opsin in the inner retina. *J Neurosci* 20(2):600–605.

Qiu X, Kumbalasiri T, Carlson SM *et al.* (2005). Induction of photosensitivity by heterologous expression of melanopsin. *Nature* 433(7027):745–749.

Roenneberg T, Daan S, Merrow M (2003). The art of entrainment. *J Biol Rhythm* 18(3):183–194.

Roenneberg T, Foster RG (1997). Twilight times: light and the circadian system. *Photochem Photobiol* 66(5):549–561.

Roseboom PH, Namboodiri MA, Zimonjic DB *et al.* (1998). Natural melatonin 'knockdown' in C57BL/6J mice: rare mechanism truncates serotonin *N*-acetyltransferase. *Brain Res Mol Brain Res* 63(1):189–197.

Ruby NF, Brennan TJ, Xie X *et al.* (2002). Role of melanopsin in circadian responses to light. *Science* 298(5601):2211–2213.

Sanyal S, De Ruiter A, Hawkins RK (1980). Development and degeneration of retina in rds mutant mice: light microscopy. *J Comp Neurol* 194(1):193–207.

Schibler U, Sassone-Corsi P (2002). A web of circadian pacemakers. *Cell* 111(7):919–922.

Sekaran S, Foster RG, Lucas RJ *et al.* (2003). Calcium imaging reveals a network of intrinsically light-sensitive inner-retinal neurons. *Curr Biol* 13(15):1290–1298.

Sekaran S, Lupi D, Jones SL *et al.* (2005). Melanopsin-dependent photoreception provides earliest light detection in the mammalian retina. *Curr Biol* 15(12):1099–1107.

Sernagor E, Eglen SJ, Wong RO (2001). Development of retinal ganglion cell structure and function. *Prog Retin Eye Res* 20(2):139–174.

Shand J, Foster RG (1999). The extraretinal photoreceptors of non-mammalian vertebrates. In: Archer SN, Djamgoz MBA, Loew ER *et al.* (eds), *Adaptive Mechanisms in the Ecology of Vision*. Kluwer Academic Publishers, Dordrecht, Netherlands, pp. 197–222.

Sollars PJ, Smeraski CA, Kaufman JD *et al.* (2003). Melanopsin and non-melanopsin expressing retinal ganglion cells innervate the hypothalamic suprachiasmatic nucleus. *Vis Neurosci* 20(6):601–610.

Soucy E, Wang Y, Nirenberg S *et al.* (1998). A novel signaling pathway from rod photoreceptors to ganglion cells in mammalian retina. *Neuron* 21(3):481–493.

Tarttelin EE, Bellingham J, Bibb LC *et al.* (2003). Expression of opsin genes early in ocular development of humans and mice. *Exp Eye Res* 76(3):393–396.

Travis GH, Brennan MB, Danielson PE *et al.* (1989). Identification of a photoreceptor-specific mRNA encoded by the gene responsible for retinal degeneration slow (rds). *Nature* 338(6210):70–73.

Travis GH, Sutcliffe JG, Bok D (1991). The retinal degeneration slow (rds) gene product is a photoreceptor disc membrane-associated glycoprotein. *Neuron* 6(1):61–70.

Tu DC, Owens LA, Anderson L *et al.* (2006). Inner retinal photoreception independent of the visual retinoid cycle. *P Natl Acad Sci USA* 103(27):10426–10431.

Tu DC, Zhang D, Demas J *et al.* (2005). Physiologic diversity and development of intrinsically photosensitive retinal ganglion cells. *Neuron* 48(6):987–999.

Viney TJ, Balint K, Hillier D *et al.* (2007). Local retinal circuits of melanopsin-containing ganglion cells identified by transsynaptic viral tracing. *Curr Biol* 17(11):981–988.

Wang Y, Macke JP, Merbs SL *et al.* (1992). A locus control region adjacent to the human red and green visual pigment genes. *Neuron* 9(3):429–440.

Yoshimura T, Ebihara S (1996). Spectral sensitivity of photoreceptors mediating phase-shifts of circadian rhythms in retinally degenerate CBA/J (rd/rd) and normal CBA/N (+/+)mice. *J Comp Physiol A* 178(6):797–802.

CHAPTER

13

The Rabbit in Cataract/IOL Surgery

Arlene Gwon

Department of Ophthalmology, University of California, Irvine

OUTLINE

INTRODUCTION

Dating as far back as the 17th century reference has been made to the ocular anatomy of the rabbit (http://www.netvet.wustl.edu; Prince, 1964). One of the first laboratory experiments using the rabbit eye as a model for lens regeneration was reported by Cocteau and D'Etoille nearly 200 years ago (Cocteau, 1827). The ocular anatomy of the rabbit is similar to that in man and, as a result, it continues to be one of the most frequently used animal models for ophthalmic research. The following is a review of the ocular anatomy of the rabbit and its use in ocular research studies.

RABBIT EYE ANATOMY: COMPARISONS AND CONTRASTS WITH THE HUMAN EYE

As compared to a rabbits overall size, the eye is relatively large and varies with age. At birth the globe is about 6mm in diameter but it grows very rapidly and at 7–10 days postnatal it is approximately two-thirds the size of the adult rabbit eye. It appears compressed in its antero-posterior dimension which measures 16–19mm while its equatorial diameter is approximately 18mm. This is in contrast to the human eye which is relatively spherical (Table 13.1). The rabbit has similar methods for changing the focal power of the eye, but much less capacity for this than man, i.e., its accommodation is limited (Prince, 1964). The accommodative power of the rabbit's eye has been found to range from 0 to approximately 2.5 diopters (D), but usually in a mature animal it is seldom more than 1.5D.

The rabbit eye possesses the same six extraocular muscles as in man (i.e., a superior, inferior, lateral and medial recti muscles, and superior and inferior oblique muscles); however, it does not display significant voluntary eye movement. In contrast to the human eye, the rabbit eye possesses an active retractor bulbi muscle, whose primary function is to pull back the eye. It arises from the apex of the orbit and runs within the muscle cone enveloping the optic nerve and posterior globe where it inserts into the sclera. Other structures present in the rabbit eye and not in the human eye include the Harder's gland, an acino-tubular gland that lies primarily on the nasal side of the orbit and functions to lubricate the nicitating membrane (or third eyelid) that slips

TABLE 13.1 Ocular dimensions in rabbit and man

	Rabbit	Man
Globe (mm)		
Anteroposterior	16–19	24
Horizontal	18–20	23.5
Vertical	17–18	
Cornea		
Diameter (mm)		
Horizontal	15	11.7
Vertical	13.5–14	10.6
Radius of curvature (mm)	7.0–7.5	7.5–8.0
Corneal curvature		
Birth	60 D	
30 Weeks	50 D	
60–80 Weeks	40–43 D	
Corneal thickness (mm)		
Center	0.3–0.4	0.5
Peripheral	0.45	0.7–1.0
Endothelial cell density (cells/mm^2)	3200	4,003–1,547 at 12–74 years
Anterior chamber		
AC depth (mm)	2.9 ± 0.36	3.5 ± 0.35 (with no refractive error)
AC diameter (mm)	Not known	12.5
Aqueous volume (ml)	0.25–3.0	0.24–0.28
AC angle	Not known	Between 15° and 40°
Pupil diameter (mm)	7 (5–11)	2.5–4.0
Crystalline lens		
Diameter (mm)	11	9–10 at 40 years
Thickness (anteroposterior) (mm)	7.6	4 at 40 years
Anterior radius (mm)	5.3	11
Posterior radius (mm)	5	6
Volume (cm^3)		90 at birth; 163 at 30–40 years; 244 at 80–90 years
Capsule thickness (μ)	3.4–51	4–23
Central anterior	30	4
Central posterior	3.4	14
Capsule bag diameter (mm)	10.4–10.7 DB 1.9–2.9 kg 11.1–11.5 NZQ 3.5–4.5 kg	10.2–10.5

Sources: Assia and Apple (1992); Bron *et al.* (1997); Davis (1929); Fatt (1978); Galand *et al.* (1984); Gwon and Gruber (2002); Kurz *et al.* (2006); Neumann and Apple (1986); Oyster (1999); Prince (1964); Tañá and Belmonte (1993); Vasavada and Singh (1998).

over the eye when the globe is retracted. As it pulls the orbit inwards the nicitating membrane is forced out from the inner canthus across the cornea by the propulsive action of the retractor bulbi muscle (Duke-Elder, 1958). The rabbit eye also has a larger lacrimal gland, unique lacrimal and aqueous drainage systems, and its cornea is devoid of a significant Bowman's membrane. The rabbit's corneal endothelium also regenerates in response to loss from injury. The rabbit does not have a true fovea centralis and has a unique bipolar and outer plexiform layer in the retina.

The conjunctiva of the rabbit eye is a mucus membrane consisting of an epithelial layer and substantia propria or stroma, and is divided into two continuous parts that aid in suspension of the eye anteriorly. The palpebral portion is firmly adhered to the posterior surface of the lids while the bulbar portion is loosely attached. Together they form the superior and inferior fornices in the fold created by their contact. The palpebral conjunctiva is approximately 40μ thick. The epidermis consists of non-keratinized squamous epithelium near the lid margin continuing on to a stratified columnar character of varying thickness through most of the conjunctiva. It contains both goblet cells and intraepithelial glands. The site of entry for the lacrimal gland and auxiliary gland ducts is the fornix. The bulbar conjunctiva is thinner than that of the palpebral, ranging from 10 to 30μ with less goblet cells. The epithelium consists of two cell layers; a row of basal cells and 1–3 rows of flatter surface cells. The substantia propria consists of a thin adenoid layer enclosing lymphocytes.

The rabbit cornea is unusually prominent and wide, transmitting almost 100% of light in the visible spectrum. It is not circular and has a horizontal diameter averaging 15 mm and a vertical diameter averaging 13.5–14.0 mm. The radius of curvature is usually between 7.0 and 7.5 mm but varies with the rabbit's age and size, constantly changing during the first 15 months of life. At birth it has a power of 60 D, decreases to 50 D by 30 weeks, and stabilizes at 40–43 D by 60–80 weeks. The corneal thickness is approximately 0.3–0.4 mm at its center and approximately 0.45 mm near the limbus. The rabbit corneal epithelium is thinner than that of man, approximately 30–40μ in thickness. It consists of a row of columnar basal cells beneath two rows of polygonal and up to six rows of wing shaped and squamous cells on the external surface. The epithelial basement membrane is continuous with that of the conjunctival epithelium and it receives tonofibrils from the basal epithelium. It is not clear if there is a Bowman's layer between the epithelium and stroma. The layer between the epithelium and the stroma is so fine (1–2μ) that it is thought of as a modified zone of the anterior stroma when present. The stroma is approximately 0.24 mm thick and consists of lamellae of collagen fibrils from 20 to30 μ in diameter that interweave with one another. Elastic fibers and flat elongated stromal cells lie between the lamellae. The Descemet's layer is 7–8 μ thick and increases with age becoming as thick as 15–22 μ in the adult. The endothelium consists of a single layer of uniform flattened hexagonal or polygonal cells averaging 20 μ in diameter (approximately 3,200/mm^2) and 3–5 μ thick. While the cells are mostly 20 μ in diameter in the adult rabbit, in young animals they may be only 5–15 μ. In the young animal, mitotic activity occurs irregularly while in the adult animal endothelial mitosis as well as amitotic activity occurs following injury. After endothelial injury, amitotic activity occurs early with mitotic activity appearing after 24 h and intensifies around the wound area during the subsequent 48–55 h.

In the rabbit eye, the anterior chamber angle is very difficult to visualize due to the bowing of the iris anteriorly. The trabecular meshwork is shallower and the 1.2–2.4 μ pore size is smaller than the 1.5–4.0 μ pore size in man. As in man, the trabecular meshwork in the rabbit eye is innervated by the ciliary nerves. The spaces of Fontana or cilioscleral sinuses lie posterior to the anterior iris pillars and communicate freely with the anterior chamber. The rabbit does not have a Schlemm's canal; however, a definite space located external to the corneo-scleral trabeculae separates it from the spaces of Fontana. It is closely associated with the trabecular meshwork and has been called the trabecular vein or canal. These collector canals are connected to the deep scleral channels which traverse the sclera to connect with the episcleral and conjunctival veins (Davis, 1929; Sheppard, 1962). The average depth of the anterior chamber is approximately 2.9 ± 0.36 mm. The average anterior chamber volume is 0.25–0.30 ml and the posterior chamber volume is 0.050–0.075 ml. The intraocular pressure of the rabbit is similar to man and generally falls between 20 and 25 mmHg. Aqueous outflow has been reported to be between 2.75 and 3.66 ml/min.

The ciliary body in the rabbit eye is very poorly developed and comparatively flat due to the scarcity of muscle fibers. The ciliary processes are well developed forming the secretory part of the eye. They arise from the anterior portion of the ciliary body and merge into the posterior surface of the iris at its base and extend to within 1 mm of the pupillary margin. The zonular fibers appear to be part of the cell membrane of the ciliary processes. A great number of the zonular fibers arise from the valleys between the ciliary processes and a diminishing number appear nearer the tips. The zonular fibers course to within 200–400 μ of the lens, break up into fibrils, and some insert into the equator of the lens capsule. Others insert anteriorly and posteriorly to the lens equator (Davis, 1929; Sheppard, 1962).

The rabbit eye pupil is usually 7 mm in diameter – contracting in light to 5 mm and dilating under emotion to 11 mm. The iris consists of three customary layers, the anterior endothelium that is continuous with that of the cornea, the stroma, and the posterior epithelium, continuous with that of the ciliary processes. The iris sphincter muscle is adjacent to the pupillary margin and extends outwards close to the epithelium. It changes from concentric to radial as it approaches the iris root (Sheppard, 1962; Prince, 1964).

The lens of the rabbit eye grows with increasing age and weight of the animal. Compared to the human lens, the rabbit lens is larger and more spherical and takes up more space in the globe. Depending upon the breed, the rabbit lens weighs between 0.142 and 0.304 g in a young animal and between 0.540 and 0.558 g in a fully mature rabbit. The anterior surface has a radius of curvature of 5.0 mm and the posterior surface radius of curvature is 5.3 mm. The anterior lens loses most of its curvature in the first 12 weeks of life, flattening up to 1.8 mm. The lens has an average anterior–posterior diameter of 7.0 mm and an equatorial diameter of 9–11 mm and has a power of approximately 10 D. It has two single line sutures, the anterior suture is vertical and the posterior is horizontal. As in man, the anterior lens capsule is thicker (10–25 μ) than the posterior capsule (4–6 μ). The rabbit lens nucleus is less sharply demarcated than that of the human lens and as in man scleroses and hardens with age. The monolayer of cuboidal anterior epithelium measures 17 μ thick and lens differentiation with elongation of lens fibers occurs at the equator. The hexagonal lens fibers form a complicated interlacing and interdigitating system which is tightly knit and capable of 1.5–2.0 D of accommodation (Prince, 1964; Kuszak *et al.*, 1991; 2000, Kuszak and Costello, 2004).

The vitreous gel of the rabbit eye is 99% water and weighs about 1.4 g. Hyaluronic acid contributes to the viscosity of the gel and can either be produced within or for the vitreous. The choroid is well developed and without a tapetum and consists of 75–80% blood with a supporting structure of collagenous and elastic connective tissue. It is attached to the retinal pigment epithelium on its inner surface and its capillary arrangement is very similar to that of man.

The optic nerve head appears as a large oval disc which is deeply cupped and lying above the posterior pole of the globe. Two broad white bands of opaque nerve fibers, the medullary rays, stream from it nasally and temporally. The retinal artery and vein enter the optic nerve ventrally and divide just before or after emerging from the nerve into nasal and temporal branches to travel along the medullated nerve fibers. The remainder of the rabbit retina is avascular and the histological structure is essentially the same as that of the human eye. Differences include a retinal pigment epithelium that is irregular in size and arrangement, unlike the regular hexagonal configuration in humans, and extremely long, thin rods and cones.

The rabbit eye sclera varies in thickness and is adjacent to the limbus. It averages 0.5 mm and thins posteriorly to 0.2 mm near the optic nerve. The sclera consists of 75% collagen, 10% other proteins including mucoproteins, 1% polysaccharides and some elastic fibers. It has a water content of about 68% (Sheppard, 1962; Prince, 1964).

USE OF THE RABBIT IN OCULAR SURGERY RESEARCH

The rabbit and its eye are ideally suited for ophthalmic research for several reasons. The animal is docile, easy to handle, comes in various sizes depending on the breed (http://www.netvet.wustl.edu) thus providing a range of sizes to work with and is economical compared to other mammals such as dogs or primates. Like other animals used in research, the United States Food and Drug Administration (US FDA), the American National Standards Institute (ANSI), and International Standards Organization/Committee European Normalization (ISO/CEN) have all implemented guidelines for the conduction of studies utilizing the rabbit model (http://www.hc-sc.gc.ca; Anon. ANSI Z80.7-2022).

Because the rabbit eye is relatively large it has proved useful for the assessment of both new technologies as well as ophthalmic surgical procedures. For example, the rabbit eye model has proved useful in the assessment of new technologies for removal of the natural crystalline lens/cataract such as surgical blades, phacoemulsification systems, intraocular lenses (IOLs), IOL insertion systems, ocular irrigating solutions, ophthalmic viscosurgical devices (OVDs) and other novel technologies. Currently, it is one of the accepted animal models for evaluating the biocompatibility of IOLs. The rabbit eye model has also been used in the evaluation of ophthalmic surgical procedures, including corneal transplantation, corneal inlays and onlays, evaluation of microkeratomes and laser refractive procedures, trabeculectomy, glaucoma shunts and endocyclophoto-coagulation, vitrectomy, proliferative vitreoretinopathy and intravitreal drug delivery. The following section reviews the use of the rabbit eye in selected types of ophthalmologic research.

Lens/Cataract Surgery

Species Selection

The New Zealand white (NZW) rabbit has been and is currently most often used in lens and cataract surgery evaluations, although white non-pigmented Rex rabbits may also be used. In the rabbit eye the visibility of ocular structures during surgery is excellent, video or still photography is easily performed through the surgical microscope or slit lamp biomicroscope.

In laser studies, the New Zealand/Dutch Belt (NZDB) or Dutch Belt (DB) pigmented rabbits are generally used because of their similarity to human ocular pigmentation. These animals are also used in studies utilizing specialized instrumentation that requires a pigmented retina such as the assessment of lens opacity (Gwon *et al.*, 1998). DB pigmented rabbits show minimal growth over time whereas the NZDB hybrid will grow and are preferred if growth of the animal or its structures is monitored. In pigmented species there is more pigment dispersion at the time of surgery and may result in pigment deposits on the IOL.

Age at Start of Study

Rabbits acquired from animal research sources are generally between 6 and 12 weeks of age. Because the rabbit eye and lens grow with increasing age and weight with an age-related reduction in postoperative inflammatory response, animals of differing age and weight are used in different types of evaluations. Specifically, younger rabbits tend to have a greater postoperative inflammatory response, which is greater than that observed in adult human eyes and characterized by a heavy fibrin reaction. As such, their response to cataract and IOL implantation is similar to that of young children. Studies have found that the fibrin formation peri- and postoperatively and posterior capsule opacification (PCO)/lens regenerative growth is inversely related to the age of the animal with early rapid lens regrowth stimulation noted directly following surgery. In very young NZW rabbits weighting 2 kg, PCO was seen as early as 6 weeks (Odrich

TABLE 13.2 Posterior capsule opacification: Time of onset

Rabbit	Age	Weight (kg)	IOL	PCO onset (weeks)	References
NZA/flemish giant	Very young	2.0	None	6	Odrich *et al.* (1985)
NZA	Young	2.5–3.5	PMMA	8	Gwon and Gruber (1994c)
NZA	Adult	3.0–4.0	Silicone	12	Gwon and Gruber (1994b)

et al., 1985). In our own unpublished studies, PCO/lens regrowth was first noted at 8 weeks postoperative in young NZW rabbits weighing 2.5–3.5 kg and at 12 weeks postoperative in adult NZW rabbits weighting 3.0–4.0 kg (Table 13.2) (Gwon and Gruber, 1994a, c).

Because of the age-related reduction in the postoperative inflammatory response and fibrin formation, older, larger rabbits are often considered more desirable in certain types of intraocular studies such as phacoemulsification capability. The 12–24 month old rabbit has a hard lens that on a 0 to 4 grading scale is similar to a +3 to +4 human cataract density/hardness. In addition, the reduction in fibrin formation seen in these older and larger animals may ameliorate the need for intraocular heparin.

Sex and Number

To date, no differences in animal response to ophthalmologic studies have been related to the sex of the rabbit used in the study. With respect to the numbers of eyes evaluated, most toxicological studies utilize a minimum of 6 eyes per group while most research studies will have 8–12 eyes per group. Bilateral ocular testing is preferable if allowed by local regulations according to the ANSI and ISO/CEN standards (http://www.hc-sc.gc.ca; Anon. ANSI Z80.7-2022).

For purposes of keeping the study groups uniform and on comparable time lines, it is preferable to have no more than 20–30 animals in any one study. Statistical differences that can be detected in paired and unpaired comparison studies are listed in Table 13.3.

Preoperative Evaluation

All the animals used in clinical research should be evaluated for general health and ocular status prior to the study. Baseline slit lamp biomicroscopy and other measurements, depending on the study variables being monitored, can be performed from 1 to 10 days preoperatively.

TABLE 13.3 Sample size estimation for rabbit studies evaluating PCO[a]

Number of rabbits	Difference in mean scores on 0–4 scale that can be detected between eyes
Paired comparisons (experimental product in one eye, control in the fellow eye	
14 (28 eyes) (number of eyes necessary?)	1.00
10 (20 eyes)	1.25
7 (14 eyes)	1.50
5 (10 eyes)	2.00
This indicates that if you are interested in detecting a mean difference between eyes (experimental versus control) of 1.5 points (e.g. mean score of 3.0 versus 1.5) then 7 rabbits (14 eyes) would be sufficient.	
Unpaired Comparisons (different rabbits used for experimental and control groups)	
20	1.00
13	1.25
10	1.50
6	2.00
This indicates that if you are interested in detecting a mean difference between groups (experimental versus control) of 2.0 points (e.g. mean score of 3.0 versus 1.0) then 6 rabbits/group (12 total) would be sufficient.	

Source: Cohen (1988).

Based on earlier research, the sample sizes were determined to detect differences in PCO and lens regrowth scores between experimental groups

[a]Paired and unpaired sample estimates based on the following assumptions – two-sided testing with alpha of 0.05, power of 0.80 and standard deviation of 1.3 for paired samples and 1.1 for unpaired samples. Estimates based on statistical power tables (Cohen, 1988). It is important to note that differing standard deviation values from other research settings could lead to very different sample size estimates.

Anesthesia

Rabbits can be anesthetized with approximately 5 mg/kg xylazine and 50 mg/kg ketamine HCl, intramuscularly.

Surgery

For optimum lens exposure, the surgical eye may be dilated with 1% cyclopentolate and 10% phenylephrine; eyelashes are trimmed; and the ocular area is disinfected with povidone iodine (Professional Disposables, Inc., Orangeburg, NY). A wire lid speculum is inserted to retract the lids and a corneal stab incision with a 1 mm sharp point blade may be made at the 1 or 2 O'clock position to accommodate a second instrument such as a phaco chopper for breaking the lens or spatula for controlling the IOL during insertion. In young rabbits, the lens can often be removed with only irrigation/aspiration and little or no phacoemulsification power, in which case this paracentesis incision is often unnecessary. Viscoelastic is then injected to fill and maintain anterior chamber depth (ACD). A corneal or limbal incision can be made at the 12 O'clock position with an appropriate size keratome. Keratomes come in a variety of sizes and the choice usually depends on the type of IOL being implanted. A 3.0–3.2 mm keratome incision is usually made to accommodate insertion of the phacoemulsification needle. However, it may be smaller when using a 21 gauge phaco needle and when implanting a one-piece lens with an injector, i.e., a 2.5–3.0 mm incision. For some three-piece IOLs a slightly larger incision may be needed. Generally hydrophobic acrylic IOLs require a larger incision than the silicone IOLs. In special cases, as when implanting an experimental dual optic lens with forceps, the incision may need to be enlarged to 4.5–5.0 mm due to its bulk (Werner *et al.*, 2004, 2006b). This is usually done just prior to IOL implantation.

Capsulorhexis Size

A viscoelastic or OVD is injected to maintain ACD following which continuous curvilinear capsulorhexis (CCC) is performed. Capsulorhexis size should be kept constant as much as possible. Capsulorhexis is usually kept smaller than the IOL optic size. The rabbit has a fair amount of posterior vitreous pressure following lens removal. By keeping the capsulorhexis smaller than the IOL optic, there is less possibility of IOL extrusion into the anterior chamber in the postoperative period. From our unpublished studies, a 7–8 mm CCC will contract to approximately 6 mm after lens extraction and IOL implantation. The final 6 mm CCC is generally sufficient to assist in retaining an IOL with a 6.5 mm optical zone in the capsular bag for long follow-up times. For IOLs with 6 mm optical zone, a final CCC of 5 mm may be preferable.

Results of one of our unpublished studies, which monitored IOL (PMMA with 6.5 mm optic) position and capsulorhexis size for 4 months, found that capsulorhexis size decreased by approximately 1.0 mm following lens extraction and IOL implantation. Over time, as PCO and lens regrowth progressed, the capsulorhexis enlarged and IOL dislocation occurred in those eyes with a 7–8 mm CCC size at the close of surgery. IOLs remained in the capsular bag in eyes with a ≤6 mm CCC at the close of surgery (Gwon and Gruber, 1994b). Similarly, Tetz *et al.* (1996a) used a 6 mm final capsulorhexis size and had no lens dislocations when followed for up to 5 months. Whereas, Hettlich *et al.* (1992) reported a 25–50% IOL dislocation rate utilizing a 7 mm final capsulorhexis size.

A reliable capsulorhexis technique for rabbits was described by Auffarth and coworkers. (Auffarth *et al.*, 1994a). This technique resulted in a stable, intact capsulorhexis and was also successfully used in a study by Kent *et al.* (1995).

Capsulorhexis size <5 mm at the close of surgery may promote anterior capsule contraction and lens regrowth. These events may interfere with the ability to view the IOL optic surface and assess PCO development. A capsulorhexis size no more than 0.5 mm smaller than the IOL optic diameter was recommended by Ravalico *et al.* (1996).

Lens/cataract Removal

Extracapsular lens extraction (ECLE) can easily be performed manually in the very young rabbit or by phacoemulsification in older rabbits with a harder nucleus. A 19 or 21 gauge phacoemulsification tip is inserted through the corneal wound and ECLE is performed by phacoemulsification and irrigation/aspiration with balanced salt solution mixed with 5% heparin and 1:100,000 epinephrine. Heparin inhibits the severe fibrin formation seen in performing surgery on young rabbits. In older animals it may not be necessary to add heparin. Epinephrine will help maintain pupil dilation during surgery. Considerable care is taken to remove all lens cortical material by diligent irrigation and aspiration. However, previous studies have shown no difference in lens regeneration/PCO rates when low vacuum suction is used to strip or polish the anterior–posterior capsule (Odrich *et al.*, 1985; Gwon *et al.*, 1992).

IOL Implantation

The corneal incision may be enlarged to accommodate the IOL being implanted. Viscoelastic, usually a cohesive hyaluronic acid that will be easily removed, is injected to deepen the anterior chamber and separate the anterior and posterior capsule. The IOL is then implanted with the desired inserter system or forceps. After implantation, the viscoelastic is removed by irrigation/aspiration. Removal of the viscoelastic is done to prevent intraocular pressure elevation in the early postoperative period. At the completion of the procedure, the corneal incision may be closed with 10–0 nylon sutures and an antibiotic, e.g. 0.25 ml (20 mg) of gentamicin, and a corticosteroid, e.g., 0.1 ml of dexamethasone (2 mg), are injected subconjunctivally.

Postoperative Medication

Postoperatively, the antibiotic prophylaxis is continued for 1–2 weeks and the anti-inflammatory corticosteroid for approximately 4–6 weeks. In accordance with ISO/CEN guidelines the postoperative use of corticosteroids should have appropriate controls to assure that "test-material-related inflammation is not masked" (http://www.hc-sc.gc.ca; Anon. ANSI Z80.7-2022). The medication may be given topically 4 times daily tapering to twice daily over the 1–1.5 month period if personnel issues are not a concern. Alternatively, 0.25 ml (20 mg) of gentamicin and 0.1 ml of dexamethasone may be given subconjunctivally every 3 days for 2 weeks for treating the initial postoperative inflammatory response to surgery.

IOL Biocompatibility

As previously mentioned, the rabbit intraocular implantation test is routinely used in the evaluation of IOL biocompatibility according to US FDA, ANSI and ISO/CEN guidelines and in the evaluation of other potential therapeutics to improve cataract/refractive lens surgery (http://www.hc-sc.gc.ca; Anon. ANSI Z80.7-2022; Tamura *et al.*, 1990; Laurell *et al.*, 1997; Norton *et al.*, 1999; Wallentin Lundberg, 2000; Scheib and Garner, 2004; Chew *et al.*, 2006; Kleinman *et al.*, 2006; Koura *et al.*, 2006). Throughout the experimental period, rabbits are observed for any abnormal clinical signs, including any abnormal ocular findings such as pain, excessive hyperemia or discharge.

With the aid of slit lamp biomicroscopy, the ocular status can be determined. The evaluation usually includes the status of the conjunctiva, cornea, anterior chamber and iris, the IOL, posterior synechiae, anterior lens regrowth/tissue ongrowth and PCO. All slit lamp findings may be graded on a scale of 0 to 4 (0 = none, 1+ = trace, 2+ = mild, 3+ = moderate, 4+ = severe) based on the methods of McDonald and Shadduck (Table 13.4) (McDonald and Shadduck, 1977).

TABLE 13.4 Anterior chamber inflammation grading scale

Cells	
0	No cells seen.
+1	1–9 cells per high power field.
+2	*Minimal*: Sparse and scattered or localized cells. 10 to 20 cells per high power field.
+3	Numerous and scattered and/or clumped cells. 20 to 30 cells per high power field.
+4	*Severe*: More than 30 cells per high power field. High concentration of cells throughout most or all of the anterior chamber, and/or clumped and cascading down the anterior lens surface.
Flare	
0	No Tyndall effect.
+1	Tyndall beam in the anterior chamber has a mild intensity.
+2	Tyndall beam in the anterior chamber has a moderate intensity.
+3	Tyndall beam is very intense. The aqueous has a white, milky appearance.
+4	Tyndall beam has severe marked intensity. Fibrin fills the anterior chamber and obscures view of the pupil.

Postoperatively, fibrin formation may be very severe in young rabbits, grade 3 to 4+. This usually resolves within 1 week with the standard postoperative corticosteroid course and without posterior synechiae formation or any sequelae. However, posterior synechiae is a frequent occurrence when studies are carried out for an extended period of time and may be treated with Nd:YAG lysis.

In general, the rabbit's postoperative inflammatory response to lens extraction and IOL implantation is mild and generally resolves by 2 weeks, the same as in humans, with current small corneal incision sizes and minimal surgical trauma when using phacoemulsification and irrigation/aspiration and foldable IOLs inserted with an injection system. However, as most rabbits are young, having behavior similar to a human child rather than adult, at the time of surgery, they may have transient fibrin formation in the early postoperative period that resolves by 1 week. With standard monofocal three-piece or one-piece IOLs the onset of posterior synechiae development may occur as early as 1–2 months postoperatively and gradually increases throughout the study making long-term studies (>6 months) problematic. The cause of this progressive posterior synechiae formation in the rabbit model has never been delineated. However, one possible explanation could be the presence of subclinical iritis related to pressure on the ciliary body from oversized IOLs with overall diameters of 13–14 mm. In studies lasting as long as 2 years, capsule bag filling lenses or silicone disc IOLs of 10–11 mm diameter progressive posterior synechiae formation was not seen. The empty rabbit capsule bag diameter has been measured at 11.1 mm in 3.5 kg NZW rabbits and 11.5 mm in the 4.5 kg rabbit in an unpublished study (Gwon and Gruber, 2002). Another possible cause is the rabbits' wound healing response to an open capsule as posterior synechiae is limited or absent in studies in lens regeneration and lens refilling.

Anterior lens regrowth has been routinely noted in rabbit IOL surgery and its onset at 6–8 weeks is not unusual. Mild IOL deposits consisting of cells/fibrin/foreign body reaction are not uncommon but may be excessive if the IOL has biocompatibility issues. The optic clarity of IOLs is easily viewed at the slit lamp and can be monitored for the presence of vacuoles, glistenings, opacification or calcification. In a study by Buchen and colleagues calcification of a hydrophilic IOL was noted at 1 month following intramuscular placement and at 4–10 months postoperative following intracapsular implantation (Buchen *et al.*, 2001). Similarly, the onset of calcification on hydrophilic acrylic IOLs was noted at 3.5–25 months postoperative in patients (Bucher, 1994; Bucher *et al.*, 1995; Foot *et al.*, 2004; Neuhann *et al.*, 2004; Werner *et al.*, 2006a).

Fibrotic type of posterior capsule haze/opacity may be noted at 1–4 weeks and gradually increases over time. Posterior capsule regenerative lens growth may be seen as early as 1–2 months depending on the age of the animal at the time of surgery with younger animals having an earlier onset. It gradually progresses over time and can be considerably thicker (graded 3 to 4+ on a scale of 0 to 4) by 6 months. The assessment of PCO deserves special consideration.

Posterior Capsule Opacification

PCO is the most common cause of decreased visual acuity following cataract extraction and IOL implantation (McDonnell *et al.*, 1983; Apple *et al.*, 1984, 1989, 1992; Nishi, 1986; Maltzman *et al.*, 1989; Menezo *et al.*, 1989; Auffarth *et al.*, 1995; Nasisse *et al.*, 1995). This opacification occurs secondary to anterior lens epithelial cell migration and myoblastic transformation contributing to wrinkling of the posterior capsule and resulting in visual distortion (Fagerholm and Philipson 1981; McDonnell *et al.*, 1983, 1984; Cobo *et al.*, 1984; Jacob, 1987; Apple *et al.*, 1992; Auffarth *et al.*, 1995). It is known to be affected by age (Moisseiev *et al.*, 1989), method of cataract extraction (Moisseiev *et al.*, 1989; Shepherd, 1989), degree of surgical trauma (Shepherd, 1989; Tan and Chee, 1993; Steinert *et al.*, 1995), capsulorhexis size (Gimbel and Neuhman, 1990; Assia *et al.*, 1991; Dangel *et al.*, 1994; Ravalico *et al.*, 1996), the amount of residual cortical tissue (Green and Boase 1989; Nishi and Nishi, 1992), polishing of the capsular bag (Green and Boase, 1989), IOL design (Irvine, 1980; Jacobi, 1985; Liesgang *et al.*, 1985; Downing, 1986; Hansen *et al.*,1988; Tetz *et al.*, 1988b; Sellman and Lindstrom, 1988; Davis and Hill 1989; Born and Ryan 1990; Lowes, 1990; Davis *et al.*, 1991; Apple *et al.*, 1992; Jaffee and Horwitz, 1992; Martin *et al.*, 1992; Ohmi and Uenoyama, 1993; Tan and Chee, 1993; Gwon and Gruber, 1994c; Auffarth *et al.*, 1995; Mamalis *et al.*, 1995; Olsen and Olson, 1995; Nishi *et al.*, 1998a; Saika, 2004; Werner *et al.*, 2004; Morrell and Pearce, 1989; Nasisse *et al.*, 1995), and IOL placement in the capsular bag (Lundgren *et al.*, 1992; Martin *et al.*, 1992).

PCO may occur in two forms that are referred to as a "fibrosis" type and a "pearl" formation. In humans, the fibrosis type usually is noted early, i.e., 2–6 months postoperatively and is related to inflammation whereas the pearl formation occurs several months to years later and is related to the migration, proliferation and transformation of lens epithelial cells (Apple *et al.*, 1992).

In vitro studies of lens epithelial cell cultures are routinely used to study the growth and behavior of these cells in response to potential therapeutic agents. Such studies are extremely helpful in evaluating pharmacological agents intended to destroy the lens cells that affect pearl formation, but have limited value in assessing fibrosis and the effects of breakdown of the blood aqueous barrier or the influence of IOL design on PCO. Thus, it is a routine procedure to evaluate PCO in any lens extraction/IOL implantation study. While *in vivo* models of proliferative PCO have been conducted in the rodent (Lois *et al.*, 2003, 2005; Call *et al.*, 2004; Medvedovic *et al.*, 2006), the rabbit model is the animal most frequently used for the study of PCO based on its similarity to the human anatomy, ease of surgery, time to development of PCO and cost effectiveness. These studies generally follow the surgical procedure described above with various modifications made depending on IOL design, potential diagnostic or therapeutic drug and/or device being evaluated.

As mentioned, PCO is known to be affected by surgical technique and inflammation/fibrin formation, IOL type and position in the capsule bag and is inversely related to the age of the animal and the size of the anterior capsulotomy/capsulorhexis. Therefore, it is helpful to have one surgeon performing all

TABLE 13.5 Posterior capsule opacification grading scales

Using direct illumination and retroillumination with pupil dilation using 1% Mydriacyl (Alcon Laboratories), the presence/absence of posterior capsule haze/opacity is graded on a 0–4 basis as follows:

None 0	Normal posterior capsule with no area of opacity. Red reflex bright.
Trace +1	Mild loss of transparency with cloudiness extending through most of posterior capsule. Red reflex mildly diminished.
Mild +2	Some loss of transparency involving the posterior capsule. Red reflex fairly bright.
Moderate +3	Moderate loss of transparency with difficulty visualizing the retina. Red reflex markedly diminished.
Severe +4	Posterior capsule very opaque with inability to view the retina. Red reflex barely visible.

The presence/absence of posterior capsule lens regrowth is graded on a 0–4 basis as follows:

None 0	Normal posterior capsule with no lens regrowth between the IOL optic and posterior capsule.
Trace +1	A thin layer of lens regrowth between the IOL optic and posterior capsule, less than 0.5 mm thick.
Mild +2	A small layer of lens regrowth between the IOL optic and posterior capsule, approximately 0.5 mm thick.
Moderate +3	Moderate lens regrowth between the IOL optic and posterior capsule, approximately 1 mm thick.
Severe +4	Lens regrowth between the IOL optic and posterior capsule is greater than IOL (>1.5 mm) thickness.

procedures, keeping the capsulotomy size constant and ensuring good lens cortical removal. Postoperatively, while inflammation/fibrin is graded on 0–4 scale, posterior synechiae and anterior lens regrowth/tissue ongrowth can be quantified by estimating percent/degree of pupillary area involvement. Adhesions between the anterior and posterior capsule create a barrier to the proliferation of lens epithelial cells and should be recorded. Posterior convex IOLs with angulated haptics appear to retard or partially inhibit PCO by placing the capsule on tension and increasing contact of the optic with the capsule (Tetz *et al.*, 1988a, 1994, 1996a; Auffarth *et al.*, 1994b; Kent, 1995; Saika, 2004). Thus, it is important to note IOL position in relation to the posterior capsule. A review of PCO and IOLs can be found in a study by Saika (2004) (Table 13.5).

Clinical Grading Based on Slit Lamp Biomicroscopy/Photography

Since the first introduction by Hansen *et al.* (1988) and Tetz *et al.* (1988b), various investigators have used multiple methods for estimating PCO formation based on slit lamp biomicroscopy and photography. Most have graded PCO directly behind the IOL optic area on a scale of 0–4.

Prior to evaluation, pupils can be dilated with 1% tropicamide and slit lamp photography can be performed with a 35 mm or a digital camera attached to a slit lamp. For direct illumination photos, photographs are taken with a diffuse slit beam placed at an approximately 30–45° angle between the observation and illumination axis with the incident light from the temporal side of the eye. The slit beam is focused on the posterior capsule and photographs are taken at 16× magnification. For retro-illumination photos, photographs are taken with a direct beam adjacent to the pupil margin, coaxial to the optical axis of the eye. The biomicroscope optics are focused on the posterior capsule and photographs are taken at 16× magnification.

A variety of methods may be used to grade the amount of PCO such as:

- The surface area of the optic involved with epithelial cell ingrowth is graded from 0 to 100% and multiplied by the thickness or density of the opacified area graded from 0 to 4 to give the relative amount of PCO (Hansen *et al.*, 1988; Tetz *et al.*, 1988a, 1994, 1996a, b). In a variation of this method, PCO is graded on a scale of 0–4 only in the 2 mm central optic zone (Irvine, 1980).
- PCO is graded on a scale of 0–3, with 0 = no PCO, 1 = PCO covering the haptic only, 2 = PCO covering part of the optic, 3 = PCO covering optic and haptic totally (Hettlich *et al.*, 1992).
- PCO is graded on a scale of 0–3 based on clarity of fundus view, with 0 = no PCO, 1 = minimal PCO, fundus visualized/clearly visible, 2 = moderate PCO, fundus partially obscured/blurred, 3 = severe PCO, fundus completely obscured/barely visible (Odrich *et al.*, 1985; Legler *et al.*, 1993).
- The degree of haze/opacity in the optic area has also been graded on a scale of 0–4 (Fig. 13.1, Table 13.4) with the thickness of the lens regrowth/regenerative tissue between the optic and posterior capsule graded separately on a 0–4 scale (Fig. 13.2; Table 13.3) (Gwon and Gruber, 1994).
- Photographic images have been graded on a scale of 0–10 in 0.5 steps with a completely clear capsule being graded as 0 and 10 being a capsule completely covered with severe, inhomogeneous, strongly light-attenuating PCO, classified as the most intense PCO possible (Sacu *et al.*, 2005).

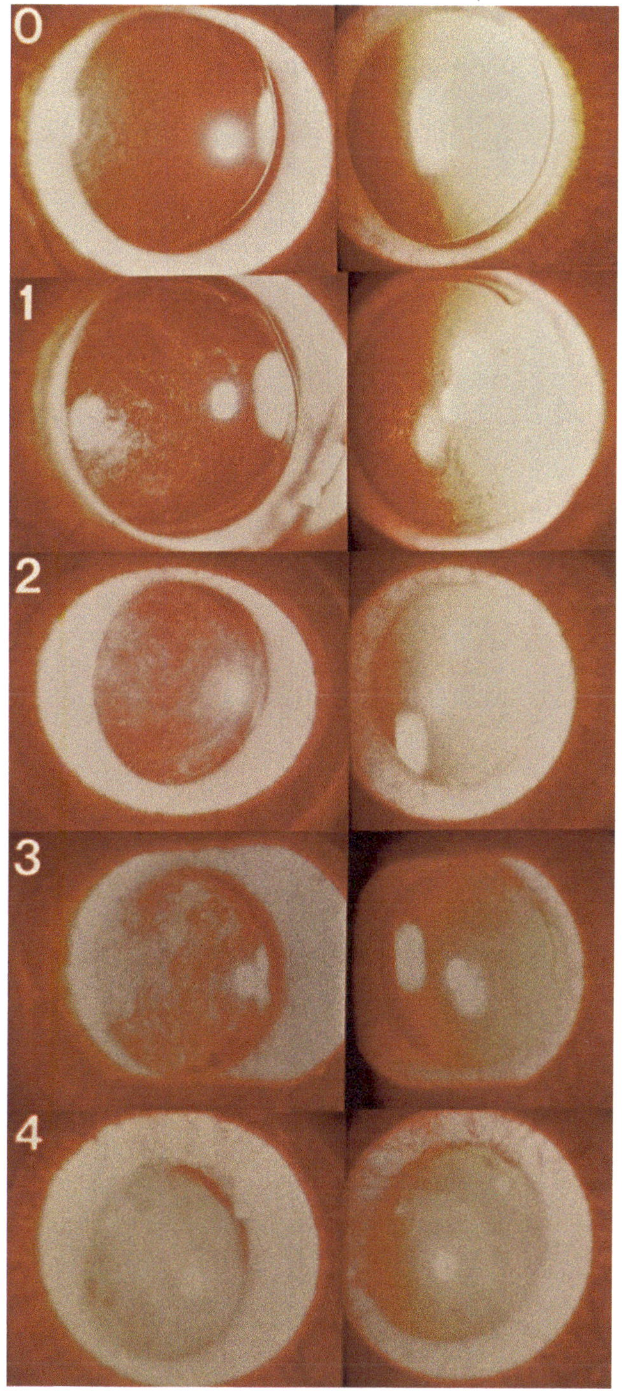

FIGURE 13.1 The presence/absence of *posterior capsule haze (degree of opacity)* is graded on a 0 to 4 basis.

Digital Image Analysis

Various investigators have used digital image analysis to objectively quantify PCO photographs and to monitor changes in the lens epithelial migration across the posterior capsule. Several investigators have used the Anterior Segment Analysis System (EAS-1000, Nidek, Inc.) to take a Scheimpflug slit image and transferred it to an image analysis computer which calculated the area densitometry (Hayashi *et al.*, 1998a,b,c, 2001 Tobari *et al.*, 1999; Wang and Woung, 2000; Hayashi, 2005). Hayashi found good correlation between the opacification density value and the patient's visual acuity (Hayashi *et al.*, 1998a,b,c, 2001; Hayashi, 2005).

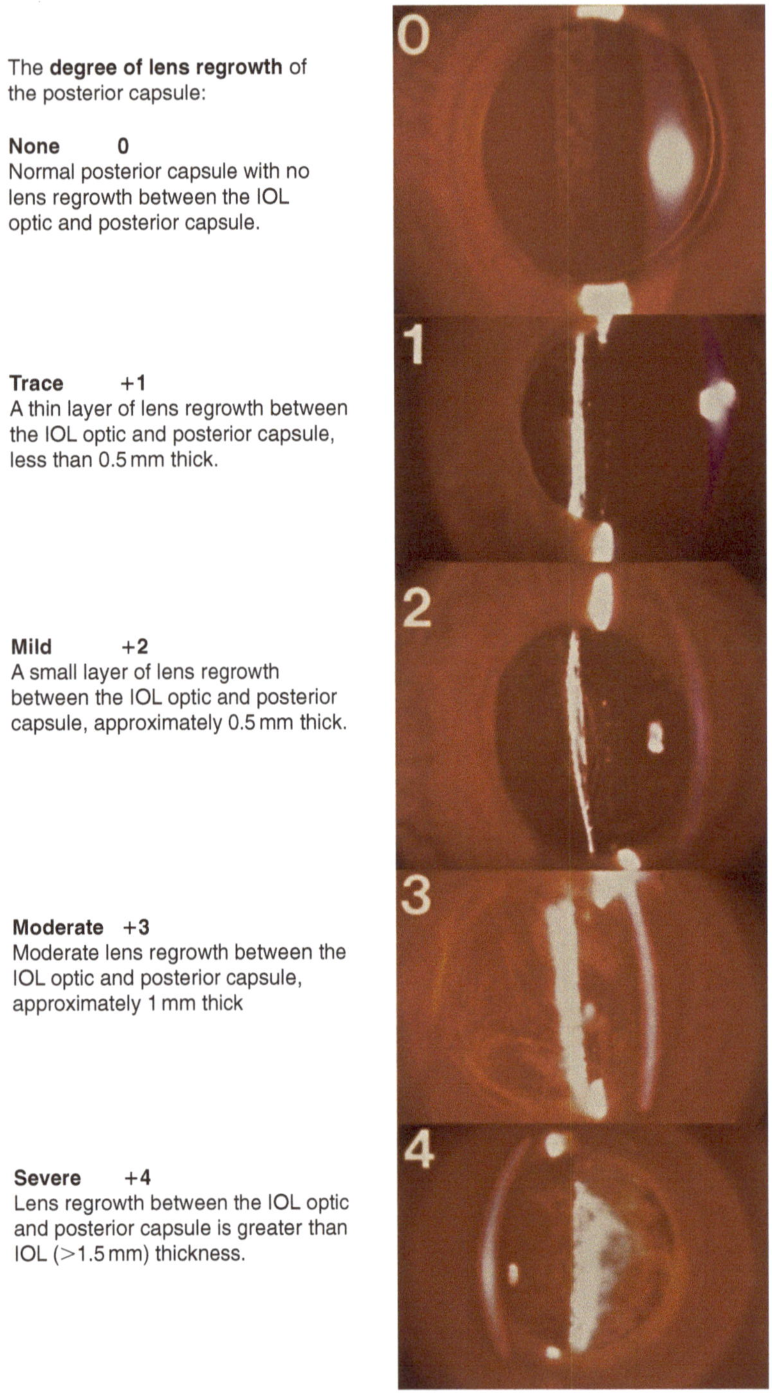

FIGURE 13.2 The presence/absence of *posterior capsule lens regrowth* is graded on a 0 to 4 basis.

Other investigators have taken standardized retro-illumination slit lamp photographs and developed computer software to quantify the area of PCO. The Evaluation of Posterior Capsule Opacification or EPCO system was first introduced by Tetz *et al.* (1997) and further evaluated in several subsequent studies (Tetz and Nimsgern, 1999; Auffarth *et al.*, 2003; Tetz and Wildeck, 2005). This commercially available system is based on the morphological assessment of PCO. The density of opacification in the area behind the IOL optic (usually 5.0–6.0 mm) is graded clinically from 0 to 4 with 0 = none, 1 = trace, 2 = mild, 3 = moderate and

4 = severe. The individual score is calculated by multiplying the density of the opacification by the fraction of capsule area involved behind the IOL optic (Tetz *et al.*, 1997; Tetz and Nimsgern, 1999; Auffarth *et al.*, 2003; Tetz and Wildeck, 2005). Nishi and coworkers (2005) applied this system to the postmortem assessment of PCO in rabbits. Following removal of the anterior capsule opacification, they took digital photographs of the posterior view and evaluated them on the EPCO system.

The POCOman system is an interactive program developed for the semi-objective assessment of PCO by researchers at King's College and St. Thomas' Hospital in London and is available free on the internet (http://www.ph.kcl.ac.uk/poco/POCOman.html). The digital retro-illumination images of the posterior capsule must be in bitmap format. The images are evaluated using pixel analysis based on texture differences. A grid, consisting of three equally spaced concentric ring segments is divided by radial lines forming 56 segments of approximately equal area. The observer marks segments with more than 50% of the area covered by PCO and grades them on a scale of 1–3 (1 = minimal texture, mild PCO, 2 = increased texture and pearls, moderate PCO, 3 = strongly textured and dark opacity, severe PCO). The program calculates the percentage area of PCO within the defined domain and a severity score ranging from 0 (totally clear) to 3 (total severe opacification) (Pande *et al.*, 1997; Ursell *et al.*, 1998; Hollick *et al.*, 1999; Barman *et al.*, 2000; Hollick *et al.*, 2000; Buehl *et al.*, 2002; Bender *et al.*, 2004; Wren *et al.*, 2005).

The AQUA (Automated Quantification of After-Cataract) system was developed at the University of Vienna in cooperation with the Technical University of Graz. It is also based on texture analysis of digital images. The program detects the capsulorhexis edge semi-automatically (computer-aided) and calculates the grade of disorder of a bitmap. This value is converted to a score between 0 and 10 (0 = clear capsule and 10 = severe PCO). The system is fully automated (there is no subjective aspect to the evaluation) and correlates well with subjective scoring of PCO (Friedman *et al.*, 1999 Findl *et al.*, 2003a; Buehl *et al.*, 2004, 2007; Sacu *et al.*, 2004a,b, 2005).

In a study by Findl *et al.* (2003b), their subjective method correlated well with the subjective EPCO system and the objective AQUA system. They found that the POCO system which assesses PCO area, did not adequately describe PCO density and includes a subjective step in the analysis in the process.

The duration of a PCO study has varied considerably throughout the literature. While slit lamp biomicroscopic/photographic grading and digital image analysis has been routine in human clinical studies, it is sometimes omitted in rabbit studies that are conducted for a short duration or supplemented with postmortem evaluation in longer duration studies. A postoperative follow-up time of 3 months in rabbits is roughly comparable with a 2–3 years follow-up in humans that is required for PCO comparison studies.

Postmortem Evaluation

At the end of the study, animals may be euthanized with an injection of sodium pentobarbital (Eutha-6, Western Medical Supply Co., Inc.) into the marginal ear vein. After euthanization, rabbit eyes can be enucleated and dissected approximately 10 mm posterior to the limbus. The IOL in the capsule bag can be viewed from the posterior aspect as described by Miyake and PCO can be graded according to the method described by Apple, commonly referred to as a "Miyake–Apple view. The amount of Soemmering formation is graded on a 0–4 basis with 0 = no PCO and 4 = severe PCO. This is similar to the clinical slit lamp grading scale and may be done separately for PCO in the optic area and in the periphery of the capsule bag. Digital photographic images can also be analyzed with appropriate software (Miyake and Miyaki, 1985; Apple *et al.*, 1989, 1990, 1992; Apple *et al.*, 1990; Apple *et al.*, 1992; Lundgren *et al.*, 1992; Zetterstrom *et al.*, 1996; Mamalis *et al.*, 1996; Chew *et al.*, 2006).

Another method used by Lundgren *et al.* (1992), is to measure the wet mass. The capsule bag can be removed and the IOL and capsule bag can be weighed separately to quantify PCO growth at the close of study. Recent studies by Tetz (Tetz *et al.*, 1996c) as well as those by others (Kugelberg *et al.*, 1997, Wallentin and Lundberg 2000; Wallentin *et al.*, 2000; Nishi, 2003) suggest a reasonable correlation between a clinical grading method and the gravimetric findings.

Standard histological and or electron microscopic examination of the eye tissue may also be performed. In recent studies, animals were euthanized as early as 2 weeks and as long as 1 year for histological determination of PCO. The shorter duration studies have generally been used to evaluate the inhibition of lens epithelial migration around a square IOL optic edge. The eyes can be fixed in 10% neutral buffered formalin, 2% glutaraldehyde or 1% osmium tetroxide and processed for standard light microscopy or electron microscopy. Tissue can then be stained with hematoxylin-eosin, periodic acid-Schiff and Masson's trichrome or other stains (Mamalis *et al.*, 1996; Saika *et al.*, 1996; Hepsen *et al.*, 1997; Nishi *et al.*, 1998a, 2004; Assia *et al.*, 1999; Werner *et al.*, 2000; Maloof *et al.*, 2005; Abdelwahab *et al.*, 2006; Matsushima *et al.*, 2006, 2007).

Accommodating IOL

The current trend in cataract and IOL placement entails the use of potentially accommodating lenses that would restore accommodation to the presbyopic eye. These lenses are often evaluated in the rabbit model for biocompatibility, ease of implantation, stability and fit in the capsule bag and interlenticular opacification and PCO (Hara *et al.*, 1992; Assia *et al.*, 1999; Pandey, 2002; Mireskandari *et al.*, 2004, 2005; Werner *et al.*, 2004, 2006b; Holmen *et al.*, 2005).

The rabbit has long been considered a poor model for studying accommodation because its zonule/ciliary muscle apparatus is poorly developed relative to that of the human (Smythe, 1958; Prince, 1964). Herbivores such as the rabbit are adapted for monitoring visual function and as such are believed to have negligible power of accommodation (Bito *et al.*, 1987). As reviewed in the anatomy section, the development of the ciliary body is almost negligible; the circular fibers seem to be missing altogether although there are many fine elastic fibers in the stromal tissue. The ciliary body is comparatively flat due to the scarcity of muscle fibers, with the thickest part being about 0.3mm (Smythe, 1958; Prince, 1964). Additionally, the few muscle fibers present are buried in dense connective tissue and they are not as granular as those in the cat and primates. Granules are known to be present in large numbers in animals having powerful accommodation. The location of the ciliary processes is also very different from that in the human. They arise from the anterior portion of the ciliary body, merge into the posterior surface of the iris at the base, and then extend to within 1mm of the pupillary margin of the iris, frequently being joined to the latter for much of their length (Prince, 1964).

The rabbit lens itself is resilient enough to permit from 1.5 to 2.5D of accommodation. Pupil dilatation and lens flattening has been reported upon stimulation of the long ciliary nerves. It is believed that the small amount of accommodation achieved by the rabbit eye is from engorgement of, or reduction of blood volume in the ciliary processes that in turn change both the diameter of the pupil and the position of the lens slightly, instead of from muscular action (Prince, 1964).

Thus, the accommodation mechanism in the rabbit is very different from that of the human or primate. With this understanding, an extensive literature search was conducted which yielded only one reference on refractive change in the phakic rabbit eye (Jungschaffer *et al.*, 1994). The investigators reported an increase in amplitude of accommodation after *in vivo* thermal treatment of the capsular bag. The pharmacologically induced amplitude of accommodation was determined to be 2.8D in pigmented rabbits by taking the difference in cycloplegic refraction (5D) after tropicamide 1% instillation and cyclospasmic refraction (2.2D) after pilocarpine 4% instillation. The effect, however, was short lived. There have been reports, however, on changes in refraction upon stimulation by a miotic agent in pseudophakic rabbit lenses that had been filled with an injectable silicone polymer (Agarwal *et al.*, 1967a; Nishi, 1989; Nishi *et al.*, 1998b).

Due to the scarcity of information on accommodation in rabbits, we conducted a series of studies to evaluate the utility of the rabbit model for evaluating potentially accommodating IOLs by monitoring changes in anterior chamber depth (ACD) using Scheimpflug imaging analysis before and after pharmacologic stimulation with carbachol 2.25% in phakic, aphakic, and pseudophakic eyes that had been implanted with a three-piece silicone IOL or a prototype silicone disc IOL (Gwon and Gruber, 1998, 2000a, b, 2002, 2004). The rabbits ranged in age from 8 weeks to 18 months at the time of testing and were 2 weeks to 15 months post lens extraction and IOL implantation.

While no appreciable response was detected in the three-piece silicone IOL eyes, ACD decreases as large as 0.6mm were detected with the silicone disc lens. The greatest change in ACD was noted 10–15 months postoperative suggesting the eye needs time to recover after IOL implant surgery to generate a significant response to carbachol stimulation. The IOL movement also appeared to depend on the age of the rabbit and the postoperative sequelae, which in turn depends on the design of the IOL and the surgical technique. Slit lamp biomicroscopy showed significantly less posterior synechia and PCO in young and adult eyes implanted with the silicone disc IOL than in adult eyes implanted with the three-piece silicone IOL and notably less than typically seen with other IOL implant studies in rabbits (Gwon and Gruber, 1998, 2000a,b, 2002, 2004).

It is noteworthy that slit lamp biomicroscopy and postmortem Miyake analysis showed that the posterior capsule lens opacification in the eyes implanted with both the three-piece silicone IOL and the silicone disc lenses diminished or disappeared over time. Assia (Assia *et al.*, 1999) similarly noted less PCO with a full-size IOL in rabbits implanted with hydrogel lenses that measured 10.0mm in diameter and 4.0 to 4.2mm thick. Other investigators have reported PCO regression in both rabbits and humans (Hollick *et al.*, 1998; Caballero *et al.*, 2000, 2001 Meacock *et al.*, 2001; Nakashima and

Yoshitomi, 2002; Neumayer *et al.*, 2005, 2006; Wolf and Findl, 2005, 2006). It is also of interest that no untoward events occurred with the silicone disc IOLs in a capsule devoid of any lens cellular growth. Thus the absence of posterior capsule lens epithelium does not appear to affect the integrity of the capsule or the ability of the lens to move axially in response to pharmacologic stimulation (Gwon and Gruber, 1998, 2000a, b, 2002, 2004).

In summary, the rabbit lens or capsular bag with and without an IOL is capable of a small axial movement upon carbachol stimulation. Whether this is related to changes in the iris and or ciliary muscle has not been determined. However, it is of interest that a 40% decrease in accommodation was reported when the iris was removed prior to accommodative stimulus in the Rhesus monkey indicating a strong iris component to accommodation in another mammal (Neider *et al.*, 1990).

The amount of movement/decrease in ACD seen in these rabbit studies is considerably less than the desired minimum 1.0 mm change required for a significant increase in diopter power with an accommodating IOL. Thus, following initial biocompatibility studies in rabbits, evaluation of novel potentially accommodating IOLs is generally performed in the primate model.

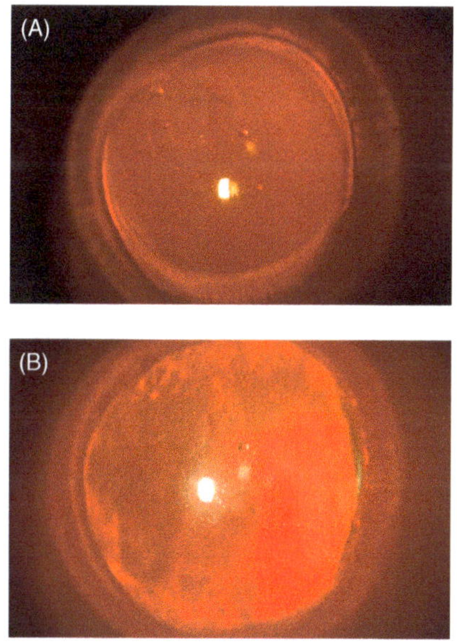

FIGURE 13.3 Injectable Polymer with Silicone Plug: (a) day 1 postoperative; (b) day 102 postoperative.

Lens Refilling

The concept of injecting a synthetic polymer to replace the natural or cataractous crystalline lens was first suggested by Kessler (1964, 1966, 1975). Since that time numerous investigators have worked on developing a suitable polymer that would have the flexibility of the natural lens and be capable of restoring accommodation in the presbyopic and/or cataractous eye. These polymers are generally liquid for easy injection into the capsule bag. Once in the bag, they polymerize to create a lens that molds to the shape of the capsule bag. Most investigators have utilized silicone polymers of low modulus, such as polydemethylsiloxane, or hydrogels, such as poly(1-hydroxy-1,3-propandiyl) and acrylamide. A recent publication by Norrby (2005) as well as by others (Agarwal *et al.*, 1967a,b; Parel *et al.*, 1981, 1986; Haefliger *et al.*, 1987; Hettlich *et al.*, 1994; Nishi *et al.*, 1998b; de Groot *et al.*, 2001; Han *et al.*, 2003; Koopmans *et al.*, 2003; Aliyar *et al.*, 2005; Kwon *et al.*, 2005; Yoo *et al.*, 2006; Wong *et al.*, 2007) have reviewed the current state of injectable polymers for lens refilling.

The standard technique for evaluation in the *in vivo* rabbit model is a modification of the endocapsular lens extraction described by Gindi (Gindi *et al.*, 1985) and Gwon (Gwon et al., 1993a). A 1 mm anterior capsulotomy is recommended to optimize the refractive result and a 2.7 or 4.5 mm silicone plug is inserted into the capsule bag and maneuvered behind the anterior capsulotomy. The silicone plug is well tolerated in both rabbits and primates and remains permanently (Fig. 13a 13b). The injectable polymer is then injected by sliding the silicone plug to facilitate placement of the cannula for delivery of the polymer. When the desired amount of silicone material has filled the capsule bag, the silicone plug is repositioned behind the capsulotomy (Tahi *et al.*, 1999, 2002; Stachs *et al.*, 2003). If a silicone plug is not available, a collagen patch can be used to seal the capsulotomy (Gwon *et al.*, 1993a). Another method suggested by Nishi and Hara in several studies is to insert an inflatable endocapsular balloon that can then be filled with the polymer (Nishi, 1989; Nishi *et al.*, 1989, 1992, 1997; Hara *et al.*, 1994).

The rabbit model continues to be the primary *in vivo* system used by investigators as they strive to resolve issues associated with this technology, including scarring and folds in the capsule, epithelial cell proliferation and secondary capsular opacification.

Lens Regeneration

Since the first description by Cocteau and D'Etoille (Cocteau, 1827), the residual lens epithelial cells that contribute to PCO have been shown to regenerate and differentiate more normally if the integrity of the lens capsule is restored following endocapsular lens extraction in rabbits (Mayer, 1832; Middlemore, 1832; Valentin, 1844; Milliot, 1872; Randolph, 1900; Sikharuldze, 1956; Stewart, 1960). Restoring the lens capsule integrity by insertion of a collagen patch at the time of surgery has enhanced the growth rate and shape/structure of the regenerated lenses (Gwon *et al.*, 1993a). Lens fiber differentiation has been shown to follow a process similar to embryological development with cellular proliferation along the anterior and posterior capsule, followed by elongation of the posterior epithelial cells, anterior migration of fiber nuclei and subsequent differentiation at the equatorial zone (Gwon *et al.*, 1990). The regenerated lenses have been shown to contain all the major crystallins (alpha, beta and gamma) in proportions similar to fetal or normal lenses (Gwon *et al.*, 1989). Regeneration is noted as early as 2–3 weeks postoperatively and capsule bag filling with regenerated lens tissue is seen at 7–10 weeks postoperatively (Table 13.1) (Gwon *et al.*, 1992). In addition, lens regeneration has been shown to occur after endocapsular extraction of a concanavalin A-induced cataract (Gwon *et al.*, 1993b). A historical review of lens regeneration in mammals can be found in a recent paper by Gwon (2006).

As an isolated organ, relatively free from systemic and vascular influence, the lens regeneration model provides a controlled environment. It is well suited for studying the process of cell growth and differentiation and as an investigative tool in research aimed at preventing secondary cataracts. While lens regeneration has been demonstrated in other mammals, including mice, cats, dogs and monkeys (Cocteau, 1827; Randolph, 1900; Agarwal *et al.*, 1964; Gwon *et al.*, 1993a; Gwon and Gruber, 1998; Shekhawat *et al.*, 2001; Gwon, 2006) it is well characterized in rabbits and a significant "starting point" for the successful regeneration of the human lens, a "natural" replacement lens for a cataract that would have the refractive and accommodative properties of the original lens.

The procedure for our lens regeneration studies is similar to that described for general cataract/lens surgery and for endocapsular lens extraction by Gindi (Gindi *et al.*, 1985) which is modified for capsule bag sealing (Gwon *et al.*, 1993a). Following routine dilation, betadine prep, and eyelash trim, a limbal incision is made at the 12:00 O'clock position with a 2.85 mm keratome. A small 1–2 mm anterior capsulotomy is performed and phacoemulsification and irrigation/aspiration of the lens is performed using balanced salt solution without heparin or epinephrine. A small amount of hyaluronic acid is injected to separate the anterior and posterior capsules and facilitate placement of a collagen patch. A collagen shield is cut freehand to approximately 2–3 times the size of the capsulotomy which varies between 1.0 and 2.0 mm. The customized collagen patch is coated with a viscoelastic and inserted into the capsule bag. A lens hook is used to maneuver the patch behind the anterior capsulotomy with at least a 1 mm overlap internally. A biodegradable viscous material such as hyaluronic acid is used to distend the capsule bag followed by a small air bubble to stabilize the patch against the capsule. The capsule bag is irrigated to remove the hyaluronic acid if using a non-hyaluronic acid scaffold (Gwon and Gruber, 2007).

Postoperative antibiotics and corticosteroids (as detailed above) are given as is routine for any cataract surgical procedure. The postoperative course is generally mild with inflammation (i.e. anterior chamber cells/flare/fibrin) usually lasting approximately 1 week with a closed capsule bag as compared to 2–3 weeks when an IOL is implanted and the anterior capsule remains open. The collagen patch usually dissolves by 2–3 weeks leaving a small linear anterior capsulotomy scar (Gwon *et al.*, 1990). Regenerated lens tissue can be seen to progress from the periphery centrally beginning at about 2–3weeks and complete filling of the capsule bag occurs as early as 4–6 weeks depending on the age of the animal and the type of scaffold implanted (Fig. 13.4).

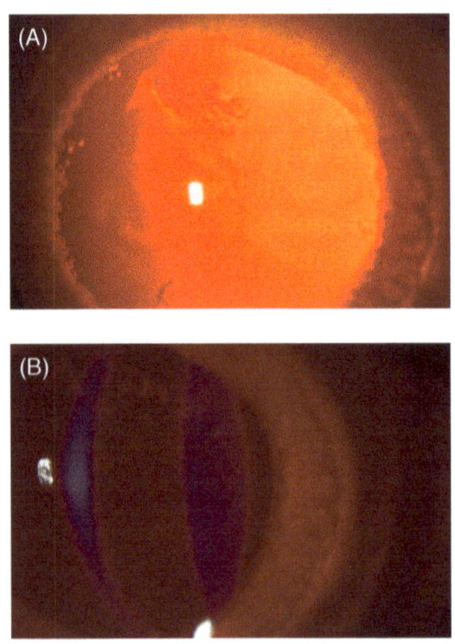

FIGURE13.4 Regenerated Lens Day 43.

SUMMARY

In summary, the rabbit is similar to man in its response to lens/cataract surgery. The postoperative inflammatory reaction, development of PCO, PCO regression and lens regenerative capability make it an excellent *in vivo* model for the assessment of new technology for the treatment of cataracts and many other ocular surgical procedural improvements. However, as the rabbit has limited accommodative ability, the primate model is better suited to the evaluation of accommodation and potentially accommodating IOLs.

REFERENCES

Abdelwahab MT, Kugelberg M, Kugelberg U, Zetterstrom C (2006). After-cataract evaluation after using balanced salt solution, distilled deionized water and fluorouracil with sealed-capsule irrigation device in eyes of 4-week-old rabbits. *J Cataract Refract Surg* 32:1955–1960.

Agarwal LP, Angra SK, Khosla PK, Tandon HD (1964). Lens regeneration in mammals II: Monkeys. *Orient Arch Ophthalmol* 2:47–59.

Agarwal LP, Narsimhan EC, Mohan M (1967a). Experimental lens refilling. *Orient Arch Ophthalmol* 5:205–212.

Agarwal LP, Narsimhan EC, Mohan M (1967b). Experimental lens refilling. *Orient Arch Ophthalmol* 5:278–280.

Aliyar HA, Hamilton PD, Ravi N (2005). Refilling of ocular lens capsule with copolymeric hydrogel containing reversible disulfide. *Biomacromolecules* 6:204–211.

Anon. ANSI Z80.7−2022, Ophthalmic Optics—Intraocular Lenses ISO 11979−5:1999, Ophthalmic implants–Intraocular lenses. Part 5: Biocompatibility. Available from: http://www.hc-sc.gc.ca. Accessed June 2007.

Apple D, Mamalis N, Loftfield K *et al.* (1984). Complications of intraocular lenses. A historical and histopathological review. *Surv Ophthalmol* 29:1–54.

Apple DJ, Kincaid MC, Mamalis N, Olson RJ (1989). *Intraocular Lenses: Evolution, Designs, Complications, and Pathology*. Williams & Wilkins, Baltimore, MD.

Apple DJ, Lim ES, Morgan RC *et al.* (1990). Preparation and study of human eyes obtained with Miyake posterior photographic technique. *Ophthalmology* 97:810–816.

Apple DJ, Solomon KD, Tetz MR *et al.* (1992). Posterior capsule opacification. *Surv Ophthalmol* 37:73–116.

Assia EI, Apple DJ (1992). Side-view analysis of the lens. I: The crystalline lens and the evacuated bag. *Arch Ophthalmol* 110:89–93.

Assia E, Apple D, Barden A *et al.* (1991). An experimental study comparing various anterior capsulectomy techniques. *Arch Ophthalmol* 109:642–647.

Assia EI, Blumenthal M, Apple DJ (1999). Effect of expandable full-size intraocular lenses on lens centration and capsule opacification in rabbits. *J Cataract Refract Surg* 25:347–356.

Auffarth GU, Wesendahl TA, Newland TJ, Apple DJ (1994a). A capsulorhexis in the rabbit eye as a model for pediatric capsulotomy. *J Cataract Refract Surg* 20:188–191.

Auffarth GU, Wesendahl TA, Assia EI, Apple DJ (1994b). The concept of modern capsular surgery. *Proc Illinois Soc Ophthalmol Otolaryngol* 1:5–25.

Auffarth GU, Wesendahl, TA, Assia EI, Apple DJ (1995). Pathophysiology of modern capsular surgery. In: Steinert RF (ed.), *Cataract Surgery: Technique, Complications, and Management*. WB Saunders, Philadelphia, PA, pp. 314–324.

Auffarth GU, Golescu A, Becker MD, Völcker HE (2003). Quantification of posterior capsule opacification with round and sharp edge intraocular lenses. *Ophthalmology* 110:772–780.

Barman SA, Hollick EJ, Boyce JF *et al.* (2000). Quantification of posterior capsule opacification in digital images after cataract surgery. *Invest Ophthalmol Vis Sci* 41:3882–3892.

Bender L, Spalton DJ, Uyanonvara B *et al.* (2004). POCOman: New system for quantifying posterior capsule opacification. *J Cataract Refract Surg* 30:2058–2063.

Bito LZ, Kaufman PL, DeRousseau CJ, Koretz J (1987). Presbyopia: An animal model and experimental approaches for the study of the mechanism of accommodation and ocular ageing. *Eye* 1:222–230.

Born C, Ryan D (1990). Effect of intraocular lens optic design on posterior capsular opacification. *J Cataract Refract Surg* 16:188–192.

Bron AJ, Tipathi RC, Tripathi BJ (1997). *Wolff's Anatomy of the Eye and Orbit*, 8th edn. Chapman and Hall Medical, London, England.

Buchen SY, Cunanan CM, Gwon A, Weinschenk, JI, III, Gruber, L, Knight PM (2001). Assessing intraocular lens calcification in an animal model. *J Cataract Refract Surg* 27:1473–1484.

Bucher P (1994). Calcification of an IOGEL 1003 lens. Foldable technology symposium. *Eur J Implant Ref Surg* 6:175.

Bucher PJM, Buchi ER, Daiker BC (1995). Dystrophic calcification of an implanted hydroxyethylmethacrylate intraocular lens. *Arch Ophthalmol* 113:1431–1435.

Buehl W, Findl O, Menapace R *et al.* (2002). Effect of an acrylic intraocular lens with a sharp posterior optic edge on posterior capsule opacification. *J Cataract Refract Surg* 28:1105–1111.

Buehl W, Menapace R, Sacu S *et al.* (2004). Effect of a silicone intraocular lens with a sharp posterior optic edge on posterior capsule opacification. *J Cataract Refract Surg* 30:1661–1667.

Buehl W, Menapace R, Findl O, Neumayer T, Bolz M, Prinz A (2007). Long-term effect of optic edge design in a silicone intraocular lens on posterior capsule opacification. *Am J Ophthalmol* 143:913–919.

Caballero A, Marin JM, Salinas M (2000). Spontaneous regression of Elschnig pearl posterior capsule opacification. *J Cataract Refract Surg* 26:779–780.

Caballero A, Garcia-Elskamp C, Losada M, Salinas M, Marin JM (2001). Natural evolution of Elschnig pearl posterior capsule opacification after posterior capsulotomy. *J Cataract Refract Surg* 27:1979–1986.

Call MK, Grogg MW, Del Rio−Tsonis K, Tsonis PA (2004). Lens regeneration in mice: Implications in cataracts. *Exper Eye Res* 78:297–299.

Chew J, Werner L, Stevens S, Hunter B, Mamalis N (2006). Evaluation of the effects of hydrodissection with antimitotics using a rabbit model of Soemmering's ring formation. *Clin Exp Ophthalmol* 34:449–456.

Cobo LM, Ohsawa E, Chandler BS *et al.* (1984). Pathogenesis of capsular opacification after extracapsular cataract extraction: An animal model. *Ophthalmology* 91:857–863.

Cocteau LL (1827). Reproduction du crystallin. *J Physiologie ExpPathologique* 1:30.

Cohen J (1988). *Statistical Power Analysis for the Behavioral Sciences*, 2nd edn. Lawrence Relbaum Associates, Hillsdale, NJ.

Dangel ME, Kirkham SM, Phipps MJ (1994). Posterior capsule opacification in extracapsular cataract extraction and the triple procedure: A comparative study. *Ophthalmic Surg* 25:82–87.

Davis FA (1929). The Anatomy and Histology of the Eye and Orbit of the Rabbit. *Trans Am Ophthalmol Soc* 27:402–441.

Davis P, Hill P (1989). Inhibition of capsule opacification by convex surface posterior three-piece all PMMA C-loop lenses: A fellow eye and same lens study. *Eur Implant Refract Surg* 1:237–240.

Davis PL, Hill P, Coffey A (1991). Convex posterior PMMA implants: Do PMMA vs prolene haptics alter capsular opacity. *Eur J Implant Refract Surg* 3:127–130.

de Groot JH, van Beijma FJ, Haitjema HJ *et al.* (2001). Injectable intraocular lens materials based upon hydrogels biomacromolecules. *Biomacromolecules* 2(3):628–634.

Downing JE (1986). Long-term discission rate after placing posterior chamber lenses with the convex surface posterior. *J Cataract Refract Surg* 12:651–654.

Duke-Elder S (1958). The eye of mammals. In: Sir Stewart , Duke-Elder (ed.), *System of Opthalmology, Volume 1: The Eye in Evolution*. Henry Kimpton, London, pp. 49–507. Chapter XV

Fagerholm PP, Philipson BT (1981). Human lens epithelium in normal and cataractous lenses. *Invest Opthalmol Vis Sci* 21:408–414.

Fatt I (1978). *Physiology of the Eye: An Introduction to the Vegetative Functions*. Butterworth Inc, Woburn, MA.

Findl O, Buehl W, Siegl H, Pinz A (2003a). Removal of reflections in the photographic assessment of PCO by fusion of digital retroillumination images. *Invest Ophthalmol Vis Sci* 44: 275–280.

Findl O, Buehl W, Menapace R *et al.* (2003b). Comparison of four methods for quantifying posterior capsule opacification. *J Cataract Refract Surg* 29:106–111.

Foot L, Werner L, Gills JP *et al.* (2004). Surface calcification of silicone plate intraocular lenses in patients with asteroid hyalosis. *Am J Ophthalmol* 137:979–987.

Friedman DS, Duncan DD, Munoz B, West SK, Schein OD (1999). Digital image capture and automated analysis of posterior capsule opacification. *Invest Ophthalmol Vis Sci* 40:1715–1726.

Galand A, Bonhomme L, Collée M (1984). Direct measurement of the capsular bag. *Am Intra-ocular Implant Soc J* 10:475–476.

Gimbel HV, Neuhman T (1990). Development, advantages and methods of the continuous circular capsulorrhexis technique. *J Cataract Refract Surg* 16:31–37.

Gindi JJ, Wan WL, Schanzlin DJ (1985). Endocapsular cataract surgery-1. Surgical technique: Cataract. *Int J of Cataract Surg* 2:5–10.

Green WT, Boase DL (1989). How clean is your capsule. *Eye* 3:678–684.

Gwon A (2006). Lens regeneration in mammals: A review. *Surv Ophthalmol* 51:51–62.

Gwon A, Gruber L (1994a). *Evaluation of Capsulorhexis Size on IOLs*. Data on File AMO#120, July 11.

Gwon A, Gruber L (1994b). *Determination of the Inflammatory Potential of SI-40 IOLs in a Rabbit Intraocular Implant Model*. Data on File AMO#119, December 15.

Gwon A, Gruber L (1994c). *An Evaluation of AMO DuraFlex vs Standard PMMA vs Froste IOL Implantation in the Rabbit Eye*. Data on File AMO#122, December 29.

Gwon A, Gruber L (1998). *An Evaluation of the Accommodative Potential in Phakic, Sphakic and Pseudophakic New Zealand white and Dutch Belt Rabbits*. Data on File AMO#205, December.

Gwon A, Gruber L (2000a). *An Evaluation of the Accommodation Potential of the Aphakic and Pseudophakic Rabbit Eye Implanted with a Flat or Angled Silicone Disc IOL (11 mm)*. Data on File AMO#183 and 235, December.

Gwon A, Gruber L (2000b). *An Evaluation of the Acccommodative Potential in Pseudophakic Silicone Disc IOL (10 mm) and Aphakic Young, Young Adult and Adult New Zealand White Rabbits*. Data on File AMO#227, December.

Gwon A, Gruber L (2002). *An Evaluation of the Rabbit Lens Capsular Gag Diameter for Implantation of IOLs and an Assessment of the Ease of Viscoelastic Removal Following Implantation of a Silicone Disc IOL*. Data on File AMO#308, January 24.

Gwon A, Gruber L (2004). An animal model for the evaluation of accommodating IOLs. *Invest Ophthalmol Vis Sci* 45(suppl 1):167.

Gwon A, Gruber L (2007). Lens regeneration with an injectable polymeric scaffold. *Invest Ophthalmol Vis Sci*(suppl):5434.

Gwon A, Enomoto H, Horowitz J, Garner M (1989). Induction of de novo synthesis of crystalline lenses in aphakic rabbits. *Exp Eye Res* 49:913–926.

Gwon AE, Gruber L, Mundwiler K (1990). A histologic study of lens regeneration in aphakic rabbits. *Invest Ophthalmol Vis Sci* 31:540–547.

Gwon AE, Jones RL, Gruber LJ, Mantras C (1992). Lens regeneration in rabbits measured by image analysis. *Invest Ophthalmol Vis Sci* 33:2279–2283.

Gwon A, Mantras C, Gruber L (1993a). Restoring lens capsule integrity enhances lens regeneration in New Zealand albino rabbits and cats. *J Cataract Refract Surg* 19:735–746.

Gwon A, Mantras C, Gruber L, Cunanan C (1993b). Lens regeneration in NZA rabbits following endocapsular cataract extraction. *Invest Ophthalmol* 34:2124–2129.

Gwon A, Kuszak J, Gruber L (1998). Intralenticulars in pigmented rabbits: Opacity lensmeter assessment. *J Cataract Refract Surg* 25:268–277.

Haefliger E, Parel JM, Fantes F *et al.* (1987). Accommodation of an endocapsular silicone lens (Phaco-Ersatz) in the nonhuman primate. *Ophthalmology* 94:471–477.

Han YK, Kwon JW, Kim JS, Cho CS, Wee WR, Lee JH (2003). In vitro and in vivo study of lens refilling with poloxamer hydrogel. *Br J Ophthalmol* 87:1399–1402.

Hansen TE, Otland N, Corydon L (1988). Posterior capsule fibrosis and intraocular lens design. *J Cataract Refract Surg* 14: 383–386.

Hara T, Hara T, Yasuda A, Mizumoto Y, Yamada Y (1992). Accommodative intraocular lens with spring action. Part 2: Fixation in the living rabbit. *Ophthalmic Surg* 23(9):632–635.

Hara T, Sakka Y, Sakanishi K, Yamada Y, Nakamae K, Hayashi F (1994). Complications associated with endocapsular balloon implantation in rabbit eyes. *J Cataract Refract Surg* 20:507–512.

Hayashi K, Hayashi H, Nakao F, Hayashi F (1998a). Reproducibility of posterior capsule opacification measurement using Scheimpflug videophotography. *J Cataract Refract Surg* 24:1632–1635.

Hayashi K, Hayashi H, Nakao F, Hayashi F (1998b). In vivo quantitative measurement of posterior capsule opacification after extracapsular cataract surgery. *Am J Ophthalmol* 125:837–843.

Hayashi H, Hayashi K, Nakao F, Hayashi F (1998c). Quantitative comparison of posterior capsule opacification after polymethacrylate, silicone and soft acrylic intraocular lens implantation. *Arch Ophthalmol* 116:1579–1582.

Hayashi K, Hayashi H (2005). Posterior capsule opacification in the presence of an intraocular lens with a sharp versus rounded optic edge. *Ophthalmology* 112:1550–1556.

Hayashi K, Hayashi H, Nakao F, Hayashi F (2001). Changes in posterior capsule opacification after poly(methyl methacrylate), silicone and acrylic intraocular lens implantation. *J Cataract Refract Surg* 27:817–824.

Hepsen IF, Bayramlar H, Gultek A, Ozen S, Tilgen F, Evereklioglu C (1997). Caffeic acid phenylethly ester to inhibit posterior capsule opacification in rabbits. *J Cataract Refract Surg* 23:1572–1576.

Hettlich HJ, Kaufman R, Harmeyer H (1992). In vitro and in vivo evaluation of a hydrophilized silicone intraocular lens. *J Cataract Refract Surg* 18:140–146.

Hettlich HJ, Lucke K, Asiyo-Vogel MN, Schulte M, Vogel A (1994). Lens refilling and endocapsular polymerization of an injectable intraocular lens: In vitro and in vivo study of potential risks and benefits. *J Cataract Refract Surg* 20:115–123.

Hollick EJ, Spalton DJ, Ursell PG, Pande MV (1998). Lens epithelial cell regression the posterior capsule with different intraocular lens materials. *Br J Ophthalmol* 82:1182–1188.

Hollick EJ, Spalton DJ, Ursell PG *et al.* (1999). The effect of polymethylmethacrylate, silicone and polyacrylic intraocular lenses on posterior capsule opacification 3 years after cataract surgery. *Ophthalmology* 106:49–55.

Hollick EJ, Spalton DJ, Ursell PG, Meacock WR, Barman S, Boyce J (2000). Posterior capsule opacification with hydrogel, polymethylmethacrylate and silicone intraocular lenses: Two--year results of a randomized prospective trial. *Am J Ophthalmol* 129:577–584.

Holmen JB, Soderberg P, Lundgren B (2005). *Prophylactic treatment of posterior capsular opacification (PCO): A comparison between perfect-capsule and PCO-guard IOVs* 46(suppl S):2858.

http://www.hc-sc.gc.ca/dhp-mps/md-im/standards normes/md_rec_stand_im_norm_lst_2_e.html#10, accessed June 21, 2007.

http://www.netvet.wustl.edu/species/rabbits/rabtmodl.txt, accessed June 21, 2007.

http://www.ph.kcl.ac.uk, accessed June 21, 2007.

Irvine A (1980). Extracapsular cataract extraction and pseudophakos implantation in primates: A clinico-pathologic study. *Tr Am Ophthalmol Soc* 2:781–807. LXXCVIII

Jacob TJC (1987). Human lens epithelial cells in culture: A quantitative evaluation of growth rate and proliferative capacity. *Exp Eye Res* 45:93–104.

Jacobi K (1985). Die behandlung der cataracta secundaria. *Fortschr Ophthalmol* 82:405–409.

Jaffee NS, Horwitz J (1992). Lens and cataract. In: *Textbook of Ophthalmology*, Vol. 3. Gower Medical Publishing, New York, NY.

Jungschaffer DA, Saber E, Zimmerman KM *et al.* (1994). Refractive changes induced by electrocautery of the rabbit anterior lens capsule. *J Cataract Refract Surg* 20:132–137.

Kent DG, Sims JCR, Apple DJ (1995). Pediatric capsulorhexis technique. *J Cataract Refract Surg* 21:236. Letter

Kessler J (1964). Experiments in refilling the lens. *Arch Ophthalmol* 71:412–417.

Kessler J (1966). Refilling the rabbit lens. Further experiments. *Arch Ophthalmol* 76:596–598.

Kessler J (1975). Lens refilling and regrowth of lens substance in the rabbit eye. *Ann Ophthalmol* 7:1059–1062.

Kleinman G, Apple DJ, Chew J *et al.* (2006). Hydrophilic acrylic intraocular lens as a drug-delivery system: Pilot study. *J Cataract Refract Surg* 32:652–654.

Koopmans SA, Terwee T, Barkhof J, Haitjema HJ, Kooijman A (2003). Polymer refilling of presbyopic human lenses in vitro restores the ability to undergo accommodative changes. *Invest Ophthalmol Vis Sci* 44:250–257.

Koura Y, Fukushima A, Nishino K *et al.* (2006). Inflammatory reaction following cataract surgery and implantation of acrylic intraocular lens in rabbits with endotoxin-induced uveitis. *Eye* 20:606–610.

Kugelberg U, Zetterstrom C, Lundgren B, Syren-Nordquist S (1997). Ocular growth in newborn rabbit eyes implanted with apoly(methyl methacrylate) or silicone intraocular lens. *J Cataract Refract Surg* 23:629–634.

Kurz S, Krummenauer F, Dumbach C, Pfeiffer N, Dick B (2006). Effect of a closed foldable equator ring on capsular bag shrinkage in cataract surgery. *J Cataract Refract Surg* 32:1615–1620.

Kuszak JR, Costello, MJ (2004). The structure of the vertebrate lens. In: Lovicu FJ, Robinson ML (eds), *Development of the Ocular Lens*. Cambridge University Press, Cambridge, UK, pp. 71–118.

Kuszak JR, Sivak JG, Weerheim JA (1991). Lens optical quality is a direct function of lens sutural architecture. *Invest Ophthalmol Vis Sci* 32(7):2119–2129.

Kuszak JR, Sivak JG, Herbert KL, Scheib S, Garner W, Graff G (2000). The relationship between rabbit lens optical quality and sutural anatomy after vitrectomy. *Exp Eye Res* 71:267–282.

Kwon JW, Han YK, Lee WJ *et al.* (2005). Biocompatibility of poloxamer hydrogel as an injectable intraocular lens: A pilot study. *J Cataract Refract Surg* 31:607–613.

Laurell CG, Wickstrom K, Zetterstrom C, Lundgren B (1997). Inflammatory response after endocapsular phacoemulsification or conventional extracapsular lens extraction in the rabbit eye. *Acta Ophtalmol Scandinavica* 75:401–404.

Legler UFC, Apple DJ, Assia EI *et al.* (1993). Inhibition of posterior capsule opacification: The effect of colchicine in a sustained drug delivery system. *J Cataract Refract Surg* 19:462–470.

Liesgang TJ, Bourne WM, Ilstrup DM (1985). Secondary surgical neodymium: YAG laser discissions. *Am J Ophthalmol* 100:510–519.

Lois N, Dawson R, McKinnon AD, Forrester JV (2003). A new model of posterior capsule opacification in rodents. *Invest Ophthalmol Vis Sci* 44:3450–3457.

Lois N, Taylor J, McKinnon AD, Forrester JV (2005). Posterior capsule opacification in mice. *Arch Ophthalmol* 123:71–77.

Lowes M (1990). The effect of posterior vaulting of intraocular lens implants on capsular opacification. *Eur J Implant Refract Surg* 2:47–52.

Lundgren B, Jonsson E, Rolfsen W (1992). Secondary cataract. An in vivo model for studies on secondary cataract in rabbits. *Acta Ophthalmol* 70(suppl 205):25–28.

Maloof AJ, Pandey SK, Neilson G, Milverton EJ (2005). Selective death of lens epithelial cells using demineralized water and Triton X-100 with perfect capsule sealed capsule irrigation. A histological study in rabbit eyes. *Arch Ophthalmol* 123:1378–1384.

Maltzman B, Haupt E, Notis C (1989). Relationship between age at time of cataract extraction and time interval before capsulotomy for opacification. *Ophthalmic Surg* 20:321–324.

Mamalis N, Crandall A, Linebarger E *et al.* (1995). Effect of intraocular lens size on posterior capsule opacification after phacoemulsification. *J Cataract Refract Surg* 21:99–102.

Mamalis N, Omar O, Veiga J, Tanner D, Pirayesh A, Fernquist D (1996). Comparison of two plate-haptic intraocular lenses in a rabbit model. *J Cataract Refract Surg* 22:1291–1295.

Martin RG, Sanders DR, Souchek J, Raanan MG, DeLuca M (1992). Effect of posterior chamber intraocular lens design and surgical placement of postoperative outcome. *J Cataract Refract Surg* 18:333–341.

Matsushima H, Iwamoto H, Mukai K, Obara Y (2006). Active oxygen processing for acrylic intraocular lenses to prevent posterior capsule opacification. *J Cataract Refract Surg* 32:1035–1040.

Matsushima H, Iwamoto H, Mukai K, Senoo T, Obara Y (2007). Prevention of posterior capsule opacification using round-edged PMMA IOL. *J Cataract Refract Surg* 33:1133–1134.

Mayer M (1832). Uber die reproduktion der krystallinse. Berlin: von Graefe und Walther. *Journal der Chirurgie und Augenheilkunde* 17:524.

McDonnell P, Zarbin M, Green W (1983). Posterior capsule opacification in pseudophakic eyes. *Ophthalmology* 90: 1548–1553.
McDonnell PJ, Stark WJ, Green RG (1984). Posterior capsular opacification: A specular microscopic study. *Ophthalmology* 9(7):853–856.
McDonald TO, Shadduck JA (1977). Eye irritation. In: Marzulli , Maibach (eds), *Advances in Modern Toxicology, Volume 4: Dermatotoxicology and Pharmacology*. Hemisphere, Washington, pp. 162–166.
Meacock WR, Spalton DJ, Hollick EJ, Barman S, Boyce JF (2001). The effect of polymethylmethacrylate and acrysof intraocular lenses on the posterior capsule in patients with large capsulorrhexis. *Japanese J Ophthalmol* 45(4):348–354.
Medvedovic M, Tomlinson CR, Call MK, Grogg M, Tsonis PA (2006). Gene expression and discovery during lens regeneration in mouse: Regulation of epithelial to mesenchymal transition and lens differentiation. *Molecular Vis* 12:422–440.
Menezo JL, Cisneros-Lanuza AL, Ferrer-Galindo E (1989). Late capsular opacification. *Eur J Implant Refract Surg* 1:123–126.
Middlemore R (1832). On the reproduction of the crystalline lens. *Lond Med Gaz* 10:344–348.
Milliot B (1872). De la regeneration du crystallin chez quelques mammiferes. *Journal de l'Anatomie et de la Physiologie (Paris)* 8:1.
Mireskandari K, Mead AL, Tuft SJ, Anderson IK, Khaw PT (2004). A model of cataract surgery, digital imaging and software analysis to investigate posterior capsule opacification in the rabbit. *Invest Ophthalmol Vis Sci* 45(suppl):171.
Mireskandari K, Jones E, Mead AL, Anderson IK, Tuft SJ, Khaw PT (2005). The efficacy of CAT-152 in the prevention of posterior capsule opacification (PCO) in the rabbit model of cataract surgery and lens implantation. *Invest Ophthalmol Vis Sci* 46(suppl):2863.
Miyake K, Miyaki C (1985). Intraoperative posterior chambers haptic fixation in the human cadaver eye. *Ophthalmic Surg* 16:230–236.
Moisseiev J, Bartov E, Schochat A, Blumenthal M (1989). Long-term study of the prevalence of capsular opacification following extracapsular cataract extraction. *J Cataract Refract Surg* 15:531–533.
Morrell AJ, Pearce JL (1989). Cataract surgery with posterior chamber lens implantation in patients aged 20–45. *Eur J Implant Refract Surg* 1:85–87.
Nakashima Y, Yoshitomi F (2002). Regression of Elschnig pearls on the posterior capsule in a pseudophakic eye. *Arch Ophthalmol* 120:397–398.
Nasisse MP, Dykstra MJ, Cobo LM (1995). Lens capsule opacification in aphakic and pseudophakic eyes. *Graefe's Arch Clin Exp Ophthalmol* 233:63–70.
Neider MW, Crawford K, Kaufman PL, Bito LZ (1990). In vivo videography of the rhesus monkey accommodative apparatus. Age-related loss of ciliary muscle response to central stimulation. *Arch Ophthalmol* 108:69–74.
Neuhann IM, Werner L, Izak AM *et al.* (2004). Late postoperative opacification of a hydrophilic acrylic (Hydrogel) intraocular lens. *Ophthalmology* 111:2094–2101.
Neumann G, Apple DJ (1986). *Pathology of the Eye*. Springer-Verlag, New York, NY.
Neumayer T, Findl O, Wolf B, Stefan S, Menapace R (2005). Long-term changes in the morphology of Elschnig pearls. *J Cataract Refract Surg* 31(11):2120–2128.
Neumayer T, Wolf B, Findl O (2006). Daily changes in the morphology of Elschnig pearls. *Am J Ophthlalmol* 141:517–523.
Nishi O (1986). Incidence of posterior capsule opacification in eyes with and without posterior chamber intraocular lenses. *J Cataract Refract Surg* 12:519–522.
Nishi O (1989). Refilling the lens of the rabbit eye after intercapsular cataract surgery using an endocapsular balloon and an anterior capsule suturing technique. *J Cataract Refract Surg* 15:450–454.
Nishi O, Nishi K (1992). Disruption of the blood–aqueous barrier by residual lens epithelial cells after intraocular lens implantation. *Ophthalmic Surg* 23:325–329.
Nishi O, Nishi K (2003). Effect of the optic size of a single-piece acrylic intraocular lens on posterior capsule opcification. *J Cataract Refract Surg* 29:348–353.
Nishi O, Hara T, Hara T, Hayashi F, Sakka Y, Iwata S (1989). Further development of experimental techniques for refilling the lens of animal eyes with a balloon. *J Cataract Refract Surg* 15:584–588.
Nishi O, Hara T, Hara T, Sakka Y, Hayashi F, Nakamae K, Yamada Y (1992). Refilling the lens with an inflatable endocapsular balloon: Surgical procedure in animal eyes. *Graefe's Arch Clin Exp Ophthalmol* 230:47–55.
Nishi O, Nakai Y, Mizumoto Y, Yamada Y (1997). Capsule opacification after refilling the capsule with an inflatable endocapsular balloon study. *J Cataract Refract Surg* 23:1548–1555.
Nishi O, Nishi K, Mano C, Ichihara M, Honda T (1998a). The inhibition of lens epithelial cell migration by a discontinuous capsular bend created by a band-shaped circular loop or a capsule-bending ring. *Ophthalmic Surg Lasers* 29:119–125.
Nishi O, Nishi K, Mano C *et al.* (1998b). Lens refilling with injectable silicone in rabbit eyes. *J Cataract Refract Surg* 24:975–982.
Nishi O, Nishi K, Osakabe Y (2004). Effect of intraocular lenses on preventing posterior capsule opacification: Design versus material. *J Cataract Refract Surg* 30:2170–2176.
Nishi O, Nishi K, Osakabe Y (2005). Evaluation of posterior capsule opacification using a new posterior view method in rabbits. Single-piece acrylic versus 3-piece acrylic intraocular lens. *J Cataract Refract Surg* 31:2369–2374.
Norrby S (2005). *Injectable Polymer, Refractive Lens Surgery*. Springer Berlin, Heidelberg. pp. 173–186
Norton JN, Kohnen T, Hackett RB, Patel A, Koch DD (1999). Ocular biocompatibility testing of intraocular lenses: A 1 year study in pseudophakic rabbit eyes. *J Cataract Refract Surg* 25:1467–1479.
Odrich M, Hall S, Worgul B *et al.* (1985). Posterior capsule opacification: Experimental analyses. *Ophthalmic Res* 17:75–84.
Ohmi S, Uenoyama K (1993). Experimental evaluation of posterior capsule opacification and intraocular lens decectration: Comparison of intraocular lenses of 12.5 mm and 14.0 mm diameter. *J Cataract Refract Surg* 19:77–82.
Olsen GM, Olson RJ (1995). Prospective study of cataract surgery, capsulotomy, and retinal detachment. *J Cataract Refract Surg* 21:136–139.
Oyster CW (1999). *The Human Eye, Structure and Function*. Sinauer Associates, Sunderland, MA.
Pande MV, Ursell PG, Spalton DJ, Heath G, Kundaiker S (1997). High-resolution digital retroillumination imaging of posterior lens capsule after cataract surgery. *J Cataract Refract Surg* 23: 1521–1527.
Pandey SK, Werner L, Apple DJ, Izak AM, Trivedi RH, McLeod SD (2002). Evaluation of an accommodative intraocular lens in human and rabbit eyes. *ARVO Annual Meeting Abstract Search and Program Planner*. Abstract No. 444.
Parel J-M, Treffers WF, Gelender H, Norton EWD (1981). Phaco-Ersatz: A new approach to cataract surgery. *Ophthalmology* 88(suppl):95.

Parel J-M, Gelender H, Treffers WF, Norton EWD (1986). Phaco-Ersatz: Cataract surgery designed to preserve accommodation. *Graefe's Arch Clin Exp Ophthalmol* 224:165–173.

Prince JH (1964). *The Rabbit in Eye Research.* Charles C. Thomas, Springfield, IL.

Randolph RL (1900). The regeneration of the crystalline lens: An experimental study. *Johns Hopkins Hospital Reports* 9:237–263.

Ravalico G, Tognetto D, Palomba MA, Busatto P, Baccara F (1996). Capsulorhexis size and posterior capsule opacification. *J Cataract Refract Surg* 22:98–103.

Sacu S, Findl O, Menapace R, Buehl W, Wirtitsch M (2004a). Comparison of posterior capsule opacification between the 1-piece and 3-piece Acrysof intraocular lenses. *Ophthalmology* 111:1840–1846.

Sacu S, Menapace R, Buehl W, Rainer G, Findl O (2004b). Effect of intraocular lens optic edge design and material on fibrotic capsule opacification and capsulorhexis contraction. *J Cataract Refract Surg* 30:1875–1882.

Sacu S, Menapace R, Findl O, Kiss B, Buehl W, Georgopoulos M (2005). Long-term efficacy of adding a sharp posterior optic edge to a three-piece silicone intraocular lenses on capsule opacification: Five-year results of a randomized study. *Am J Ophthalmol* 139:696–703.

Saika S (2004). Relationship between posterior capsule opacification and intraocular lens biocompatibility. *Prog Ret Eye Res* 23:283–305.

Saika S, Ohmi S, Kanagawa R *et al.* (1996). Lens epithelial cell outgrowth and matrix formation on intraocular lenses in rabbit eyes. *J Cataract Refract Surg* 22:835–840.

Saika S, Ohmi S, Tanaka S, Ohnishi Y, Yamanaka A, Ooshima A (1997a). Light and scanning electron microscopy of rabbit lens capsules with intraocular lenses. *J Cataract Refract Surg* 23:787–794.

Saika S, Tanaka S, Ohmi S *et al.* (1997b). Deposition of extracellular matrix on intraocular lenses in rabbits: An immunohistochemical and transmission electron microsopic study. *Graefe's Arch Clin Exp Ophthalmol* 235:241–247.

Scheib SA, Garner WH (2004). Anti-inflammatory effects of topical ocular Maxidex® administration to rabbits following vitrectomy or lensectomy. *Exp Eye Res* 79:893–902.

Sellman TR, Lindstrom RL (1988). Effect of a plano-convex posterior chamber lens on capsular opacification from Elschnig pearl formation. *J Cataract Refract Surg* 14:68–72.

Shekhawat DS, Jangir OP, Prakash A, Pawan S (2001). Lens regeneration in mice under the influence of vitamin A. *J Biosci Indian Acad Sci* 26(5):571–576.

Shepherd JR (1989). Capsular opacification associated with silicone implants. *J Cataract Refract Surg* 15:448–450.

Sheppard B (1962). The anatomy and histology of the normal rabbit eye with special reference to the ciliary zone. *Arch Ophthalmol* 67:87–100.

Sikharuldze TA (1956). Exchange of crystallin lens in rabbits by embryonic skin ectoderm. *Bull Acad Sci Georg SSR* 14:337.

Smythe RH (1958). *Veterinary Ophthalmology*, 2nd edn. Bailliere Tindall and Cox, London.

Stachs O, Schneider H. Stave J, Schmitz KP, Terwee T, Guthoff R (2003). The anterior segment in rabbit eyes after lens refilling with injectable silicone polymer. *ARVO Annual Meeting Abstract Search and Program Planner*. Abstract No. 273.

Steinert RF, Bayliss B, Brint SF, Giamporcaro JE, Hunkeler JD (1995). Long-term clinical results of the AMO® PhacoFlex® Model SI18 intraocular lens. *J Cataract Refract Surg* 21:331–338.

Stewart DS (1960). Further observations on regenerated crystalline lenses in rabbits, with special reference to their refractive qualities. *Trans Ophthalmol Soc UK* 80:357.

Tahi H, Chapon P, Hamaoui M, Lee W, Holden B, Parel JM (1999). Restoring accommodation: surgical technique and preliminary evaluation in rabbits. *Ophthalmic Technologies Conference 9.* San Jose, CA, 1999–01–23 Journal: *SPIE proceedings series*, 3591 267–269.

Tahi H, Fernandez V, Chapon P, Lee W, Hamaoui M, Parel J-M M (2002). Mini-capsulorhexis valve, a new device to control capsular bag refilling procedures. *ARVO Annual Meeting Abstract Search and Program Planner*. Abstract No. 400.

Tamura M, Kanagawa R, Saika S, Ohmi S, Nakao T, Kinoshita K (1990). Comparison of the cellular response on intraocular lenses implanted in rabbit eyes with and without extracapsular lens extraction. *J Cataract Refract Surg* 16:746–750.

Tan DTH, Chee SP (1993). Early central posterior capsular fibrosis in sulcus-fixated biconvex intraocular lenses. *J Cataract Refract Surg* 19:471–480.

Tañá P, Belmonte J (1993). Elasticity of the capsulorhexis and delivery of the nucleus. *Eur J Implant Ref Surg* 5:103–108.

Tetz MR, Nimsgern C (1999). Posterior capsule opacification. Part 2: Clinical findings. *J Cataract Refract Surg* 25:1662–1674.

Tetz MR, Wildeck A (2005). Evaluating and defining the sharpness of intraocular lenses. Part 1: Influence of optic design on the growth of the lens epithelial cells in vitro. *J Cataract Refract Surg* 31:2172–2179.

Tetz M, Imkamp E, Hansen S *et al.* (1988a). Experimentelle studie zur hinterkapseltrübung und optischen dezentrierung verschiedener hinterkammerlinsen nach intrakapsulärer implantation. *Fortschr Ophthalmol* 85:682–688.

Tetz M, O'Morchoe DJC, Gwin T *et al.* (1988b). Posterior capsular opacification and intraocular lens decentration. Part II: Experimental findings on a prototype circular intraocular lens design. *J Cataract Refract Surg* 14:614–623.

Tetz MR, Lehrer I, Klein U, Völcker HE (1994). Cataracta secundaria bei diabetes mellitus. In: Pham DT, Wollensack J, Rochels R, Hartmann C (eds), *8 Kongreß der Deutschen Gesellschaft für Intraokularlinsen Implantation (DGII)*. Springer, Berlin, Heidelberg, New York, pp. S398–sS406.

Tetz MR, Sperker M, Blum M, Auffarth GU, Völcker HE (1996a). Klinische nachstarbewertung in pseudophaken augen: methodik und reproduzierbarkeit. *Der Ophthalmologe* 93:33–37.

Tetz MR, Sperker M, Blum M, Auffarth GU, Völcker HE (1996b). Evaluation of a photographic image analysis system for PCO scoring. *J Cataract Refract Surgery*(special issue):27–30.

Tetz MR, Ries MW, Lucas C, Stricker H, Volcker HE (1996c). Inhibition of posterior capsule opacification by an intraocular-lens-bound sustained drug delivery system: An experimental animal study literature review. *J Cataract Refract Surg* 22:1070–1078.

Tetz MR, Auffarth GU, Sperker M, Blum M, Völcker HE (1997). Photographic image analysis system of posterior capsule opacification. *J Cataract Refract Surg* 23:1515–1520.

Tobari I, Iwaki Y, Miyake K (1999). Effect of tranilast eyedrops in preventing posterior capsule opacification: Preliminary report. *J Cataract Refract Surg* 25:1394–1399.

Ursell PG, Spalton DJ, Pande MV *et al.* (1998). Relationship between intraocular lens biomaterials and posterior capsule opacification. *J Cataract Refract Surg* 24:352–360.

Valentin G (1844). Mikroskopische Uuntersuchung zweier wiedererzeugter krystallinsen des kaninchens. *Zeitschrift f Ration Medicin Bd* 1:227–237.

Vasavada A, Singh R (1998). Relationship between lens and capsular bag size. *J Cataract Refract Surg* 24:547–551.

Wallentin N, Lundberg C (2000). Steroid and anti-CD18 treatment have no effect on after-cataract formation following surgery in rabbits. *Curr Eye Res* 20(5):384–393.

Wallentin N, Lundgren B, Lundberg C (2000). Lack of correlation between intraocular inflammation and after-cataract formation in the rabbit eye. *J Cataract Refract Surg* 26:1389–1397.

Wang MC, Woung LC (2000). Digital retroilluminated photography to analyze posterior capsule opacification in eyes with intraocular lenses. *J Cataract Refract Surg* 26:56–61.

Werner L, Mamalis N, Pandey S *et al.* (2000). Posterior capsule opacification in rabbit eyes implanted with hydrophilic acrylic intraocular lenses with enhanced square edge. *J Cataract Refract Surg* 30(11):2403–2409.

Werner L, Pandey SK, Izak AM *et al.* (2004). Capsular bag opacification after experimental implantation of a new accommodating intraocular lens in rabbit eyes. *J Cataract Refract Surg* 30:1114–1123.

Werner L, Hunter B, Stevens S, Chew JJL, Mamalis N (2006a). Role of silicon contamination on calcification of hydrophilic acrylic intraocular lenses. *Am J Ophthalmol* 141:35–43.

Werner L, Mamalis N, Stevens S, Hunter B, Chew JJL, Vargas LG (2006b). Intralenticular opacification: Dual-optic versus piggy-back intraocular lenses. *J Cataract Refract Surg* 32:655–661.

Wolf B, Findl O (2005). Short-term changes in the morphology of posterior capsule opacification. *J Cataract Refract Surg* 31:962–968.

Wong KH, Koopmans SA, Terwee T, Kooijman AC (2007). Changes in spherical aberration after lens refilling with silicone oil. *Invest Ophthalmol Vis Sci* 48:1261–1267.

Wren SME, Spalton DJ, Jose R, Boyce J, Heatly C (2005). Factors that influence the development of posterior capsule opacification with a polyacrylic intraocular lens. *Am J Ophthalmol* 139:691–695.

Yoo MK, Choi YJ, Lee JH, Wee WR, Cho CS (2006). Injectable intraocular lens using hydrogels. *J Drug Del Sci Tech* 17(1):81–85.

Zetterstrom C, Kugelberg U, Lundgren B, Syren-Nordquist S (1996). After-cataract formation in newborn rabbits implanted with intraocular lenses. *J Cataract Refract Surg* 22:85–88.

CHAPTER

14

The Primate in Cataract/IOL Surgery

Arlene Gwon
Department of Ophthalmology, University of California, Irvine

OUTLINE

INTRODUCTION

Non-human primates are generally well suited to the study of human disease due to their evolutionary similarity to humans. The Rhesus (*Macaca mulatta*) and cynomologus (*Acaca fasciularis*) monkeys are the two species most commonly used in ocular research. But due to their prohibitive cost, difficulty in handling and limited availability, the primate has been used infrequently in cataract surgery to study endoophthalmitis, viscoelastics, and intraocular lens designs. However, they are particularly helpful in the study of accommodation and potentially accommodating intraocular lenses (IOLs).

LENS/CATARACT SURGERY

The primate eye is very similar to the human eye though approximately 20% smaller. The lens diameter of the cynomologus monkey ranges from 7.0 to 8.0 mm and that of the Rhesus monkey has been measured at 7.5–9.6 mm and lens thickness ranges from 2.84 to 4.42 mm (Nishi *et al.*, 1993; Manns *et al.*, 2007). Thus, IOLs must be made slightly smaller (overall diameter of 8–10 mm) than the 10–13 mm IOLs designed for humans, which have a capsular bag diameter of 10.2–10.5 mm (Galand *et al.*, 1984; Assia and Apple, 1992; Tañá and Belmonte, 1993; Vasavada and Singh, 1998) to accommodate the size of the lens in the animal model. However, this has not been a problem when performing studies with an injectable polymer in lens refilling experiments or when studying lens regeneration when the anterior capsulotomy is sealed.

The cataract surgical procedure is similar to that described for rabbit and humans. A 5–6 mm capsulorhexis is made and phacoemulsification and irrigation/aspiration of the natural lens/cataract are followed by insertion of the IOL. While some investigators have performed surgery without supplemental heparin in the balanced salt irrigation solution, Lambert *et al.* have noted a significant decrease in the post-operative fibrin reaction if heparin is added (Lambert *et al.*, 1995, 1999; Lambert, 1997; Lambert and Grossniklaus, 1997).

In studies of lens refilling and lens regeneration, a small 1–2mm capsulorhexis and endocapsular lens extraction are performed by the method described in separate studies by Gindi *et al.* (1985) and Gwon *et al.* (1993). A modification of the standard endocapsular procedure is provided by Nishi *et al.* (1992) for insertion of an endocapsular balloon. Following injection of the synthetic polymer or scaffold, the anterior capsulotomy may be closed with a silicone plug or collagen patch (Gwon *et al.*, 1993; Koopmans *et al.*, 2006).

In these studies antibiotic and corticosteroids are given at the end of surgery, but the post-operative regimen has been varied. Because of the difficulty in administration of medication to the primate, some investigators have given no post-operative medications and others have applied topical or subconjunctival antibiotic and corticosteroids for approximately 2 weeks (Lundgren *et al.*, 1992; Lambert *et al.*, 1995, 1999; Lambert, 1997; Lambert and Grossniklaus, 1997; Koopmans *et al.*, 2006).

Post-operative eye examinations are most often performed under general anesthesia thus limiting their frequency. A visually significant fibrinous reaction occurs at 1 week and lasts as long as 18 weeks. In infant primates, a pupillary membrane formation occurs which may extend onto both the anterior and posterior PMMA IOL optic surface. These membranes are generally resistant to medical and surgical therapy, but have responded to Nd:YAG membranectomy after repeated treatments. The addition of heparin to the irrigation solution or implantation of a heparin surface modified IOL is associated with less severe papillary membrane formation (Lundgren *et al.*, 1992; Lambert *et al.*, 1995, 1999; Lambert, 1997; Lambert and Grossniklaus, 1997). Membrane formation has not been reported in older primates when the anterior capsulotomy is sealed in lens refilling and lens regeneration studies (Agarwal *et al.*, 1964; Haefliger *et al.*, 1987; Nishi *et al.*, 1992, 1993; Haefliger and Parel, 1994; Sakka *et al.*, 1996; Hashizoe *et al.*, 1998; Nishi and Nishi, 1998; Koopmans *et al.*, 2006).

ACCOMMODATION

Primates are generally believed to be the best and only appropriate animal model for studies on accommodation. The accommodation in Rhesus and cynomologus monkeys, particularly Rhesus monkeys, has been extensively investigated (Bito *et al.*, 1982, 1987; Neider *et al.*, 1990; Glasser and Kaufman, 1999). While the primate and human share many similarities in the key aspects of accommodation, there are a few differences. These similarities and differences are highlighted below.

The Similarities:

- The lens continues to grow through adulthood (Bito *et al.*, 1987).
- Accommodation is the only non-productive function that is completely lost well before the end of the lifespan (Bito *et al.*, 1987).
- Accommodation is age-dependent and the ability to accommodate will be completely lost eventually (Bito *et al.*, 1982).
- Accommodation is associated with an increase in lenticular axial thickness and anterior chamber shallowing by the same amount. The location of the posterior capsular surface relative to the posterior cornea remains unchanged (Bito *et al.*, 1982).
- Rhesus monkeys undergo an age-related decrease in pharmacologically induced accommodation highly comparable to the age-dependent decrease in physiologically induced accommodation in humans (Bito *et al.*, 1982; Neider *et al.*, 1990).

The Differences:

- The lifespan of the Rhesus monkey is only one-third that of the human (Bito *et al.*, 1987).
- Rhesus monkeys possess a greater capacity for accommodation and have a much closer near point than humans (Bito *et al.*, 1982). Under carbachol stimulation, the accommodation amplitude of a juvenile Rhesus monkey is greater than 30D (Bito *et al.*, 1987), whereas the maximum accommodation in the adolescent human is approximately 14D (Borish, 1970).
- In humans, the age-dependent change in the anterior surface is almost 10 times that of the posterior surface resulting in the same curvature of the anterior and posterior surfaces by age 75–80. Initially, the anterior surface in the primate is much less sharply curved than the posterior surface, and this difference is maintained throughout the primate's lifetime (Bito *et al.*, 1987; Koretz *et al.*, 1987).
- The average resting refraction under ketamine is −5D in primates (Bito *et al.*, 1982) versus −1 to 2D in humans.
- The Rhesus lens decreases in thickness by 0.1mm/year during youth (5–6 years) and then increases by 0.014mm (Koretz *et al.*, 1987), whereas in the human the lens thickness remains unchanged in the pre-adult and then increases by 0.023mm/year between ages 20 and 60. (Brown, 1974).

Additionally, the iris has been shown to play a significant role in pharmacologically stimulated accommodation in the Rhesus monkey (Crawford *et al.*, 1990). The amplitude of accommodation was found to be 40% less after total iridectomy, as well as a decrease in anterior chamber shallowing and lens thickening. This difference was not seen with submaximal accommodation induced by intramuscular pilocarpine infusion or maximum accommodation induced by mid-brain stimulation. It is believed that the iris sphincter muscle pulls the ciliary body farther forward and inward than does maximum ciliary muscle contraction alone, allowing additional lens rounding and additional accommodation power (Crawford *et al.*, 1990). The role of the iris in accommodation in the human has not been investigated.

In lens refilling experiments, accommodative amplitude studies have been conducted during the initial post-operative period. A decrease in the anterior chamber of depth 0.5 mm for up to 4 years post-operative with a silicone polymeric lens was noted by Haefliger and Parel (1994), as much as 4.6 D accommodative amplitude by Nishi *et al.* (Nishi *et al.*, 1993; Nishi and Nishi, 1998), and up to 6.3 D by Koopmans *et al.* (2006). However, the development of capsule fibrosis and posterior capsule opacification remain problematic in bringing this technology forward.

In lens regeneration experiments, Agarwal *et al.* showed lens regeneration filling the capsule bag by 24 weeks, but with irregular optical quality in Rhesus monkeys (Agarwal *et al.*, 1964; Gwon, 2006).

SUMMARY

Based on the literature available to date, the primates, particularly Rhesus monkeys, are by far the best animal study model on accommodation and potentially accommodating IOLs or injectable polymers. The amplitude and mechanism of accommodation in all other animal species are different enough from humans making data obtained using any of these other animals irrelevant. If an alternate study model must be used for feasibility assessment of a potentially accommodating IOL, the rabbit is the best choice due to its small size, low cost, and ready availability. The best animal model for assessing efficacy of any accommodating intraocular lens is the non-human primate.

REFERENCES

Agarwal LP, Angra SK, Khosla PK, Tandon HD (1964). Lens regeneration in mammals. II: Monkeys. *Orient Arch Ophthalmol* 2:47–59.

Assia EI, Apple DJ (1992). Side-view analysis of the lens. I. The crystalline lens and the evacuated bag. *Arch Ophthalmol* 110:89–93.

Bito LZ, Kaufman PL, DeRousseau CJ, Koretz J (1987). Presbyopia: an animal model and experimental approaches for the study of the mechanism of accommodation and ocular ageing. *Eye* 1:222–230.

Bito LZ, DeRousseau CJ, Kaufman PL, Bito JW (1982). Age-dependent loss of accommodative amplitude in rhesus monkeys: an animal model for presbyopia. *Invest Ophthalmol Vis Sci* Jul 23:23–31.

Borish IM (1970). Accommodation and presbyopia. In: Borish IM (ed.), *Clinical Refraction*, 3rd edn. The Professional Press, Inc, Chicago, IL, p. 169.

Brown N (1974). The change in lens curvature with age. *Exp Eye Res* 19:175–184.

Crawford KS, Kaufman PL, Bito LZ (1990). The role of the iris in accommodation of Rhesus monkeys. *Invest Ophthalmol Vis Sci* 31(10):2185–2190.

Galand A, Bonhomme L, Collée M (1984). Direct measurement of the capsular bag. *Am Intra-Ocular Implan Soc J* 10:475–476.

Gindi JJ, Wan WL, Schanzlin DJ (1985). Endocapsular cataract surgery.1. Surgical technique. Cataract. *Int J Cataract Surg* 2:5–10.

Glasser A, Kaufman PL (1999). The mechanism of accommodation in primates. *Ophthalmology* 106:863–872.

Gwon A (2006). Lens regeneration in mammals: a review. *Surv Ophthalmol* 51:51–62.

Gwon A, Mantras C, Gruber L (1993). Restoring lens capsule integrity enhances lens regeneration in New Zealand albino rabbits and cats. *J Cataract Refract Surg* 19:735–746.

Haefliger E, Parel J-M (1994). Accommodation of an endocapsular silicone lens (phaco-ersatz) in the aging rhesus monkey. *J Refract Corneal Surg* 10:550–555.

Haefliger E, Parel J-M, Fantes F *et al.* (1987). Accommodation of an endocapsular silicone lens (Phaco-Ersatz) in the nonhuman primate. *Ophthalmology* 94:471–477.

Hashizoe M, Hara T, Ogura Y, Sakanishi K, Honda T, Hara T (1998). Equator ring efficacy in maintaining capsular bag integrity and transparency after cataract removal in monkey eyes. *Graefes Arch Clin Exp Ophthalmol* 236(5):375–379.

Koopmans SA, Terwee T, Glasser A et al. (2006). Accommodative lens refilling in rhesus monkeys. *Invest Ophthalmol Vis Sci* 47(7):2976–2984.

Koretz JF, Neider MW, Kaufman PL, Bertasso AM, DeRousseau CJ, Bito LZ (1987). Slit-lamp studies of the rhesus monkey eye. I. Survey of the anterior segment. *Exp Eye Res* 44:307–318.

Lambert SR (1997). Monkey model of neonatal monocular pseudophakia. *Sem Ophthalmol* 12:81–88.

Lambert SR, Aiyer A, Grossniklaus H (1999). Infantile lensectomy and intraocular lens implantation with long-term follow up in a monkey model. *J Pediatr Ophth Strab* 36:271–278.

Lambert SR, Grossniklaus HE (1997). Intraocular lens implantation in infant monkeys: clinical and histopathological findings. *J Cataract Refract Surg* 23:605–611.

Lambert SR, Fernandez A, Grossniklaus H, Drews-Botsch C, Eggers H, Boothe RG (1995). Neonatal lensectomy and intraocular lens implantation: effect in Rhesus monkeys. *Invest Ophthalmol Vis Sci* 36:300–310.

Lundgren B, Selen G, Spangberg M, Hafstand A (1992). Fibrinous reaction on implanted intraocular lenses. *J Cataract Refract Surg* 18:236–239.

Manns F, Parel JM, Denham D *et al.* (2007). Optomechanical response of human and monkey lenses in a lens stretcher. *Invest Ophthalmol Vis Sci* 48:3260–3268.

Neider MW, Crawford K, Kaufman PL, Bito LZ (1990). In vivo videography of the rhesus monkey accommodative apparatus.

Age-related loss of ciliary muscle response to central stimulation. *Arch Ophthalmol* 108:69–74.
Nishi O, Nishi K (1998). Accommodation amplitude after lens refilling with injectable silicone by sealing the capsule with a plug in primates. *Arch Ophthalmol* 116(10):1358–1361.
Nishi O, Nakai Y, Yamada Y, Mizumoto Y (1993). Amplitudes of accommodation of primate lenses refilled with two types of inflatable endocapsular balloons. *Arch Ophthalmol* 111:1677–1684.
Nishi O, Hara T, Hara T, Sakka Y, Hayashi F, Nakamae K, Yamada Y (1992). Refilling the lens with an inflatable endocapsular balloon: surgical procedure in animal eyes. *Graefes Arch Clin Exp Ophthalmol* 230:47–55.
Sakka Y, Hara T, Yamada Y, Hara T, Hayashi F (1996). Accommodation in primate eyes after implantation of refilled endocapsular balloon. *Am J Ophthalmol* 121:210–212.
Tañá P, Belmonte J (1993). Elasticity of the capsulorhexis and delivery of the nucleus. *Eur J Implant Refract Surg* 5:103–108.
Vasavada A, Singh R (1998). Relationship between lens and capsular bag size. *J Cataract Refract Surg* 24:547–551.

Index

Kolibri 360 GmbH
Ferdinand-Jühlke-Straße 7
99095 Erfurt, Deutschland
produktsicherheit@kolibri360.de